S0-BWT-443

Lecture Notes in Physics

Edited by J. Ehlers, München, K. Hepp, Zürich and H. A. Weidenmüller, Heidelberg

37

Trends in Elementary Particle Theory

Springer-Verlag
Berlin · Heidelberg · New York

Lecture Notes in Physics

Edited by J. Ehlers, München, K. Hepp, Zürich and
H. A. Weidenmüller, Heidelberg
Managing Editor: W. Beiglböck, Heidelberg

37

Trends in Elementary Particle Theory

International Summer Institute on
Theoretical Physics in Bonn 1974

Edited by H. Rollnik and K. Dietz

Springer-Verlag
Berlin · Heidelberg · New York 1975

Editors
Prof. Horst Rollnik
Prof. Klaus Dietz
Physikalisches Institut
der Universität Bonn
Nußallee 12
53 Bonn/BRD

Library of Congress Cataloging in Publication Data

International Summer Institute on Theoretical Physics,
6th, Bonn, 1974.
Trends in elementary particle theory.

(Lecture notes in physics ; 37)
Bibliography: p.
Includes index.
1. Particles (Nuclear physics)--Congresses.
I. Dietz, Klaus, 1934- II. Rollnik, Horst,
1931- III. Title. IV. Series.
QC793.I556 1974 539.7'21 75-8826

ISBN 3-540-07160-1 Springer-Verlag Berlin · Heidelberg · New York
ISBN 0-387-07160-1 Springer-Verlag New York · Heidelberg · Berlin

Offsetprinting and bookbinding: Julius Beltz, Hemsbach/Bergstr.

PREFACE

This Volume of the Lecture Notes in Physics contains the proceedings of the International Summer Institute organized by the University of Bonn in 1974. This Institute was the sixth in the series of Summerschools held by universities of the Federal Republic of Germany.

The aim of the lectures was to present "Trends in Elementary Particle Theory". Special emphasis was given to the following subjects:

Unified Field Theories
Parton Models
Field Theory and Statistical Mechanics
Colliding Beam Physics.

The responsibility for the final preparation of the manuscripts for printing was in the hand of the editors.

We gratefully acknowledge the assistance of Dr. R. Meyer, H. Schwilden and H. Horstmeier and of our secretaries Frau C. Al-Haidary and Frau R. Fuchs.

The Institute was sponsored by the NATO Advanced Study Institute Programme and supported by the Bundesministerium für Forschung und Technologie in Bonn and the Ministerium für Forschung und Wissenschaft des Landes Nordrhein-Westfalen in Düsseldorf.

Finally we thank all the lecturers, participants and members of the Bonn Physics Institute whose efforts made the Summer Institute so successful and enjoyable.

January 1975

K. Dietz
H. Rollnik

TABLE OF CONTENTS

NEW DEVELOPMENTS IN QUANTUM FIELD THEORY

RENORMALIZATION OF GAUGE THEORIES

J. Zinn-Justin

Centre d'Etudes Nucléaires de Saclay, Gif-sur-Yvette, France

I. THE CHANGE OF VARIABLES IN THE FUNCTIONAL INTEGRATION.

The functional formulation of quantum field theory through the Feynman path integral is widely used in order to derive algebraic properties of the perturbation series expansion, like for example Ward Identities. But because the path integral is in general not well defined in the topological sense (in the sense of a limit), a question is often raised: how rigorous are the results obtained by this method and is it necessary to verify them explicitly on the Feynman graph expansion. The answer to this question is of course that the results obtained by functional integration are exactly as rigorous as those derived from manipulations on the perturbation series, because the path integral has a perfectly well defined algebraic meaning. In order to clarify this point we shall give in this chapter a few justifications of the results derived by this method, using pure algebraic techniques.

1. Definition of the path integral

Let Z(J) be the generating functional of the complete Green functions $Z_N(x_1,\ldots,x_N)$ defined by a path integral:

$$Z(J) = N^{-1}\int [d\varphi] \exp[-S(\varphi) + \int dx\, J(x)\varphi(x)] \tag{1}$$

with $S(\phi)$ given by:

$$S(\varphi) - \frac{1}{2}\int K(x,y)\varphi(x)\varphi(y)\,dx\,dy + V(\varphi) \tag{2}$$

where $V(\phi)$ is a formal power series in ϕ. For convenience the normalization factor N is chosen so that Z(0) = 1 and the conventions are those of an euclidian field theory. The Green functions are given by:

$$Z_N(x_1,\ldots,x_N) = \frac{\delta}{\delta J(x_1)} \cdots \frac{\delta}{\delta J(x_N)} Z(J)\Big|_{J=0}$$

We shall define Z(J) by a pure algebraic formula which is also used to calculate Z(J) in perturbation series.

First formally:

$$Z(J) = N^{-1} \exp - V\left[\frac{\delta}{\delta J}\right] \int [d\varphi] \exp - \frac{1}{2} \int dx\, dy\, \varphi(x) \varphi(y) K(x;y) + \int J(x) \varphi(x)\, dx \tag{3}$$

The integral reduces now to a simple gaussian integral. We are able to compute such an integral for an arbitrary number of variables:

$$\int dx_1 \ldots dx_n \exp(-\tfrac{1}{2} x_i x_j K_{ij} + b_i x_i) = (2\pi)^{n/2} (\det K)^{-1/2} \exp \tfrac{1}{2} b_i K_{ij}^{(-1)} b_j \tag{4}$$

This result can trivially be generalized to the continuous case:

$$Z(J) = N'^{-1} \exp - V\left[\frac{\delta}{\delta J}\right] \exp \frac{1}{2} \int J(x) \Delta(x;y) J(y)\, dx\, dy \tag{5}$$

where we have introduced the propagator $\Delta(x;y)$

$$\int dz\, \Delta(x;z) K(z;y) = \delta(x-y) \tag{6}$$

Now the expression (5) which generates the whole perturbation expansion is perfectly well defined provided it has been, if needed, properly regularized.

It is our rigorous algebraic definition of the path integral.

2. Uniqueness

In calculating the integral (1) we have separated in the exponent of the integrand a special part composed of a quadratic and linear form. Let us show that if we choose a different quadratic and linear form the result of the integration is not changed.

Let us separate K(x,y) in two parts:

$$K(x;y) = K_1(x,y) + K_2(x,y) \tag{7}$$

and write now equation (1) in the form:

$$Z(J) = N^{-1} \int [d\varphi] \exp[-\tfrac{1}{2} \int \varphi(x) K_1(x;y) \varphi(y)\, dx\, dy + \int [J(x) + L(x)] \varphi(x)\, dx$$

$$-\tfrac{1}{2} \int \varphi(x) K_2(x;y) \varphi(y)\, dx\, dy - \int L(x) \varphi(x)\, dx - V(\varphi)] \tag{8}$$

Instead of the expression (5) we now obtain:

$$Z(J) = N_1^{-1} \exp - V[\tfrac{\delta}{\delta J}] - \tfrac{1}{2} \int \frac{\delta}{\delta J(x)} K_2(x;y) \frac{\delta}{\delta J(y)}\, dx\, dy - \int L(x) \frac{\delta}{\delta J(x)}\, dx$$

$$\exp \tfrac{1}{2} \int (J(x) + L(x)) K_1^{(-1)}(x;y)(J(y) + L(y))\, dx\, dy \tag{9}$$

The operator $\exp - \int L(x) \frac{\delta}{\delta J}(x)\, dx$ transforms $J(x) + L(x)$ in $J(x)$. Therefore to prove the identity of expressions (5) and (9) it is sufficient to compute $Z_0(J)$:

$$Z_0(J) = \exp -\tfrac{1}{2} \int \frac{\delta}{\delta J(x)} K_2(x;y) \frac{\delta}{\delta J(y)}\, dx\, dy \; \exp \tfrac{1}{2} \int J(x) K_1^{(-1)}(x;y) J(y)\, dx\, dy \tag{10}$$

Let us compute $\frac{\delta}{\delta J(z)} Z_0(J)$. Commuting $J(x)$ with the differential operator leads to

$$\frac{\delta}{\delta J(z)} Z_0(J) = -\int (K_1^{(-1)} K_2)(x;y) \frac{\delta}{\delta J(y)}\, dy\, Z_0(J) + \int dy\, K_1^{(-1)}(z;y) J(y)\, Z_0(J) \tag{11}$$

which can also be written:

$$\frac{\delta}{\delta J(z)} Z_0(J) = \int [K_1 + K_2]^{(-1)}(z;y) J(y)\, dy\, Z_0(J) \tag{12}$$

whose solution is:

$$Z_0(J) = Z_0(0) \exp \tfrac{1}{2} \int J(x) K^{(-1)}(x;y) J(y)\, dx\, dy \tag{13}$$

Therefore the expressions (5) and (9) are identical.

Now usually, one uses the path integral (1) to derive some algebraic properties of the perturbation series expansion. For example one integrates by part, or changes variables in the integral. We have to show that the resulting identities are true for the function Z(J) defined by equation (5). It is immediate to see that the problem of a linear change of variables reduces to the problem of the separation of a quadratic part in the exponent of the integrand which is already solved. Therefore we shall only consider non linear change of variables. The change of variables will always be given under the form of a formal power series. It has to be canonical i.e. the coefficient of the linear term in the power series has to be non-singular.

3. Integration by part and equation of motion

The integration by part formula relies on the fact that the integral of a total derivative vanishes. This yields the identity:

$$\int [d\varphi] \left[\frac{\delta S}{\delta \varphi(x)} - J(x) \right] \exp\left[-S(\varphi) + \int J(x)\varphi(x) \right] = 0 \tag{14}$$

which can be formally rewritten as an equation for Z(J):

$$\left\{ \frac{\delta S}{\delta \varphi(x)} \left[\frac{\delta}{\delta J} \right] - J(x) \right\} Z(J) = 0 \tag{15}$$

Equation (15) is the equation of motion for Z(J). We have to show that Z(J) defined by equation (5) satisfies indeed this equation. Expressing $S(\phi)$ in terms of $V(\phi)$ we obtain the identity:

$$\left\{ \int K(x;y) \frac{\delta}{\delta J(y)} dy + \frac{\delta}{\delta \varphi(x)} V\left[\frac{\delta}{\delta J} \right] - J(x) \right\} \exp - V\left[\frac{\delta}{\delta J} \right] \exp \frac{1}{2} \int dx\, dy\, J(x) J(y) \Delta(x;y) = 0 \tag{16}$$

In order to derive equation (16) we use the commutation relation:

$$\left[J(x) , \exp - V\left[\frac{\delta}{\delta J} \right] \right] = \frac{\delta V}{\delta \varphi(x)} \left[\frac{\delta}{\delta J} \right] \exp - V\left[\frac{\delta}{\delta J} \right] \tag{17}$$

Equation (16) reduces then to the trivial identity:

$$\left[\int K(x;y) \frac{\delta}{\delta J(y)} dy - J(x) \right] \exp \frac{1}{2} \int dx\, dy\, J(x) J(y) \Delta(x;y) = 0 \tag{18}$$

which follows from the definition of $\Delta(x,y)$.

4. Infinitesimal change of variables

Let us now show that an integral is left invariant by an infinitesimal change of variables:

$$\varphi(x) = \chi(x) + \varepsilon F(x,\chi) \tag{19}$$

where $F(x, \chi)$ is an arbitrary functional of χ :

$$F(x,\chi) = \sum \frac{1}{n!} \int dy_1 \dots dy_n \, F^{(n)}(x; y_1, \dots, y_n) \, \chi(y_1) \dots \chi(y_n) \tag{20}$$

After such a change of variables the integral (1) becomes at first order in ε:

$$\begin{aligned} Z(J) = \int [d\chi] \Big[1 + \varepsilon \int \frac{\delta F(x,\chi)}{\delta \chi(x)} dx\Big] \Big\{1 - \varepsilon \int F(x,\chi) \frac{\delta S}{\delta \chi(x)} dx \\ + \varepsilon \int J(x) F(x,\chi) dx \Big\} \exp\Big[- S(\varphi) + \int J(x) \varphi(x) dx\Big] \end{aligned} \tag{21}$$

Expressing that $Z(J)$ is independent of ε yields the following identity:

$$\Big\{ \int dx \Big[J(x) F(x, \tfrac{\delta}{\delta J}) - F(x, \tfrac{\delta}{\delta J}) \frac{\delta S}{\delta \chi(x)} \Big[\frac{\delta}{\delta J}\Big] + \frac{\delta F(x, \frac{\delta}{\delta J})}{\delta \chi(x)} \Big\} Z(J) = 0 \tag{22}$$

We have now to show that this identity is satisfied by $Z(J)$ using equation (5). But actually equation (22) is directly a consequence of the equation of motion (15). Let us act with $F(x; \frac{\delta}{\delta J})$ on equation (15):

$$\int dx \, F(x, \tfrac{\delta}{\delta J}) \Big\{ \frac{\delta S}{\delta \chi(x)} \Big[\frac{\delta}{\delta J}\Big] - J(x) \Big\} Z(J) = 0 \tag{23}$$

We now use the commutation relation:

$$F(x; \tfrac{\delta}{\delta J}) J(x) = J(x) F(x; \tfrac{\delta}{\delta J}) + \frac{\delta F}{\delta \chi(x)}(x; \tfrac{\delta}{\delta J}) \tag{24}$$

This transforms equation (23) in equation (22).

The equation (22) has therefore been derived by pure algebraic methods. This equation implies all the Ward identities derived both in the case of global and local gauge symmetries. Nevertheless, for completeness, we shall show that the consequences of a finite change of variables can also be derived in a pure algebraic way.

5. Finite change of variables

If we perform now a finite change of variables of the same form as previously:

$$\varphi(x) = \chi(x) + F(x;\chi) \tag{25}$$

The integral giving Z(J) becomes

$$Z(J) = \int [d\chi] \det\left[\delta(x-y) + \frac{\delta F}{\delta\chi(y)}(x;\chi)\right] \exp\left\{-\frac{1}{2}\int dx\, dy\, \chi(x) K(x,y) \chi(y) - V'(\chi) + \int dx\, J(x)[\chi(x) + F(x;\chi)]\right\} \tag{26}$$

where $V'(\chi)$ is given by:

$$V'(\chi) = V(\chi + F(x;\chi)) + \frac{1}{2}\int dx\, dy\, [\chi(x) + F(x;\chi)] K(x,y) [\chi(y) + F(y,\chi)] - \frac{1}{2}\int dx\, dy\, \chi(x) K(x;y) \chi(y) \tag{27}$$

We therefore obtain the following new expression for Z(J):

$$Z(J) = :\exp\int J(x) F(x;\tfrac{\delta}{\delta J})\, dx : \exp - V'(\tfrac{\delta}{\delta J}) \det\left[\delta(x-y) + \frac{\delta F}{\delta\chi(y)}(x;\tfrac{\delta}{\delta J})\right] \exp\frac{1}{2}\int dx\, dy\, J(x) \Delta(x,y) J(y) \tag{28}$$

where: $\exp \int J(x)\, F(x; \frac{\delta}{\delta J})\, dx$: stands for:

$$:\exp\int J(x) F(x;\tfrac{\delta}{\delta J})\, dx : = \sum_0^\infty \frac{1}{n!}\int dx_1 \ldots dx_n\, J(x_1)\ldots J(x_n)\, F(x_1;\tfrac{\delta}{\delta J})\ldots F(x_n;\tfrac{\delta}{\delta J})$$

and det(1+M) is defined by its power series expansion in the variable M, through the formula

$$\det(1+M) = \exp \operatorname{tr} \ln(1+M)$$

In order to prove equation (28) we have to derive a few simpler identities.

a) Let us first show:

$$(: \exp \int J(x) F(x; \frac{\delta}{\delta J}) :) \left[\frac{\delta}{\delta J(y)} + F(y; \frac{\delta}{\delta J}) \right] = \frac{\delta}{\delta J(y)} : \exp \int J(x) F(x; \frac{\delta}{\delta J}) dx \tag{29}$$

This identity can be written explicitly in the following form:

$$\sum_{n=0}^{\infty} \frac{1}{n!} \int dx_1 \ldots dx_n \, J(x_1) \ldots J(x_n) F(x_1; \frac{\delta}{\delta J}) \ldots F(x_n; \frac{\delta}{\delta J}) F(y; \frac{\delta}{\delta J})$$

$$= \sum_{n=0}^{\infty} \frac{1}{n!} \int dx_1 \ldots dx_n \left[\frac{\delta}{\delta J(y)} , J(x_1) \ldots J(x_n) \right] F(x_1; \frac{\delta}{\delta J}) \ldots F(x_n; \frac{\delta}{\delta J}) \tag{30}$$

and the evaluation of the commutator $\left[\frac{\delta}{\delta J(y)}, J(x_1) \ldots J(x_n) \right]$ leads to the result.

b) Another useful identity is:

$$\exp - \int F(x; \frac{\delta}{\delta J}) K(x;y) \frac{\delta}{\delta J(y)} dx\, dy \, \exp \frac{1}{2} \int J(x) \Delta(x,y) J(y) \, dx\, dy$$

$$= \exp \frac{1}{2} \int F(x; \frac{\delta}{\delta J}) K(x;y) F(y; \frac{\delta}{\delta J}) dx\, dy \, \vdots \exp - \int F(x; \frac{\delta}{\delta J}) J(x) dx \, \vdots \tag{31}$$

where we have defined:

$$\vdots \exp - \int F(x; \frac{\delta}{\delta J}) J(x) dx \, \vdots = \sum_{n=0}^{\infty} \frac{(-1)^n}{n!} \int dx_1 \ldots dx_n \, F(x_1; \frac{\delta}{\delta J}) \ldots$$

$$F(x_n; \frac{\delta}{\delta J}) J(x_1) \ldots J(x_n) \tag{32}$$

To prove identity (31) it is sufficient to remark that the operator $\exp - \int F(x;\frac{\delta}{\delta J}) K(x,y) \frac{\delta}{\delta J(y)} dx\, dy$ would transform J in J - KF if F would be independent of $\frac{\delta}{\delta J}$. In order to obtain the same algebraic result, one has simply to avoid to commute any function $F(x;\frac{\delta}{\delta J})$ with J.

c) Repeated use of identity (29) transforms (28) in:

$$Z(J) = \exp - V[\tfrac{\delta}{\delta J}] : \exp \int J(x)\, F(x;\tfrac{\delta}{\delta J})\, dx : \det[\delta(x-y) + \frac{\delta F}{\delta \chi(y)}(x;\tfrac{\delta}{\delta J})]$$

$$\exp\{-\tfrac{1}{2}\int dx\, dy\, [\tfrac{\delta}{\delta J(x)} K(x,y) F(y;\tfrac{\delta}{\delta J}) + F(x;\tfrac{\delta}{\delta J}) K(x,y) \tfrac{\delta}{\delta J(y)} \tag{33}$$

$$+ 2 F(x;\tfrac{\delta}{\delta J}) K(x,y) F(y;\tfrac{\delta}{\delta J})]\} \exp \tfrac{1}{2} \int dx\, dy\, J(x)\, \Delta(x,y)\, J(y)$$

Identity (31) transforms then equation (33) in:

$$Z(J) = \exp - V[\tfrac{\delta}{\delta J}] : \exp \int J(x)\, F(x;\tfrac{\delta}{\delta J})\, dx : \det[\delta(x-y) + \frac{\delta F(x;\tfrac{\delta}{\delta J})}{\delta \chi(y)}] \tag{34}$$

$$\vdots \exp - \int F(x;\tfrac{\delta}{\delta J})\, J(x)\, dx \vdots \exp \tfrac{1}{2} \int J(x)\, \Delta(x,y)\, J(y)\, dx\, dy$$

Therefore, in order to achieve the derivation of equation (28) we have to prove the identity between differential operators:

$$A(F) = : \exp \int J(x)\, F(x;\tfrac{\delta}{\delta J})\, dx : \det[\delta(x-y) + \frac{\delta F(x;\tfrac{\delta}{\delta J})}{\delta \chi(y)}] \tag{35}$$

$$\vdots \exp - \int F(x;\tfrac{\delta}{\delta J})\, J(x)\, dx \vdots \equiv 1$$

This can be done in two steps.

d) Let us first remark that:

$$\vdots \exp - \int F(x;\tfrac{\delta}{\delta J})\, J(x)\, dx \vdots \frac{\delta}{\delta J(y)} = [\frac{\delta}{\delta J(y)} + F(y;\tfrac{\delta}{\delta J})] \vdots \exp - \int F(x;\tfrac{\delta}{\delta J})\, J(x)\, dx \tag{36}$$

which can be derived exactly in the same way as the identity (29). Then the identities (29) and (36) lead to:

$$A(F) \frac{\delta}{\delta J(x)} = \frac{\delta}{\delta J(x)} A(F) \tag{37}$$

Equation (37) shows that A(F) is a power series in $\frac{\delta}{\delta J(x)}$ with coefficients independent of J. To prove identity (35) we shall act with A(F) on the functional $\exp \int J(x)\, \phi(x) dx$. A(F) can then be written as a power series in $\phi(x)$ of the form:

$$A(F) = A_-(F)\,A_+(F) \tag{38}$$

where we have defined:

$$A_+(F) = \sum_{n=0}^{\infty} \frac{1}{n!} \int dx_1 \ldots dx_n \, F(x_1;\varphi) \ldots F(x_n;\varphi) \frac{\delta}{\delta\varphi(x_1)} \cdots \frac{\delta}{\delta\varphi(x_n)} \tag{39a}$$

$$A_-(F) = \sum_{n=0}^{\infty} \frac{(-1)^n}{n!} \int dx_1 \ldots dx_n \frac{\delta}{\delta\varphi(x_1)} \cdots \frac{\delta}{\delta\varphi(x_n)} F(x_1,\varphi) \ldots F(x_n,\varphi) \tag{39b}$$

$$\det\left[\delta(x-y) + \frac{\delta F(x,\varphi)}{\delta\varphi(y)}\right]$$

Property (37) means that A(F) commutes with any functional of $\phi(x)$. It is therefore sufficient to show that A(F) acting on the functional 1 gives 1 which clearly reduces to:

$$A_-(F)\,1 = 1 \tag{40}$$

or $A_-(F)$ considered as a function of ϕ is identical to 1.

One can first for example prove the composition law for $A_-(F)$

$$A_-(F_1)\,A_-(F_2) = A_-(F_2 + F_1(F_2 + \varphi))$$

This reduces the proof, in the case of an arbitrary number of discrete variables to the proof of the property for one variable which is very simple. It can also be used for a continuous set of variables.

6. Integral on fermions

In order to treat fermions in the same way as bosons, one has to introduce the notion of derivatives and integrals on elements of an antisymmetric algebra. An introduction to this method can be found in Berezin's book "Methods of second quantization".

We shall therefore only list, without proof, a few properties.

Let $\{x_i\}$ be a set of generators of such an antisymmetric algebra. The derivative with respect to such an x_i is defined in the same way as in the case of commuting quantities up to a question of sign. Let us call D such a derivation. The derivative of a product is given by:

$$D(AB) = A\,D(B) + D(A)\,P(B) \tag{41}$$

where P(B) is obtained from B by replacing all the x_i by $-x_i$.

Integration of fermions is identical to derivation. By definition:

$$\int dx_i \equiv \frac{\delta}{\delta x_i} \tag{42}$$

With this definition, it is easy to prove the two main properties of the integration on fermions:

- the value of the gaussian integral

$$\int \prod_i dx_i \exp(x_i a_{ij} x_j) = (\det A)^{1/2} \tag{43}$$

for an antisymmetric matrix A;

- the rule of change of variables: in a change of variables, the measure $\prod_i dx_i$ to be multiplied by the inverse of the jacobian of the transformation.

Let us now make a last remark: If we introduce sources for the $\{x_i\}$ anticommuting variables, they will necessarily belong to the same antisymmetric algebra as the $\{x_i\}$.

REFERENCES

For the three chapters, see E. S. Abers and B.W. Lee, Phys. Rep. 9C, 1 (1973) and references therein contained.

References for chapter I.

Berezin - The method of second quantization. Academic Press N.Y. (1966)

J. Schwinger, Proceedings of the National Academy of Sciences 37,452 (1951)

A. Vassiliev - Cargèse lectures 239 (1970). D. Bessis ed. Gordon and Breach

Proofs, similar to those given in this chapter have also been given by A. A. Slavnov (unpublished).

II. RENORMALIZATION OF GLOBAL SYMMETRIES

1. Construction of the symmetric theory

We shall now consider a set of fields ϕ_i transforming under a real representation of a compact group G. The infinitesimal form of the transformation is:

$$\delta\varphi_i(x) = t^\alpha_{ij}\,\omega_\alpha\,\varphi_j(x) \qquad \text{with} \quad t^\alpha_{ij} = -t^\alpha_{ji} \tag{1}$$

where the ω_α parametrize the transformation.

Let us consider a lagrangian $\mathcal{L}$ invariant under G. The action $S(\varphi)$, integral of the lagrangian density $\mathcal{L}$, satisfies therefore:

$$\int dx\, \frac{\delta S(\varphi)}{\delta\varphi_i(x)}\, t^\alpha_{ij}\,\varphi_j(x) = 0 \tag{2}$$

The generating functional of the complete Green's functions $Z(J)$, corresponding to lagrangian $\mathcal{L}$ is given by:

$$Z(J) = \int [d\varphi] \exp i \int dx\, [\mathcal{L}(\varphi) + J_i(x)\varphi_i(x)] \tag{3}$$

Let us perform in integral (3) the infinitesimal change of variables:

$$\varphi_i = \varphi'_i + t^\alpha_{ij}\,\omega_\alpha\,\varphi'_j \tag{4}$$

As a consequence of the invariance of the lagrangian $\mathcal{L}$ the functional $Z(J)$ satisfies:

$$Z(J_j) = Z(J_j + J_i t^\alpha_{ij}\,\omega_\alpha)$$

which yields the equation:

$$\int dx\, J_i(x)\, t^\alpha_{ij} \frac{\delta}{\delta J_j(x)} Z(J) = 0 \tag{5}$$

The generating functional of the connected Green functions $W(J)$ given in terms of $Z(J)$ by:

$$W(J) = i \ln Z(J) \tag{6}$$

satisfies the same equation.

The renormalization conditions are expressed on the generating functional of the one-particle irreducible (1PI) Green's functions $\Gamma(\chi)$ which is the Legendre transform of $W(J)$:

$$\begin{cases} \Gamma(\chi) + W(J) + \int J_i(x) \chi_i(x)\, dx = 0 \\ \chi_i(x) = -\dfrac{\partial W}{\partial J_i(x)} \end{cases} \tag{7}$$

It satisfies the equation:

$$\int dx \frac{\delta \Gamma}{\delta \chi_i(x)} t_{ij}^{\alpha} \chi_j(x) = 0 \tag{8}$$

The action $S(\phi)$ and the functional $\Gamma(\chi)$ satisfy the same equation. Equation (8) is actually only valid if the lagrangian $\mathcal{L}(\phi)$ has been regularized, to make finite the perturbation series, in a symmetric way, i. e. in such a way that the regularized action satisfies also equation (2). This is always possible for a renormalizable lagrangian invariant under a global symmetry. The ω_α being constants independent of the space time point x, $\partial_\mu \phi_i(x)$ transforms as $\phi_i(x)$. Therefore it is always possible to replace the term $\partial_\mu \phi_i(x)^2$ in the lagrangian by a regularized term:

$$-\int dx \left[\partial_\mu \varphi_i(x) \right]^2 \longrightarrow \int dx\, \varphi_i(x) \left[\Box + \alpha \frac{\Box^2}{\Lambda^2} + \beta \frac{\Box^3}{\Lambda^4} + \cdots \right] \varphi_i(x)$$

which will make the theory finite without breaking the symmetry. This argument would also apply if we would consider fermions instead of bosons as we do here for simplicity.

The problem is now to construct a renormalized symmetric theory by adjusting the coefficients in the lagrangian, or by adding counter-terms, in such a way that the perturbation series has a finite limit when the cut-off becomes infinite.

We shall consider the loopwise expansion of $\Gamma(\chi)$:

$$\Gamma(\chi) = \sum_{n=0}^{\infty} \Gamma_n(\chi) \tag{9}$$

It is well known that the tree approximation of $\Gamma(\phi)$, $\Gamma_0(\phi)$ is simply the action:

$$\Gamma_0(\chi) = S(\varphi) \tag{10}$$

Each term in the loop expansion satisfies equation (8) and in particular $\Gamma_1(\phi)$. In the large Λ limit, $\Gamma_1(\phi)$ has an infinite part which is a local polynomial in χ of canonical dimension 4. Let us call $-\delta S_1(\chi)$ this infinite part. It clearly satisfies also equation (8). If we add to the action $S(\chi)$, $\delta S_1(\chi)$, then the loop expansion will be finite up to order 1. Furthermore the new action $S_1(\chi)$:

$$S_1(\chi) = S(\chi) + \delta S_1(\chi) \tag{11}$$

will still be a renormalizable action satisfying equation (2).

This argument can easily be generalized, in order to prove by induction that one can construct a renormalized symmetric theory, by adding to $S(\chi)$ a set of symmetric counter terms. Indeed assume that we have added to $S(\phi)$ the counter terms which make the theory finite up to order n and that $S_n(\phi)$ still satisfies equation (2). Then the divergencies of $\Gamma_{n+1}(\phi)$ will only come from the superficial divergencies of the diagrams, all divergent subgraphs being already renormalized Therefore the divergent part $-\delta S_{n+1}(\chi)$ of $\Gamma_{n+1}(\phi)$ will be a local polynomial of dimension 4. Because $S_n(\phi)$ still satisfies equation (2) $\delta S_{n+1}(\chi)$ will satisfy equation also and so will the renormalized action to order (n+1) $S_{n+1}(\phi) = S_n(\phi) + \delta S_{n+1}(\phi)$ which makes $\Gamma_{n+1}(\chi)$ finite. The resulting action will have the form of the most general local polynomial of dimension 4 satisfying equation (2).

2. Linear symmetry breaking

We shall break the symmetry of the lagrangian, by adding to the lagrangian a term linear in the field. The new action is:

$$S(\varphi) = S_{sym}(\varphi) + \int dx\, C_i\, \varphi_i(x) \tag{12}$$

In these conditions the fields ϕ_i develop a non zero expectation value v_i and one has to translate ϕ_i:

$$\varphi_i = \varphi_i' + v_i \quad , \quad \langle \varphi_i' \rangle = 0 \tag{13}$$

The new generating functional is given by:

$$Z(J) = \int [d\varphi] \exp i \int [\mathcal{L}_{sym}(\varphi_i + v_i) + (J_i(x) + C_i)\, \varphi_i(x)]\, dx \tag{14}$$

where v_i is given by:

$$v_i = - \frac{\partial W_{sym}}{\partial J_i} \Big|_{J_i = C_i} \tag{15}$$

The new generating functional W(J) is then:

$$W(J) = W_{sym}(J+C) - W_{sym}(C) + v_i \int J_i(x)\, dx \tag{16}$$

and the 1PI functional $\Gamma(\varphi)$:

$$\Gamma(\chi) = \Gamma_{sym}(v_i + \chi_i) + C_i \int \chi_i(x)\, dx - \Gamma_{sym}(v_i) \tag{17}$$

with v_i given by

$$C_i = - \frac{\delta \Gamma_{sym}(v_i)}{\delta \chi} \tag{18}$$

Two remarks are in order:

The generating functional of the non symmetric theory can be entirely expressed, in a finite way, in terms of the symmetric functional. Therefore if we have taken for $S_{sym}(\phi)$ the renormalized action with all the counter terms, $\Gamma(\phi)$ is finite. Furthermore it satisfies the analog of equation (8):

$$\int dx \left[\frac{\delta \Gamma}{\delta \chi_i(x)} - C_i \right] t^{\alpha}_{ij} \left[\chi_j(x) + v_j \right] = 0 \tag{19}$$

In the same way as equation (8) implied reciprocally the symmetry of the lagrangian, equation (19) implies that the Green's functions are constructed from a lagrangian containing a linear term breaking the symmetry. Equation (19) contains all the Ward-Takahashi identities of the broken symmetry and can be used to renormalize the theory.

Spontaneous linear symmetry breaking

If the symmetric lagrangian $\mathcal{L}_{sym}$ is such that the equation:

$$\frac{\delta \mathcal{L}}{\delta \varphi_i(x)} \Big|_{\varphi_i(x) = v_i} = 0 \tag{20}$$

has a non vanishing solution, then the field ϕ_i acquires a spontaneous expectation value, and the symmetry is spontaneously broken. This occurs if the mass term in $\mathcal{L}$ is negative. This situation can be considered as a non trivial limit of the linear symmetry breaking when C_i goes to zero.

The 1PI generating functional satisfies then the limit of

equation (19):

$$\int dx \frac{\delta \Gamma}{\delta \chi(x)} A^{\alpha}_{ij} [\chi_j(x) + v_j] = 0 \tag{20}$$

In particular if we take the derivative of this equation with respect to χ and set $\chi = 0$ we obtain:

$$\int dx \frac{\delta^2 \Gamma}{\delta \chi_i(x) \delta \chi_k(x)} t^{\alpha}_{ij} v_j = 0 \tag{21}$$

If we call $\Gamma^{(2)}_{ij}$ (p) the inverse propagator in momentum space we have:

$$\Gamma^{(2)}_{ki}(p=0) t^{\alpha}_{ij} v_j = 0 \tag{22}$$

which is the expression of the Goldstone theorem in this framework. Indeed $\Gamma^{(2)}_{ij}$ (p=0) has as many eigenvectors corresponding to the eigenvalue zero, as there exist linearly independent vectors of the form $t^{\alpha}_{ij} v_j$. If the Lie algebra of the group G has N generators, and the little group of the vector v_i has m infinitesimal generators, $\Gamma^{(2)}_{ij}$ (p=0) has N - m eigenvectors associated to the eigenvalue zero, which correspond to massless Goldstone particles.

Large momenta behaviour of the Green's functions

Equation (19) shows directly, through Weinberg's theorem, that the 1.P.I. Green functions $\Gamma^{(n)}(p_i)$ are symmetric for large momenta. It is even possible, using equation (19) combined with Wilson's short distance expansion to characterize the subleading terms of the $\Gamma^{(2)}(p)$ The fact that the $\Gamma^{(n)}(p_i)$ are symmetric in this limit illustrates the fact that the breaking of symmetry is "soft".

3. Quadratic symmetry breaking

Let us now study the effect of adding to the symmetric lagrangian a term quadratic in the fields which breaks the symmetry:

$$\mathcal{S}(\varphi) = \mathcal{S}_{sym}(\varphi) - \frac{1}{2} C_{ij} \int dx\, \varphi_i(x) \varphi_j(x) \tag{23}$$

If one expands the new generating functional Z(J) in terms of C_{ij}, the term linear in C_{ij} will consist in symmetric Green's functions with an insertion of $\int dx\, \phi_i(x) \phi_j(x)$, which are not finite. Therefore certainly new renormalizations are needed. Let us write the Ward identities of this broken symmetry by expressing that the integral giving Z(J) is invariant through the infinitesimal change of variables (4):

$$\varphi_i = \varphi_i' + t_{ij}^{\alpha} \omega_{\alpha} \varphi_j'$$

We obtain:

$$0 = \int [d\varphi] \, \delta [S_{sym}(\varphi) - \tfrac{1}{2} C_{ij} \int dx \, \varphi_i(x) \varphi_j(x) + \int J_i(x) \varphi_i(x) \, dx]$$

$$\exp i [S(\varphi) + \int J_i(x) \varphi_i(x) \, dx]$$

which gives an equation for Z(J):

$$\int dx \, [J_i(x) t_{ij}^{\alpha} \frac{\delta}{\delta J_i(x)} + C_{ij} t_{ik}^{\alpha} \frac{\delta}{\delta J_k(x)} \frac{\delta}{\delta J_i(x)}] Z(J) = 0 \qquad (24)$$

The new feature of this W-T. identity is that it is not a relation between Green's functions, but between Green's functions and vertex functions of the vertex $\int dx \ \phi_i(x) \ \phi_j(x)$. One cannot therefore avoid the introduction of an arbitrary number of insertions of this product.

a) <u>Insertion of the vertex $\phi_i \ \phi_j$.</u>

In order to do this we shall introduce a source term for $\phi_i \ \phi_j$ and consider:

$$Z(J,L) = \int [d\varphi] \exp i \int dx \, [\mathcal{L}_{sym}(\varphi) + J_i(x) \varphi_i(x) - \tfrac{1}{2} L_{ij}(x) \varphi_i(x) \varphi_j(x)] \qquad (25)$$

Now the W.T. identities for Z(J,L) takes the form of a <u>linear first order</u> differential equation:

$$\int dx \, [J_i(x) \, t_{ij}^{\alpha} \frac{\delta}{\delta J_j(x)} + 2 L_{ik}(x) t_{kj}^{\alpha} \frac{\delta}{\delta L_{ij}(x)}] Z(J,L) = 0 \qquad (26)$$

The same equation applies to W(J,L).

As before we have to write the equation satisfied by the 1PI generating functional $\Gamma(\chi,L)$. Remarking that:

$$\frac{\delta W}{\delta L} + \frac{\delta \Gamma}{\delta L} = 0 \tag{27}$$

we obtain an equation for Γ :

$$\frac{\delta \Gamma}{\delta \chi_i} t^{\alpha}_{ij} \chi_j - 2 L_{ik} t^{\alpha}_{kj} \frac{\delta \Gamma}{\delta L_{ij}} = 0 \tag{28}$$

Again this equation holds between regularized Green's functions. We write the loop expansion of $\Gamma(\chi,L)$ and study the large cut-off limit of the various terms of the expansion. Power counting arguments show that besides the divergencies already discussed new divergent terms appear of the form:

$$\Gamma_{div}(\chi,L) = A^{(1)}_{ij} L_{ij} + A^{(2)}_{i_1 j_1, i_2 j_2} L_{i_1 j_1} L_{i_2 j_2}$$

$$+ B^{(1)}_{ij,k} L_{ij} \varphi_k + B^{(2)}_{ij,kl} L_{ij} \varphi_k \varphi_l \tag{29}$$

They will be cancelled by adding order by order counter terms of the same form to the lagrangian. The divergent terms of Γ satisfy equation (28). Therefore the complete action with all the counter terms, will be the most general local polynomial of degree 4 in ϕ and L (L has dimension 2) which satisfy equation (28).

$$\int dx \left\{ \frac{\delta S(\varphi,L)}{\delta \varphi_i(x)} t^{\alpha}_{ij} \varphi_j(x) - 2 L_{ik}(x) t^{\alpha}_{kj} \frac{\delta S(\varphi,L)}{\delta L_{ij}(x)} \right\} = 0 \tag{30}$$

which means that $S(\phi,L)$ is invariant under the transformation:

$$\begin{cases} \delta \varphi_i = t^{\alpha}_{ij} \varphi_j \omega_\alpha \\ \delta L_{ij} = [L_{ik} t^{\alpha}_{kj} + L_{jk} t^{\alpha}_{ik}] \omega_\alpha \end{cases} \tag{31}$$

In this form, it is clear that if $S(\phi,L)$ satisfies equation (29), $\Gamma(\chi,L)$ will still satisfy equation (28).

We are now in position to discuss the quadratic symmetry breaking.

b) Quadratic symmetry breaking

If we now give to $L_{ij}(x)$ a constant value C_{ij}, the action $S(\phi, L_{ij} = C_{ij})$ will be the action of a symmetric theory with quadratic symmetry breaking.

We see that, if the symmetry does not forbid terms of the form $L_{ij}\ \phi_k$, the quadratic breaking term induces also a linear breaking of the symmetry. This is a general feature: A breaking term of a given degree induces in general all possible symmetry breaking terms of lower degree.

Furthermore the W.T. identities of the broken symmetry take now the form of a coupled system of equations between Green's functions and functions with an arbitrary number of $\int dx\ \phi_i(x)\ \phi_j(x)$ insertions. For the renormalization procedure, only the insertions of one or two such fields have to be considered:

$$\int dx \left\{ \frac{\delta \Gamma(\chi, L)}{\delta \chi_i(x)} t^\alpha_{ij} \chi_j(x) - 2 C_{ik} t^\alpha_{kj} \frac{\delta \Gamma(\chi, L)}{\delta L_{ij}(x)} \right\}\bigg|_{L=C} = 0 \tag{31a}$$

$$\int dy \frac{\delta}{\delta L_{mn}(y)} \int dx \left\{ \frac{\delta \Gamma(\chi, L)}{\delta \chi_i(x)} t^\alpha_{ij} \chi_j(x) - 2 L_{ik}(x) t^\alpha_{kj} \frac{\delta \Gamma(\chi, L)}{\delta L_{ij}(x)} \right\}\bigg|_{L=C} = 0 \tag{31b}$$

From this system one can derive all the renormalization conditions. These results can be easily generalized to breaking terms of dimension 3.

The rule is to introduce source terms for all the composite fields breaking the symmetry, and write the linear first order differential equation which expresses the symmetry. Then it is easy to show that the renormalized lagrangian with all its counter terms will be the most general local polynomial of dimension 4 in the field and in the sources for the composite fields (the dimension of the sources is $4 - \delta$, where δ is the dimension of the composite field). It is not very useful to consider a breaking term of dimension 4. The corresponding source has dimension zero and therefore the complete lagrangian will contain monomials of arbitrary degree in this source. The final theory in this case will simply be characterized by the symmetry group G' of the complete interaction, consisting of the symmetric part and the breaking terms.

REFERENCES

References for chapter II.

The generating functional for 1.P.I. Green's functions (proper vertices) and its construction by Legendre transform have been introduced in:

C. De Dominicis, J.Math. Phys. 4, 255 (1963)
C. De Dominicis and P. C. Martin, J. Math. Phys. 5, 14, 31 (1964)
G. Jona-Lasinio, Nuov. Cim. 34, 1790 (1964)

The W.T. identities have been discussed in:

B. W. Lee, Chiral Dynamics, Document in modern Physics, Gordon and Breach NY (1972) and references therein contained.

K. Symanzik, Renormalization of theories with broken symmetry, Cargèse lectures (1970) D. Bessis ed. Gordon and Breach.

III. GAUGE SYMMETRIES

Up to now we have only considered global symmetries, i.e. symmetries whose parameters are independent of the space-time variable. We shall now study the renormalization of local symmetries. In discussing these symmetries we shall use highly condensed notations. For this reason we shall call the set of all fields A_i. The index i stands for all attributes of the fields. For example A_i will contain the gauge field $B_\mu^\alpha(x)$ and i stands for the point x, the Lorentz μ and the group index α. A_i denotes both the gauge and matter fields.

The infinitesimal local gauge transformation may be written as:

$$\delta A_i = (\Lambda_j^\alpha + t_{ij}^\alpha A_j)\,\omega_\alpha \tag{1}$$

where $\omega_\alpha = \omega_\alpha(x)$ is a space-time dependent parameter of a compact Lie group G. The inhomogeneous term Λ_i^α in (1) is of the form

$$\Lambda_i^\alpha = \left[\frac{1}{g}\right]^{\alpha\beta} \partial_\mu \delta(x-y) \quad \text{for } A_i = B_\mu^\alpha(y) \tag{2}$$

where $g_{\alpha\beta} = g_\alpha \, \delta_{\alpha\beta}$ is the coupling constant matrix, and is a constant for scalar fields.

As in the previous chapter t_{ij}^α is real antisymmetric. The group structure implies the identity:

$$t_{ik}^\alpha (t_{kj}^\beta A_j + \Lambda_k^\beta) - t_{ik}^\beta (t_{kj}^\alpha A_j + \Lambda_k^\alpha) = f_{\alpha\beta\gamma}(t_{ij}^\gamma A_j + \Lambda_i^\gamma) \tag{3}$$

where $f_{\alpha\beta\gamma}$ are the completely antisymmetric structure constants of the Lie algebra of the group G.

Introducing the notation:

$$D_i^\alpha(A) = \Lambda_i^\alpha + t_{ij}^\alpha A_j \tag{4}$$

equation (3) can be rewritten:

$$\frac{\partial D_i^\alpha}{\partial A_k} D_k^\beta - \frac{\partial D_i^\beta}{\partial A_k} D_k^\alpha = f_{\alpha\beta\gamma} D_i^\gamma \tag{5}$$

A lagrangian $\mathcal{L}$ invariant under a gauge transformation (1), will satisfy the equation:

$$D_i^\alpha (A) \frac{\delta \mathcal{L}}{\delta A_i} = 0 \tag{6}$$

1. Quantization

The quantization of a gauge invariant lagrangian is performed by introducing a gauge fixing function $F_\alpha(A)$. The Green's functions of the gauge theory can be derived from the functional integral:

$$Z(J) = \int \det(M) [dA_i] \exp i [\mathcal{L}(A) - \tfrac{1}{2} F_\alpha(A)^2 + J_i A_i] \tag{7}$$

where $M_{\alpha\beta}$ is the variation of $F_\alpha(A)$ with respect to a gauge transformation:

$$M_{\alpha\beta}(A) = \frac{\delta F_\alpha(A)}{\delta A_i} D_i^\beta (A) \tag{8}$$

On equation (7) it is not completely obvious that the theory, constructed in this way, is renormalizable when $\mathcal{L}(A) - \frac{1}{2} F_\alpha(A)^2$ is of dimension 4, because det M generates a non local interaction. It is therefore convenient to use the identity:

$$\det(M) = \int [dC][d\bar{C}] \exp i \bar{C}_\alpha M_{\alpha\beta} C_\beta \tag{9}$$

where C_α and $\bar{C}_\beta$ are anticommuting fields corresponding to scalar fermions, and called generally Faddeev-Popov ghosts. The functional Z(J) can then be written in a way which shows that the theory is renormalizable by simple power counting:

$$Z(J) = \int [dA_i][d\bar{C}][dC] \exp i [\mathcal{L}(A, C, \bar{C}) + J_i A_i] \tag{10}$$

with

$$\mathcal{L}(A, C, \bar{C}) = \mathcal{L}(A) - \tfrac{1}{2} F_\alpha^2(A) + \bar{C}_\alpha M_{\alpha\beta} C_\beta \tag{11}$$

Notice that $M_{\alpha\beta}$ is in general not hermitian so that the ghost-line is orientable.

2. The Ward-Takahashi identities

The usual derivation of the W.T. identitites uses a fundamental property of the measure $\det M [dA_i]$. It is an invariant measure for the non linear gauge transformation of type (1) but with ω_α function of A:

$$\omega_\alpha = M^{(-1)}_{\alpha\beta}(A)\,\lambda_\beta \tag{12}$$

This transformation has the property of translating the gauge fixed term:

$$\delta[F_\alpha(A)] = \lambda_\alpha \tag{13}$$

Performing on integral (7) an infinitesimal change of variables of form (12) and expressing that the result is independent of λ_α yield the W.T. identities for Z(J):

$$\left\{ F_\alpha\left[\frac{1}{i}\frac{\delta}{\delta J}\right] - J_i D_i^\beta\left(\frac{1}{i}\frac{\delta}{\delta J}\right) M^{(-1)}_{\beta\alpha}\left(\frac{1}{i}\frac{\delta}{\delta J}\right)\right\} Z(J) = 0 \tag{14}$$

It is convenient to define:

$$Z_{\alpha\beta}(J) = M^{(-1)}_{\alpha\beta}\left(\frac{1}{i}\frac{\delta}{\delta J}\right) Z(J) \tag{15}$$

$Z_{\alpha\beta}(J)$ is the ghost propagator in presence of external vector fields.

Then the W.T. can be rewritten:

$$F_\alpha\left[\frac{1}{i}\frac{\delta}{\delta J}\right] Z(J) - J_i D_i^\beta\left[\frac{1}{i}\frac{\delta}{\delta J}\right] Z_{\beta\alpha}(J) = 0 \tag{16a}$$

$$M_{\alpha\gamma}\left(\frac{1}{i}\frac{\delta}{\delta J}\right) Z_{\gamma\beta}(J) = \delta_{\alpha\beta}\, Z(J) \tag{16b}$$

The second equation is simply the equation of motion of the $\bar{C}$ ghost field. The discussion of the renormalization of gauge theories is based on these two equations. After the application of an invariant regularization to the theory (for example the dimensional regularization), these equations imply the form of the counter-terms. Actually the renormalization conditions apply to the 1PI Green's functions. Unfortunately it is not easy to derive from these equations, W.T. identitites for the 1PI generating functional. So in a first step, the identities (16) have been written for all the Green's functions which are superficially divergent, and decomposed "by hand" in equations for 1PI Green's functions. More recently Lee has been able to transform directly equations (16) in equations for 1PI generating functional in the case of linear gauge fixing functions $F_\alpha(A)$. Using a method introduced by Becchi, Rouet and Stora we shall give a simpler derivation of these identities first in the case of linear gauges, which can be then easily generalized to quadratic gauges, and write these identities in a form which simplifies the discussion of the renormalization.

3. Renormalization of gauge theories with linear gauge functions

We shall first consider the case of linear gauge functions:

$$F_\alpha(A) = F_{\alpha i} A_i \tag{17}$$

Power counting would allow us to use also gauge fixing functions quadratic in the fields. Therefore in section (4) we shall generalize the method to this case.

a) Other derivation of the W.T. identities

The arguments which lead to the W.T. identities (16) were purely geometrical and therefore did not involve the ghost fields which only appear in the interpretation of these identities in terms of ordinary Green's functions. But from the point of view of the renormalization theory, the ghost fields have to be considered exactly on the same level as the gauge and matter fields. Therefore a question arises: is it possible to derive the W.T. identities as consequence of a symmetry of the effective lagrangian $\mathcal{L}(A,C,\bar{C})$ introduced in formula (11)? Surprisingly enough such a symmetry exists. It is a global symmetry (the parameters do not depend on space time) of anticommuting type which we shall describe below and which is actually suggested by examining the equation (16a). It is defined by:

$$\delta A_i = D_i^\alpha C_\alpha \delta\lambda \tag{18a}$$

$$\delta C_\alpha = -\frac{1}{2} f_{\alpha\beta\gamma} C_\beta C_\gamma \delta\lambda \tag{18b}$$

$$\delta \bar{C}_\alpha = - F_\alpha(A) \delta\lambda \tag{18c}$$

where $\delta\lambda$ is an anticommuting constant.

First $\mathcal{L}(A)$ is clearly invariant because (18a) is a special gauge transformation. Let us now show:

$$\delta [D_i^\alpha C_\alpha] = 0 \tag{19}$$

Indeed the variation of $D_i^\alpha C_\alpha$ reads:

$$\delta [D_i^\alpha C_\alpha] = \frac{\delta D_i^\alpha}{\delta A_j} D_j^\beta C_\beta \, \delta\lambda \, C_\alpha - \frac{1}{2} D_i^\gamma f_{\gamma\alpha\beta} C_\alpha C_\beta \, \delta\lambda$$

C_α and C_β anticommute, the coefficient of $C_\alpha C_\beta$ can therefore be antisymmetrized in $(\alpha\,\beta)$, and one then sees that it vanishes as a consequence of equation (5).

Equation (19) also implies that $M_{\alpha\beta}\, C_\beta$ is invariant:

$$M_{\alpha\beta} C_\beta = \frac{\delta F_\alpha}{\delta A_i} D_i^\beta C_\beta$$

therefore we have:

$$\delta [M_{\alpha\beta} C_\beta] = \delta \left[\frac{\delta F_\alpha}{\delta A_i} \right] D_i^\beta C_\beta$$

we compute now the variation of $\frac{\delta F_\alpha}{\delta A_i}$:

$$\delta \left[\frac{\delta F_\alpha}{\delta A_i} \right] = \frac{\delta^2 F_\alpha}{\delta A_i \, \delta A_j} D_j^\gamma C_\gamma \, \delta\lambda$$

This leads to the result:

$$\delta [M_{\alpha\beta} C_\beta] = - \frac{\delta^2 F_\alpha}{\delta A_i \, \delta A_j} D_i^\beta D_j^\gamma C_\beta C_\gamma \, \delta\lambda \tag{20}$$

The coefficient of $C_\beta\, C_\gamma$ is symmetric in β and γ and therefore the right handside vanishes:

$$\delta [M_{\alpha\beta} C_\beta] = 0 \tag{21}$$

Let us now compute the variation of the non gauge invariant part of the effective lagrangian:

$$\delta [\bar{C}_\alpha M_{\alpha\beta} C_\beta - \frac{1}{2} F_\alpha^2(A)] = F_\alpha(A) M_{\alpha\beta} C_\beta \, \delta\lambda - F_\alpha(A) \frac{\delta F_\alpha}{\delta A_i} D_i^\beta C_\beta \, \delta\lambda$$

The r.h.s. vanishes by definition of $M_{\alpha\beta}$.

Therefore the complete lagrangian is invariant:

$$\delta[\mathcal{L}(A,C,\bar{C})] = 0 \tag{22}$$

It is easy also to verify that the measure $dA_i\ dC\ d\bar{C}$ is an invariant measure for this transformation.

Let us now first show that this invariance property implies equation (16a). Let us consider the integral:

$$0 = \int [dA_i][dC][d\bar{C}]\ \bar{C}_\alpha \exp i[\mathcal{L}(A,C,\bar{C}) + J_i A_i]$$

Let us now perform a change of variables corresponding to a transformation (18) and let us write the coefficient of $\delta\lambda$:

$$\int [dA_i][dC][d\bar{C}] \{-F_\alpha(A) + \bar{C}_\alpha J_i D_i^\beta C_\beta\} \exp i[\mathcal{L}(A,C,\bar{C}) + J_i A_i] = 0 \tag{23}$$

This is exactly equation (16a). But to consider only this equation is not consistent with the ideas that we have exposed in chapter II. Indeed we tried to argue that one should introduce sources for all fields which appear in the W.T. identities. Therefore we shall introduce sources for the fields C_α and $\bar{C}_\alpha$ as well as for the composite fields $D_i^\alpha C_\alpha$ and $f_{\alpha\beta\gamma}\ C_\beta\ C_\gamma$ which will now be generated by the transformation (18). In general one has also to add a source for the field $F_\alpha(A)$ except if $F_\alpha(A)$ is linear in the field A_i. For simplicity we shall consider first the latter case. We shall define:

$$F_\alpha(A) = F_{\alpha i} A_i \tag{24}$$

Let us therefore consider the generating functional:

$$\begin{aligned} Z(J,\eta,\bar{\eta},K,L) = \int [dA][dC][d\bar{C}] \exp i[\mathcal{L}(A,C,\bar{C}) + J_i A_i + \bar{\eta}_\alpha C_\alpha + \\ + \bar{C}_\alpha \eta_\alpha + K_i D_i^\alpha(A) C_\alpha - \tfrac{1}{2} L_\alpha f_{\alpha\beta\gamma} C_\beta C_\gamma \end{aligned} \tag{25}$$

In this expression, η_α, $\bar{\eta}_\alpha$ and K_i are anticommuting sources. Thanks to the invariance of $\mathcal{L}(A,C,\bar{C})$ with respect to the transformations (18) we can now write generalized W.T. identities for this functional. We perform a change of variables corresponding to a transformation (18) and write as usual that the integral is invariant. We first remark that:

$$\delta[f_{\alpha\beta\gamma} C_\alpha C_\gamma] = 0 \tag{26}$$

as a consequence of the Jacobi identity for $f_{\alpha\beta\gamma}$. We therefore obtain:

$$\left[J_i \frac{\delta}{\delta K_i} + \bar{\eta}_\alpha \frac{\delta}{\delta L_\alpha} + \eta_\alpha F_\alpha \left(\frac{\delta}{\delta J}\right)\right] Z(J, \eta, \bar{\eta}, K, L) = 0 \tag{27}$$

As before, together with the W.T. identities, we shall use the equation of motion of the $\bar{C}$ field, obtained by the change of variable $\bar{C}$ in $\bar{C} + \delta\bar{C}$ in the integral:

$$\left[M_{\alpha\beta}\left(\frac{1}{i}\frac{\delta}{\delta J}\right) C_\beta + \eta_\alpha \right] Z(J, \eta, \bar{\eta}, K, L) = 0 \tag{28}$$

But remarking that:

$$M_{\alpha\beta} C_\beta = F_{\alpha i} D_i C_\beta \tag{29}$$

we can rewrite equation (28):

$$\left[\frac{1}{i} F_{\alpha i} \frac{\delta}{\delta K_i} + \eta_\alpha \right] Z(J, \eta, \bar{\eta}, K, L) = 0 \tag{30}$$

Equations (27) and (30) are now first order linear differential equations. It is very easy to write the corresponding identities for the generating functional of connected Green's functions and then for the 1.P.I. generating functional.

b) But before doing this, let us make <u>a few remarks:</u>

Let us introduce the notation:

$$S(A, C, \bar{C}, K, L) = \mathcal{L}(A, C, \bar{C}) + K_i D_i^\alpha(A) C_\alpha - \frac{1}{2} L_\alpha f_{\alpha\beta\gamma} C_\beta C_\gamma \tag{31}$$

The invariance of S with respect to the symmetry can be expressed by the equation:

$$D_i^\alpha(A) C_\alpha \frac{\delta S}{\delta A_i} - \frac{1}{2} f_{\alpha\beta\gamma} C_\beta C_\gamma \frac{\delta S}{\delta C_\alpha} - F_\alpha(A) \frac{\delta S}{\delta \bar{C}_\alpha} = 0 \tag{32}$$

But by definition we have:

$$D_i^\alpha(A) C_\alpha = \frac{\delta S}{\delta K_i} \tag{33}$$

$$-\frac{1}{2} f_{\alpha\beta\gamma} C_\beta C_\gamma = \frac{\delta S}{\delta L_\alpha} \tag{34}$$

Therefore equation (32) can be rewritten:

$$\frac{\delta S}{\delta K_i}\frac{\delta S}{\delta A_i} + \frac{\delta S}{\delta L_\alpha}\frac{\delta S}{\delta C_\alpha} - F_\alpha(A)\frac{\delta S}{\delta \bar{C}_\alpha} = 0 \tag{35}$$

Furthermore one has the identity:

$$\frac{\delta S}{\delta \bar{C}_\alpha} + F_{\alpha i}\frac{\delta S}{\delta K_i} = 0 \tag{36}$$

It is clear that equation (36) implies equation (30). Indeed from the ghost field equation we have:

$$\int [dA_i][dC][d\bar{C}]\left[\frac{\delta S}{\delta \bar{C}_\alpha} - \eta_\alpha\right] \exp i\left[S + J_i A_i + \bar{\eta}_\alpha C_\alpha + \bar{C}_\alpha \eta_\alpha\right] \tag{37}$$

which together with equation (36) implies:

$$\int [dA_i][dC][d\bar{C}]\left[F_{\alpha i}\frac{\delta S}{\delta K_i} + \eta_\alpha\right] \exp i\left[S + J_i A_i + \bar{\eta}_\alpha C_\alpha + \bar{C}_\alpha \eta_\alpha\right] = 0 \tag{38}$$

which is an equivalent way of writing equation (30). Let us now show what we can derive from equation (35). We shall perform a change of variables corresponding to the transformation:

$$\delta A_i = \frac{\delta S}{\delta K_i}\delta\lambda \tag{39a}$$

$$\delta C_\alpha = \frac{\delta S}{\delta L_\alpha}\delta\lambda \tag{39b}$$

$$\delta \bar{C}_\alpha = -F_\alpha(A)\delta\lambda \tag{39c}$$

Equation (35) tells us that Z is invariant by such a change of variables, provided the measure is also invariant which requires:

$$\left\{\begin{aligned} &\frac{\delta^2 S}{\delta C_\alpha \delta L_\alpha} = 0 \\ &\frac{\delta^2 S}{\delta A_i \delta K_i} = 0 \end{aligned}\right. \tag{40}$$

This change of variables leads to the equation:

$$\int \left[\frac{\delta S}{\delta K_i} J_i + \frac{\delta S}{\delta L_\alpha}\eta_\alpha + F_\alpha(A)\eta_\alpha\right] \exp i\left[S + J_i A_i + \bar{\eta}_\alpha C_\alpha + \bar{C}_\alpha \eta_\alpha\right] = 0 \tag{41}$$

which yields immediately equation (27).

c) W. T. identities for the 1.P.I. generating functional

Let us first introduce the generating functional W for the connected Green's functions:

$$W = i \ln Z \tag{42}$$

W satisfies equation (27). Equation (30) becomes:

$$F_{\alpha i} \frac{\delta W}{\delta K_i} = \eta_\alpha \tag{43}$$

We now introduce the generating functional of the 1.P.I. Green's functions:

$$W(J,\eta,\bar\eta,K,L) + \Gamma(A,C,\bar C,K,L) + J_i A_i + \bar\eta_\alpha C_\alpha + \bar C_\alpha \eta_\alpha = 0 \tag{44}$$

with

$$\left\{\begin{array}{l} A_i = -\dfrac{\delta W}{\delta J_i} \\[2ex] C_\alpha = \dfrac{\delta W}{\delta \bar\eta_\alpha} \\[2ex] \bar C_\alpha = -\dfrac{\delta W}{\delta \eta_\alpha} \end{array}\right. \Longleftrightarrow \left\{\begin{array}{l} J_i = -\dfrac{\delta \Gamma}{\delta A_i} \\[2ex] \bar\eta_\alpha = -\dfrac{\delta \Gamma}{\delta C_\alpha} \\[2ex] \eta_\alpha = \dfrac{\delta \Gamma}{\delta \bar C_\alpha} \end{array}\right. \tag{45}$$

Remarking that if W and Γ depend on parameters Q, not involved in the Legendre transformations, like K and L in this case, they satisfy:

$$\frac{\delta W}{\delta Q} + \frac{\delta \Gamma}{\delta Q} = 0 \tag{46}$$

We can derive from the equations satisfied by W, the two equations for Γ:

$$\frac{\delta \Gamma}{\delta A_i}\frac{\delta \Gamma}{\delta K_i} + \frac{\delta \Gamma}{\delta C_\alpha}\frac{\delta \Gamma}{\delta L_\alpha} - F_\alpha(A)\frac{\delta \Gamma}{\delta \bar C_\alpha} = 0 \tag{47}$$

$$\Gamma_{\alpha i}\frac{\delta \Gamma}{\delta K_i} + \frac{\delta \Gamma}{\delta \bar C_\alpha} = 0 \tag{48}$$

We shall now discuss the renormalization of the gauge theories in terms of these two equations.

We have already seen that $S(A,C,\bar{C},K,L)$ satisfies these two equations. Moreover, and this is the important point, the corresponding equations for S imply, up to a trace condition coming from the measure, the equations (47) and (48). We are now again in a situation very similar to those considered previously in the case of ordinary global symmetries.

One can be surprised that equation (47) does not make any explicit reference to the original gauge symmetry, in the sense that the structure constants for example are not present. This comes from the fact that the transformation laws contain explicitly the coupling constant and will therefore be renormalized.

Before going to the proof of the renormalization, let us introduce an auxiliary functional $\tilde{\Gamma}$:

$$\tilde{\Gamma}(A,C,\bar{C},K,L) = \Gamma(A,C,\bar{C},K,L) + \frac{1}{2} F_\alpha(A)^2 \tag{49}$$

It is easy to see that $\tilde{\Gamma}$ satisfies the two equations:

$$\left\{\begin{array}{l} \dfrac{\delta\tilde{\Gamma}}{\delta A_i}\dfrac{\delta\tilde{\Gamma}}{\delta K_i} + \dfrac{\delta\tilde{\Gamma}}{\delta C_\alpha}\dfrac{\delta\tilde{\Gamma}}{\delta L_\alpha} = 0 \qquad (50) \\[2ex] F_{\alpha i}\dfrac{\delta\tilde{\Gamma}}{\delta K_i} + \dfrac{\delta\tilde{\Gamma}}{\delta \bar{C}_\alpha} = 0 \qquad (51) \end{array}\right.$$

d) <u>Renormalization</u>

Let us now briefly explain how the theory can be renormalized with the use of the equations (50) and (51). Let us expand loopwise $\tilde{\Gamma}$ as we did in the previous chapter:

$$\tilde{\Gamma} = \sum_{n=0}^{\infty} \tilde{\Gamma}_n \tag{52}$$

We have:

$$\tilde{\Gamma}_0(A,C,\bar{C},K,L) = \tilde{S}(A,C,\bar{C},K,L) \tag{53}$$

where we have introduced the notation:

$$\tilde{S} = S + \frac{1}{2} F_\alpha(A)^2 \tag{54}$$

Before writing the consequences of the W.T. identities for $\tilde{\Gamma}$ let us introduce the symbolic notation:

$$\frac{\delta\tilde{\Gamma}}{\delta A_i}\frac{\delta\tilde{\Gamma}}{\delta K_i}+\frac{\delta\tilde{\Gamma}}{\delta c_\alpha}\frac{\delta\tilde{\Gamma}}{\delta L_\alpha}=\tilde{\Gamma}*\tilde{\Gamma} \tag{55}$$

The first divergencies appear in the one-loop functional $\tilde{\Gamma}_1$. From equation (50,51), the divergent part $\tilde{\Gamma}_{1,div.}$ of $\tilde{\Gamma}_1$ satisfies:

$$F_{\alpha i}\frac{\delta\tilde{\Gamma}_{1,div.}}{\delta K_i}+\frac{\delta\tilde{\Gamma}_{1,div.}}{\delta\bar{c}_\alpha}=0 \tag{56}$$

$$\tilde{\Gamma}_{1,div.}*\tilde{S}+\tilde{S}*\tilde{\Gamma}_{1,div.}=0 \tag{57}$$

In order to obtain a finite one-loop functional, we have to add the counter term $-\tilde{\Gamma}_{1,div}$ to $\tilde{S}$:

$$\tilde{S}_1=\tilde{S}-\tilde{\Gamma}_{1,div.} \tag{58}$$

But equations (56) and (57) are exactly the equations which imply that one can choose the one loop counterterm in such a way that $\tilde{S}_1$ satisfies at the one-loop order equations (50) and (51). By adding then to $\tilde{S}_1$ higher terms, which do not modify the one-loop functional, one can construct a renormalized lagrangian which satisfies exactly equations (50) and (51) and leads to a finite one-loop functional $\tilde{\Gamma}_1$.

We shall now proceed by induction. We shall assume that we have constructed a renormalized $\tilde{S}_n$ which satisfies equations (50) and (51) and generates a finite perturbation series up to order n. Then the corresponding $\tilde{\Gamma}$ will still satisfy equations (50) and (51). $\tilde{\Gamma}_1$ up to $\tilde{\Gamma}_n$ will be finite and the divergencies of $\tilde{\Gamma}_{n+1}$ will only come from the over all divergencies of the diagrams, all the subdiagrams having already been renormalized. The divergent part of $\tilde{\Gamma}_{n+1}$ will therefore be a local polynomial of degree four in A,C,$\bar{C}$,K and L which satisfies the consequences of equations (50) and (51). At order n+1 these equations yield:

$$F_{\alpha i}\frac{\delta\tilde{\Gamma}_{n+1,div}}{\delta K_i}+\frac{\delta\tilde{\Gamma}_{n+1,div}}{\delta\bar{c}_\alpha}=0 \tag{59}$$

$$\tilde{\Gamma}_{n+1,div.} * \tilde{S} + \tilde{S} * \tilde{\Gamma}_{n+1,div.} = -\tilde{\Gamma}_n * \tilde{\Gamma}_1 - \tilde{\Gamma}_1 * \tilde{\Gamma}_n \tag{60}$$

$$- \tilde{\Gamma}_{n-1} * \tilde{\Gamma}_2 \quad \ldots \quad \text{-finite part } (\tilde{\Gamma}_{n+1} * S + S * \tilde{\Gamma}_{n+1})$$

In the r.h.s. of equation (60) all terms are finite by hypothesis, therefore the divergent part $\tilde{\Gamma}_{n+1,div}$ of $\tilde{\Gamma}_{n+1}$ satisfies:

$$\tilde{\Gamma}_{n+1,div.} * \tilde{S} + \tilde{S} * \Gamma_{n+1,div.} = 0 \tag{61}$$

But equations (59) and (61) tell us that the divergencies of Γ_{n+1} can be removed by adding to $\tilde{S}$ a counter-term such that the renormalized lagrangian $\tilde{S}$ at order n + 1 still satisfies the equations (50) and (51). Therefore the result is proved by induction. $\tilde{S}_{ren}$ is the most general local polynomial of degree 4 which satisfies equations (50) and (51).

2. The renormalized Lagrangian

The renormalized lagrangian $\tilde{S}_r$ (without the gauge fixing term) satisfies from the previous analysis:

$$\begin{cases} F_{\alpha i} \dfrac{\delta \tilde{S}_r}{\delta K_i} + \dfrac{\delta \tilde{S}_r}{\delta \bar{C}_\alpha} = 0 \\[2ex] \dfrac{\delta \tilde{S}_r}{\delta A_i} \dfrac{\delta \tilde{S}_r}{\delta K_i} + \dfrac{\delta \tilde{S}_r}{\delta C_\alpha} \dfrac{\delta \tilde{S}_r}{\delta L_\alpha} = 0 \end{cases}$$

By power counting the dimension of K_i and L_α is two, the dimension of A_i, C, $\bar{C}$ one. Furthermore ghost number conservation tells us that K_i has to go with a factor C_α, and L_α with a factor C_β C_γ. The requirement that $\tilde{S}_r$ should be of dimension four then allows us to write the decomposition:

$$\tilde{S}_r = D_i^{\alpha(r)}(A_i) K_i C_\alpha - \frac{1}{2} f^{(r)}_{\alpha\beta\gamma} L_\alpha C_\beta C_\gamma + \mathcal{L}^{(r)}(A, C, \bar{C}) \tag{62}$$

where by power counting $D^{(r)}$ has to be of dimension one and therefore linear in A, and $f^{(r)}_{\alpha\beta\gamma}$ has to be a constant.

If we now use the first equation, we obtain:

$$\tilde{S}_r = -\frac{1}{2} f^{(r)}_{\alpha\beta\gamma} L_\alpha C_\beta C_\gamma + D_i^{\alpha(r)}(A) K_i C_\alpha + \bar{C}_\alpha F_{\alpha i} D_i^{\beta(r)} C_\beta + \mathcal{L}^{(r)}(A) \quad (63)$$

We then use this decomposition in the second equation and identify the coefficients of L_α, K_i and the part function of A only:

$$f^{(r)}_{\rho\alpha\gamma} f^{(r)}_{\alpha\delta\varepsilon} C_\gamma C_\delta C_\varepsilon = 0 \quad (64a)$$

$$\left[\frac{\delta D_i^\alpha}{\delta A_j} D_j^\beta - \frac{1}{2} D_i^\gamma f^{(r)}_{\gamma\alpha\beta} \right] C_\alpha C_\beta = 0 \quad (64b)$$

$$D_i^{\alpha(r)}(A) \frac{\delta \mathcal{L}^{(r)}}{\delta A_i} = 0 \quad (64c)$$

The first equation, due to the anticommutation of the C_α, shows that the $f^{(r)}_{\alpha\beta\gamma}$ satisfy the Jacobi identity. The second equation shows that $D^{(r)}(A)$ satisfies the same equation as D(A) and the last one proves that $\mathcal{L}(A)$ is gauge invariant.

We have now still to show that the symmetry group and the representations are the same as in the original lagrangian. This will, in particular, insure that the measure is invariant under the transformations (39). The conditions (40) indeed give:

$$\begin{cases} f^{(r)}_{\alpha\alpha\beta} = 0 \\ \dfrac{\delta D_i^{\alpha(r)}}{\delta A_i} = 0 \end{cases} \quad (65)$$

which will be automatically satisfied if the symmetry group and the representations have not changed. The simple continuity argument,

$$\begin{cases} f^{(r)}_{\alpha\beta\gamma} = f_{\alpha\beta\gamma} + f^{(1)}_{\alpha\beta\gamma} + \dots \\ t^{\alpha(r)}_{ij} = t^\alpha_{ij} + \dots \end{cases} \quad (66)$$

is sufficient in the simplest cases to prove the result. But, if for example the symmetry group contains abelian factors, the continuity argument is not quite sufficient. In the case of a linear gauge fixing function one can, following Lee, give a different argument. One can first consider the so called "R-gauges", i.e. gauges in which the gauge fixing term $F_\alpha^2(A)$ does not break the global symmetry. Then the W.T. identities of the global symmetry, symmetry in which the group elements do not depend on space-time, tell us that $S^{(r)}$ has also the same global symmetry and therefore the group and the representations have not changed.

One then remarks that in a linear gauge fixing function $F_\alpha(A)$ the part which has dimension two comes only from $\partial_\mu B_\mu^a(x)$ and gives a globally symmetric term. One can therefore write:

$$F_\alpha(A)^2 = \frac{1}{\xi}\int [\partial_\mu B_\mu^a(x)]^2 dx + \Delta(A) \tag{67}$$

where $\Delta(A)$ contains only terms of dimension less than four. If we now first renormalize the lagrangian containing only the symmetric part of the gauge fixing term, and then treat $\Delta(A)$ as a perturbation, power counting will show that the counter-terms needed to renormalize an arbitrary number of insertions of $\Delta(A)$ are of dimension three at most. Therefore the part of dimension four of $S^{(r)}$ will not be modified and this part fixes completely the group structure. This completes the proof for all linear gauge fixing functions. Unfortunately this last part of the argument cannot be generalized to gauge fixing function containing a non symmetric quadratic part.

4. Quadratic gauge functions

Let us now briefly show the modification which appear in the case of gauge fixing function containing quadratic terms in the fields. $F_\alpha(A)$ is now a composite operator and we have to introduce a source R_α for this operator. We shall therefore consider the generating functional:

$$Z(J,\eta,\bar{\eta},K,L,R) = \int [dA][dC][d\bar{C}] \exp i\,[S(A,C,\bar{C},K,L,R) + J_i A_i + \bar{\eta}_\alpha C_\alpha + \bar{C}_\alpha \eta_\alpha] \tag{67}$$

where S is now defined by:

$$S = \mathcal{L}(A,C,\bar{C}) + K_i D_i^\alpha(A) C_\alpha - \frac{1}{2} L_\alpha f_{\alpha\beta\gamma} C_\beta C_\gamma - R_\alpha F_\alpha(A) \tag{68}$$

We remark that under the transformation (18) $F_\alpha(A)$ transforms like:

$$\varepsilon\,[F_\alpha(A)] = -\frac{\delta S}{\delta \bar{C}_\alpha}\,\delta\lambda \tag{69}$$

Therefore performing a change of variables of type (18) and using the equation of motion of $\bar{C}$ leads to the W.T. identity:

$$[J_i \frac{\delta}{\delta K_i} + \bar{\eta}_\alpha \frac{\delta}{\delta L_\alpha} - \eta_\alpha \frac{\delta}{\delta R_\alpha} + i\eta_\alpha R_\alpha] Z = 0 \tag{70}$$

The equation of motion of $\bar{C}$ has this time already been used and is no longer useful as a separate equation. Again introducing the functional W defined by equation (42), we obtain:

$$[J_i \frac{\delta}{\delta K_i} + \bar{\eta}_\alpha \frac{\delta}{\delta L_\alpha} - \eta_\alpha \frac{\delta}{\delta R_\alpha}] W - \eta_\alpha R_\alpha = 0 \tag{71}$$

Performing then the Legendre transformation (44,45) yields an equation for Γ:

$$\frac{\delta\Gamma}{\delta A_i}\frac{\delta\Gamma}{\delta K_i} + \frac{\delta\Gamma}{\delta C_\alpha}\frac{\delta\Gamma}{\delta L_\alpha} + \frac{\delta\Gamma}{\delta \bar{C}_\alpha}\left(\frac{\delta\Gamma}{\delta R_\alpha} - R_\alpha\right) = 0 \tag{72}$$

Furthermore it is easy to verify that if S satisfies equation (72), as it does before renormalization, then equations (70-72) are satisfied if the measure is invariant for the transformation:

$$\begin{aligned} \delta A_i &= \frac{\delta S}{\delta K_i}\,\delta\lambda \\ \delta C_\alpha &= \frac{\delta S}{\delta L_\alpha}\,\delta\lambda \\ \delta \bar{C}_\alpha &= \frac{\delta S}{\delta R_\alpha}\,\delta\lambda \end{aligned} \tag{73}$$

Arguments similar to those given in the previous section show then that the renormalized effective lagrangian $S^{(r)}$ is the most general polynomial of dimension four satisfying:

$$\frac{\delta S^{(r)}}{\delta A_i}\frac{\delta S^{(r)}}{\delta K_i} + \frac{\delta S^{(r)}}{\delta C_\alpha}\frac{\delta S^{(r)}}{\delta L_\alpha} + \frac{\delta S^{(r)}}{\delta \bar{C}_\alpha}\left(\frac{\delta S^{(r)}}{\delta R_\alpha} - R_\alpha\right) = 0 \tag{74}$$

Remarking that R_α is of dimension two, we can write:

$$S^{(r)}(AC\bar{C}KLR) = \frac{1}{2} a_{\alpha\beta} R_\alpha R_\beta - F_\alpha^{(r)}(AC\bar{C})R_\alpha + S^{(r)}(AC\bar{C}KL) \tag{75}$$

Equation (74) yields then three equations. From the coefficient of R we obtain:

$$(a_{\gamma\beta} - \delta_{\gamma\beta}) \frac{\delta F_\alpha^{(r)}}{\delta \bar{C}_\beta} (A C \bar{C}) + \alpha \leftrightarrow \gamma = 0 \tag{76}$$

The coefficient of R then yields:

$$(a_{\alpha\beta} - \delta_{\alpha\beta}) \frac{\delta S^{(r)}}{\delta \bar{C}_\beta} - \frac{\delta F_\alpha^{(r)}}{\delta A_i} \frac{\delta S^{(r)}}{\delta K_i} - \frac{\delta F_\alpha^{(r)}}{\delta C_\beta} \frac{\delta S^{(r)}}{\delta L_\beta} + \frac{\delta F_\alpha^{(r)}}{\delta \bar{C}_\beta} F_\beta^{(r)} = 0 \tag{77}$$

The part independent of R gives the last equation:

$$\frac{\delta S^{(r)}}{\delta A_i} \frac{\delta S^{(r)}}{\delta K_i} + \frac{\delta S^{(r)}}{\delta C_\alpha} \frac{\delta S^{(r)}}{\delta L_\alpha} - F_\alpha^{(r)}(A C \bar{C}) \frac{\delta S^{(r)}}{\delta \bar{C}_\alpha} = 0 \tag{78}$$

The complete solution of these equations is rather long to write, but two comments are in order. The gauge fixing terms becomes:

$$-\frac{1}{2} F_\alpha^{(r)}(A)(1-a)^{-1}_{\alpha\beta} F_\beta^{(r)}(A) \tag{79}$$

the form of the $\bar{C}$ C term is modified, and a $\bar{C}$ $\bar{C}$ C C term appears, both related to $\bar{C}$ C dependence of F_α(A C C).

This proves the renormalizability of the quadratic gauge up to the point already mentioned in the preceeding section: when the quadratic part of F(A) is not symmetric under the space-time independent tran formations of the group, the equations above are not always sufficient to prove that the renormalized transformations are equivalent to the initial one, and therefore that the measure is invariant.

5. Generalized Landau gauge

The special gauges corresponding to an integration with a δ-function can in general be obtained as a limit of the gauges studied before. It is nevertheless some times useful to study them directly. We shall do it here for completeness. Consider the generating functiona

$$Z(J) = \int [dA_i]\, \delta[F_\alpha(A)]\, [dC][d\bar{C}] \exp i [\mathcal{L}(A) + \bar{C}_\alpha M_{\alpha\beta} C_\beta + J_i A_i] \tag{80}$$

It is necessary, in order to be able to write an effective lagrangian, to replace the δ-function by an integral representation:

$$Z(J) = \int [dA][dQ][dC][d\bar{C}] \exp i [\mathcal{L}(A) + Q_\alpha F_\alpha(A) + \bar{C}_\alpha M_{\alpha\beta} C_\beta + J_i A_i] \quad (81)$$

The lagrangian $\mathcal{L}$ (A,C,$\bar{C}$,Q) defined by

$$\mathcal{L}(A,C,\bar{C},Q) = \mathcal{L}(A) + Q_\alpha F_\alpha(A) + \bar{C}_\alpha M_{\alpha\beta} C_\beta \quad (82)$$

is now invariant under the transformation:

$$\begin{cases} \delta A_i = D_i^\alpha(A) C_\alpha \delta\lambda \\ \delta C_\alpha = -\frac{1}{2} f_{\alpha\beta\gamma} C_\beta C_\gamma \delta\lambda \\ \delta \bar{C}_\alpha = Q_\alpha \delta\lambda \end{cases} \quad (83)$$

In order to renormalize the theory, one has to introduce the generating functional

$$Z(J,\eta,\bar{\eta},K,L,R) = \int [dA][dC][d\bar{C}][dQ] \exp i [S + J_i A_i + \bar{\eta}_\alpha C_\alpha + \bar{C}_\alpha \eta_\alpha + R_\alpha Q_\alpha] \quad (84)$$

where we have defined:

$$S = \mathcal{L}(A,C,\bar{C},Q) + K_i D_i^\alpha C_\alpha - \frac{1}{2} f_{\alpha\beta\gamma} L_\alpha C_\beta C_\gamma$$

We have here assumed that $F_\alpha(A)$ was linear in A. In this case the renormalization follows from three equations.

The W.T. identity can be written:

$$\left[J_i \frac{\delta}{\delta K_i} + \bar{\eta}_\alpha \frac{\delta}{\delta L_\alpha} - \eta_\alpha \frac{\delta}{\delta R_\alpha} \right] Z = 0 \quad (85)$$

We have again the $\bar{C}$ equation of motion:

$$\left[\frac{1}{i} F_{\alpha i} \frac{\delta}{\delta K_i} + \eta_\alpha \right] Z = 0 \quad (86)$$

But now we have to take in account the Q-equation of motion:

$$\left[F_\alpha \left(\frac{1}{i}\frac{\delta}{\delta J}\right) + R_\alpha \right] Z = 0 \tag{87}$$

The Legendre transformation has now to be performed with respect to J, η, $\bar{\eta}$ and R. The 1PI functional Γ satisfies then the equations:

$$\frac{\delta\Gamma}{\delta A_i}\frac{\delta\Gamma}{\delta K_i} + \frac{\delta\Gamma}{\delta C_\alpha}\frac{\delta\Gamma}{\delta L_\alpha} + \frac{\delta\Gamma}{\delta\bar{C}_\alpha} Q_\alpha = 0 \tag{88}$$

$$F_{\alpha i}\frac{\delta}{\delta K_i}\Gamma + \frac{\delta\Gamma}{\delta\bar{C}_\alpha} = 0 \tag{89}$$

$$\frac{\delta\Gamma}{\delta Q_\alpha} = F_\alpha(A) \tag{90}$$

We define now:

$$\tilde{\Gamma} = \Gamma - Q_\alpha F_\alpha(A) \tag{91}$$

The functional $\tilde{\Gamma}$ satisfies exactly the same equation as in section (3).

REFERENCES

References for chapter III

Introduction of non abelian gauge theories

C.N. Yang and R. Mills, Phys. Rev. 96, 191 (1954)
R. Utiyama, Phys. Rev. 101, 1597 (1956)
M. Gell-Mann and S. Glashow, Ann. Phys. (N.Y.) 15, 437 (1961)

The quantization has been discussed in:

R.P. Feynman, Acta Phys. Polonica 26, 697 (1963)
V.N. Popov and L.D. Faddeev "Perturbation Theory for gauge invariant fields", Kiev ITP report (unpublished)
L.D. Faddeev and V.N. Popov; Phys. Lett. 25B, 29 (1967)
B. De Witt Phys. Rev. 162, 1195, 1239 (1967)
S. Mandelstam. Phys. Rev. 175, 1580 (1968)
E.S. Fradkin and I.V. Tuytin, Phys. Lett. 30B, 562 (1969), Phys. Rev. D2, 2841 (1970)
M.T. Veltman, Nucl. Phys. B21, 288 (1970)
G. 't Hooft, Nucl. Phys. B33, 173 (1971)

The Ward-Takahashi identities were first discussed in:

J.C. Ward, Phys. Rev. 78, 1824 (1950)
Y. Takahashi, Nuov. Cim. 6, 370 (1957)

The W.T. identities for the non-abelian gauge theories were first discussed in:

A.A. Slavnov, Theor. and Math. Phys. 10, 152 (1972)
English translation: Theor. and Math. Phys. 10, 99 (1972)
J.C. Taylor, Nucl. Phys. 10, 99 (1971)

The dimensional regularization for gauge theories was introduced in:

G. 't Hooft and M. Veltman, Nucl. Phys. B44, 189 (1972)

The renormalization of gauge-theories for Green's functions was studied in:

B.W. Lee and J. Zinn-Justin, Phys. Rev. D5, 3121, 3137 (1972) and D7, 1049 (1973)
G. 't Hooft and M.T. Veltman, Nucl. Phys. B50, 318 (1972)

W.T. identities for the 1PI generating functional have been obtained in:

B.W. Lee, Phys. Lett 46B, 214 (1974) and Phys. Rev.D9, 933 (1974)

The "super-symmetry" of gauge theories and its use in the derivation of W.T. identities have been outlined in:

C. Becchi, A. Rouet and S. Stora, Renormalization of the Higgs-Kibble model (Marseilles preprint,to be published)

This method seems useful in the study of the renormalization of gauge invariant operators

H. Stern-Kluberg and J.B. Zuber, Ward identities and some clues to the renormalization of gauge invariant operators, Saclay preprint D. Ph-T/74-56 (to be published)

UNIFIED MODELS OF ELECTROMAGNETIC AND WEAK INTERACTIONS*

H. Pietschmann

Institut für Theoretische Physik, Universität Wien
and
Institut für Hochenergiephysik, Österr. Akademie d. Wissenschaften
Vienna, Austria

1. Introduction

The seventeenth International Conference on High Energy Physics in London 1974 was spirited by an almost euphoric mood of the community of elementary particle physicists. This was due to the great impact on our understanding derived from the new insights in field theory. In turn, this was stimulated by the developments in non-Abelian gauge theories. The discovery and firm establishment of the existence of weak neutral currents has greatly supported the view that gauge theories might have some bearance on physical reality. Indeed, one can now think of possible unification of <u>all</u> interactions without leaving the grounds of scientific seriousness.

In our lectures, we shall be concerned with the model aspect of the new theories, in other words we shall work closely to experiments. Nobody has yet set up a realistic model of unification of all interactions. But some models which unify weak and electromagnetic interactions have considerable chance to survive experimental tests, possibly with modifications. Therefore we do not touch upon the interesting question of "gauging" strong interactions also but restrict to the weak and electromagnetic interactions. The aspect of renormalizability is treated in the lectures of Zinn-Justin so that we can concentrate on the model aspect, i.e. experimental predictions and the like.

There is a vast variety of different models in the literature and we shall not attempt at any completeness. Rather we pick a few examples and try to be explicit in all calculations. The reader who has followed through the computational steps in one particular case should have no problem to do the same calculation in other cases.

---

* Supported by "Fond zur Förderung der wissenschaftlichen Forschung in Österreich"

(A superficial knowledge of all models does of course not bring about the ability to compute any of them!)

None of the models is supported by experiments beyond doubt. Therefore, it is still advisable to analyse the new findings from a phenomenological point of view, not being biased by a particular choice of model. Once the phenomenological parameters are known empirically, they can be compared to the predictions of the various models. We shall follow this path before we enter into the description of special models.

2. The "Classical" Weak Interaction

Before we go on to explore newly discovered phenomena, let us recall in brief the rather astounding knowledge on weak interaction we already had before gauge models came along. We shall call this part the "classical" weak interactions. It can be described by the interaction of a current with itself, in the so-called current-current (CC) form

$$\mathcal{L}_{CC} = \frac{G}{\sqrt{2}} \mathcal{J}_\lambda^+(x) \mathcal{J}^{\lambda -}(x) \tag{1}$$

with

$$(\mathcal{J}_\lambda^+)^\dagger = \mathcal{J}_\lambda^- \tag{2}$$

The total weak current is given by

$$\mathcal{J}_\lambda^+(x) = \sum_\ell \bar{\nu}_\ell(x) \gamma_\lambda (1+\gamma_5) \ell(x) + \\ + \cos\vartheta \, j_\lambda^+(x) + \sin\vartheta \, J_\lambda^+(x) \tag{3}$$

where the sum goes over electronic and muonic type leptons and ϑ is the Cabibbo angle.

The strangeness conserving and strangeness changing hadronic currents can be expressed in terms of quark fields $q(x)$ by

$$j_\lambda^+(x) = \bar{q}(x) \gamma_\lambda (1+\gamma_5) \frac{\lambda_1 + i\lambda_2}{2} q(x) \tag{4a}$$

$$J_\lambda^+(x) = \bar{q}(x) \gamma_\lambda (1+\gamma_5) \frac{\lambda_4 + i\lambda_5}{2} q(x) \tag{4b}$$

where λ_i are the familiar 3 x 3 Gell-Mann matrices of SU(3).

The weak current has some intimate relations to the electromagnetic current. This was noticed rather early and found its first rigorous formulation in the conserved vector current hypothesis. Later on, current algebra deepened this relationship even further. In fact, the weak current can be constructed out of the electromagnetic current. Let us write the electromagnetic current as

$$j_\mu(x) = \sum_\ell \bar{\ell}(x)\gamma_\mu \ell(x) + j_\mu^v(x) + j_\mu^s(x) \tag{5}$$

where the isovector and the isoscalar parts are given by

$$j_\mu^v(x) = \bar{q}(x)\gamma_\mu \frac{\lambda_3}{2} q(x) \tag{6a}$$

$$j_\mu^s(x) = \bar{q}(x)\gamma_\mu \frac{\lambda_8}{2\sqrt{3}} q(x) \tag{6b}$$

The fact that we write the hadronic currents in terms of quarks already presupposes and expresses certain selection rules. For example, there are no isotensor currents and the like. Throughout our lectures we shall use quark fields only to express transformation properties of currents. We shall not go into the interesting questions whether quarks exist as real particles, whether they are confined, imprisoned or anything like that.

The "classical" weak current can be obtained out of the electromagnetic current by the following 5 steps:

i) exchange one of the leptons by its neutrino in each of the 2 leptonic parts

ii) drop the isoscalar part

iii) rotate the isovector part in isospin-space, i.e. replace

$$\frac{\lambda_3}{2} \longrightarrow \frac{\lambda_1 + i\lambda_2}{2}$$

iv) rotate in the plane of SU(3) space spanned by ($\lambda_1 + i\lambda_2$) and ($\lambda_4 + i\lambda_5$) around the Cabibbo angle ϑ, i.e. replace

$$(\lambda_1 + i\lambda_2) \longrightarrow \cos\vartheta(\lambda_1 + i\lambda_2) + \sin\vartheta(\lambda_4 + i\lambda_5)$$

v) replace γ_λ by $\gamma_\lambda(1+\gamma_5)$.

Notice that the matrices λ_6 and λ_7 do not show up at all in the currents! They belong to $\Delta Q=0$ with $|\Delta S| = 1$ transforming parts and it is well established experimentally[1] that there are no neutral strangeness changing currents as can be seen from table 1. The rates for the 2 observed modes agree well with predictions from combined weak and

$$B(K^{\pm} \to \pi^{\pm} e^{+}e^{-}) = (2.3 \pm 0.8) \cdot 10^{-7}$$

$$B(K^{\pm} \to \pi^{\pm} \mu^{+}\mu^{-}) < 2.4 \cdot 10^{-6}$$

$$B(K^{\pm} \to \pi^{\pm} \nu \bar{\nu}) < 5.6 \cdot 10^{-7}$$

$$B(\Sigma^{+} \to p e^{+}e^{-}) < 7 \cdot 10^{-6}$$

$$B(K_{L}^{o} \to \mu^{+}\mu^{-}) = (1.2 {}^{+0.8}_{-0.4}) \cdot 10^{-8}$$

Table 1 : Absence of neutral strangeness changing currents. (B are branching ratios to total decay rates)

ΔI \ ΔQ	+1	0	-1
0	×	j^s	×
$\frac{1}{2}$	J^+	×	J^-
1	j^+	j^v	j^-

Table 2: $\Delta I, \Delta Q$ properties of currents

ΔS \ ΔQ	+1	0	-1
+1	J^+	×	×
0	j^+	j^s, j^v	j^-
-1	×	×	J^-

Table 3: $\Delta Q, \Delta S$ properties of currents.

electromagnetic transitions.

The symmetry structure of these currents is summarized in tables 2 and 3. In the entries to these tables, we have dropped vector indices and space-time dependence of the currents to facilitate reading. The crosses in tables 2 and 3 indicate that not all possible combinations are realized with currents existing in nature.

In equ. (1) we have written down the current-current form of the "classical" weak interactions. What about the intermediate boson? In fact, the intermediate boson will be a prerequisite without which gauge models cannot be conceived. From now on, we shall always assume its existence in spite of the fact that it has not yet been observed and lower mass limits are of the order of $M_W > 10$ GeV[2)]. Why, then, did we not use the intermediate vector boson (IVB) form of the "classical" weak interaction to begin with? The answer is very simple: At low energies, the 2 forms are completely equivalent. Let us write the IVB-Lagrangian

$$\mathcal{L}_{IVB} = g_{sw}\left(W^\lambda \mathcal{J}^+_\lambda + h.c.\right) \tag{7}$$

where g_{sw} is the "semi-weak" coupling constant.
Since no process with real intermediate bosons has yet been observed, we can invert the equation of motion for the IVB (W-boson)

$$\partial^\nu\left[\partial_\nu W_\mu(x) - \partial_\mu W_\nu(x)\right] + M_W^2 W_\mu(x) = g_{sw}\mathcal{J}_\mu(x) \tag{8}$$

to obtain in momentum space

$$W_\lambda = -g_{sw}\frac{g_{\lambda\nu} - \frac{k_\lambda k_\nu}{M_W^2}}{k^2 - M_W^2}\mathcal{J}^\nu \tag{9}$$

Thus the effective interaction of 2 weak currents with each other, mediated by a W-boson becomes

$$-g_{sw}^2\mathcal{J}^+_\lambda\frac{g^{\lambda\nu} - \frac{k^\lambda k^\nu}{M_W^2}}{k^2 - M_W^2}\mathcal{J}^-_\nu = \frac{g_{sw}^2}{M_W^2}\mathcal{J}^+_\lambda\mathcal{J}^{\lambda -} + O\left(\frac{k^2}{M_W^2}\right) \tag{10}$$

A comparison with eq. (1) gives the famous relation

$$\frac{g_{sw}^2}{M_W^2} = \frac{G}{\sqrt{2}} \tag{11}$$

and we have complete equivalence in the energy range where terms of order k^2/M_W^2 can be neglected.

3. Neutral Weak Currents

Rather early in the history of weak interactions, the existence of weak neutral currents was considered[3]. However, their absence in strangeness changing transitions became quite apparent (table 1!) and strangeness conserving neutral weak currents could not be reached and tested in decay experiments.

Through the pioneering work of the Gargamelle-collaboration (Aachen,Brussels, CERN, Paris, Milano, Orsay, London) and subsequent confirmation in other laboratories, the existence of weak neutral currents in strangeness conserving transitions seems now firmly established. For a discussion of the experiments, we refer to the literature[2]. Here, we just notice their existence and the fact that they have been observed through neutral neutrino currents (these observations rule out a complete class of gauge models!).

We shall base our phenomenological analysis on the following assumptions[4]:

i) Weak neutral currents are of vector nature also (i.e. combinations of V and A). They are mediated by the neutral boson Z_μ.

ii) Weak neutral hadronic currents can be constructed out of quarks in much the same way as in eqs. (3) and (5).

From the second assumption it follows, that the weak neutral hadronic current is a sum of isoscalar and isovector contribution, which we write as

$$j_\lambda^{n} = j_\lambda^{n,s} + j_\lambda^{n,v} \tag{12}$$

From the first assumption, it follows that each of these contributions in turn consists of a vector and an axial-vector part, so that we have

$$j_\lambda^{n,s} = j_\lambda^{n,s,V} + j_\lambda^{n,s,A} \tag{13a}$$

$$j_\lambda^{n,v} = j_\lambda^{n,v,V} + j_\lambda^{n,v,A} \tag{13b}$$

When we collect all pieces to write down the weak neutral hadronic current it is better to explicitly take out certain weight parameters α_i in order to keep the definition of the currents free of model-dependent parameters.

$$j_\lambda^{n} = \alpha_1 j_\lambda^{n,s,V} + \alpha_2 j_\lambda^{n,s,A} + \alpha_3 j_\lambda^{n,v,V} + \alpha_4 j_\lambda^{n,v,A} \tag{14}$$

This is a very general ansatz for the weak neutral hadronic current. In order to achieve a reasonable analysis of experiments, we have to invoke further assumptions.In doing so we shall follow the spirit of Pais and Treiman[4]. First we note that the 2 vector parts of eqs (13) have the same transformation properties as the 2 pieces of the electromagnetic current defined in eqs. (6). Recalling that we have taken out a numerical parameter , we shall assume them to be identical, i.e.

$$j_\lambda^{n,s,V} = j_\lambda^{s} \tag{15a}$$

$$j_\lambda^{n,v,V} = j_\lambda^{v} \tag{15b}$$

The isovector axial part of eq (13b) transforms as the neutral member of an isotriplet whose other components are given by eq (4a) and its hermitian conjugate. Thus we assume

$$j_\lambda^{n,v,A} = \bar{q}(x)\,\gamma_\lambda\gamma_5\,\frac{\lambda_3}{2}\,q(x) \equiv j_\lambda^{A} \tag{16}$$

The remaining piece, $j_\lambda^{n,s,A}$, has no analog among "classical" current We shall therefore drop it from our phenomenological analysis. This is a rather drastic assumption which is very worth-while to be tested experimentally. (Suggestions for tests are given in reference 4)). It will turn out that this piece does not, in fact, show up in the gauge models we are going to discuss. It is now quite natural to assume further that the 3 pieces do not occur in arbitrary combinations but rather as the electromagnetic current and the isospin rotated weak current so that we are left with only 2 free parameters.

$$\begin{aligned} j_\mu^{n} &= \rho\, j_\mu + \lambda\,(j_\mu^{v} + j_\mu^{A}) = \rho\,(j_\mu^{s} + j_\mu^{v}) + \lambda\,(j_\mu^{v} + j_\mu^{A}) = \\ &= \rho\, j_\mu^{s} + \lambda\, j_\mu^{A} + (\rho+\lambda)\, j_\mu^{v} \end{aligned} \tag{17}$$

The parameters ρ and λ will take on special values in each particular gauge model.

As soon as we extend our analysis to include strange particles, isospin rotations are no longer sufficient to generate weak currents. We have to generalize to SU(3) rotations. They will, of course, also yield isoscalar pieces composed of strange particles as will be shown explicitly towards the end of section 4. However, as far as neutral current experiments are concerned, we can neglect "associated production" contributions and shall thus restrict our phenomenological analysis to the isospin subgroup. The general case is discussed in reference 4.

So far we have dealt with the hadronic part of the weak neutral current. Let us now turn to its leptonic part. Here, we shall follow Llewellyn Smith and Nanopoulos[5]. The weak neutral leptonic current can in general contain 4 pieces: A neutrino contribution and a lepton contribution of both electronic and muonic type. Each piece consists of a vector part and an axial vector part. Because of electron-muon universality, the weight factors should be identical for electronic and muonic pieces of the same nature. Because of the definite helicity of the neutrino ("2-component neutrino assumption"), vector and axial vector part of the neutrino contributions have to have the same weight factor also. This leaves us with 3 free parameters which we arrange in the following manner:

$$\ell_\lambda^n(x) = \sum_\ell \left\{ g_1 \bar{\nu}_\ell(x) \gamma_\lambda (1+\gamma_5) \nu_\ell(x) + \bar{\ell}(x) \gamma_\lambda \left(g_L \frac{1+\gamma_5}{2} + g_R \frac{1-\gamma_5}{2} \right) \ell(x) \right\} \tag{18}$$

where the sum is as usual over muonic and electronic type leptons.

The total weak neutral current can now be written with 5 free parameters

$$\begin{aligned} \mathcal{J}_\lambda^n(x) &= \ell_\lambda^n(x) + j_\lambda^n(x) = \\ &= \sum_\ell \left\{ g_1 \bar{\nu}_\ell(x) \gamma_\lambda (1+\gamma_5) \nu_\ell(x) + \bar{\ell}(x) \gamma_\lambda \left(g_L \frac{1+\gamma_5}{2} + g_R \frac{1-\gamma_5}{2} \right) \ell(x) \right\} + \\ &\quad + \rho\, j_\lambda^s(x) + \lambda\, j_\lambda^A(x) + (\rho+\lambda)\, j_\lambda^v(x) \end{aligned} \tag{19}$$

The 5 parameters take specific values in each particular gauge model. On the other hand, processes where weak neutral currents play a role can be computed from eq. (19), giving experimental information on the

5 parameters. In turn, this information can be compared to the various theoretical predictions, hopefully ruling out one or the other. At the same time, it narrows down the allowed region for the parameters (such as mixing angles) of the surviving models. This is, in our opinion, the most unbiased way to test for particular gauge models.

4. Gauge Models

Let us now turn to the construction of actual gauge models. We have already said in the beginning of section 3 that a full class of gauge models is ruled out by the experimental detection of weak neutral currents. Moreover, heavy leptons have not yet been found and the experimental lower limit for their mass is presently of the order of 8 GeV[6]. With these findings in view, it is best to concentrate on the "minimal" model, the model which is constructed with the minimal number of newly predicted particles; it is the Salam-Weinberg model[7].

In setting up the Salam-Weinberg model, we shall simplify matters by forgetting about muonic type of leptons. Their addition in the final Lagrangian is a trivial matter. First we shall concentrate on the leptons. In weak interactions, all particles couple with their left-handed chirality projection and we thus define a left-handed doublet

$$L(x) = \frac{1+\gamma_5}{2}\begin{pmatrix}\nu_e\\ e\end{pmatrix} \tag{20}$$

This doublet spans a space called "weak isospin space". It is easy to construct the charged weak leptonic current of eq (3) by means of this doublet

$$\ell_\lambda^{\pm}(x) = 2\,\bar{L}(x)\,\gamma_\lambda\,\tau^{\pm}\,L(x) \tag{21}$$

where $\tau^{\pm}$ are the usual Pauli-matrices acting in weak isospin space.

Since we are out to unify weak and electromagnetic interactions, we also have to construct the electromagnetic current out of the doublet (20). To do this, still need the right-handed components of the electron, which we consequently define as a singlet

$$R(x) = \frac{1-\gamma_5}{2}\,e(x) \tag{22}$$

The electromagnetic current can now be written as

$$l_\lambda(x) = \bar{e}(x)\gamma_\lambda e(x) = \frac{1}{2}(\bar{L}\gamma_\lambda L - \bar{L}\gamma_\lambda \tau_3 L) + \bar{R}\gamma_\lambda R \tag{23}$$

We see that it contains the third component of the weak isotriplet whose other 2 components are given by eq (21). In addition, it contains a weak isosinglet whose composition is completely specified.

In a gauge model, we want to couple our currents in a gauge invariant way. This leads naturally to a triplet plus a singlet of vector gauge bosons. Thus the underlying gauge group structure is SU(2) ⊗ U(1). The gauge invariant interaction Lagrangian is

$$\mathcal{L}_1(x) = \frac{g}{2}\bar{L}\gamma_\lambda\vec{\tau}L\vec{A}^\lambda + g'\left[\frac{1}{2}\bar{L}\gamma_\lambda L + \bar{R}\gamma_\lambda R\right]B^\lambda \tag{24}$$

In writing down eq (24) we have made use of the freedom of different coupling constants for triplet and singlet interaction. Consequently, none of the gauge bosons is coupled to the electromagnetic current, eq (23). But a certain mixture of B^λ and the neutral triplet boson $A^3{}_\lambda$ is! Since the gauge group has to be broken in some way because of the considerable mass differences of the bosons,the physical vector bosons can indeed be such mixtures. Hence we mix A^3 and B by the ansatz

$$\begin{aligned} A^3_\mu &= \cos\varphi\, Z_\mu - \sin\varphi\, A_\mu \\ B_\mu &= \sin\varphi\, Z_\mu + \cos\varphi\, A_\mu \end{aligned} \tag{25}$$

where the "Salam-Weinberg mixing angle" φ is defined by the requirement that A_μ be the physical photon.

Indeed, if we insert eq (25) in the Lagrangian (21) we obtain for the interesting neutral part, i.e. the part containing the neutral vector bosons

$$\begin{aligned} \mathcal{L}_1(x) = \text{charged part } + \\ +\left[g'\cos\varphi\left(\frac{1}{2}\bar{L}\gamma_\mu L + \bar{R}\gamma_\mu R\right) - \frac{g}{2}\sin\varphi\,\bar{L}\gamma_\mu\tau^3 L\right]A^\mu + \\ +\left[g'\sin\varphi\left(\frac{1}{2}\bar{L}\gamma_\mu L + \bar{R}\gamma_\mu R\right) + \frac{g}{2}\cos\varphi\,\bar{L}\gamma_\mu\tau^3 L\right]Z^\mu \end{aligned} \tag{26}$$

A comparison with eq (23) together with the requirement that A^μ be the physical photon immediately yields

$$e = g\sin\varphi = g'\cos\varphi \tag{27}$$

or

$$tg\ \varphi = \frac{g'}{g} \tag{28}$$

so that the mixing angle can be interpreted as the arctangens of the ratio of singlet and triplet coupling constant.

Much more interesting is of course the predicted coupling of the new neutral boson Z^{μ}. With a little algebra, the "neutral current" can be recast into the form

$$g' \sin\varphi \left(\frac{1}{2}\bar{L}\gamma_\mu L + \bar{R}\gamma_\mu R\right) + \frac{g}{2}\cos\varphi\, \bar{L}\gamma_\mu \tau^3 L =$$

$$= \frac{g}{4\cos\varphi}\left\{\bar{\nu}\gamma_\mu(1+\gamma_5)\nu - \bar{e}\gamma_\mu(C_V+\gamma_5)e\right\} \tag{29}$$

with

$$C_V = 1 - 4\sin^2\varphi \tag{30}$$

Notice in passing that the special value $\sin\varphi = 1/2$ opens the interesti possibility that neutrinos are coupled to the neutral weak current with V - A, (to the electromagnetic current not at all of course), whereas charged leptons are coupled to the electromagnetic current with pure V and to the neutral weak current with pure A.

Eventually, we shall of course discuss the coupling to hadrons also. It is a difficult endeavour because of the absence of neutral weak current with $\Delta S \neq 0$. For the moment, let us neglect strange particles altogether. This is a fair approximation because of the smallness of the Cabibbo angle. If we do this and interest ourselves in the coupling of the physical nucleons, we can simply take over eq (29) with few adjustments. This is because the structure of the weak neutral current was determined solely by the structure of the electromagnetic current and the triplet interaction (21). Both are very similar for nucleons. Thus we get, for the coupling of nucleons to the Z-boson, the following current

$$\frac{1}{2}\left\{\bar{N}\gamma_\mu(1+\gamma_5)N - \bar{P}\gamma_\mu(C_V+\gamma_5)P\right\} =$$

$$= \frac{1}{2}\left[\bar{N}\gamma_\mu(1+\gamma_5)N - \bar{P}\gamma_\mu(1+\gamma_5)P\right] + 2\sin^2\varphi\, \bar{P}\gamma_\mu P \tag{31}$$

where C_V is given by eq (30) and P and N refer to proton and neutron respectively.

The first term of eq (31) is just the rotated isospin current whereas the second term is proportional to the electromagnetic current of nucleons. Comparing eqs. (29) and (31) with eq (19) allows us to determine the 5 parameters of the phenomenological ansatz of section 3, for the Salam-Weinberg model in the following way:

$$\begin{aligned} g_1 &= \frac{1}{2} \\ g_L &= 2\sin^2\varphi - 1 \\ g_R &= 2\sin^2\varphi \\ \lambda &= 1 \\ \rho &= -2\sin^2\varphi \end{aligned} \tag{32}$$

The current (31) can be used to predict a parity violating potential between electrons and an atomic nucleus. This provides a possible means to test the Salam-Weinberg model with atomic physics[8]. Because of the large mass of the Z-boson, the potential in x-space is well approximated by a δ-function and we can write for the coupling of electrons to nucleons

$$\left(\frac{g^2}{4\cos\varphi}\right)^2 \bar{e}\gamma_\lambda(C_V+\gamma_5)e\,\frac{1}{M_Z^2}\,\delta^{(3)}(\vec{x})\left\{\bar{n}\gamma^\lambda(1+\gamma_5)n - \bar{p}\gamma^\lambda(C_V+\gamma_5)p\right\} \tag{33}$$

where C_V is given by eq. (30). To obtain the potential of orbiting electrons, we can take the static approximation for the nucleons and neglect their spin contributions. In doing so we replace

$$\bar{u}_r\gamma_\mu u_{r'} \longrightarrow \delta_{\mu 0}\,\delta_{rr'} \tag{34a}$$

$$\bar{u}_r\gamma_\mu\gamma_5 u_{r'} \longrightarrow -\delta_{\mu k}\langle\sigma_k\rangle_{rr'} \tag{34b}$$

Eqs. (34) are, of course, well-known from nuclear β-decay. Since we neglect nuclear spin contributions, only the time-component of the vector part has to be kept. For the parity violating part of the potential, it hence suffices to consider the time-component of the electrons axial vector. Elementary Dirac-algebra yields

$$\bar{u}_2\gamma_0\gamma_5 u_1 = \frac{1}{2m_e}\left\{(\vec{\sigma}\vec{p})_1 + (\vec{\sigma}\vec{p})_2\right\} \tag{35}$$

It remains to collect the numerical factors. We shall show subsequently that we can equate

$$\frac{g^2}{16\cos^2\varphi M_Z^2} = \frac{g^2}{16 M_W^2} = \frac{G}{2\sqrt{2}} \tag{36}$$

so that the parity violating potential for an electron orbiting around a nucleus of Z protons and N neutrons becomes

$$V_{p.v.} = \frac{G}{4\sqrt{2} m_e}\left\{\vec{\sigma}\vec{p}\,\delta^{(3)}(\vec{x}) + \delta^{(3)}(\vec{r})\,\vec{\sigma}\vec{p}\right\} Q_W(Z,N) \tag{37}$$

with

$$Q_W(Z,N) = C_V Z - N = (1-4\sin^2\varphi)Z - N \tag{38}$$

It is a challenge to try whether one can obtain information on neutral currents or even details of a gauge model by spectroscopy of atomic transitions[8].

After all this phenomenology let us now turn to the crucial problem of masses of intermediate bosons. No mass term for vector boson can be used in the Lagrangian because it would break the gauge symmetry and thus render the theory non-renormalizable. On the other hand, we know that the weak vector bosons have rather high masses (supposing that they exist at all). To circumvent this apparent contradiction, let us first study a simple model due to Higgs[9]. We consider the Lagrangia of a complex scalar field ϕ with quartic self-interaction and in interaction with the (electromagnetic) vector field A_μ. It is given by

$$\mathcal{L}_H = \frac{1}{4}(F_{\mu\nu})^2 + |(\partial_\mu - ieA_\mu)\phi|^2 - \mu^2|\phi|^2 - h|\phi|^4 \tag{39}$$

with

$$F_{\mu\nu} = \partial_\mu A_\nu - \partial_\nu A_\mu \tag{40}$$

This is a perfectly renormalizable model. It is invariant under the usual gauge transformations

$$A_\mu \to A_\mu + \partial_\mu \Lambda \tag{41a}$$

$$\phi \to \phi\, e^{ie\Lambda} \tag{41b}$$

$$\phi^* \to \phi^* e^{-ie\Lambda} \tag{41c}$$

μ is the mass of the scalar field and μ^2 has to be positive, of course. But let us now consider what kind of changes the model undergoes if we allow μ^2 to become negative. In this case, the "classical potential"

$$V(\phi) = \mu^2 \phi^2 + h\,\phi^4 \tag{42}$$

acquires minima away from the origin. The situation is shown in Fig. 1.

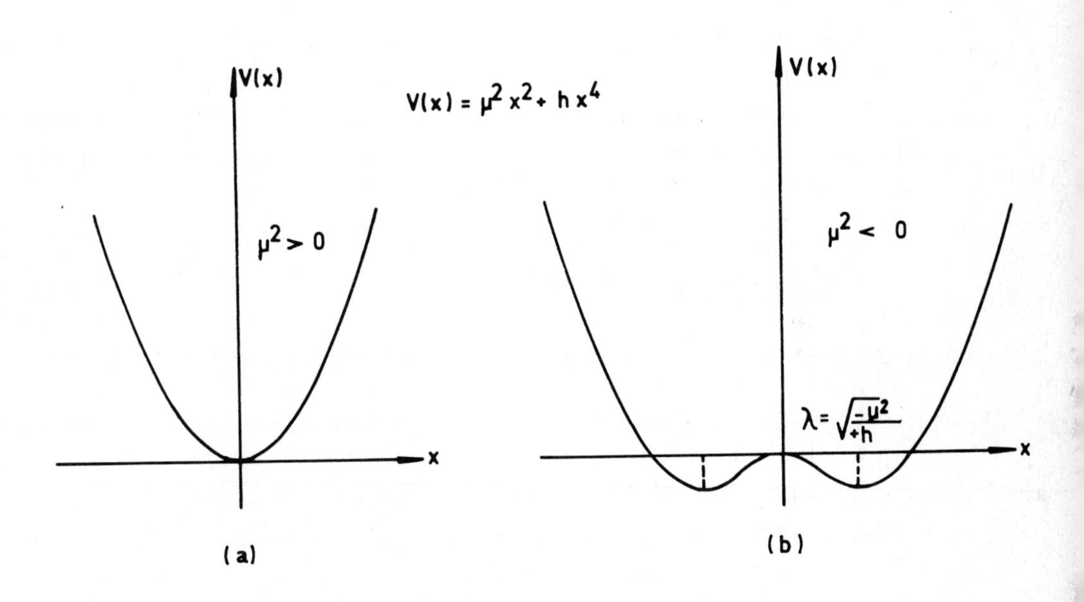

Fig. 1
The "Classical Potential" $\mu^2\Phi^2 + h\Phi^4$ for positive and negative μ^2.

Quantum mechanically, this corresponds to the situation where ϕ has a non-vanishing vacuum expectation value. Let us write

$$\langle 0|\,\phi\,|0\rangle = \frac{1}{\sqrt{2}}\,\lambda \tag{43}$$

λ can always be chosen real by fixing a phase. We shall show presently, that 2 possibilities arise: Either λ vanishes (which is the normal case) or λ takes the value of the classical potential minimum, i.e.

$$\lambda = \sqrt{\frac{-\mu^2}{h}} \qquad\qquad \mu^2 < 0 \tag{44}$$

To see this, we want to use real fields, i.e. we have to explicitly decompose the complex field ϕ . This can be done in 2 ways: We can separate real and imaginary part (the so-called R-formalism) or we can separate absolute value and phase (the so-called U-formalism). It is by no means trivial that the 2 procedures are equivalent. However, in perturbation theory, their equivalence has been shown by G.t'Hooft[10] who also showed that the renormalizability of the model is not spoiled by assuming eq (44).

Since we are mainly concerned with model building, i.e. experimental consequences, we shall choose the U-formalism . The R-formalism is linear and therefore more suitable for mathematical investigations of renormalizability for example. But it contains ghost-fields which are eliminated in the U-formalism . Let us thus write

$$\phi = \frac{1}{\sqrt{2}}(\lambda + \chi)\, e^{\frac{i}{\lambda}\vartheta} \tag{45}$$

where χ and ϑ are real fields. By means of choosing the gauge

$$\Lambda = -\frac{1}{e\lambda}\vartheta \tag{46}$$

the "phase field" ϑ can be eliminated entirely. The question is, where does the corresponding degree of freedom go? It turns out that it is taken up by the vector particle which acquires mass. To see this, insert

$$\phi = \frac{1}{\sqrt{2}}(\lambda + \chi) \tag{47}$$

into eq (39). It becomes

$$\mathcal{L}_H = \frac{1}{4}(F_{\mu\nu})^2 + \frac{1}{2}(\partial_\mu \chi)^2 - \lambda\chi(\mu^2 + h\lambda^2) + \\ + \frac{1}{2}e^2\lambda^2(A_\mu)^2 - h\lambda^2\chi^2 \text{ + coupling terms.} \tag{48}$$

We have omitted to write out the coupling terms because they are only interesting in a realistic case. Observe that the term linear in χ drops out in the 2 cases described before and in eq (44) corroborating the statement. Both the vector particle and the "Higgs scalar" have acquired mass

$$m_A = e\lambda \tag{49a}$$

$$m_\chi^2 = 2h\lambda^2 \tag{49b}$$

This is exactly a mechanism serving for the purpose of attributing mass to vector mesons. We shall therefore adapt it to the more realistic case of the Salam-Weinberg model. Since we are here dealing with a non abelian gauge group, the Higgs field will be more complicated. Let us introduce a doublet of complex fields

$$\phi = \begin{pmatrix} \phi^+ \\ \phi^0 \end{pmatrix} \tag{50}$$

In accord with eq (39), its self-interaction is assumed to be

$$\mathcal{L}_2 = -\mu^2 |\phi|^2 - h|\phi|^4 \tag{51}$$

where

$$|\phi|^2 = |\phi^+|^2 + |\phi^0|^2 \tag{52}$$

The doublet ϕ has to be coupled in a gauge invariant way to the 4 vector mesons introduced in eq (24). This is achieved by

$$\mathcal{L}_3 = \left| \left(\partial_\mu - i\frac{g}{2}\vec{\tau}\vec{A}_\mu + i\frac{g'}{2}B_\mu\right)\phi \right|^2 \tag{53}$$

We have to be cautious not to give a mass to all 4 vector mesons because the physical photon has to remain massless. Let us consider the neutral part of the Lagrangian (53). Recalling eq (25), we can write

$$\left(-i\frac{g}{2}\tau_3 A_\mu^3 + i\frac{g'}{2}B_\mu\right)\phi = \frac{ig}{2\cos\varphi}\begin{pmatrix} (A_\mu \sin 2\varphi - Z_\mu \cos 2\varphi)\phi^+ \\ Z_\mu \phi^0 \end{pmatrix} \tag{54}$$

The photon does not couple to the neutral Higgs field. In the U-formalism we write in analogy to eq (45)

$$\begin{pmatrix} \phi^+ \\ \phi^0 \end{pmatrix} = e^{\frac{i}{\lambda}\vec{\Theta}\vec{\tau}} \begin{pmatrix} 0 \\ \frac{1}{\sqrt{2}}(\lambda + \chi) \end{pmatrix} \tag{55}$$

All 3 "phase fields" $\vec{\Theta}$ can be "gauged away" leaving

$$\begin{pmatrix} \phi^+ \\ \phi^0 \end{pmatrix} \rightarrow \frac{1}{\sqrt{2}} \begin{pmatrix} 0 \\ \lambda + \chi \end{pmatrix} \tag{56}$$

so that the photon does not acquire mass because it is never multiplied into λ. Indeed, since we have eliminated 3 "ghosts", i.e. 3 degrees of freedom from the Higgs system, only 3 of the 4 vector particles can add another degree of freedom by gaining mass.

The free Lagrangian for the Higgs fields is, of course, contained in eq (53). For fermions and vector bosons we have to add

$$\mathcal{L}_0 = \frac{1}{4} \vec{A}_{\mu\nu} \vec{A}^{\mu\nu} + \frac{1}{4} B_{\mu\nu} B^{\mu\nu} + i \bar{R} \not{\partial} R + i \bar{L} \not{\partial} L \tag{57}$$

where

$$\vec{A}_{\mu\nu} = \partial_\mu \vec{A}_\nu - \partial_\nu \vec{A}_\mu + g [\vec{A}_\mu \times \vec{A}_\nu] \tag{58a}$$

$$B_{\mu\nu} = \partial_\mu B_\nu - \partial_\nu B_\mu \tag{58b}$$

No mass terms are included in eq (57). We have just described how the vector mesons acquire mass. For the fermions, we have to add a mass term in a gauge invariant way. Also, it has to be such that the neutrino remains mass-less. This can be done in the following way:

$$\mathcal{L}_4 = \frac{\sqrt{2}}{\lambda} m_e \left(\bar{R} \phi^\dagger L + \bar{L} \phi R \right) \tag{59}$$

Because of the requirement of gauge invariance, eq (59) automatically yields a coupling of the Higgs boson to charged leptons. Its strength is proportional to the lepton mass and therefore 200 times bigger for muons than it is for electrons! At what level this induces deviations from electron - muon universality is still an open question.

To write down the complete Lagrangian for the leptonic part of the Salam-Weinberg model we simply collect eqs (24), (51), (53), (57) and (59), i.e.

$$\mathcal{L}_{SW} = \sum_{i=0}^{4} \mathcal{L}_i \tag{60}$$

To invoke the Higgs mechanism , we then perform the replacement (56). Finally, we add the Lagrangian for the muonic type of leptons and obtain

$$
\begin{aligned}
\mathcal{L}_{SW} = \mathcal{L}_0 &+ \sum_{\ell} \Big\{ \Big[\frac{g}{2\sqrt{2}} \bar{\ell} \gamma_\lambda (1+\gamma_5) \nu_\ell W^\lambda + h.c. \Big] + e \bar{\ell} \gamma_\mu \ell A^\mu + \\
&+ \frac{g}{4\cos\varphi} \Big[\bar{\nu}_\ell \gamma_\lambda (1+\gamma_5) \nu_\ell - \bar{\ell} \gamma_\lambda (C_V + \gamma_5) \ell \Big] Z^\lambda \Big\} - \\
&- h\lambda^2 \chi^2 - h\lambda \chi^3 - \frac{h}{4} \chi^4 + \frac{1}{2} (\partial_\mu \chi)^2 + \\
&+ \frac{g^2}{4} \Big(W_\mu^\dagger W^\mu + \frac{1}{2\cos^2\varphi} Z_\mu Z^\mu \Big) (\lambda^2 + 2\lambda\chi + \chi^2) + \\
&+ \sum_{\ell} m_\ell \Big(1 + \frac{1}{\lambda} \chi \Big) \bar{\ell} \ell
\end{aligned}
\tag{61}
$$

where the sum goes over electronic and muonic type of leptons. C_V is defined in eq (30).

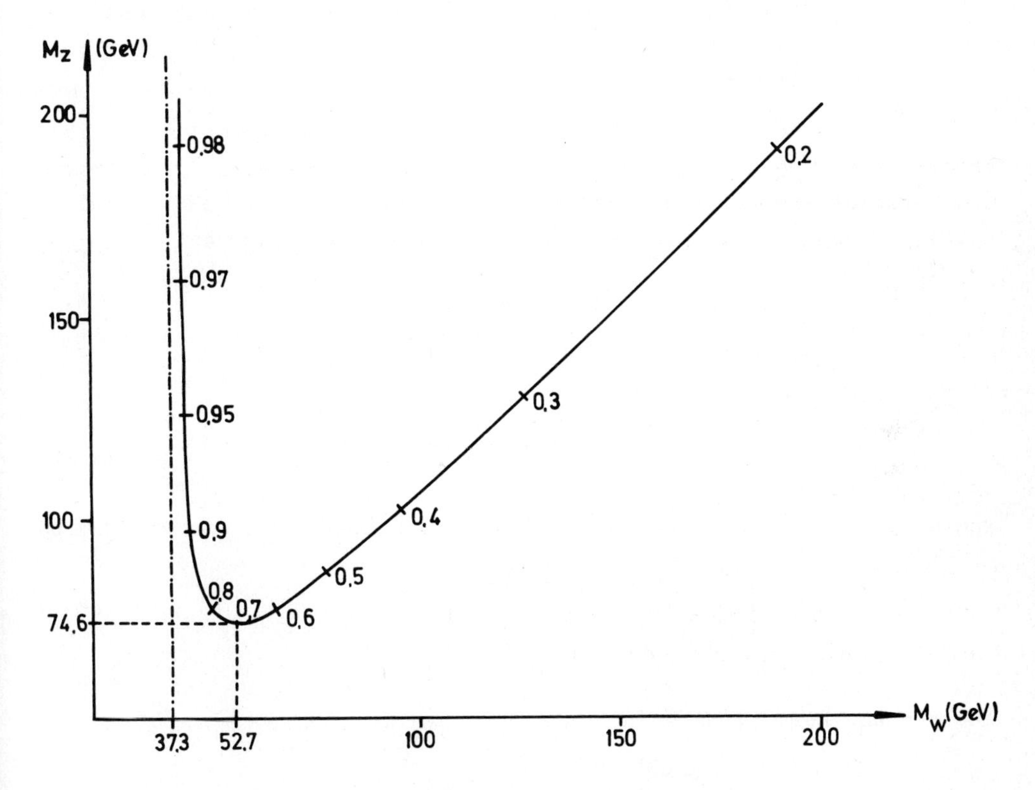

Fig. 2

Masses of the intermediate bosons as predicted by the Salam-Weinberg model. The parameter is sin ϕ.

We can now read off eq (61) several important relations:

$$\frac{G}{\sqrt{2}} = \frac{g^2}{8M_W^2} \qquad (62a) \qquad\qquad M_Z = \frac{M_W}{\cos\varphi} \qquad (62c)$$

$$M_W = \frac{g\lambda}{2} \qquad (62b) \qquad\qquad m_\chi = \lambda\sqrt{2h} \qquad (62d)$$

Combining eqs (62a), (62b) and (27) gives

$$\frac{G}{\sqrt{2}} = \frac{1}{2\lambda^2} = \frac{e^2}{8M_W^2 \sin^2\varphi} \tag{63}$$

proving eq (36). Eq (62c) allows for a new interpretation of the mixing angle. Its cosine is just the ratio of charged to neutral intermediate boson masses.

We have already mentioned the problem of including hadrons in the model. Table 1 shows the absence of neutral strangeness changing currents. How can we then incorporate hadrons if they do not couple to the Z-boson? All we know is that strangeness changing currents do not couple to the Z-boson. In fact, the coupling of strangeness conserving currents is a consequence of experimental findings described by D.C. Cundy in his talk[2].

To overcome the problem, let us first try to formulate it in a precise way. The standard weak hadronic currents expressed in terms of quark fields are written down in eqs (4). Omitting all inessential details, the currents can be combined in the following way:

$$\bar{q}(x)\, C_+\, q(x) = \bar{p}(x)\, n_C(x) \tag{64}$$

with

$$n_C(x) = n(x)\cos\vartheta + \lambda(x)\sin\vartheta \tag{65}$$

where p, n and λ are the quark fields. The matrix C_+ is defined by

$$C_+ = \begin{pmatrix} 0 & \cos\vartheta & \sin\vartheta \\ 0 & 0 & 0 \\ 0 & 0 & 0 \end{pmatrix} \tag{66}$$

and for the hermitian conjugate current we use

$$C_- = (C_+)^T \tag{67}$$

The requirement of gauge invariance necessitates the coupling of a neutral current containing

$$C_3 = [C_+, C_-] = \begin{pmatrix} 1 & 0 & 0 \\ 0 & -\cos^2\vartheta & -\frac{1}{2}\sin 2\vartheta \\ 0 & -\frac{1}{2}\sin 2\vartheta & -\sin^2\vartheta \end{pmatrix} \tag{68}$$

Written out fully, this current reads

$$\bar{q} C_3 q = \bar{p}p - \cos^2\vartheta\, \bar{n}n - \sin^2\vartheta\, \bar{\lambda}\lambda - \frac{1}{2}\sin 2\vartheta (\bar{n}\lambda + \bar{\lambda}n) \tag{69}$$

The last term in this current is a neutral strangeness changing piece and therefore unacceptable.

To circumvent this problem, we follow Glashow, Iliopoulos and Maiani[11]. Again, their suggestion is the "minimal" one in the sense that they get away with the smallest number of new particles. In addition to the 3 well-known quarks, they predict a new one. It carries the quantum numbers of the proton quark but is distinguished from it by a new quantum number, called "charm".
The quarks form a quartet now

$$q' = \begin{pmatrix} p' \\ p \\ n \\ \lambda \end{pmatrix} \tag{70}$$

where p' is the charmed quark. The current (64) is generalized to

$$\bar{q}' C_+' q' = \bar{p}' \lambda_c + \bar{p} n_c \tag{71}$$

with

$$C_+' = \begin{pmatrix} 0 & A \\ 0 & 0 \end{pmatrix} \tag{72}$$

and

$$A = \begin{pmatrix} -\sin\vartheta & \cos\vartheta \\ \cos\vartheta & \sin\vartheta \end{pmatrix} \tag{73}$$

n_c is defined in eq (65) and

$$\lambda_c = -n\sin\vartheta + \lambda\cos\vartheta \tag{74}$$

Eq (68) generalizes to

$$C_3' = [C_+', C_-'] = \begin{pmatrix} I & 0 \\ 0 & -I \end{pmatrix} \tag{75}$$

so that there are no off-diagonal matrix elements for the neutral current. In other words, there are no neutral strangeness changing transitions. The price was, as often, the introduction of a new quantum number, the charm.

Let us work out this idea a bit more detailed. In analogy to the lepton case, we define 2 left-handed doublets

$$L_n = \frac{1+\gamma_5}{2} \begin{pmatrix} p \\ n_c \end{pmatrix} \tag{76a}$$

$$L_\lambda = \frac{1+\gamma_5}{2} \begin{pmatrix} p' \\ \lambda_c \end{pmatrix} \tag{76b}$$

The charged weak hadronic current of eq (71) can then be written as

$$\begin{aligned} &\bar{p}\gamma_\sigma(1+\gamma_5)n_c + \bar{p}'\gamma_\sigma(1+\gamma_5)\lambda_c = \\ &= 2(\bar{L}_n\gamma_\sigma\tau^+L_n + \bar{L}_\lambda\gamma_\sigma\tau^+L_\lambda) \end{aligned} \tag{77}$$

so that the triplet interaction is fixed to be

$$\frac{g}{2}(\bar{L}_n\vec{\tau}\gamma_\mu L_n + \bar{L}_\lambda\vec{\tau}\gamma_\mu L_\lambda)\vec{A}_\mu \tag{78}$$

In order to find the singlet interaction we follow beaten tracks and define 4 right-handed singlets

$$R_p = \frac{1-\gamma_5}{2}p \tag{79a}$$

$$R_n = \frac{1-\gamma_5}{2}n \tag{79b}$$

$$R_\lambda = \frac{1-\gamma_5}{2}\lambda \tag{79c}$$

$$R_{p'} = \frac{1-\gamma_5}{2}p' \tag{79d}$$

The electromagnetic current can then be expressed through left-handed doublets and right-handed singlets in the following way

$$
\begin{aligned}
j_\mu^{e.m.} &= \frac{2}{3}\bar{p}\gamma_\mu p + \frac{2}{3}\bar{p}'\gamma_\mu p - \frac{1}{3}\bar{n}\gamma_\mu n - \frac{1}{3}\bar{\lambda}\gamma_\mu\lambda = \\
&= \frac{1}{2}\bar{L}_n\tau_3\gamma_\mu L_n + \frac{1}{6}\bar{L}_n\gamma_\mu L_n + \frac{2}{3}\bar{R}_p\gamma_\mu R_p - \frac{1}{3}\bar{R}_n\gamma_\mu R_n + \\
&+ \frac{1}{2}\bar{L}_\lambda\tau_3\gamma_\mu L_\lambda + \frac{1}{6}\bar{L}_\lambda\gamma_\mu L_\lambda + \frac{2}{3}\bar{R}_{p'}\gamma_\mu R_{p'} - \frac{1}{3}\bar{R}_\lambda\gamma_\mu R_\lambda
\end{aligned} \tag{80}
$$

This fixes the interaction Lagrangian in accord with our general prescription which led to eq (24).

$$
\begin{aligned}
\mathcal{L}_{GIM} &= \frac{g}{2}\left(\bar{L}_n\vec{\tau}\gamma_\mu L_n + \bar{L}_\lambda\vec{\tau}\gamma_\mu L_\lambda\right)\vec{A}^\mu + \\
&+ g'\Big(\frac{1}{6}\bar{L}_n\gamma_\mu L_n + \frac{2}{3}\bar{R}_p\gamma_\mu R_p - \frac{1}{3}\bar{R}_n\gamma_\mu R_n + \\
&+ \frac{1}{6}\bar{L}_\lambda\gamma_\mu L_\lambda + \frac{2}{3}\bar{R}_{p'}\gamma_\mu R_{p'} - \frac{1}{3}\bar{R}_\lambda\gamma_\mu R_\lambda\Big)B^\mu
\end{aligned} \tag{81}
$$

Just as in the case of the leptons, we are of course interested in the coupling of the physical vector bosons, i.e. in the measurable neutral currents. To bring this out, we have to invoke the mixing of eq (25) again. After some algebra, we obtain

$$
\begin{aligned}
\mathcal{L}_{GIM} &= \text{charged part} + e A^\mu j_\mu^{e.m.} + \\
&+ \frac{g}{4\cos\varphi} Z^\mu\Big\{\bar{p}\gamma_\mu p\left(1-\frac{8}{3}\sin^2\varphi\right) + \bar{p}\gamma_\mu\gamma_5 p - \\
&- \bar{n}\gamma_\mu n\left(1-\frac{4}{3}\sin^2\varphi\right) - \bar{n}\gamma_\mu\gamma_5 n + \\
&+ \bar{p}'\gamma_\mu p'\left(1-\frac{8}{3}\sin^2\varphi\right) + \bar{p}'\gamma_\mu\gamma_5 p' - \\
&- \bar{\lambda}\gamma_\mu\lambda\left(1-\frac{4}{3}\sin^2\varphi\right) - \bar{\lambda}\gamma_\mu\gamma_5\lambda
\end{aligned} \tag{82}
$$

The charged weak hadronic current is given in eq (77). Using the definitions eqs (65) and (74), it can be spelled out more explicitly in the following way

$$2(\bar{L}_n \gamma_\mu \tau^+ L_n + \bar{L}_\lambda \gamma_\mu \tau^+ L_\lambda) =$$
$$= \cos\vartheta \left[\bar{p}\gamma_\mu(1+\gamma_5)n + \bar{p}'\gamma_\mu(1+\gamma_5)\lambda\right] + \tag{83}$$
$$+ \sin\vartheta \left[\bar{p}\gamma_\mu(1+\gamma_5)\lambda - \bar{p}'\gamma_\mu(1+\gamma_5)n\right]$$

It is seen, that charmed particles decay into strange particles predominantly, because their decay into non-strange particles is suppressed by a factor $\sin^2\vartheta$. On the other hand, it seems that scattering processes are well described by "valence quark dominance", that is to say that presence of strange (and charmed) quarks (plus their antiquarks) in ordinary targets is negligible. In this case, the creation probability (by weak processes) for charmed particles is suppressed by a factor $\sin^2\vartheta$.

Before we conclude this section, we should compare the explicit computations we have performed so far with one alternative at least. In this way, we shall see how difficult it is indeed to deviate from the "minimal" assumptions without predicting too many new particles. It might be bothersome to invent the new quantum number "charm". An alternative has been suggested in the model of Bég and Zee[12]. Before we sketch this model let us recall yet another complication. The number of quarks has to be triplicated by introducing "color" as a new quantum number. There are at least 3 reasons to do so: Firstly, the decay of the π^0-meson[13], secondly, quark statistics[13] and thirdly, the problem of triangle anomalies[14]. This leaves us with 12 quarks in the model described above. However, the 3 sets of quarks with different color are treated in a symmetric way. Bég and Zee relax this symmetry, but they get away with 9 quarks only. The prize in the lepton system is that they loose lepton-hadron universality. To restore it, they predict a heavy neutral lepton which mixes with the neutrino. More specifically, they define 2 left-handed doublets, one right-handed doublet and 2 left-handed singlets in the following way

$$L^+ = \frac{1+\gamma_5}{2}\begin{pmatrix} \frac{1}{3}\nu + \frac{\sqrt{8}}{3}E^0 \\ e \end{pmatrix} ; \qquad L^- = \frac{1+\gamma_5}{2}\begin{pmatrix} \frac{1}{3}\nu - \frac{\sqrt{8}}{3}E^0 \\ e \end{pmatrix} \tag{84a}$$

$$R = \frac{1-\gamma_5}{2} \begin{pmatrix} E^0 \\ e \end{pmatrix} \qquad (84b)$$

$$S^+ = \frac{1+\gamma_5}{2} \left(\frac{\sqrt{8}}{3} \nu + \frac{1}{3} E^0 \right) ; \quad S^- = \frac{1+\gamma_5}{2} \left(-\frac{\sqrt{8}}{3} \nu + \frac{1}{3} E^0 \right) \qquad (84c)$$

where E^0 is the heavy neutral lepton.

The reason, they get away without a charmed quark is that the neutral current has the following simple form:

$$\mathcal{L}^{neutral}_{BZ} = \frac{g}{2\cos\varphi} Z_\mu \left[\frac{1}{18} \sum_\ell \bar{\nu}_\ell \gamma^\mu (1+\gamma_5) \nu_\ell + (1-\sin^2\varphi)\, j^\mu_{el.magn.} \right] \qquad (85)$$

Hence the 5 parameters of the phenomenological description eq (19) are, in the case of the Bég-Zee model

$$\begin{aligned} g_1 &= \frac{1}{18} \\ g_L = g_R &= 1 - 2\sin^2\varphi \\ \lambda &= 0 \\ \rho &= 1 - 2\sin^2\varphi \end{aligned} \qquad (86)$$

The Higgs-system is also more complicated than in the Salam-Weinberg model. On the other hand, because of the mixing of eqs (84), the lower limit for the W-meson mass is 13 GeV.

The last question, we want to touch upon is naturally that of experimental distinction of various models. Clearly, this is a difficult task. Let us just hint at some possibilities. Take total cross-sections for neutral current neutrino (and anti-neutrino) scattering. They can be expressed by the 3 well-known structure functions in the scaling region

$$\sigma^{(\nu,\bar{\nu})}_{tot} = \frac{G^2}{\pi} E_\nu M \int_0^1 dx \left\{ \frac{1}{3} x F_1(x) + \frac{1}{2} F_2(x) \mp \frac{1}{3} x F_3(x) \right\} \qquad (87)$$

The upper sign refers to ν-scattering, the lower to $\bar{\nu}$. The structure function F_3 stems from V-A interference. Thus it is absent in the electromagnetic case. For an isoscalar target, because of

$$F_i^{\nu p} = F_i^{\bar{\nu} n} \qquad i = 1,2,3 \tag{88}$$

we can drop all indices from the structure functions. In this case, the whole difference in total cross-sections for neutrinos and anti-neutrinos stems from V-A interference. In the model of Bég and Zee, there is no such interference for neutral currents and the total cross-sections have to be equal. This should be testable rather easily and provides one way to support or rule out a model. Needless to say that there are many others which is one reason why the discovery of neutral currents in experiments and of gauge models in theory was such an appreciable stimulus for elementary particle physics.

REFERENCES

0. Notation and definitions follow
   H. Pietschmann, Formulae and Results in Weak Interactions (Springer Verlag Wien - New York, 1974).
1. K. Kleinknecht, Proc. XVII. Int. Conf. High Energy Physics, London (1974)
2. D.C. Cundy, Proc. XVII. Int. Conf. High Energy Physics, London (1974)
3. T.D. Lee and C.N. Yang, Phys. Rev. 119, 1410 (1960)
4. A. Pais and S. Treiman, Phys. Rev. D9, 1459 (1974)
5. C.H. Llewellyn Smith and D.V. Nanopoulos CERN preprint
6. B.C. Barish, Proc.XVII. Int. Conf. High Energy Physics London (1974)
7. A. Salam and J.C. Ward, Phys. Lett. 13, 168 (1964)
   A. Salam, Proc. 8th Nobel Symposium (ed. N. Svartholm, Almqvist and Wiksell, Stockholm 1968)
   S. Weinberg, Phys. Rev. Lett. 19, 1264 (1967) and 27, 1688 (1971)
8. M. A. Bouchiat and C.C. Bouchiat, Phys. Lett. 48B, 111 (1974)
9. P.W. Higgs, Phys. Rev. Lett. 13, 508 (1964) and Phys. Rev. 145, 1156 (1966)
   T.W.B. Kibble, Phys. Rev. 155, 1554 (1967)
10. G.'t Hooft, Nucl. Phys. B33, 173 (1971) and B35, 167 (1971)
    G.'t Hooft and M. Veltman, "Diagrammar" CERN 73-9 (1973)

11. S.L. Glashow, J. Iliopoulos, L. Maiani, Phys. Rev. D2, 1285 (1970)

12. M.A.B. Bég and A. Zee, Phys. Rev. Lett. 30, 675 (1973) and Phys. Rev. D8, 1460 (1973)

13. M. Gell-Mann, Acta Phys. Austriaca Suppl. 9, 733 (1972)

14. C. Bouchiat, J. Iliopoulos, Ph. Meyer, Phys. Lett. 38B,519 (1972)

OSTERWALDER-SCHRADER POSITIVITY IN CONFORMAL INVARIANT QUANTUM FIELD THEORY*

by

G. Mack

Institut für Theoretische Physik der Universität Bern,
Switzerland
Institute for Advanced Study, Princeton, New Jersey 08540

In the present notes we will describe an attempt to gain some insight into the nature of the axiomatic positivity constraint in local quantum field theory. It has recently become clear that it is advantageous to look at quantum field theory in terms of its Euclidean Green functions, also called Schwinger functions. As Osterwalder and Schrader [2], and also Glaser [3], have pointed out, spectrum condition and positivity of the Wightman functions (= vacuum expectation values of products of local fields) are equivalent, modulo other axioms, to a new type of positivity condition which must be satisfied by the Euclidean Green functions. We call this the OS-positivity. After some explanation of the preceding remarks in Sec. 1 we will restrict our attention to exactly conformal invariant quantum field theories. Such theories will hopefully describe the short distance behavior of more realistic theories, as was explained elsewhere [1]. In any case, they are interesting as a laboratory, because they can be analyzed to a remarkable extent by nonperturbative technique. The present note will further exemplify this, for it turns out that in conformal invariant theories the OS-positivity condition can be analyzed by group theoretical methods and that surprisingly simple sufficient conditions can be found for its validity. We think that there is a good chance that these conditions are also necessary; it will be explained in Sec. 5 why that is so.

The work reported here is still in progress at the time of this writing and is published here for the first time. It is the purpose of

---

* The present notes present the content of part of the lectures which the author presented at the Bonn summer school 1974. The rest of the material covered there can be found in Reference 1.

these notes to get the readers interested in this new development at an early stage, and we will concentrate here on the main ideas. We are confident however that the analysis can be made rigorous and complete by working out all the technical details (such as growth properties of Q^{χ}- functions in χ , equivalence relations at integer points, etc.).

1. Euclidean Green Functions

Let us consider a local quantum field theory which satisfies the usual postulates [4] (Wightman axioms): locality, spectrum condition, positivity, Poincaré invariance, uniqueness of the vacuum and temperedness (i.e. some distribution theoretic properties).

For simplicity consider a theory of one hermitian scalar fundamental field $\phi(x)$. Thus we are given a Hilbert space $\mathcal{H}$ of physical states, a vacuum Ω in $\mathcal{H}$ and a field $\phi(x)$ which becomes an operator in $\mathcal{H}$ after smearing with test functions.

Poincaré invariance means that there exists a representation of the Poincaré group by unitary operators $U(a,\Lambda)$ such that

$$U(a,\Lambda)\,\phi(x)\,U(a,\Lambda)^{-1} = \phi(\Lambda x + a)\,; \quad U(a,\Lambda)\Omega = \Omega$$

with $\{a,\Lambda\}$ standing for a Lorentz transformation by Λ followed by a translation by a.

Locality says that the commutator $[\phi(x),\phi(y)] = 0$ if $(x-y)^2 < 0$.

Positivity is the statement that all nonzero state vectors have positive norm, so that $(\Psi,\Psi) \geq 0$ in general. This is true if $\mathcal{H}$ is a Hilbert space.

Spectrum condition requires that the spectrum of the Hamiltonian (= generator of time translations) is positive semidefinite. It follows that

$$\int d^4x\, e^{-ipx}\, U(x)\Psi = 0 \qquad \text{unless} \qquad p \in \bar{V}_+$$

with $\bar{V}_+$ the closed forward lightcone and $U(x) = U(x,1)$.

In addition there are distribution theoretic requirements as we said, we shall not go into them (see [14]).

Consider now state vectors obtained by applying the field to the vacuum

$$\Psi(x_1 \ldots x_n) = \Phi(x_1) \ldots \Phi(x_n)\, \Omega \tag{1.1}$$

they should belong to $\mathcal{H}$ after smearing with test functions. The Wightman functions are defined by

$$W(x_1 \ldots x_n) = (\Omega, \Phi(x_1) \ldots \Phi(x_n)\Omega) = (\Omega, \Psi(x_1 \ldots x_n)) \tag{1.2a}$$

Because of hermiticity of the field $\Phi(x)$, one has the more general relation

$$(\Psi(x_1' \ldots x_m'), \Psi(x_1 \ldots x_n)) = W_{m+n}(x_m' \ldots x_1', x_1 \ldots x_n) \tag{1.2b}$$

Moreover, by translation invariance.

$$\Psi(x_1 \ldots x_n) = U(x_1)\Phi(0)U(x_2 - x_1)\Phi(0) \ldots U(x_n - x_{n-1})\Phi(0)\,\Omega$$

Consider now the Fourier transform $\tilde{\Psi}(p;q_1 \ldots q_{n-1})$ of this, considered as a function of x_1; $x_2 - x_1, \ldots x_n - x_{n-1}$. Clearly, because of the spectr condition as stated above,

$$\tilde{\Psi}(p;q_1 \ldots q_{n-1}) = 0 \text{ unless } p \in \bar{V}_+,\ q_i \in \bar{V}_+ \quad (i=1 \ldots n-1) \tag{1.3}$$

The inverse Fourier transform gives back

$$\Psi(x_1 \ldots x_n) = (2\pi)^{-4n} \int dp\, dq_1 \ldots dq_{n-1}\, \tilde{\Psi}(p;q_1 \ldots q_{n-1}) \exp i\{p x_1 + \Sigma q_i (x_{i+1} \tag{1.4}$$

We will now pause for a moment to recall the notion of vector valued holomorphic function. Let $\mathcal{H}$ a normed space so that for every nonzero element $a \in \mathcal{H}$ its norm $\|a\| > 0$ is defined and positive. For a Hilbert space $\mathcal{H}$, $\|a\| = (a,a)^{1/2}$. A mapping f of a domain $D \subset \mathbb{C}$ into $\mathcal{H}$ is called a vector valued holomorphic function on D if for every z_0, f(z) can be expanded in a power series around z_0 with a non-zero radius ϱ of absolute convergence. That is $f(z) = \sum_{n=0}^{\infty} a_n z^n$ with $a_n \in \mathcal{H}$, $\Sigma \|a_n\| z^n < \infty$ whenever $|z-z_0| < \varrho$.

Such functions share all the properties of complex holomorphic functions. In the standard text book in analysis, ref. [10] the theory of holomorphic functions is developed right away for functions with values in an arbitrary normed space; the notion generalizes readily to functions of several variables.

Consider now the generalization of expression (1.4) to complex arguments z_1, viz

$$\psi(z_1 \ldots z_n) = (2\pi)^{-4n} \int dp\, dq_1 \ldots dq_{n-1}\, \hat{\psi}(p; q_1 \ldots q_{n-1})\, \exp i\{pz_1 + \sum q_i(z_{i+1} - z_i)\}$$

Because of the support property (1.3) of $\hat{\psi}$

ψ $(z_1 \ldots z_n)$ is defined and holomorphic for arguments $z_j = x_j + iy_j$

with $y_1 \in \bar{V}_+$ and $y_{j+1} - y_j \in \bar{V}_+$ for all $j = 1 \ldots n-1$

(1.5)

Points of special interest are the socalled Euclidean points, viz $x_j^0 = 0$ and $\underline{y}_j = 0$ (imaginary time and real space coordinates). Define the "Euclidean state vectors" ψ^E,

$$\psi^E(\vec{x}_1 \ldots \vec{x}_n) = \psi(z_1 \ldots z_n) \qquad \text{for } z = (ix_k^4, \underline{x}_k); \quad \vec{x} = (\underline{x}, x^4)$$

defined and real analytic for $x_n^4 > x_{n-1}^4 > \ldots x_1^4 > 0$ (1.6)

We are now ready to define the Euclidean Green functions $G(\vec{x}_1 \ldots \vec{x}_n)$, viz

$$G_n(\vec{x}_1 \ldots \vec{x}_n) = (\Omega, \psi^E(\vec{x}_1 \ldots \vec{x}_n)) \tag{1.7}$$

in analogy with (1.2a). As defined here and throughout this paper, G_n is always the full disconnected Green function, this must be kept in mind. To start with, it is defined for arguments as specified in (1.6). However, the restriction $x_1^4 > 0$ is unnecessary because G_n depends on its arguments only through their differences. Indeed the same is true for the Wightman function W_n by translation invariance, and G_n is the analytic continuation of the Wightman function W_n because ψ^E is the analytic continuation of ψ $(x_1 \ldots x_n)$.

Let us now introduce the Euclidean time reversal operator Θ, which reverses x^4,

$$\Theta(\underline{x}, x^4) = (\underline{x}, -x^4) \tag{1.8}$$

Θ is really a complex conjugation of the complex variable z, because $z = (ix^4, \underline{x})$ implies $\bar{z} = (-ix^4, \underline{x}) = (i\Theta x^4, \Theta\underline{x})$.

Consider now the scalar product of two Euclidean state vectors. We find from Eq. (1.2b) by antianalytic continuation in the first m arguments and analytic continuation in the last n arguments that

$$(\psi^E(\vec{x}_1'\ldots\vec{x}_m'),\psi^E(\vec{x}_1\ldots\vec{x}_n)) = G_{m+n}(\Theta\vec{x}_m'\ldots\Theta\vec{x}_1'\,\vec{x}_1\ldots\vec{x}_n) \tag{1.9}$$

for arguments

$$\Theta x_m'^4 < \Theta x_{m-1}'^4 < \ldots < \Theta x_1'^4 < 0 < x_1^4 < \ldots < x_n^4$$

Suppose now that there is given a finite sequence (f) of test functions $f_0 \in \mathbb{C}$, $f_1(x_1)\ldots f_N(x_1\ldots x_N)$ with support in the domain of definition of ψ^E, i.e. $0 < x_1^4 < \ldots < x_k^4$ or else $f_k(\vec{x}_1\ldots\vec{x}_k) = 0$. Then

$$\psi^E(f) \equiv \sum_k \int d^{4k}x\, f_k(\vec{x}_1\ldots\vec{x}_k)\,\psi^E(\vec{x}_1\ldots\vec{x}_k) \tag{1.10}$$

is an element of the Hilbert space $\mathcal{H}$ of physical states and thus must have nonnegative norm $(\psi^E(f),\psi^E(f)) \geqslant 0$. By (1.9) this norm is expressible in terms of the Euclidean Green functions. Thus

$$\sum_{k,l} G_{k+l}(\Theta f_k^* \times f_l) \geqslant 0 \tag{E.2}$$

where

$$(\Theta f_k^* \times g_l)(\vec{x}_1'\ldots\vec{x}_k'\,\vec{x}_1\ldots\vec{x}_l) \equiv \bar{f}_k(\Theta\vec{x}_k'\ldots\Theta\vec{x}_1')\,g_l(\vec{x}_1\ldots\vec{x}_l)$$

and $G_n(h) = \int d^{4n}x\, h(\vec{x}_1\ldots\vec{x}_n)\, G_n(\vec{x}_1\ldots\vec{x}_n)$, integration being over Euclidean space. Inequality (E.2) is the OS-positivity condition, it is required to hold for finite sequences of test functions $f_k(x_1\ldots x_k)$ that vanish unless $0 < x_1^4 < \ldots < x_k^4$.

In the special case that only $f_2 \not\equiv 0$, inequality (E.2) reads explicitly

$$\int dx_1\ldots dx_4\, \bar{f}_2(\Theta\vec{x}_2,\Theta\vec{x}_1)\, G_4(\vec{x}_1\vec{x}_2\vec{x}_3\vec{x}_4)\, f_2(\vec{x}_3,\vec{x}_4) \geqslant 0 \tag{1.11}$$

So far our discussion has followed ref. 2.

Glaser has pointed out [3] that locality and the edge of the wedge theorem can be used to further extend the domain of definition and analyticity of the state vectors $\psi(z_1\ldots z_n)$ and, therefore, $\psi^E(x_1\ldots x_2)$:

Let π a permutation of 1...n and consider the vector

$$\psi_{\pi}(z_1 \ldots z_n) = \psi(z_{\pi 1} \ldots z_{\pi n}) = \text{analyt. cont. of } \phi(x_{\pi 1}) \ldots \phi(x_{\pi n})\Omega \quad (1.12)$$

This is holomorphic for $y_{\pi 1} \in \overline{V}_+$, $y_{\pi\, j+1} - y_{\pi\, j} \in \overline{V}_+$
Locality tells us that on Minkowski space

$$\psi_{\pi}(x_1 \ldots x_n) = \psi(x_1 \ldots x_n) \quad \text{for all } \pi \text{ if all } (x_i - x_j)^2 < 0 \quad (1.13)$$

Thus, the boundary values of the holomorphic functions $\psi(z_1 \ldots z_n)$ agree on a real neighborhood (they are known to be vector-valued distributions). The edge of the wedge theorem asserts that then the functions ψ_{π} are in fact analytic continuations of one and the same holomorphic function ψ $(z_1 \ldots z_n)$, this function is thus defined on the union of the original domains of definition of the original ψ_{π} and is symmetric in its arguments there by (1.13). Specializing to Euclidean arguments this contains the union over π of the sets of arguments with $0 < x^4_{\pi 1} < \ldots < x^4_{\pi n}$, i.e. n-tuples of Euclidean arguments with non-coinciding positive times $x^4_i > 0$, $x^4_i \neq x^4_j$ $(i \neq j)$. It can be shown [5] that the restriction to noncoinciding times can be weakened to non-coinciding arguments. Summing up:

> state vectors $\psi^E(\vec{x}_1 \ldots \vec{x}_n)$ in $\mathcal{H}$ are defined and (real) analytic for Euclidean arguments $\vec{x}_1 \ldots \vec{x}_n$ such that $\vec{x}_i \neq \vec{x}_j$ for $i \neq j$ and all $x^4_i > 0$. They are symmetric in their arguments, viz. $\psi^E(\vec{x}_{\pi 1} \ldots \vec{x}_{\pi n}) = \psi^E(\vec{x}_1 \ldots \vec{x}_n)$ for all permutations π.

Using this and the fact that the Green functions depend only on differences of their arguments, we see that the Euclidean Green functions $G_n(\vec{x}_1 \ldots \vec{x}_n)$ are defined by Eq. (1.7) for arbitrary noncoinciding arguments $\vec{x}_1 \ldots \vec{x}_n$. By symmetry of ψ^E,

$$G_n(\vec{x}_{\pi 1} \ldots \vec{x}_{\pi n}) = G_n(\vec{x}_1 \ldots \vec{x}_n) \quad \text{for all permutations } \pi \quad \text{(E.3)}$$

Finally it is known that the Euclidean Green functions so defined are invariant under the Euclidean Poincaré-group, viz

$$G_n(m\vec{x}_1 + \vec{a} \ldots m\vec{x}_n + \vec{a}) = G_n(\vec{x}_1 \ldots \vec{x}_n) \quad \text{(E.1)}$$

for arbitrary 4-rotations $m \in SO(4)$.

Osterwalder and Schrader have shown [2] that postulates (E.1),

(E.2) and (E.3) as stated above together with a standard cluster property and a rather involved distribution theoretic axiom are sufficient conditions to guarantee that the Euclidean Green functions G_n (n=0,1...) determine a local quantum field theory satisfying the Wightman axioms.

Lastly it may be worthwhile pointing out that positivity (E.2) will actually hold for more general finite sequences of Schwartz test functions than stated there; it suffices that they vanish with all their derivatives when $\vec{x}_i = \vec{x}_j$ for some $i \neq j$, or some $x_i^4 \leq 0$. This follows immediately from assertion (1.14).

2. Conformal Invariance

From now on we shall restrict our attention to quantum field theories whose Euclidean Green functions are exactly conformal invariant The Euclidean conformal group is $\mathcal{G} \simeq SO_\ell(5,1)$, it is compounded from [1]

4-rotations $\vec{x} \to m\,\vec{x}$, $m \in SO(4)$
translations $\vec{x} \to t_c \vec{x} = \vec{x} + \vec{c}$
dilations $\vec{x} \to \rho\vec{x}$, $\rho > 0$
special conformal transformations $\vec{x} \to R t_c R^{-1}$
where $R\vec{x} = \dfrac{\vec{x}}{x^2}$.

We will assume that our theory is parity invariant, it follows then that the Euclidean Green functions are invariant under Euclidean time reversal Θ . It is known that ΘR is an element of the identity component of the conformal group $\mathcal{G}$, thus parity invariance plus conformal invariance implies R-invariance.
Conversely R-invariance alone implies full conformal invariance in a Poincaré invariant theory, because special conformal transformations, 4-rotations and translations generate the whole group $\mathcal{G}$.

Let d the dimension of the fundamental field in the theory (the dimension is a new quantum number) then the requirement of R-invariance for the Euclidean Green functions is

$$G_n(\vec{x}_1 \ldots \vec{x}_n) = (x_1^2 \ldots x_n^2)^{-d}\, G_n(R\vec{x}_1 \ldots R\vec{x}_n)$$

One knows from Wilson's work [6] that the dimension d can in general be noninteger and is dynamically determined. Positivity of the 2-point function requires however that $d \geqslant 1$ resp. $d \geqslant \frac{1}{2}D-1$ in a world with D space time dimensions. In order not to have to distinguish between several cases we will assume that in fact $\frac{1}{2}D-1 \leq d \leq \frac{1}{2}D$. It is interesting to consider theories in an arbitrary even number D of space time dimensions, the considerations of Sec. 1 generalize immediately to this case.

3. Conformal Partial Wave Expansion

Given a conformal invariant Euclidean n-point Green function, graphically represented by a bubble with n legs, we can select two if its arguments and decompose it into terms corresponding to the exchange of definite conformal quantum numbers flowing between the selected pair of legs and the remaining ones. This is called the (Euclidean) conformal partial wave expansion [1].

Its virtues are first of all these: The connected Green functions in a Lagrangian quantum field theory are known to satisfy dynamical equations which amount to a coupled set of infinitely many nonlinear integral equations. All these integral equations are "solved" simultaneously by the conformal partial wave expansion, i.e. they are converted from integral equations to algebraic constraints. These algebraic constraints amount to demanding presence of certain simple poles in the partial waves qua analytic functions of the continuous conformal quantum number ("dimension") δ, with factorizing residues.

All this has been derived and discussed in detail in earlier lectures by the author which are already published [1]. We will therefore only review here very briefly the formulae which we will need later on.

The conformal quantum numbers (Casimir invariants of SO(5,1) resp. SO(D+1,1) D = # of space time dimensions) are $\chi = [\ell, \delta]$, with ℓ an SO(4)-spin (resp. SO(D)-spin) and δ a complex number, the "dimension". We will only be interested here in completely symmetric traceless tensor representations of SO(4), they can be characterized by their rank ℓ. Thus ℓ will be a nonnegative integer from now on.

There are several series of unitary representations of SO(D+1,1). If the space time dimension D is even, they are [7]

<u>identity representation</u> (1-dimensional)
<u>principal series</u>: ℓ arbitrary, $\delta = \frac{1}{2}D + i\sigma$, $-\infty < \sigma < \infty$
<u>supplementary series</u>: includes in particular $\ell = 0$, $0 < \delta < D$ real
<u>exceptional series:</u> associated with certain "integer points" (integer δ)

The exceptional series does never appear in the decomposition of the Euclidean Green functions for $D > 2$, and for $D=2$ it is equivalent to principal series representations [9] .

More generally, there exist (Banach-space) representations of the Euclidean conformal group $\mathcal{G}$ for arbitrary $\chi = [\ell, \delta]$, δ complex. They contain the unitary ones as special cases (resp. irreducible parts thereof for the exceptional series) and are constructed as induced representations as follows [1,8] .

The representation space $\mathcal{E}^{\chi}$ (or a dense subspace thereof) consists of functions $\varphi_{\alpha}(x)$ on Euclidean space $\{\vec{x}\}$, $\alpha = (\alpha_1 \ldots \alpha_\ell)$ being tensor indices, viz $\alpha_i = 1 \ldots 4$ resp. D. The little group H of $\vec{x} = 0$ consists of Euclidean 4-rotations $m \in M$, dilations $a \in A$ and special conformal transformations $n \in N$, viz $H = MAN$. The transformation law of functions $\varphi \in \mathcal{E}^{\chi}$ under $\Lambda \in \mathcal{G}$ is

$$(T(\Lambda)\varphi)_{\alpha}(\vec{x}) = D^{\chi}_{\alpha\beta}(h)\varphi_{\beta}(\Lambda^{-1}\vec{x}) \quad \text{(sum over } \beta\text{)} \tag{3.1}$$

Herein the little group element $h \in H$ depends on $\vec{x}$ and Λ and is given b

$$h = t_x^{-1} \Lambda t_{x'} \; , \quad \vec{x}' = \Lambda^{-1}\vec{x},$$

t_x the translation taking 0 to $\vec{x}$. The inducing representation D^{χ} is given by

$$D^{\chi}_{\alpha\beta}(man) = |a|^{-\delta} D^{\ell}_{\alpha\beta}(m)$$

if a is a dilation by $|a|$. D^{ℓ} is the completely symmetric ℓ-th rank tensor representation of $M \simeq SO(4)$, note that all $D^{\ell}_{\alpha\beta}(m)$ are therefore real. This representation can be extended to a representation of the group O(4) which is obtained by adjoining to M the reflection by Θ , so that also $D^{\ell}(\Theta)$ is defined. Representations $\chi = [\ell, \delta]$ and $-\chi = [\ell, D-\delta$ are equivalent except at integer points. In addition there are further "partial equivalences" at integer points which are very important but cannot be discussed here (see refs. 9 and 16).

Since we will be interested in discussing positivity, we will need to consider the full disconnected Green functions as we did througł out Sec. 1. In terms of the connected Green functions

disc. = + + etc. (3.2)

We will for simplicity consider a theory of one hermitian scalar field ϕ, e.g. a ϕ^3-theory. [The considerations can be extended to ϕ^4-theory with only small changes by introducing also the field $\phi^2(x)$ right from the start, i.e. a theory of two hermitian scalar fields having different dimensions.] The 2-point function is specified up to normalization by the dimension d of the field ϕ; we will fix its normalization by [1]

$$\text{———} = G(\vec{x}_1\vec{x}_2) = (2\pi)^{-\frac{1}{2}D}\,\Gamma(d)\,(\tfrac{1}{2}|\vec{x}_1-\vec{x}_2|^2)^{-d}/\Gamma(\tfrac{1}{2}D-d)$$

We will now write down the conformal partial wave expansion for the full disconnected 4-point Green function

$$\text{[diagram: disc, legs } \vec{x}_1,\vec{x}_2,\vec{x}_3,\vec{x}_4] = \sum\!\!\!\!\!\!\int d\chi\,[1+g(\chi)]\ \text{[diagram: exchange of } \chi] + \text{Id.} \tag{3.3}$$

$$= \sum\!\!\!\!\!\!\int d\chi\,[1+g(\chi)]\int dx\,\Gamma^{\chi}(\vec{x}_1\vec{x}_2|\vec{x})\,\Gamma^{\tilde{\chi}}(\vec{x}_3\vec{x}_4|\vec{x}) + \text{Id.}$$

Id. stands for a contribution belonging to χ = identity representation: it is given explicitly by the very first term on the right hand side (r.h.s) of Eq. (3.2), viz. Id. = $G_2(x_1x_2)\,G_2(x_3x_4)$.

The integration $\sum\!\!\!\!\!\int d\chi$ runs over a certain subset of the unitary representations of $\mathcal{G}$, we will come back to this below. All the dynamical information on the disconnected Greenfunction is in the partial wave amplitudes $1+g(\chi)$. It is written in this way to agree with the notation of ref.[1]. All the rest are kinematical factors determined by group theory.

Explicitly, the Clebsch Gordan kernels

$$\text{[diagram: } \vec{x}_1,\vec{x}_2 \to \vec{x},\alpha;\ \chi] = \Gamma^{\chi}_{\alpha_1\ldots\alpha_\ell}(\vec{x}_1\vec{x}_2|\vec{x})$$

$$= N(\chi)\,(\tfrac{1}{2}x_{12}^2)^{-d+\frac{1}{2}\delta-\frac{1}{2}\ell}\,(\tfrac{1}{2}x_{13}^2\,\tfrac{1}{2}x_{23}^2)^{-\frac{1}{2}(\delta-\ell)}\,(\hat{x}_{\alpha_1}\cdots\hat{x}_{\alpha_\ell}-\text{traces}) \tag{3.4}$$

with $x_{ij} = \vec{x}_i - \vec{x}_j$, $\hat{x}_\alpha = \frac{(x_{13})\alpha}{\frac{1}{2}x_{13}^2} - \frac{(x_{23})\alpha}{\frac{1}{2}x_{23}^2} = \nabla_{3\alpha} \ln(x_{23}^2/x_{13}^2)$

and for D = 2h

$$N(\chi) = (2\pi)^{-h} \left\{ \frac{\Gamma(-h+d+\frac{1}{2}\delta+\frac{1}{2}l)\Gamma(d-\frac{1}{2}\delta+\frac{1}{2}l)\Gamma(\frac{1}{2}\delta+\frac{1}{2}l)^2}{\Gamma(2h-d-\frac{1}{2}\delta+\frac{1}{2}l)\Gamma(h-d+\frac{1}{2}\delta+\frac{1}{2}l)\Gamma(h-\frac{1}{2}\delta+\frac{1}{2}l)^2} \right\}^{1/2}$$

Because of the equivalence of representations χ and $-\chi$, the partial waves can, and will be required to satisfy a symmetry relation

$$g(\chi) = g(-\chi) \tag{3.5}$$

Our choice of overall factors is such that [11] the Clebsch Gordan kernels for representations χ and $-\chi$ are related by

$$\Gamma_\beta^\chi(\vec{x}_1\vec{x}_2|\vec{x}) = \int dx' \, \Gamma_\alpha^{-\chi}(\vec{x}_1\vec{x}_2|\vec{x}')\Delta_{\alpha\beta}^\chi(\vec{x}',\vec{x}) \tag{3.6}$$

with intertwining kernel

$$-\Delta_{\alpha\beta}^\chi(\vec{x},\vec{x}') = n(\chi)(\tfrac{1}{2}x^2)^{-\delta}\{g_{\alpha_1\beta_1}(\vec{x})\dots g_{\alpha_l\beta_l}(\vec{x}) - \text{traces}\}$$

with

$$g_{\alpha\beta}(\vec{x}) = -\delta_{\alpha\beta} + 2x_\alpha x_\beta/\vec{x}^2$$

$$n(\chi) = (2\pi)^{-h}\frac{\Gamma(\delta+l)\Gamma(2h-\delta-1)}{\Gamma(h-\delta)\Gamma(2h-\delta+l-1)} \tag{3.7}$$

Since (3.5) is true for all χ, hence also for $-\chi$, it follows that $\Delta^{-\chi}$ is the inverse of Δ^χ in the convolution sense. The graphical notation used e. g. in (3.3) takes all this into account if we picture $\Gamma^{-\chi}$ as a bubble with a short wiggly line. Inserting expression (3.6) for Γ^χ into the rhs. of the second equality (3.3), a more symmetrical expression results, which involves however two x-integratio

We will not now write expansions for the higher n-point function connected or otherwise, they were given in ref. 1. The expansion for th connected 4-point Green function is obtained from (3.3) by substituting

$g(\chi)$ for $1+g(\chi)$ and omitting the contribution from the identity representation.

The dynamical integral equation [12] for Green functions mentioned before imply [1] for $g(\chi)$ that it should have a simple pole in δ for $\ell = 0$ at $\delta = d$, shortly: a pole at $\chi = [0,d]$. Its residue must be positive and fixes the square of the coupling constant. As a result, the integration over representations in (3.3) can be deformed to path integrals which run as follows:

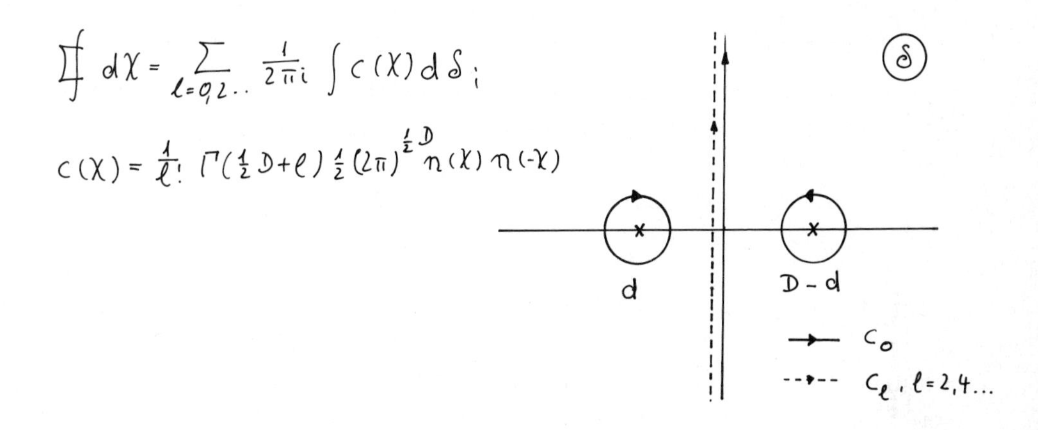

Fig. 1 paths of δ-integration

The Plancherel measure $c(\chi)$ is a polynomial. Note that the pole of $g(\chi)$ at $\chi = \chi_0 = [0,d]$ is accompanied by a brother at $-\chi_0$ by symmetry (3.5).

One finds in addition that $g(\chi)$ should also have a pole at $\chi = [2,D]$, this is required by the existence of a stress energy tensor.

Throughout this paper we will restrict our attention to theories in which the partial waves $g(\chi)$ are meromorphic function of δ for all ℓ with only simple poles. There is a good physical reason, for this assumption is necessary to have validity of operator product expansions with only a discrete number of (composite) fields appearing in them. This has also been discussed ref. 1. Further light on this analyticity assumption will be shed by our later considerations of positivity, where we will need the hypothesis that $g(\chi)$ are holomorphic in a cut δ-plane.

4. A Semigroup and Its Contractive Representations

We will now introduce a maximal noncommutative semigroup S contained in the Euclidean conformal group $\mathcal{G}$. It is defined to consist of those conformal transformations Λ in $\mathcal{G}$ which leave invariant the halfspace with positive Euclidean time x^4 resp. x^D:

$$\Lambda \in S \subset \mathcal{G} \text{ iff } 0 < x^4 < \infty \quad \text{implies} \quad 0 < (\Lambda x)^4 < \infty \tag{4.1}$$

and similarly for D space time dimensions. Henceforth such generalization will be left to the reader. In the 6-dimensional language, S consists of pseudo-orthogonal transformations of positive lightlike 6-vectors ξ which leave invariant the halfspace $\xi^4 > 0$.

Evidently S contains a subgroup $U \simeq SO_e(4.1)$ which consists of those pseudo-rotations which leave invariant ξ^4. It also contains the 1-parameter semigroup of pseudorotations $b_\tau(\tau \geqslant 0)$ in the 4-6 direction. This is seen as follows: Introduce hyperbolic coordinates, viz

$$\xi^k = r \cdot e^k \ (k = 1235)\ ;\ \xi^4 = r \sinh \sigma\ ;\ \xi^6 = r \cosh \sigma \tag{4.2}$$

Evidently $\xi^4 > 0$ if and only if $\sigma > 0$. Thus the halfspace $\xi^4 > 0$ is left invariant by the pseudorotations b_τ with $\tau \geqslant 0$ which translate the hyperpolic coordinate σ to $\sigma + \tau$. The generator H of this 1-parameter semigroup will be of special interest, it is defined by

$$b_\tau = e^{-H\tau} \tag{4.3}$$

The interior S^o of S is also a semigroup and so is $S^\sim = S^o \cup U$. It follows from the work of Toller and collaborators [13] that

$$\Lambda \in S^\sim \text{ iff } \Lambda = u_1 b_\tau u_2 \quad \text{with } u_i \in U \ (i=1,2),\ \tau \geqslant 0 \tag{4.4}$$

We observe that if Λ is in S resp. $S^\sim$ then so is $\bar{\Lambda} \equiv \Theta \Lambda^{-1} \Theta$.
Consider now any of the (possibly nonunitary) induced representations of the Euclidean conformal group $\mathcal{G}$ which were described in the last section, with representation space $\mathcal{E}^\chi$ consisting of functions $\varphi_\alpha(x)$ We restrict this representation to a representation of the semigroup $S^\sim \subset \mathcal{G}$.

One notes then that there exists a subspace $\mathcal{E}^\chi_+$ of $\mathcal{E}^\chi$ which consists of those functions which vanish in a halfspace

$$\mathcal{E}^{\chi}_{+} : \quad \varphi_{\alpha}(x) = 0 \quad \text{if } x^4 < 0 \tag{4.5}$$

It is evidently invariant under the action of the semigroup S.

It was discovered by Toller et al. that there also exists an invariant complement (or almost ~). That is another invariant subspace $\mathcal{E}^{\chi}_{-}$ such that the direct sum $\mathcal{E}^{\chi}_{+} \oplus \mathcal{E}^{\chi}_{-} = \mathcal{E}^{\circ\chi}$ is dense in the (Banach) space $\mathcal{E}^{\chi}$.

This decomposition corresponds to a split of functions $\varphi_{\alpha}(x) \in \mathcal{E}^{\circ\chi}$ as follows

$$\varphi_{\alpha}(\vec{x}) = \varphi^{+}_{\alpha}(\vec{x}) + \int dx' \, \Delta^{\chi}_{\alpha\beta}(\vec{x},\vec{x}') \, \varphi^{-}_{\beta}(\vec{x}') \tag{4.6}$$

$$\text{with} \quad \varphi^{\pm}_{\alpha}(\vec{x}) = 0 \quad \text{if } x^4 < 0$$

Note that the intertwining operator Δ^{χ} is involved, thus the possibility of the split must be related to the equivalence of representations χ and $-\chi$ of $\mathcal{G}$. We will only sketch an argument for the possibility of the split: Suppose that φ^{-} is already known, then obviously it is trivial to find also φ^{+}. Now φ^{-} is determined entirely by the values of $\varphi_{\alpha}(x)$ for $x^4 < 0$ and can be determined by solving the integral equation obtained from (4.6) by setting $x^4 < 0$ so that the first term on the rhs. is absent. This integral equation can be solved by partial wave expansion over the group $U \simeq SO(4.1)$. This follows from the fact that U acts transitively on the halfplane $x^4 > 0$. Restrictions on the functions φ that can be split come from the requirement that the U-partial waves of φ^{-} are well behaved at infinity qua functions of the continuous Casimir-invariant, this restricts the behaviour of φ for $x^4 < 0$ (in particular φ must be real analytic there).

We will next introduce a bilinear form on the representation space $\mathcal{E}^{-\chi}_{+}$ of $\tilde{S}$ by

$$(\varphi, \psi)_{\chi} = \Gamma(\tfrac{1}{2}D - \delta) \int dx\, dx' \, \bar{\varphi}_{\alpha}(\vec{x}) \, D^{\ell}_{\alpha\beta}(\Theta) \, \Delta^{\chi}_{\beta\gamma}(\Theta\vec{x},\vec{x}') \, \psi_{\gamma}(\vec{x}') \tag{4.7}$$

We ask ourselves for which χ the bilinear form (4.7) defines a positive semi-definite scalar product. For suitable choice of phase in $D^{\ell}(\Theta)$ the answer is given by the following

Lemma: If $\chi = [l, \delta]$ with either $l = 0$, $\delta > \frac{1}{2}D-1$ or $l > 0$, $\delta > D-2+l$, then $(\varphi, \varphi)_\chi \geqslant 0$, and $\| T(\Lambda) \| \leq 1$ for all $\Lambda \in S^\sim$ in the operator norm induced by the scalar product $(,)_\chi$.

Moreover, one checks by a straightforward computation that for real δ

$$(\varphi, T(\Lambda)\psi) = (T(\bar{\Lambda})\varphi, \psi) \quad \text{for } \Lambda \in S, \ \bar{\Lambda} = \Theta \Lambda^{-1} \Theta \tag{4.8}$$

It follows from this that the subspace of zero-norm vectors of $\mathcal{E}_+^{-\chi}$ is invariant under $S^\sim$.

Thus, for χ such as are specified in the lemma, we may complete the representation space $\mathcal{E}_+^{-\chi}$ to a Hilbert space $\mathcal{H}_+^{-\chi}$ after dividing out the invariant subspace of zero norm vectors, and we are supplied by a contractive representation of $S^\sim$ acting in this Hilbert space $\mathcal{H}_+^{-\chi}$, and satisfying a pseudo-hermiticity condition (4.8) which implies in particular that U is represented unitarily and b_τ is represented by selfadjoint contraction operators so that $H \geqslant 0$, selfadjoint. The last assertion holds because $\Theta H \Theta = -H$ and $\Theta u \Theta = u$ for $u \in U$. The operator H is called the conformal Hamiltonian for reasons explained elsewhere [5] .

We will not prove the lemma but make it understandable by mentioning the following theorem which was stated and proven by Lüscher and the author in Appendix C of ref. [5] .

<u>Theorem</u>: Let T a continuous representation of $S^\sim$ by contraction operators in a Hilbert space, satisfying the condition (4.8), viz $T(\Lambda) = T(\bar{\Lambda})^*$. Then T can be analytically continued to a unitary representation of the universal (∞-sheeted) covering group $\mathcal{G}^*$ of the Minkowskian conformal group $SO(4,2)/Z_2$ resp. $SO(D,2)/Z_2$.

Continuity is satisfied if $\| T(\Lambda)-1 \| \to 0$ when $\Lambda \to 1$ through values in $S^\sim$. It can be shown to be satisfied for the representations considered so far. Taking lemma and theorem together we see that we end up with a class of unitary representations of $\mathcal{G}^*$. They ought to be equivalent to the known analytic representations of $\mathcal{G}^*$ studied by Rühl [14] .

5. Positivity of the 4-Point Function

Let us start by rewriting the conformal partial wave expansion (3.3) of the 4-point function. We consider the Clebsch Gordan kernel $\Gamma_\alpha^{-\chi}(\vec{x}_3\vec{x}_4 \mid \vec{x})$ as a function of $\vec{x}$ and split in the manner of Eq. (4.6), viz.

$$\Gamma_\alpha^{\chi}(\vec{x}_3\vec{x}_4 \mid \vec{x}) = Q_\alpha^{\chi}(\vec{x}_3\vec{x}_4 \mid \vec{x}) + \int dx' \, \Delta^{\chi}_{\alpha\beta}(\vec{x},\vec{x}') \, Q_\beta^{-\chi}(\vec{x}_3\vec{x}_4 \mid \vec{x}') \quad (5.1a)$$

$$\text{for } x_3^4, x_4^4 > 0 \quad \text{with } Q_\alpha^{\pm\chi}(.. \mid \vec{x}) = 0 \text{ if } x^4 < 0 .$$

The notation takes into account the symmetry property (3.6) of Γ^χ under $\chi \to -\chi$. Because of Θ-invariance of Δ^χ and Γ^χ it follows that also

$$\Gamma_\alpha^{\chi}(\vec{x}_1\vec{x}_2 \mid \vec{x}) = Q^{\chi}_{\Theta\alpha}(\Theta\vec{x}_1,\Theta\vec{x}_2 \mid \Theta\vec{x}) + \int dx' \, \Delta^{\chi}_{\alpha\beta}(\vec{x},\vec{x}') \, Q^{-\chi}_{\Theta\beta}(\Theta\vec{x}_1,\Theta\vec{x}_2 \mid \Theta\vec{x}') \quad (5.1b)$$

$$\text{for } x_1^4, x_2^4 > 0 \quad \text{with } Q^{\pm\chi}_{\Theta\alpha} \equiv D^{\ell}_{\alpha\beta}(\Theta) \, Q_\beta^{\pm\chi}$$

The split (5.1a) may be performed in the manner sketched in Sec. 4. The Q^χ unfortunately turn out not to be good functions of $\vec{x}$, even though their U-partial waves are well defined, $\Gamma_\alpha^{\chi}(\vec{x}_3\vec{x}_4 \mid \vec{x})$ being a smooth function of $\vec{x}$ for $x^4 < 0$, x_3^4, $x_4^4 > 0$. In the present note we will for simplicity ignore this complication, and proceed heuristically.*

We will use a graphical notation

$$Q_\alpha^{-\chi}(\vec{x}_1\vec{x}_2 \mid \vec{x}) = \quad \text{if } x_1^4, x_2^4 > 0$$

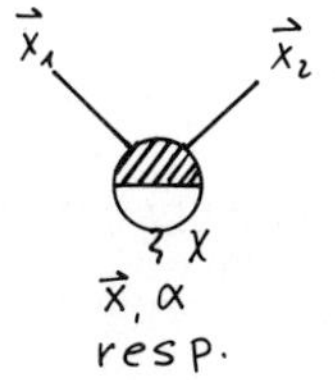

$$= Q^{-\chi}_{\Theta\alpha}(\Theta\vec{x}_1, \Theta\vec{x}_2 \mid \Theta\vec{x}) \quad \text{if } x_1^4, x_2^4 < 0 \quad (5.2)$$

---

* A related problem is that the Hilbert spaces $\mathcal{H}_+^\chi$ of Sec.4 do not anymore consist of equivalence classes of functions after completion in the norm, because Cauchy sequences need not converge in any function space topology. Both problems can be overcome by working with U-partial waves throughout.

Let us now consider the full disconnected Euclidean Green function $G_4(\vec{x}_1\vec{x}_2\vec{x}_3\vec{x}_4)$ for x_1^4, $x_2^4 < 0$; x_3^4, $x_4^4 > 0$. After inserting the split (5.1a) for one of the Γ^χ-kernels in the expansion (3.3), one has two terms. They can however be grouped together again by a change of χ -integration variable, using the symmetry property $g(\chi) = g(-\chi)$. Next one can use the identity

$\vec{x}_3$, $\vec{x}_4$, $\vec{x}_1$, $\vec{x}_2$, χ [diagram] = [diagram] $\vec{x}_3$, $\vec{x}_4$, $\vec{x}_1$, $\vec{x}_2$, χ

for $x_1^4, x_2^4 < 0;\ x_3^4, x_4^4 > 0$ (5.3)

This follows from the split (5.1b) and the support properties of $Q^{-\chi}$.

As a result we get the new expansion

[diagram: disc., $\vec{x}_1$, $\vec{x}_2$, $\vec{x}_3$, $\vec{x}_4$] = [diagram] $+ 2 \not\!\!\int d\chi\,[1+g(\chi)]$ [diagram: χ]

$$= G_2(\vec{x}_1\vec{x}_2)\,G_2(\vec{x}_3\vec{x}_4) + 2 \not\!\!\int d\chi\,[1+g(\chi)] \int dx\, dx'\, Q_\alpha^{-\chi}(\theta\vec{x}_1\,\theta\vec{x}_2\,|\,\vec{x})\, \mathcal{D}^{\ell}_{\alpha\beta}(\theta)\, \Delta^{\chi}_{\beta\gamma}(\vec{x},\vec{x}')\, Q_\gamma^{-\chi}(\vec{x}_3\vec{x}_4\,|\,\vec{x}')$$

Given a function $f_2(\vec{x}_1,\vec{x}_2)$ satisfying support properties stated after (E.2) of Sec. 1, let us define

$$\varphi_\alpha(\vec{x}) = \{N(\chi)N(-\chi)\}^{-\frac{1}{2}} \int dx_1\, dx_2\, f_2(\vec{x}_1\vec{x}_2)\, Q_\alpha^{-\chi}(\vec{x}_1\vec{x}_2\,|\,\vec{x})$$

$\varphi'_\alpha(\vec{x})$ = same with $\bar{f}_2$ in place of f_2.

Of course these functions depend on χ ($\varphi_\alpha \in \mathcal{E}_+^{-\chi}$ if $Q^{-\chi}$ were a good function).

We may now inspect the expression (1.11) which ought to be positive. Comparing with the definition (4.7) of the bilinear form $(,)_\chi$ and recalling that G_4 is symmetric in its arguments we find

$$G_4(\theta f_2^* \times f_2) = 2 \sum\!\!\!\!\!\!\int d\chi\, M(\chi)[1+g(\chi)](\bar{\varphi}', \varphi)_\chi + \mathrm{Id}. \tag{5.4a}$$

where $$M(\chi) = N(\chi) N(-\chi)\, \Gamma(\tfrac{1}{2}D-\delta)^{-1} \tag{5.4b}$$

and Id. stands for a contribution from the identity representation which is automatically positive by itself. The normalization factors N(.) where given in Eq. (3.4) and f.

We will now try to deform the path of the χ -integration in such a way that the assertion of the lemma becomes applicable.

Let us assume that the partial waves $g(\chi)$ satisfy growth conditions for $|\delta| \to \infty$ such that the path of the δ-integration can be closed to the right in Fig. 1. We also assume temporarily that there are no poles of the integrand in (5.4a) inside this closed path which come from the factors $M(\chi)(\bar{\varphi}', \varphi)_\chi$. Lastly, we observe that $\{N(\chi)N(-\chi)\}^{-1/2}\,\Gamma^\chi$ and Δ^χ are real for real δ , and so is therefore $\{N(\chi)N(-\chi)\}^{-1/2}\,Q^{-\chi}$ by its definition; it follows that $\bar{\varphi}' = \varphi$ for real δ . Suppose that $g(\chi)$ has a pole at $\chi = \chi_a = [l_a, \delta_a]$, (i.e. a pole in δ at δ_a for $l = l_a$). We define*

$$\underset{\chi_a}{\mathrm{res}_*}\,[1+g(\chi)] \equiv -c(\chi_a)\, M(\chi_a) \underset{\delta=\delta_a}{\mathrm{res}}\,[1+g(l_a, \delta_a)] \tag{5.5}$$

Note that the definition depends on the dimension d of the fundamental field through $M(\chi)$.

By Cauchy's theorem we have then

$$G_4(\theta f_2^* \times f_2) = 2 \sum_{\text{poles}} \mathrm{res}_*\,[1+g(\chi)]\,(\varphi, \varphi)_{\chi_a} + \mathrm{Id}' \tag{5.6}$$

---

* In the second paper of ref. 1 a factor $M(\chi)$ is missing in the statement of positivity constraints.

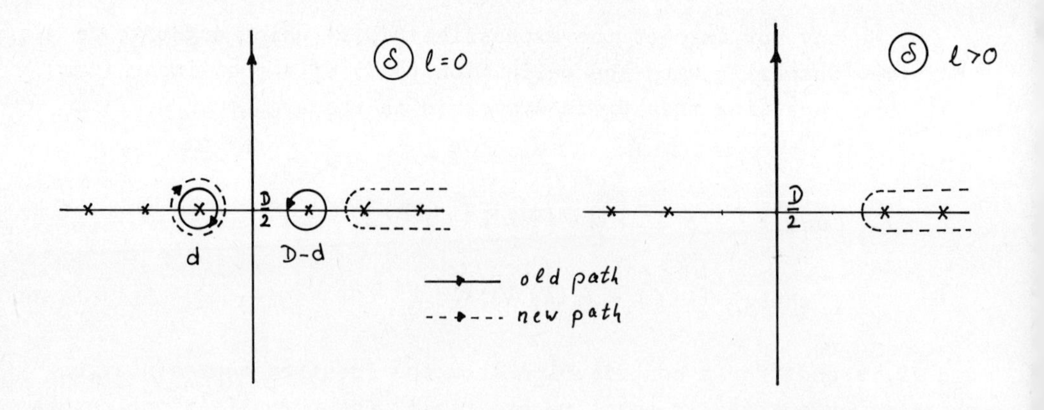

Fig. 2 Deformation of paths of δ-integrations.

with summation running over the pole of $g(\chi)$ at $\chi_o = [0,d]$ plus all poles with $\delta > \frac{1}{2}$ D except the pole at $-\chi_o$. We see that the expression is manifestly positive if all the residues of the aforementioned poles are positive and these poles are positioned at real $\delta_2 > D-2+\ell_a$ if $\ell_a > 0$, for then the hypotheses of the lemma in Sec. 4 are met.

We have so far disregarded possible singularities of the factors $M(\chi)(\bar{\varphi}', \varphi)_\chi$. Preliminary computations indicate that one must anticipate poles of two types: i) the poles of $N(\chi)N(-\chi)$ at $\delta = 2d+\ell+2n$, n = 0, 1,... ii) poles at certain integer points (δ integer). In order that such poles do not ruin positivity, partial waves $g(\chi)$ must satisfy additional kinematical constraints. Todorov and collaborators have recently shown [9] that in addition to (3.5) $g(\chi)$ must in any case satisfy further equalities relating its values at partially equivalent integer points, viz

$$g(\chi) = g(\chi') \text{ if } \chi = [\ell, D+\ell+n-1]\,;\; \chi' = [\ell+n, D+\ell-1], n = 2,4,...$$

It is hoped that these constraints will lead to cancellation of the contributions from integer point poles. Concerning poles i) it seems attractive, though not really necessary, to demand that they are cancelled by zeroes of $1 + g(\chi)$, so that expression (5.6) is valid without extra terms.

Summing up, positivity of the 4-point function will hold, if the partial waves $g(\chi)$ fulfill conditions of the following type as a function of $\chi = [\ell, \delta]$; $\ell = 0, 2 \ldots$

1. $g(\chi)$ is a meromorphic function of δ for each ℓ, with poles

only at real δ satisfying $|\delta - \frac{1}{2} D| > |\frac{1}{2} D-2+\ell|$ if $\ell > 0$. The residues of the pole at $\chi_0 = [0,d]$ and of all poles with $\delta > \frac{1}{2}$ D apart from $-\chi_0$ must be positive.

2. $g(\chi)$ satisfies growth conditions as $|\delta| \to \infty$ ($\text{Re}\,\delta > \frac{1}{2}$ D) so that the path of integration in (5.4a) can be closed to the right.

3. $g(\chi)$ must satisfy certain kinematical constraints related to the existence of kinematical poles of $M(\chi)(.,.)_\chi$, cp. text.

One can also look at the result in another way. The above conditions are imposed in order that the sum in (5.6) converges and consists of a sum of positive terms. That is

$$G_4(\Theta f_2^* \times f_2) = \sum_a r_a (\varphi,\varphi)_{\chi_a} + \text{ld}' \quad , \qquad \text{with } r_a > 0 \text{ and}$$

$$\chi_a = [\ell_a, \delta_a] \,, \quad \delta_a > \tfrac{1}{2} D - 1 \qquad \text{for } \ell = 0 \text{ and } \delta_a > D-2+\ell \text{ otherwise.}$$

We assume here and in the following that $1+g(\chi) = 0$ at the poles of $N(\chi)N(-\chi)$.

Because of the Cauchy-Schwartz inequality, an expansion of this type must then also be convergent if we smear with arbitrary test function $\Theta f_2^* \times g_2$ instead of $\Theta f_2^* \times f_2$, and thus in the distribution theoretic sense

$$G_4(\vec{x}_1 \ldots \vec{x}_4) = \sum_a r_a \; \chi_a + \text{ld}' = \sum_a r_a \; \chi_a + \text{ld}' \qquad (5.7)$$

The second equality follows from (5.3). We have thus arrived at an expansion of the type suggested in ref. [15] and Eq. (9.3) of ref.[1] except that it has not been shown that the Q^χ used here are the same (in some sense) as those used in ref.[1].

The results of Osterwalder and Schrader imply that the expansion

(5.7) remains valid, i.e. convergent when we analytically continue in the external arguments $\vec{x}_1 \ldots \vec{x}_4$ to Minkowski space through values $x_1{}^4 < x_2{}^4 < 0 < x_3{}^4 < x_4{}^4$ (x^4 = imaginary part of time) and we will in this way obtain an expansion of the Wightman function $W(x_1 x_2 x_3 x_4)$ We expect, in view of the theorem cited in Sec. 4, that if amounts to a partial wave expansion on the simply connected covering $\mathcal{G}^*$ of the Minkowskian conformal group $SO_e(4,2)/Z_2$.

$$W_4(x_1 x_2 x_3 x_4) = \sum_a r_a \int_{\substack{\text{Eucl. space} \\ x^4 > 0}} d^4\vec{x}\, \Gamma^{\chi_a}(\vec{z}_1 \vec{z}_2 | \vec{x})\, Q^{-\chi_a}(\vec{z}_3 \vec{z}_4 | \vec{x}) \tag{5.8}$$

$$\vec{z}_i = (\underset{\sim}{z}_i, z_i^4) = (\underset{\sim}{x}_i, -ix_i^0 + \varepsilon_i), \quad \varepsilon_1 < \varepsilon_2 < 0 < \varepsilon_3 < \varepsilon_4 \quad \text{infinitesimal}$$

Note that there is still an integration over half of Euclidean space involved as the formula stands now.

If the above interpretation is correct, one may even be able to show that the conditions for positivity of the 4-point function mentioned above are not only sufficient but also necessary ones. Absence of cuts in $g(\chi)$ needs a separate argument though, cp. Sec. 3, and also the precise form of the kinematical constraints is open to further study.

6. Generalization to Arbitrary n-point Functions

So far we have only investigated a special consequence of the positivity condition which involves the 4-point function alone. We now want to come to the general case.

Since we are presently interested in sufficient conditions for positivity, we will proceed by Ansatz.

Considering an m+n - point Green function, we try a decomposition into terms which correspond to exchange of definite conformal quantum numbers as follows

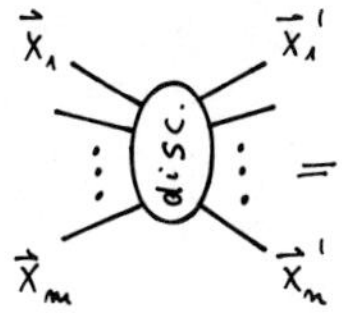

$= \sum_r \int\!\!\!\!\!\sum d\chi$

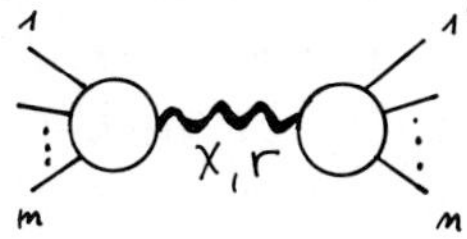

$$\equiv \sum_r \int dx\, dx'\, G_\alpha^{-\chi,r}(\vec{x}\,|\,\vec{x}_1 \ldots \vec{x}_m)\, \tilde{g}_r(\chi)\, \Delta^\chi_{\alpha\beta}(\vec{x},\vec{x}')\, G_\beta^{-\chi,r}(\vec{x}'\,|\,\vec{x}_1' \ldots \vec{x}_n') \tag{6.1}$$

We have introduced here a new notation: the χ -integration is now supposed to include also a contribution from the identity representation (We leave it to the reader to work out the necessary convections for such χ), and a fat wiggly propagator

$$= \tilde{g}_r(\chi)\, \Delta^\chi_{\alpha\beta}(x,x')$$

r are some sort of internal quantum numbers. The $G^{-\chi,r}$ are conformal invariant, their transformation law as functions of the first argument is specified by $-\chi$ and given by Eq. (3.1), similarly for the other arguments ($\ell = 0, \delta = d$ there).

Our Ansatz consists in demanding that an expansion of the form (6.1) is valid with an integrand that factorizes. That is, for arbitrary number of arguments $x_1' \ldots x_n'$, the factor $G_\alpha^{-\chi,r}(\vec{x}\,|\vec{x}_1 \ldots \vec{x}_m)$ is always the same, and similarly for the second factor.

In case no summation over internal quantum numbers is necessary, we put

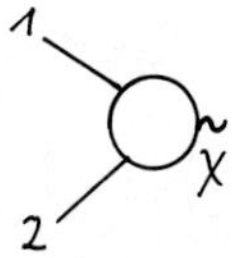

$$= G_\alpha^{-\chi}(\vec{x}\,|\,\vec{x}_1,\vec{x}_2) = \vec{\Gamma}_\alpha^{-\chi}(\vec{x}_1,\vec{x}_2\,|\,\vec{x}) = \tag{6.2}$$

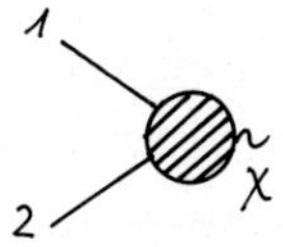

so that

$$\tilde{g}(\chi) = 1 + g(\chi)$$

For m=2, partial wave expansions of the form (6.1) were already considered in ref. 1, and we know from the discussion given there that

the partial waves for arbitrary n = 2,3,... should share the poles of $g(\chi)$. For these poles are in correspondence with the local fields (including composite ones) in the theory, and their positions indicate tensor character and dimension of such fields.
For this reason we have pulled out a factor $1+g(\chi)$ resp $\tilde{g}_r(\chi)$ in Eq. (6.1), it is supposed to contain all the "dynamical" poles of the integrand, while the factors $G^{-\chi,r}$ will be assumed to be holomorphic (apart from the possibility of certain kinematical singularities. They need a separate discussion much as in the case of the 4-point function).

Let there be given a finite sequence of test functions f_o, $f_1(\vec{x}_1),\ldots f_N(\vec{x}_1\ldots\vec{x}_v)$ with support properties as stated after (E.2) in Sec. 2. We define

$$\psi_\alpha^r(\vec{x}) = \sum_k \int dx_1\ldots dx_k\, f_k(\vec{x}_1\ldots\vec{x}_k)\, G_\alpha^{-\chi,r}(\vec{x}|\vec{x}_1\ldots\vec{x}_k) \tag{6.3}$$

and

$$\psi_\alpha'^r(\vec{x}) = \quad \text{same with } \bar{f}_k \text{ in place of } f_k.$$

Of course they depend on χ. With this notation,

$$\sum_{k,l} G_{k+l}(\Theta f_k^* \times f_l) = \sum_r \not{\int} d\chi\, \tilde{g}_r(\chi) \int dx\,dx'\, \psi_\alpha'^r(\Theta\vec{x})\, \Delta^\chi_{\alpha\beta}(\vec{x},\vec{x}')\, \psi_\beta^r(\vec{x}') \tag{6.4}$$

assuming Θ-invariance of $G^{-\chi,r}$, i.e. parity invariance of the theory.

We will now split the functions ψ, ψ' in the manner described in Eq. (4.6), viz.

$$\psi_\alpha^r(\vec{x}) = \varphi_\alpha^{+,r}(x) + \int dx'\, \Delta^\chi_{\alpha\beta}(x,x')\, \varphi_\beta^{-,r}(x') \tag{6.5}$$

and similarly for ψ'. Because of equivalence of representations χ and $-\chi$ of the Euclidean conformal group, partial waves $G^{\chi,r}$ will (or can be required to) share a symmetry property analog to (3.6). As a consequence the same is true for ψ, ψ' and so $\varphi^{+,r}$ at χ is the same as $\varphi^{-,r}$ at $-\chi$.

Inserting the split (6.5) into (6.4) and simplifying the result with the help of symmetry and support properties as in Sec. 5, we end up with

$$\sum_{k,l} G_{k+l}(\Theta f_k^* \times f_l) = \sum_r 2 \sum\!\!\!\!\!\!\int d\chi\, \hat{g}_r(\chi)\, M(\chi)\, (\bar{\varphi}'^r, \varphi^r)_\chi \qquad (6.6)$$

where $\varphi^r \equiv \varphi^{r,+}$ and $\varphi'^r \equiv \varphi'^{r,+}$

This equation is identical in appearance to Eq. (5.4a) of Sec. 5.

The further analysis then proceeds as in the case of the 4-point function. This results in some extra conditions in addition to those already satisfied by Ansatz or assumption stated above. They are of the same types as were stated in Sec. 5, i.e. the poles of $\tilde{g}_r(\chi)$ should be in permissible parts of the real δ-axis and have positive residues if $\delta > \frac{1}{2}$ D, $\chi \neq -\chi_0$. In addition there are growth conditions and kinematical constraints. However, in contrast to the Γ^χ, the partial waves $G^{\chi,r}$ are not in general completely determined by kinematics. Thus, the growth conditions will be conditions not only on $\hat{g}_r(\chi)$ but they also involve the partial waves $G^{-\chi,r}$. Also the kinematical constraints relating $G^{-\chi}$ at partially equivalent integer points will look more complicated; we will not give the explicit expressions here they are however implicit in ref. 9. A detailed study of the growth conditions has not yet been carried out at the time of this writing. Also the connection between the expansion (5.7) resp. (5.8) and operator product expansion [17] deserves further study.

ACKNOWLEDGEMENTS

The author is grateful to Professor I. Todorov for helpful comments and for communicating the results of ref. 9. He wishes to thank Professors K. Dietz, H. A. Kastrup and H. Rollnik for the invitation to lecture at this summer school. Thanks are also due to Professor Kaysen for hospitality at the Institute for Advanced Study where the manuscript was written.

REFERENCES

1. Mack, G: J. de Physique 34, Colloque C-1 (Suppl. au No. 10) 99 (1973) and in: Renormalization and Invariance in Quantum Field Theorie, E. R. Caianiello (ed.), New York: Plenum Press 1974.

2. Osterwalder, K., Schrader, R.: Commun. Math. Phys. 31, 83 (1973) and in preparation.
Osterwalder, K.: in: Constructive Quantum Field Theory, G. Velo and A.S. Wightman (Eds.) Lecture Notes in Physics 25, Springer Verlag Heidelberg 1973.

3. Glaser, V.: CERN prepring TH 1706 (1973), to appear in Commun. Math. Phys.

4. Streater, R.F., Wightman, A.S.: PCT, Spin and Statistics and all that. New York: W.A. Benjamin 1964.

5. Lüscher, M.: Mack, G.: Global conformal invariance in quantum field theory, preprint Bern (Aug. 1974), to be submitted to Commun. Math. Phys.

6. Wilson, K.: Phys. Rev. 179, 1499 (1969).

7. Hirai, P.: Proc. Japan. Acad. 38, 83, 258 (1962), 42, 323 (1965)

8. Warner, G.: Harmonic Analysis on Semi-simple Lie Groups, Vol. 1. Heidelberg: Springer Verlag 1972.

9. Dobrev, V.K., Petkova, V.B., Petrova, S.G., Todorov, I.T.: On exceptional integer points in the representation space of the pseudo orthogonal group $O^{\uparrow}(2h+1,1)$ (in preparation).

10. Dieudonné, F.: Foundations of modern analysis. New York: Academic Press 1969.

11. Dobrev, V.K., Mack, G., Petkova,V.B., Petrova, S.G., Todorov,I.T., On Clebsch Gordan expansion for O(2h+1,1). JINR preprint E2-7977, Dubna 1974.

12. Symanzik, K.: in: Lectures on High Energy Physics, Jaksic (ed.), Zagreb 1961, New York: Gordon and Breach 1965.

13. Ferrara, S., Mattioli, G., Rossi, G., Toller,M.: Nucl. Phys. B53, 366 (1973).

14. Rühl, W.: Commun. Math. Phys. 30, 287 (1973).

15. Polyakov, A.M., Non-Hamiltonian approach to the quantum field theory at small distances, preprint Chernogolovka.
See also: Migdal, A.A., 4-dimensional soluble models of conformal field theory, preprint Chernogolovka 1972.

16. Todorov, I.T.: Conformal expansions for Euclidean Green functions. Trieste: Report IC/74/67 (1974).

17. Ferrara, S., Gatto, R., Grillo, A.F.: Lettere Nuovo Cimento 2, 1363 (1971); A12; 952 (1972); Nucl. Phys. B34, 349 (1971). Ferrara, S., Gatto, R., Grillo, A.F., Parisi, G.: Nucl. Phys. B49, 77 (1972); Schroer, B., Swieca, J.A. (unpublished).

QUANTUM GRAVITY

G. t'Hooft

University of Utrecht, The Netherlands

1. Introduction

The gravitational force is by far the weakest elementary interaction between particles. It is so weak that only collective forces between large quantities of matter are observable at present, and it is elementary because it appears to obey a new symmetry principle in nature: the invariance under general coordinate transformations.

Ever since the invention of quantum mechanics and general relativity, physicists have tried to "quantize gravity"[1], and the first thing they realized is that the theory contains natural units of length (L), time (T) and mass (M). If

$$\kappa = 6.67 \cdot 10^{-11} \ \mathrm{m}^3/\mathrm{kg\ sec}^2$$

is the gravitational constant, then

$$L = \sqrt{\kappa\hbar/c^3} = 1.616 \cdot 10^{-35} \ \mathrm{m}$$

$$T = \sqrt{\kappa\hbar/c^5} = 5.39 \cdot 10^{-44} \ \mathrm{sec}$$

$$\text{and} \quad M = \sqrt{\hbar c/\kappa} = 1.221 \cdot 10^{28} \ \mathrm{eV}/c^2 = 2.177 \cdot 10^{-5} \ \mathrm{g}$$

But then the theory contains a number of obstacles.First there are the conceptual difficulties: the meaning of space and time in Einstein's general relativity as arbitrary coordinates, is very different from that of space and time in quantum mechanics. The metric tensor $g_{\mu\nu}$, which used to be always fixed and flat in quantum field theory, now becomes a local dynamical variable.

Advances have been made, from different directions[2,3,4], to devise a language to formulate quantum gravity, but then the next problem arises: the theory contains essential infinities such that a field theorist would say: it is not renormalizable. This problem, discussed in detail in section 12, may be very serious. It may very well imply that there exists no well determined, logical, way to combine gravity with quantum mechanics from first principles. And then one is led to the question: should gravity be quantized at all? After all, such quantum effects would be small, too small perhaps to be ever measurable. Perhaps the truth is very different, both from quantum theory and from general relativity.

Whatever one should, or should not do, our present picture of what happens at a length scale L and a time scale T is incomplete, and we would like to improve it. We claim that it is very worthwhile to try and improve our picture step by step, as a perturbation expansion in κ. In the following it is shown how to apply the techniques of gauge field theory and gain some remarkable results.

The sections 2-4 deal with the conventional theory of general relativity, seen from the viewpoint of a gauge field theorist. In section 5 it is indicated how quantization could be carried out in principle, but in practice we need a more sophisticated formalism to ease calculations.

This formalism, the background field method[2,5)], is explained in section 6-11 . In these sections we mainly discuss gauge theories, and gravity is hardly mentioned; gravity is just a special case here.

Back to gravity in section 12, where we discuss numerical results. It is shown there why only pure gravity is finite up to the one-loop corrections.

2. Gauge Transformations

The underlying principle of the theory of general relativity is invariance under general coordinate transformations,

$$x'^{\mu} = f^{\mu}(x) \quad . \tag{2.1}$$

It is sufficient to consider infinitesimal transformations,

$$x'^{\mu} = x^{\mu} + \eta^{\mu}(x) \text{ , } \eta \text{ infinitesimal.} \tag{2.2}$$

Or, in other words, a function $A(x)$ is transformed into

$$A'(x) = A(x + \eta(x)) = A(x) + \eta^{\lambda}(x)\partial_{\lambda}A(x) \quad . \tag{2.3}$$

If A does not undergo any other change, then it is called a scalar. We call the transformation (2.3) simply a gauge transformation, generated by the (infinitesimal) gauge function $\eta^\lambda(x)$, to be compared with Yang-Mills isospin transformations, generated by gauge functions $\Lambda^a(x)$.

For the derivative of A(x) we have

$$\partial_\mu A'(x) = \partial_\mu A(x) + \eta^\lambda{}_{,\mu}\partial_\lambda A(x) + \eta^\lambda\partial_\lambda\partial_\mu A(x) \ , \tag{2.4}$$

where $\eta^\lambda{}_{,\mu}$ stands for $\partial_\mu\eta^\lambda$, the usual convention. Any object A_μ transforming the same way, i. e.

$$A'_\mu(x) = A_\mu(x) + \eta^\lambda{}_{,\mu}A_\lambda(x) + \eta^\lambda\partial_\lambda A_\mu(x) \ , \tag{2.5}$$

will be called a covector. We shall also have contravectors $B^\mu(x)$ (note that the distinction is made by putting the index upstairs), which transform like

$$B^{\mu\prime}(x) = B^\mu(x) - \eta^\mu{}_{,\lambda}B^\lambda(x) + \eta^\lambda\partial_\lambda B^\mu(x) \ , \tag{2.6}$$

by construction such that

$$A_\mu(x)\ B^\mu(x)$$

transforms as a scalar. Similarly, one may have tensors with an arbitrary number of upper and lower indices.

Finally, there will be density functions $\omega(x)$ that transform like

$$\omega'(x) = \omega(x) + \partial_\lambda\ [\eta^\lambda(x)\ \omega(x)] \ . \tag{2.7}$$

They enable us to write integrals of scalars

$$\int\omega(x)\ A(x)\ d_4(x) \ ,$$

which are completely invariant under local gauge transformations (under certain boundary conditions).

For the construction of a complete gauge theory it is of importance that the gauge transformations form a group. Of course they do, and hence we have a Jacobi identity. Let u(i) be the gauge transformations generated by $\eta^\mu(i,x)$. Then if

$$[u(1),\ u(2)] = u(3) \ ,$$

then

$$\eta^{\mu}(3,x) = \eta^{\lambda}(2,x)\,\partial_{\lambda}\eta^{\mu}(1,x) - \eta^{\lambda}(1,x)\partial_{\lambda}\eta^{\mu}(2,x). \qquad (2.8)$$

3. The Metric Tensor

In much the same way as in a gauge field theory[6], we ask for a dynamical field that fixes the gauge of the vacuum by having a non-vanishing vacuum expectation value. (Contrary to the Yang-Mills case it seems to be impossible to construct a reasonable "symmetric" theory.) To this end we choose a two-index field, $g_{\mu\nu}(x)$, which is symmetric in its indices,

$$g_{\mu\nu} = g_{\nu\mu} , \qquad (3.1)$$

and its vacuum expectation value is

$$\langle g_{\mu\nu}(x) \rangle_0 = \delta_{\mu\nu} \qquad (3.2)$$

(our metric corresponds to a purely imaginary time coordinate).

With $g_{\mu\nu}$, or its inverse, $g^{\mu\nu}$, we can now define lengths and time-intervals at each point in space-time:

$$|\ell|^2 = g_{\mu\nu}\ell^{\mu}\ell^{\nu} ,$$

and $g_{\mu\nu}$ can be used to raise or lower indices:

$$A^{\mu} = g^{\mu\nu}A_{\nu} , \quad A_{\nu} = g_{\nu\mu}A^{\mu} , \text{ etc.} \qquad (3.3)$$

Just as in the Yang-Mills case, we can now define covariant derivatives:

$D_{\mu}A = \partial_{\mu}A$ (the derivative of a scalar transforms as a vector),

$$\begin{aligned} D_{\mu}A_{\nu} &= \partial_{\mu}A_{\nu} - \Gamma^{\alpha}_{\mu\nu}A_{\alpha} , \\ D_{\mu}B^{\nu} &= \partial_{\mu}B^{\nu} + \Gamma^{\nu}_{\mu\alpha}B^{\alpha} . \end{aligned} \qquad (3.4)$$

The field $\Gamma^{\alpha}_{\mu\nu}$ is called the Christoffel symbol and is yet to be defined. First we write down how it should transform under a gauge transformation, such that the above-defined covariant derivatives be real tensors:

$$\Gamma^{\lambda\prime}_{\mu\nu} = \Gamma^{\lambda}_{\mu\nu} + \text{(ordinary terms for 3-index tensor)} + \partial_{\mu}\partial_{\nu}\eta^{\lambda}. \qquad (3.5)$$

We see that no harm is done by making the restriction that

$$\Gamma^{\lambda}_{\mu\nu} = \Gamma^{\lambda}_{\nu\mu} , \qquad (3.6)$$

because the symmetric part of Γ only is enough to make (3.4) covariant.

We now define the field Γ by requiring

$$D_\lambda g_{\mu\nu} = 0 \quad , \tag{3.7}$$

(from which follows: $D_\lambda g^{\mu\nu} = 0$). We see that we have exactly the right number of equations. By writing Eq. (3.7) in full we find that it is easy to solve

$$\Gamma^\lambda{}_{\mu\nu} = \frac{1}{2} g^{\lambda\alpha}(\partial_\mu g_{\alpha\nu} + \partial_\nu g_{\alpha\mu} - \partial_\alpha g_{\mu\nu}) . \tag{3.8}$$

Note that $\Gamma^\lambda_{\mu\nu}$ is not a covariant tensor.

The covariant derivative may be used just like ordinary derivatives when acting on a product:

$$D(XY) = (DX)Y + XDY \quad , \tag{3.9}$$

but two covariant differentiations do not necessarily commute:

$$D_\mu D_\nu A_\alpha \neq D_\nu D_\mu A_\alpha \quad . \tag{3.10}$$

Instead, we have:

$$D_\mu D_\nu A_\alpha - D_\nu D_\mu A_\alpha = R^\beta{}_{\alpha\nu\mu} A_\beta \tag{3.11}$$

with

$$R^\beta{}_{\alpha\nu\mu} = \Gamma^\beta{}_{\alpha\mu,\nu} - \Gamma^\beta{}_{\alpha\nu,\mu} + \Gamma^\beta{}_{\tau\nu}\Gamma^\tau{}_{\alpha\mu} - \Gamma^\beta{}_{\tau\mu}\Gamma^\tau{}_{\alpha\nu} \tag{3.12}$$

The comma denotes ordinary differentiation.
Since the l.h.s. of eq. (3.11) is clearly covariant and A_β is an arbitrary vector, $R^\beta{}_{\alpha\nu\mu}$ transforms as an ordinary tensor, in contrast with $\Gamma^\beta{}_{\alpha\mu}$. It is called the Riemann or curvature tensor (see the standard text books). Indices can be raised or lowered following (3.3). Without putting in any further dynamical equation, one finds the following identities,

$$\begin{aligned} &R_{\alpha\beta\gamma\delta} = R_{\gamma\delta\alpha\beta} = - R_{\alpha\beta\delta\gamma} , \\ &R_{\alpha\beta\gamma\delta} + R_{\alpha\gamma\delta\beta} + R_{\alpha\delta\beta\gamma} = 0 , \\ &R_{\alpha\beta\gamma\delta;\mu} + R_{\alpha\beta\delta\mu;\gamma} + R_{\alpha\beta\mu\gamma;\delta} = 0 . \end{aligned} \tag{3.13}$$

The semicolon denotes covariant differentiation. Further, we define

$$R_{\mu\nu} = R^{\alpha}{}_{\mu\nu\alpha} \quad , \quad R = R_{\mu\nu} g^{\mu\nu} \quad ,$$
$$G_{\mu\nu} = R_{\mu\nu} - \tfrac{1}{2} R g_{\mu\nu} \quad , \tag{3.14}$$

which satisfy, according to (3.13):

$$R_{\mu\nu} = R_{\nu\mu} \quad , \qquad D_\mu G_{\mu\nu} = 0 \ . \tag{3.15}$$

The metric tensor also enables us to define a density function [see (2.7)],

$$\omega(x) = \sqrt{\det\,(g_{\mu\nu}(x))} \ . \tag{3.16}$$

In the quantum theory we shall encounter a fundamental problem: instead of $g_{\mu\nu}$ we could go over to a new metric $g'_{\mu\nu}$ with, for instance,

$$g'_{\mu\nu} = f_1(R) g_{\mu\nu} + f_2(R) R_{\mu\nu} \ . \tag{3.17}$$

So in a curved space there is some arbitrariness in the choice of metric (Section 12).We bypass this problem here.

4. Dynamics

The question now is whether we can make the fields $g_{\mu\nu}$ propagate. Indeed we can, because we can construct a gauge invariant action integral

$$S = - \frac{c^2}{16\pi\kappa} \int \omega \; R \; d_4 x \quad , \tag{4.1}$$

where κ is to be identified with the usual gravitational constant:

$$\kappa = 6{,}67 \cdot 10^{-11} \ m^3 \ kg^{-1} \ sec^{-2} \ . \tag{4.2}$$

For simplicity we shall take the units in which

$$\frac{c^2}{16\pi\kappa} = 1 \tag{4.3}$$

At a later stage one could put κ back in the expressions to find that the expansion in numbers of closed loops will correspond to an expansion with respect to κ.

One can also add other fields in the Lagrangian, for instance

$$\mathcal{L} = \omega \{- R - \frac{1}{2} g^{\mu\nu}\partial_\mu\phi\partial_\nu\phi - \frac{1}{2} m^2\phi^2 \} , \qquad (4.4)$$

where ϕ is a scalar field. We shall not repeat here the usual arguments to show that variation of the Lagrangian (4.4) really leads to the familiar gravitational interactions between masses, and to unfamiliar interactions between objects with a great velocity ("gravitational magnetism"). The equation for the gravitational field will be

$$G_{\mu\nu} = - \frac{1}{2} T_{\mu\nu} , \qquad (4.5)$$

where $T_{\mu\nu}$ is the usual energy-momentum tensor (Einstein's equation). The action (4.1) has much in common with the action

$$- \frac{1}{4} G^a_{\mu\nu} \; G^a_{\mu\nu}$$

in Yang-Mills theories. As we shall indicate in the next section, a massless graviton with helicity $\pm$ 2 will propagate. Notice that we have been led to Einstein's theory of gravity almost automatically. It seems to be the simplest choice if we ask for a theory with invariance under general coordinate transformations.

5. Quantization

The first thing we must do is make a shift

$$g_{\mu\nu} = \delta_{\mu\nu} + A_{\mu\nu} , \qquad (5.1)$$

and consider $A_{\mu\nu}$ as the quantum fields. Here the problem mentioned in the introduction presents itself: what if we start with

$$g^{\mu\nu} = \delta^{\mu\nu} + B^{\mu\nu} \; ? \qquad (5.2)$$

The answer is that as long as we take as our elementary field any local function of the $g_{\mu\nu}$, the obtained physical amplitudes will be the same. The transformation from one function (for example, $g_{\mu\nu}$) to the other (for example, $g^{\mu\nu}$) will be accompanied by a Jacobian, or closed loops of fictitious particles (see the Zinn-Justin lectures on gauge theories)

But the propagators of these particles are constants or pure polynomials in k, because the transformation is local. If we now turn on the dimensional regularization procedure, which has to be used in order to get gauge invariant results, then the integrals over polynomials,

$$\int d^n k \;\; \mathrm{Pol}\,(k) \;,$$

vanish[7]. This is the reason why it makes no difference whether we start from Eq. (5.1) or Eq. (5.2). Non-local functions of $g_{\mu\nu}$ are not allowed. These non-local transformations would give rise to fictitious particles that do contribute in the cutting rules[8], and they are outlawed once unitarity has been established for the choice (5.1) or (5.2). By choosing a convenient gauge, comparable with the Coulomb gauge in QED, it is indeed not difficult to establish that the theory is unitary, in a Hilbert space with massless particles with helicity $\pm$ 2.

We can work out the bilinear part of the Lagrangian in Eq. (4.1) in terms of the fields $A_{\mu\nu}$:

$$\mathcal{L} = -\tfrac{1}{4}(\partial_\mu A_{\alpha\beta})^2 + \tfrac{1}{8}(\partial_\mu A_{\alpha\alpha})^2 + \tfrac{1}{2} L_\mu^2 + \text{higher orders,} \tag{5.3}$$

with

$$L_\mu = \tfrac{1}{2}\,\partial_\mu A_{\alpha\alpha} - \partial_\alpha A_{\mu\alpha} \;. \tag{5.4}$$

For practical calculations it seems to be convenient to choose the gauge

$$\mathcal{L}^c = -\tfrac{1}{2} L_\mu^2 \tag{5.5}$$

The Lagrangian in this gauge, $\mathcal{L} + \mathcal{L}^c$, has as a kinetic term

$$-\,\partial_\mu A_{\alpha\beta}\; W_{\alpha\beta|\gamma\delta}\;\; \partial_\mu A_{\gamma\delta} \tag{5.6}$$

where W is a matrix built from δ-functions.
The propagator is then

$$\frac{1}{k^2 - i\varepsilon}\,(W^{-1})_{\alpha\beta|\gamma\delta} \quad . \tag{5.7}$$

Just in order to show the divergent character of the complications involved, we show here the Faddeev-Popov ghost[8] for this gauge, obtained by subjecting L_μ to an infinitesimal gauge transformation:

$$\mathcal{L}^{F.-P} = - \partial_\alpha \phi^*_\mu \partial_\alpha \phi^\mu + \phi^*_\mu \left[A_{\lambda\alpha,\alpha} \phi^\lambda_{,\mu} + A_{\mu\lambda}\phi^\lambda_{,\alpha\alpha} + A_{\mu\lambda,\alpha}\phi^\lambda_{,\alpha} + A_{\mu\alpha,\alpha\lambda}\phi^\lambda + A_{\mu\alpha,\lambda}\phi^\lambda_{,\alpha} - \tfrac{1}{2}A_{\alpha\alpha,\lambda}\phi^\lambda_{,\mu} - \tfrac{1}{2}A_{\alpha\alpha,\mu\lambda}\phi^\lambda - A_{\lambda\alpha,\mu}\phi^\lambda_{,\alpha} \right] . \qquad (5.8)$$

The Lagrangian (4.1), expanded in powers of $A_{\mu\nu}$, with the gauge-fixing term (5.5) and the ghost term (5.8), form a perfect quantum theory. It is, however, more complicated than necessary, because gauge invariance is given up right in the beginning by adding the bad terms (5.5) and (5.8). The background field method, discussed in the next sections, is much more elegant because gauge invariance is exploited in all stages of the calculations, thus simplifying things a lot.

6. A Prelude for the Background Field Technique: Gauge Invariant Source Insertions

The methods described in this and the following sections are not only suitable for quantum gravity, but have a very wide applicability, in particular in gauge theories, for instance for calculations of renormalization group coefficients. First, it is convenient to introduce the concept of a gauge invariant source insertion. This is an artificial term in the Lagrangian of the form

$$J(x)\, R(x) \ ,$$

where $J(x)$ is a c-number source function and $R(x)$ is some gauge invariant combination of fields, containing a linear part and quadratic or higher order corrections. Let us give some examples:

i) In a gauge theory with Higgs mechanism:

$$J_\mu \, \phi^* \cdot D_\mu \phi \ , \qquad (6.1)$$

where ϕ is the Higgs field and $D_\mu \phi$ is its covariant derivative. We take $\partial_\mu J_\mu = 0$.

If

$$< \phi > = F \ ;$$

$$\phi = F + \psi \ ,$$

then (6.1) becomes

$$J_\mu \left[gF^* A_\mu F \;+\; g\,\psi^* A_\mu F \;+\; gF^* A_\mu \psi \;+\; \psi^* D_\mu \psi \right] \qquad (6.2)$$

The first term emits or absorbs single neutral vector particles. The other terms are higher order corrections, emitting two or three particles at once. In general one can find for all physical particles a similar gauge-invariant source that produces them predominantly and one by one.

ii) In a pure gauge theory (in momentum representation):

$$J^a_\mu(k)\; A^a_\mu(k) + g\; J^{ab}_{\mu\nu}\,(p,q)\; A^a_\mu(p)\; A^b_\nu(q) + \ldots \qquad (6.3)$$

where $k_\mu J^a_\mu(k) = 0$. Integration over k, p and/or q is understood. The higher terms are determined by requiring gauge invariance under

$$A^{a\prime}_\mu(x) = A^a_\mu(x) + g\, f_{abc}\, \Lambda^b(x)\, A^c_\mu(x) - \partial_\mu\, \Lambda^a(x)\ .$$

This implies

$$ip_\mu (J^{bc}_{\mu\nu}(p,q) + J^{cb}_{\nu\mu}\,(q,p)) = J^a_\nu\,(p+q)\; f_{abc}\ , \text{ etc.} \qquad (6.4)$$

A source insertion that satisfies these conditions can be written in a closed form, in terms of an antisymmetric tensor source $J^a_{\mu\nu}(x)$:

$$J^a_{\mu\nu}(x)\;\; T \left[\exp \int_{\text{path}}^{x} g\, A_\lambda(x')\, dx'^\lambda\right]_{ab}\; G^b_{\mu\nu}(x) \quad , \qquad (6.5)$$

where $G^a_{\mu\nu} = \partial_\mu A^a_\nu - \partial_\nu A^a_\mu + g\, f_{abc}\, A^b_\mu\, A^c_\nu$, and A_λ stands for the matrix

$$A^{ac}_\lambda = f_{abc} A^b_\lambda\ .$$

The integral is along a path from infinity to x. The symbol T stands for time ordering along the path. Expanding (6.5) gives

$$J^a_{\mu\nu}(x) \left[G^a_{\mu\nu}(x) + g\, f_{abc} \int_{\text{path}}^{x} A^b_\lambda(x')\, dx'\; G^c_{\mu\nu}(x) + \ldots\right] \qquad (6.6)$$

iii) In gravity one can do a similar thing:
a source $J^{\mu\nu}(x)$ satisfying

$$J^{\mu\nu} = J^{\nu\mu}\ ; \qquad \partial_\mu\, J^{\mu\nu} = 0 \qquad (6.7)$$

can be coupled to $A_{\mu\nu}$ in eq. (5.1), and one can add higher order corrections to restore gauge invariance. The details are not very relevant for what follows.

The amplitude with which a gauge invariant source emits a single particle, obtains higher order corrections, see Fig. 1.

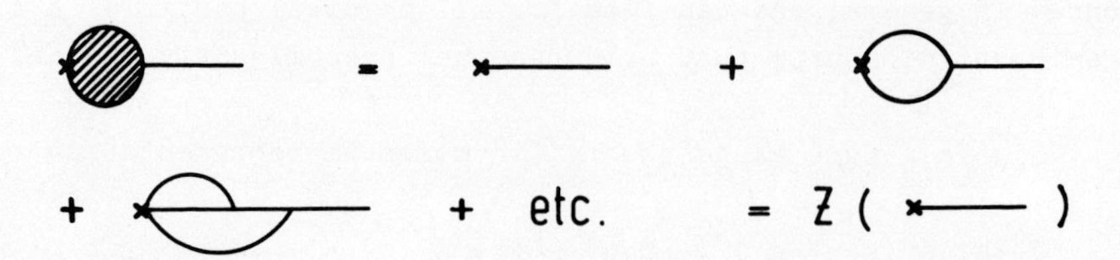

Fig. 1

A source can emit a single particle in different ways

The S-matrix can now readily be obtained in a gauge-independent way by considering vacuum-vacuum transitions in the presence of a gauge-invariant source. Then the external legs are amputated and put on mass-shell. From Fig. 1 it will be clear that in practice one can just as well calculate the amputated Green's functions directly and multiply them with some renormalization factor Z. This factor Z however may be gauge-dependent. The correct factor can be obtained by normalizing the imaginary part of the two-point function[8].

7. The Background Field Method

The background field method, useful for gauge theories, is practically indispensible for quantum gravity. Let the full Lagrangian be

$$\mathcal{L}(A) = \mathcal{L}^{inv}(A) + J\,R(A) - \frac{1}{2}\,C^2(A) + \mathcal{L}^{F.-P} \quad , \tag{7.1}$$

where $\mathcal{L}^{inv}(A)$ is the complete gauge invariant Lagrangian for all fields A_i. As described in the previous section, R(A) is a gauge invariant field combination, and J is a c-number source function. C(A) is the gauge-fixing function and $\mathcal{L}^{F.-P}$ describes the associated Feynman-deWitt Faddeev-Popov ghost[1,2,8]. All irrelevant indices have been suppressed. The gauge transformation law will be written as

$$A_i' = A_i + s^a_{ij}\,\Lambda_a\,A_j + t^a_i\,\Lambda^a \qquad (7.2)$$

where $\Lambda^a(x)$ is the infinitesimal generator. t and s are coefficients built from numbers and the space-time derivative ∂_μ.

We now introduce the notion of a classical field A^{cl}, which is a function of the c-number sources J. Usually[2,5] one defines its J dependence by requiring that A^{cl} satisfies the classical equations of motion. This is sufficient as long as we are interested in diagrams with at most one closed loop. In these notes we shall make that restriction also, but it must be kept in mind that for applications of these techniques at still higher orders it will be more convenient to add quantum corrections to the equation of motion*. At this stage we require A^{cl} also to be in the gauge

$$C(A^{cl}) = 0 \ . \qquad (7.3)$$

Next, we perform a shift:

$$A = A^{cl} + \phi \ , \qquad (7.4)$$

where now ϕ is the new quantum field, and we rewrite the Lagrangian in terms of ϕ :

$$\mathcal{L}(A^{cl}+\phi) = \mathcal{L}_0(A^{cl},J) + \mathcal{L}_1(\phi,J,A^{cl}) + \mathcal{L}_2(\phi,J,A^{cl}) - \tfrac{1}{2}C^2(A^{cl}+\phi)$$
$$+ \mathcal{L}^{F.-P.} + \mathcal{O}(\phi^3). \qquad (7.5)$$

Here $\mathcal{L}_1$ is linear in ϕ; $\mathcal{L}_2$ is quadratic in ϕ and $\mathcal{O}(\phi^3)$ is of higher order in ϕ.

Now, since A^{cl} satisfies the equation of motion

$$\frac{\delta\mathcal{L}}{\delta A^{cl}} = 0 \qquad (7.6)$$

and $C(A^{cl}) = 0$, all terms linear in ϕ cancel:

$$\mathcal{L}_1(\phi) = 0 \ . \qquad (7.7)$$

---

* This way one can avoid the so-called Feynman baskets[1,2]. We leave the details to a possible future publication.

So the source J is not coupled to terms linear in ϕ, and there are no vertices with only one ϕ-line. Therefore, the ϕ-lines can only go around in loops, and if we confine ourselves to one-loop diagrams then we can neglect the terms $\mathcal{O}(\phi^3)$. One loop diagrams now only consist of a ϕ-loop, with bilinear insertions of classical sources depending on A^{cl} and J. But remember that A^{cl} is a function of J, which can be obtained by solving the classical equations of motion by iteration. This iteration process corresponds to adding all possible trees to the single loop. Thus we reproduce the original Feynman rules, with the only change that loop lines are called ϕ-lines, and tree lines are called A^{cl}-lines.

8. The Background Field Gauge

The relevant part of the Lagrangian (7.5) is

$$\mathcal{L}(\phi) = \mathcal{L}'_{inv}(\phi, J, A^{cl}) - \tfrac{1}{2}C^2(A^{cl}+\phi) + \mathcal{L}^{F.-P.} . \tag{8.1}$$

This is just an ordinary gauge field Lagrangian where $\mathcal{L}'_{inv} = \mathcal{L}_2 + \mathcal{O}(\phi^3)$ is invariant under what we shall call gauge transformations of type Q:

$$\begin{aligned} \phi'_i &= \phi_i + s^a_{ij}\,\Lambda^a(A^{cl}_j + \phi_j) + t^a_i\,\Lambda^a \ ; \\ A^{cl\prime} &= A^{cl} \end{aligned} \tag{8.2}$$

The gauge is fixed by the function $C(A^{cl} + \phi)$.

Now there is another invariance of $\mathcal{L}'_{inv}$, also broken by this C term. We call this a gauge transformation of type C:

$$\begin{aligned} \phi'_i &= \phi_i + s^a_{ij}\,\Lambda^a\,\phi_j \ ; \\ A^{cl\prime}_i &= A^{cl}_i + s^a_{ij}\Lambda^a A^{cl}_j + t^a_i\,\Lambda^a \ . \end{aligned} \tag{8.3}$$

The power of the present formulation is that we can go over to a different gauge function $C(A^{cl},\phi)$ which breaks the Q-gauge invariance (as is necessary) but preserves gauge invariance of type C. For example, if ϕ^a_μ is the quantum part of the vector field A^a_μ, then the choice

$$C_a = D_\mu \phi^a_\mu \ , \tag{8.4}$$

where D_μ is the covariant derivative in the classical (C) sence:

$$D_\mu \phi^a_\mu = \partial_\mu \phi^a_\mu + g\, f_{abc}\, A^{cl}_{\mu b}\, \phi^c_\mu \quad ,$$

clearly preserves C-invariance. Such a gauge is also possible in the case of gravity. If

$$g_{\mu\nu} = g^{cl}_{\mu\nu} + A_{\mu\nu} \quad ,$$

we can take

$$C_\alpha = \sqrt[4]{g^{cl}} \cdot t^{\alpha\mu} \left(\tfrac{1}{2} g^{\kappa\lambda}_{cl} D_\mu A_{\kappa\lambda} - g^{\kappa\lambda}_{cl} D_\kappa A_{\mu\lambda}\right) , \qquad (8.5)$$

where

$$t^{\alpha\mu} t^{\alpha\nu} = g^{\mu\nu}_{cl} \quad .$$

Again, D_μ is the covariant derivative in the classical sence.

The one-loop (irreducible) vertex functions obtained in this gauge will be C-invariant also. It follows that they satisfy not only the Slavnov identities that describe the Q gauge symmetry, but in addition the much simpler Ward identities which are the direct generalizations of those in quantum electrodynamics.

Of course, the new Feynman rules in this background field gauge are independent of our original choice of the gauge C(A) in eq. (7.1). This implies that we can now also drop the gauge condition (7.3) for the classical fields since it can be replaced by another, arbitrary, gauge condition.

9. A Simple Example of the Background Field Gauge: Pure Yang-Mills Fields

Although we are mainly interested in quantum gravity, it is much more instructive to illustrate our methods in simpler field theories. Let us consider pure Yang-Mills fields. The invariant Lagrangian is

$$\mathcal{L}^{inv} = -\frac{1}{4} G^a_{\mu\nu} G^a_{\mu\nu} , \qquad (9.1)$$

with $$G^a_{\mu\nu} = \partial_\mu A^a_\nu - \partial_\nu A^a_\mu + g\, f_{abc} A^b_\mu A^c_\nu \quad .$$

The gauge invariance is

$$A^{a'}_{\mu} = A^{a}_{\mu} + g f_{abc}\Lambda^{b}A^{c}_{\mu} - \partial_{\mu}\Lambda^{a} . \tag{9.2}$$

We leave aside the gauge invariance sources (sect. 6).

We shift

$$\begin{aligned} A^{a}_{\mu} &= A^{cl}_{\mu a} + \phi^{a}_{\mu} , \\ G^{a}_{\mu\nu} &= G^{a\ cl}_{\mu\nu} + D_{\mu}\phi^{a}_{\nu} - D_{\nu}\phi^{a}_{\mu} + g f_{abc}\phi^{b}_{\mu}\phi^{c}_{\nu} , \\ &(\text{where } D = \partial + g f A^{cl}) \end{aligned} \tag{9.3}$$

Using

$$(D_{\mu}D_{\nu})^{ac} - (D_{\nu}D_{\mu})^{ac} = g f_{abc} G^{b\ cl}_{\mu\nu} , \tag{9.4}$$

we get

$$\begin{aligned} \mathcal{L}^{inv} = \mathcal{L}^{inv}(A^{cl}) &- \frac{1}{2}(D_{\mu}\phi_{\nu})^{2} + \frac{1}{2}(D_{\mu}\phi_{\mu})^{2} \\ &- g G^{c\ cl}_{\mu\nu} f_{abc}\phi^{a}_{\mu}\phi^{b}_{\nu} + \mathcal{O}(\phi^{3}) \\ &+ \text{total derivative.} \end{aligned} \tag{9.5}$$

A convenient background field gauge is

$$\mathcal{L}^{C} = -\frac{1}{2}C^{2} = -\frac{1}{2}(D_{\mu}\phi^{a}_{\mu})^{2} . \tag{9.6}$$

The ghost Lagrangian is then

$$\mathcal{L}^{F.-P.} = - D_{\mu}\psi^{*}_{a}D_{\mu}\psi_{a} , \tag{9.7}$$

up to irrelevant interactions with ϕ^{a}_{μ}.
Of course the ghost is also C-invariant. Note that C-invariance permits us to write the interactions of ϕ^{a}_{μ} and ψ_{a} in a very condensed way. It is this feature that prevents overpopulation of indices in the case of gravity.

The one-loop infinities can be subtracted by a C-invariant counte term in the Lagrangian. The only candidate is

$$\alpha \cdot G^{a\ cl}_{\mu\nu} G^{a\ cl}_{\mu\nu} .$$

The index α simultaneously governs the infinities of the two, three and four point vertices, and is therefore directly proportional to the Callan - Symanzik β-function[9]. To find this β-function one therefore only needs to investigate the two point function, contrary to the conventional formulation where also three point functions had to be calculated[10] in order to eliminate the Callan-Symanzik γ-function.

10. A Master Formula for all One-Loop Infinities

From the preceeding it will be clear that any one-loop amplitude* can be obtained from a Lagrangian bilinear in a set of fields ϕ_i. One can then rearrange the coefficients in such a way that

$$\mathcal{L} = \sqrt{g}\,\{-\tfrac{1}{2}(\partial_\mu\phi_i + N^{ij}_\mu\phi_j)g^{\mu\nu}(\partial_\nu\phi_i + N^{ik}_\nu\phi_k) + \tfrac{1}{2}\phi_i X_{ij}\phi_j\} , \qquad (10.1)$$

where N, g and X are arbitrary functions of space time x. Further,

$$N^{ij}_\mu = - N^{ji}_\mu ; \qquad X_{ij} = X_{ji} .$$

In dealing with gravity[3] we found it very convenient first to calculate the infinities of this general Lagrangian, in terms of N, g and X . Of course, the background metric is allowed to be curved. Afterwards one may substitute the details such as: the way N and X depend on the background fields; and the fact that the indices i, j actually stand for pairs of Lorentz-indices. In gauge theories the objects N are mostly background gauge fields and in gravity the N contain the Christoffel symbols Γ.

In ref. 3), we used dimensional regularization and considered the poles at $n \to 4$ (n is the number of space-time dimensions). From power counting arguments one easily deduces that the residues of these poles can consist only of a limited number of terms. This number is even more restricted if we use the observation that, whatever N, g or X are, there is a C-gauge symmetry:

$\mathcal{L}$ is invariant under

---

* provided a Feynman-like background field gauge is chosen.

$$\phi'(x) = \phi(x) + \Lambda(x)\phi(x)$$

$$N'_\mu(x) = N_\mu - \partial_\mu\Lambda + \Lambda N_\mu - N_\mu\Lambda \qquad (10.2)$$

$$X'(x) = X + \Lambda X - X\Lambda$$

where Λ is an infinitesimal, antisymmetric matrix.

A straightforward calculation yields, that all one-loop infinitie as $n \to 4$ are absorbed by the counter-Lagrangian[3)]

$$\Delta\mathcal{L} = \frac{1}{8\pi^2(n-4)} \sqrt{g}\, \mathrm{Tr}\,\{ \frac{1}{24} Y^{\mu\nu}Y_{\mu\nu} + \frac{1}{4} X^2 - \frac{1}{12} RX$$

$$+ \frac{1}{120} R_{\mu\nu}R^{\mu\nu}\, I + \frac{1}{240} R^2\, I \} \qquad (10.3)$$

where

$$Y_{\mu\nu} = \partial_\mu N_\nu - \partial_\nu N_\mu + N_\mu N_\nu - N_\nu N_\mu \qquad (10.4)$$

and

Tr I = number of fields.
$R_{\mu\nu}$ and R are the Riemann curvature tensors for the background metric.
Raising and lowering indices by use of the background $g^{\mu\nu}$ is understood.

The formula (10.3) can also be used to calculate a similar master formula in the case of Fermions[9)].

11. Substituting Equations of Motion.

In some cases the resulting formula (10.3) is not the end of the story. One may still make use of the information that the background fields are not arbitrary, but satisfy equations of motion. If, for instance, one finds a contribution in $\Delta\mathcal{L}$ proportional to

$$(D_\mu A^{cl})^2 \ , \qquad (11.1)$$

and the equation of motion is

$$D^2 A^{cl} = V(A^{cl}) \ , \qquad (11.2)$$

then one may replace (10.1) by

$$- A^{cl} \, V(A^{cl}) \, . \tag{11.3}$$

Addition of infinitesimal terms in the Lagrangian that vanish if the equation of motion is fulfilled, corresponds exactly to making an infinitesimal field renormalization: if the equation of motion is

$$\frac{\delta \mathcal{L}}{\delta A} = 0 \quad , \tag{11.4}$$

then the terms in question must be

$$\varepsilon \cdot B \cdot \frac{\delta \mathcal{L}}{\delta A} \quad , \tag{11.5}$$

and we have

$$\mathcal{L}(A + \varepsilon B) = \mathcal{L}(A) + \varepsilon \, B \, \frac{\delta \mathcal{L}}{\delta A} \, . \tag{11.6}$$

$A' = A + \varepsilon B$ is a field renormalization.
Note that all terms in $\Delta\mathcal{L}$ are always infinitesimal, because we neglect everything that comes from two-loop diagrams.

12. Some Numerical Results. Conclusions.

The master formula (10.3) has been applied to calculate the infinity structure in different cases. For pure gravitation we used[3] the gauge (8.5) and found from the gravitons

$$\Delta \mathcal{L} = \frac{1}{\varepsilon} \sqrt{g} \left(\frac{7}{24} R^2 + \frac{7}{12} R_{\mu\nu} R^{\mu\nu} \right) \quad , \tag{12.1}$$

where

$$\frac{1}{\varepsilon} = 1/8\pi^2(n-4) \, ,$$

and from the ghosts

$$\Delta \mathcal{L} = \frac{1}{\varepsilon} \sqrt{g} \left(- \frac{17}{60} R^2 - \frac{7}{30} R_{\mu\nu} R^{\mu\nu} \right) \tag{12.2}$$

Here $g_{\mu\nu}$ and $R_{\mu\nu}$ are the classical metric and curvature .(At this level it is never necessary to consider quantum fields in the counter terms. We are also not interested in the renormalization of the source terms). From power counting one would expect a third possible term of the form

$$\sqrt{g}\ R_{\alpha\beta\gamma\delta}\ R^{\alpha\beta\gamma\delta} \quad . \tag{12.3}$$

It indeed occurs, but we made use of the identity[2,3,11)]

$$\sqrt{g}\ (R_{\alpha\beta\gamma\delta}\ R^{\alpha\beta\gamma\delta} - 4\ R_{\mu\nu}R^{\mu\nu} + R^2)$$

$$= \text{ total derivative } \quad , \tag{12.4}$$

which implies that terms of the form (12.3) can be eliminated.

So all together we have

$$\Delta\mathcal{L} = \frac{1}{\varepsilon}\sqrt{g}\ (\frac{1}{120}\ R^2 + \frac{7}{20}\ R_{\mu\nu}R^{\mu\nu}) \ . \tag{12.5}$$

However, we still have the equations of motion of the background field (see previous section), which is

$$R_{\mu\nu} = 0 \quad ; \quad R = 0 \tag{12.6}$$

This means that the infinity vanishes in pure gravity. It is interesting to observe that the field renormalization that corresponds to the elimination of this infinity (see previous section) is of an unusual type:

$$g_{\mu\nu} \rightarrow g_{\mu\nu} + \alpha\ R_{\mu\nu} + \beta\ R\ g_{\mu\nu} \quad . \tag{12.7}$$

This was the reason for our remark in the end of sect. 3. Note that eq. (12.4) was essential for this result.

Next we studied pure gravity with a (massless) Klein-Gordon field ϕ. The classical equations of motion are

$$\begin{aligned} D_\mu D^\mu \phi &= 0 \ ; \\ R_{\mu\nu} &= -\frac{1}{2}\ D_\mu\phi D_\nu\phi \\ R &= -\frac{1}{2}\ (D\phi)^2 \end{aligned} \tag{12.8}$$

There is one type of counterterm which is allowed by power counting and cannot be eliminated by substitution of the equations of motion:

$$\Delta\mathcal{L} = \frac{\sqrt{g}}{\varepsilon} \cdot \frac{203}{80} \cdot R^2 \tag{12.9}$$

The next thing that has been tried is pure gravity with in addition

Maxwell fields[12].
The Lagrangian is

$$\mathcal{L} = \sqrt{g}\,(-R - \frac{1}{4} F_{\mu\nu}F_{\alpha\beta}\, g^{\mu\alpha}g^{\nu\beta}).$$

$$F_{\mu\nu} = \partial_\mu A_\nu - \partial_\nu A_\mu \,. \tag{12.10}$$

The equations of motion are

$$G_{\mu\nu} \equiv R_{\mu\nu} - \frac{1}{2} g_{\mu\nu} R = -\frac{1}{2} T_{\mu\nu} \;;$$

$$T_{\mu\nu} \equiv F_{\mu\alpha}F^{\alpha}{}_{\nu} - \frac{1}{4} g_{\mu\nu} F^{\alpha\beta}F_{\alpha\beta} \,,$$

$$D_\alpha F^{\alpha\beta} = 0 \,,$$

$$D_\alpha F_{\beta\gamma} + D_\beta F_{\gamma\alpha} + D_\gamma F_{\alpha\beta} = 0 \,. \tag{12.11}$$

From which

$$R = \frac{1}{2} T^{\alpha}{}_{\alpha} = 0 \,.$$

In principle one may expect

$$\Delta\mathcal{L} = \frac{1}{\varepsilon}\sqrt{g}\,[\,\alpha_1 R_{\mu\nu}R^{\mu\nu} + \alpha_2 (F_{\alpha\beta}F^{\alpha\beta})^2 + \alpha_3 R_{\mu\nu\alpha\beta} F^{\mu\nu}F^{\alpha\beta}] \,. \tag{12.12}$$

Explicit calculation shows however:

$$\Delta\mathcal{L} = \frac{1}{\varepsilon}\sqrt{g}\cdot\frac{137}{60} R_{\mu\nu}R^{\mu\nu}. \tag{12.13}$$

So we see that there are some cancellations:

$$\alpha_2 = \alpha_3 = 0 \,. \tag{12.14}$$

Indeed, they occur in a miraculous way during the calculation and have not yet been explained.
In the coupled Einstein-Yang-Mills system [13] these cancellations persist and many new cancellations occur. Starting from the obvious generalization of the Abelian Lagrangian (12.10), one finds

$$\Delta\mathcal{L} = \frac{1}{\varepsilon}\sqrt{g}\,\{[\frac{137}{60} + \frac{r-1}{10}] R_{\mu\nu}R^{\mu\nu} + f^2 C_2 \frac{11}{12} F_{\alpha\beta}F^{\alpha\beta}\} \tag{12.15}$$

where f is the gauge coupling constant,

$$C_2\delta_{ab} = f_{apq}f_{bpq} \quad ,$$

$$C_2 r = f_{apq}f_{apq} \qquad (12.16)$$

Five other coefficients each happen to vanish. The second term in (12.15) is of the renormalizable type. It renormalizes the gauge coupling constant and fixes the Callan-Symanzik β-function.

Finally, also the combined Einstein-Dirac system has been investigated and nonrenormalizable infinities have been found also[14)].

The fact that in all these systems where matter in some form is added to pure gravity infinities of the nonrenormalizable dimension survive really means that these theories cannot be renormalized in the perturbation expansion. In the case of pure gravity the infinities have been shown to be non-physical up to the one-loop level. No calculations have been performed to investigate renormalizability in the order of two loops. The calculations of Nieuwenhuizen and Deser show that "miraculous" cancellations often occur . Perhaps this is an indication of a new sort of symmetry that we are not aware of. Investigation of this symmetry could reveal new renormalizable models with gravity.

Even so, a renormalized perturbation expansion would only be a small step forward. At very small distances the gravitational effects must be large, because of the dimension of the gravitational constant, so the expansion would break down at small distances anyhow. We have the impression that not only a better mathematical analysis is needed, but also new physics. What we learned (see eq. (12.7) and the remarks in the end of sect.3) is that in such a theory the metric tensor might not at all be such a fundamental concept. In any case, its definition is not unambiguous.

REFERENCES

1. R. P. Feynman, Acta Phys. Polon. 24, 697 (1963)
   S. Mandelstam, Phys. Rev. 175, 1580, 1604 (1968)
   E. S. Fradkin and I. V. Tyntin, Phys. Rev. D2, 2841 (1970)
   L. D. Faddeev and V. N. Popov, Phys. Letters 25B, 29 (1967)

2. B. S. DeWitt, in Relativity, Groups and Topology, Summerschool of Theor. Physics, Les Houches, France, 1963 (Gordon and Breach, New York, London)
B. S. DeWitt, Phys. Rev. 162, 1195, 1239 (1967)

3. G.'t Hooft and M. Veltman, CERN preprint TH 1723 (1973), to be publ. in Annales de l'Institut Henri Poincaré

4. D. Christodoulou, proceedings of the Academy of Athens 47 (20 January 1972)
D. Christodoulou, CERN preprint TH 1894 (1974)

5. J. Honerkamp, Nucl. Phys. B48, 269 (1972);
J. Honerkamp, Proc. Marseille Conf.,19-23 June 1972, CERN preprint TH 1558

6. G. 't Hooft, Nucl. Phys. B35, 167 (1971)

7. G. 't Hooft and M. Veltman, Nucl. Phys. B44, 189 (1972)

8. G. 't Hooft and M. Veltman, DIAGRAMMAR, CERN report 73-9 (1973)

9. G. 't Hooft, Nucl. Phys. B61, 455 (1973)
G. 't Hooft, Nucl. Phys. B62, 444 (1973)

10. H. D. Politzer, Phys. Rev. Letters 30, 1346 (1973)
D. J. Gross and F. Wilczek, Phys. Rev. Letters 30, 1343 (1973)

11. R. Bach, Math. Z. 9, 110 (1921)
C. Lanczos, Ann. Math. 39, 842 (1938)

12. S. Deser and P. van Nieuwenhuizen, Phys.Rev.Letters ,32,245(1974)
Brandeis preprint (1974), to be publ. in Phys. Rev.

13. S. Deser, Hung-Sheng Tsao and P. van Nieuwenhuizen, Brandeis preprint (June 1974),to be publ. in Phys. Rev.

14. S. Deser and P. Van Nieuwenhuizen, Brandeis preprint (1974), to be publ. in Phys. Rev.

QUANTUM FIELD THEORY AND STATISTICAL MECHANICS

F. Jegerlehner

Freie Universität Berlin, Germany

1. Introduction

In the last decade the ideas of Migdal and Polyakov[1] on one hand and Kadanoffs[2] work on the other hand led to Wilsons[3] renormalization group approach and the Wilson-Fischer ε-expansion[4]. These two steps renewed the interest of many physicists in critical phenomena and second order phase transitions[5]. The origin of the methods which contributed so much to the understanding of the statistical mechanics of phase transitions is quantum field theory, which on the other hand benefited a lot from the applications to phase transitions. Many of the relevant new ideas in the area of quantum field theory and the theory of phase transitions developed parallel in the two fields so the ideas of scaling of operator product expansions etc. The aim of the present lectures is a discussion of critical behaviour directly in renormalized field theory. But first I will briefly discuss some characteristic properties of second order phase transitions and give a heuristic understanding how the relation between quantum field theory and statistical mechanics near criticality comes about.
We will then turn to renormalized quantum field theory in $4-\varepsilon$ dimensions, calculate critical indices, and introduce a (pre -) scaling parametrization which will turn out to be most appropriate for a discussion of the scaling behaviour. We then investigate the structure of corrections to scaling for the thermodynamical quantities and the correlation functions.

2. Critical Phenomena[2] [5]

a. The thermodynamic quantities

We first introduce the relevant thermodynamic quantities for the study of second order phase transitions. For definiteness we start from a ferromagnetic Lenz-Ising system.

On a D dimensional lattice G_a in configuration space, with lattice spacing a, there is associated a discrete classical spin $\sigma_{\vec{n}} = \pm 1$ to each lattice point $\vec{n}$ (labeled by integers). The spins interact with its nearest neighbours (n.n.) only. Parallel spins are attractive with energy -K, a spin parallel to an external magnetic field H has energy -H; for antiparallel spins the energy is K and H. Accordingly the Lenz-Ising Hamiltonian reads

$$\mathcal{H} = -K \sum_{\substack{\vec{n}\,\vec{m} \\ n.n.}} \sigma_{\vec{n}} \sigma_{\vec{m}} - H \sum_{\vec{n}} \sigma_{\vec{n}} \tag{2.1}$$

For a finite system with N spins we obtain in the usual way thermodynamical quantities from the partition function

$$Z_N = \sum_{\substack{all \\ conf.}} \exp(-\beta \mathcal{H}) \; ; \quad \beta = \frac{1}{kT} \tag{2.2}$$

For the free energy density

$$f(K, H) = \frac{F_N}{N} \; ; \quad F_N = -\ln Z_N \tag{2.3}$$

the spin correlation functions

$$\langle \sigma_{\vec{n}} \rangle = Z_N^{-1} \sum_{conf.} \sigma_{\vec{n}} \, e^{-\beta \mathcal{H}}$$

$$\langle \sigma_{\vec{n}} \sigma_{\vec{m}} \rangle = Z_N^{-1} \sum_{conf.} \sigma_{\vec{n}} \sigma_{\vec{m}} \, e^{-\beta \mathcal{H}} \qquad etc. \tag{2.4}$$

and the energy correlation functions

$$\langle E_{\vec{n}} \rangle = Z_N^{-1} \sum_{conf.} E_{\vec{n}} \, e^{-\beta \mathcal{H}}$$

$$\langle E_{\vec{n}} E_{\vec{m}} \rangle = Z_N^{-1} \sum_{conf.} E_{\vec{n}} E_{\vec{m}} \, e^{-\beta \mathcal{H}} \qquad etc.$$

$$E_{\vec{n}} \doteq \sum_{\substack{\vec{m} \\ n.n.}} \sigma_{\vec{n}} \sigma_{\vec{m}}$$

we will always take the thermodynamic limit $N \to \infty$.
From the free energy density f we define the thermodynamical quantities ($k = \beta K$; $h = \beta H$):

$$M = \frac{\partial f}{\partial h} = \langle \sigma_{\vec{n}} \rangle = \langle \sigma_0 \rangle \qquad \text{magnetization density}$$

$$E = \frac{\partial f}{\partial k} = \langle E_{\vec{n}} \rangle = \langle E_0 \rangle \qquad \text{energy density}$$

$$\begin{aligned} \chi &= -\frac{\partial^2 f}{\partial h^2} = \sum_{\vec{n}} \langle \sigma_{\vec{n}} \sigma_0 \rangle^{conn} \qquad \text{susceptibility} \\ &= \sum_{\vec{n}} \{ \langle \sigma_{\vec{n}} \sigma_0 \rangle - \langle \sigma_{\vec{n}} \rangle \langle \sigma_0 \rangle \} \end{aligned} \tag{2.5}$$

$$\begin{aligned} C &= -\frac{\partial^2 f}{\partial k^2} = \sum_{\vec{n}} \langle E_{\vec{n}} E_0 \rangle^{conn} \qquad \text{specific heat.} \\ &= \sum_{\vec{n}} \{ \langle E_{\vec{n}} E_0 \rangle - \langle E_{\vec{n}} \rangle \langle E_0 \rangle \} \end{aligned}$$

Here we have taken into account translation invariance.

b) Second order phase transitions

The 2nd order phase transitions are related to the spin fluctuations in the system. For large distances the spin correlations away from the critical point show the Ornstein-Zernicke exponential fall off

$$\langle \sigma_{\vec{n}} \sigma_0 \rangle \underset{|\vec{x}| \gg a}{\simeq} \frac{e^{-|\vec{x}|/\xi}}{|\vec{x}|} \quad ; \quad \vec{x} = \vec{n}\, a \tag{2.6}$$

This relation defines the correlation length ξ (the most important parameter in the study of 2. order phase transitions).
The phase diagram for a ferromagnetic system is depicted in Fig. 1

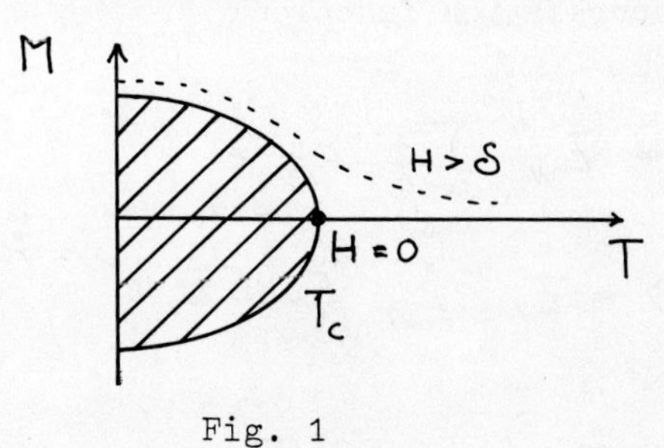

Fig. 1

For $H \neq 0$ all T and $H = 0$, $T > T_c$ the thermodynamic functions are analytic functions of T and H due to finite ξ , which means that the physics actually takes place in a finite box of size $L \gtrsim \xi$.

For $H = 0$, $T < T_c$ the spins are aligned and a spontaneous magnetization

$$< \sigma > = \pm M(T)$$

occurs. The state is not longer uniquely defined as a function of T. A first order phase-transition takes place as H changes sign and the system jumps from M to -M. As $T \to T_c$ from below the first order transition disappears. This is the critical point of a second order phase transition. It is necessarily a point of non analyticity as $M(H,T) = 0$ for $T > T_c$ and $\neq 0$ for $T < T_c$.
What happens is that for $H = 0$, $T < T_c$ there is a net magnetization, in the z direction say, and clusters of spins pointing in the wrong direction of maximal size ξ (with a clustering down to microscopic scale). To turn the spins it costs energy and therefore macroscopically the system is in a stable state.
As $\xi \to \infty$ ($T \to T_c$, $T \leq T_c$) criticality is approached; the difference in magnetization approaches zero and together with it the energy cost per area of producing a region of wrong phase. This is the region of large scale weak fluctuations in magnetization. The physics is then no longer determined by what's happening in a finite box. We are faced with a system of infinitly many degrees of freedom. With the critical behaviour there are associated characteristic singularities which are caused by correlations over infinite distances in space i.e. the fundamental reason is the divergence of ξ at the critical point

$$\xi \propto t^{-\nu} \qquad \text{at } H=0 \quad \text{as } t' - \frac{T-T_c}{T_c} \to 0 \tag{2.7}$$

The correlation functions then behave as

$$\begin{aligned} < \sigma_{\vec{x}} \sigma_0 > &\simeq \frac{const.}{|\vec{x}|^{2d_\sigma}} \\ & \qquad\qquad ; \quad |\vec{x}| \to \infty \\ < E_{\vec{x}} E_0 > &\simeq \frac{const.}{|\vec{x}|^{2d_E}} \end{aligned} \tag{2.8}$$

causing divergent thermodynamic quantities (infinite sums over the densities). The singularities may be parametrized by power laws as calculations from the Lenz-Ising model and the mean field theory (qualitatively correct picture) as well as experiments confirm. The exponents are the critical indices defined by:

$$M|_H \simeq \begin{cases} 0 \\ (-t)^{\beta} \end{cases}$$

$$\chi|_H \simeq \begin{cases} t^{-\gamma} \\ (-t)^{-\gamma'} \end{cases} \qquad t \to \begin{cases} +0 \\ -0 \end{cases} \tag{2.9}$$

$$C|_H \simeq \begin{cases} t^{-\alpha} \\ (-t)^{-\alpha'} \end{cases}$$

$$M|_{t=0} \simeq \pm |H|^{1/\delta} \quad ; \quad H \to \pm 0$$

c. The Kadanoff picture of critical behaviour[2)]

What is an appropriate theory of critical phenomena? Kadanoff had the idea that the critical system can be reduced to the consideratio of the physics contained in a finite box. Kadanoff's block spin picture can be roughly described as follows: The microscopic theory is described by cells of size a^D As $T \simeq T_c$ (i.e. $\xi \gg a$) a coarser division of the system into cells should give a good approximation to the macroscopic properties of the system. Hence one obtaines a new description of the system by forming block spins i.e. cells of size L^D ($a < L \ll \xi$); within these cells the spins are strongly correlated and behave essentially as one big spin with nearest neighbour interaction.
By forming the big spin

$$\tilde{\sigma}_i = \sum_{cell_i} \sigma_{i'} \tag{2.10}$$

one actually averages out the non relevant degrees of freedom."Renormalizing" the big spins to $\pm$ 1 one gets an equivalent description of the system

$$H\!\!\!H_a \doteq \beta \mathcal{H}_a = -k \sum_{n.n.} \sigma_{\vec{n}} \sigma_{\vec{m}} - h \sum \sigma_{\vec{n}} \qquad ; \vec{n} \in G_a$$

by the Hamiltonian

$$H\!\!\!H_L \doteq \beta \mathcal{H}_L = -\tilde{k} \sum_{n.n.} \sigma_{\vec{n}} \sigma_{\vec{m}} - \tilde{h} \sum \sigma_{\vec{n}} \qquad ; \vec{n} \in G_L \tag{2.11}$$

with $\tilde{k} = (L/a)^y k$ and $\tilde{h} = (L/a)^x h$ (2.12)

For exactly aligned spins in each cell

$$y = y_o = D - 1 \qquad \text{and} \qquad x = x_o = D \tag{2.13}$$

The crucial point is that the spins are not exactly lined up due to fluctuations down to microscopic scale and therefore the coefficients x and y have not the values (2.13), they merely have to be considered as unknown parameters*. It will be one of the main goals of a theory of critical phenomena to explain and calculate these indices. In a precise formulation of the block spin picture the "average" (2.10) has to be done actually in the partition function. This will be discussed in detail by Wegner in his lectures. The transformation

$$\mathbb{H}_a \longrightarrow \mathbb{H}_L = T_{L/a}\, \mathbb{H}_a \tag{2.14}$$

is called a renormalization group (RG) transformation[3]. It has the semigroup property.
For $\xi \gg L > a$ we expect the physics described by $\mathbb{H}_L$ to be essentially unchanged

$$\mathbb{H}_L \simeq \mathbb{H}_a$$

At the critical point the physics is expected to be independent of the cell size such that

$$\mathbb{H}_L^{(o)} = \mathbb{H}_a^{(o)} = \mathbb{H}^* \qquad \forall L \quad \text{as } H = 0;\ T = T_c \tag{2.15}$$

i. e. we have a fixed point of the above transformation $T_{L/a}$.

d. Fixed point properties[3)6)]

As the system deviates from criticality $\xi \neq \infty$ the Hamiltonian may be viewed as consisting of a critical part $\mathbb{H}^*$ and a remainder

$$\mathbb{H} = \mathbb{H}^* + \delta\mathbb{H}$$

i.e.

$$\mathbb{H} = -k_c \sum \sigma_{\vec{n}} \sigma_{\vec{m}} - (k - k_c) \sum \sigma_{\vec{n}} \sigma_{\vec{m}} - h \sum \sigma_{\vec{n}} = \mathbb{H}^* + \sum_i h_i\, O_i \tag{2.16}$$

---

*(Note that the homogeneous Ansatz (2.12) is assumed to make sense only near criticality.)

with $k - k_c$ proportional to the reduced temperatur t. Hence h_i are the parameters ("fields") which describe the deviation from criticality and the O_i's are the conjugate operators. For infinitesimal

$$\delta H = \sum_i h_i O_i$$

and the O_i's choosen (if possible) diagonal under $T_{L/a}$ it follows:

$$T_{L/a} O_i = (L/a)^{y_i} O_i$$

$$\delta H_L = T_{L/a} \, \delta H_a = \sum_i h_i (L/a)^{y_i} O_i = \sum_i \tilde{h}_i O_i \tag{2.17}$$

According to whether $h_i \rightarrow \tilde{h}_i$ is increasing or decreasing the eigen-operators (and the conjugate fields) are classified:

$$\begin{array}{ll} y_i > 0 & \text{relevant} \\ y_i < 0 & \text{irrelevant} \\ y_i = 0 & \text{marginal} \end{array} \tag{2.18}$$

If the relevant fields are zero we call $H = H^{(0)}$ critical.
We have

$$T_{L/a} H^{(0)} \longrightarrow H^{*} ; \quad L \rightarrow \infty \tag{2.19}$$

under suitable behaviour of the marginal fields. As we will see in our field theoretical treatment the marginal operators alone determine the fixed point (if any) properties of H .
When (2.17) can be realized globally ($\forall L$) by a suitable choice of the parametrization of non marginal fields we call this a parametrization in terms of global scaling fields. These fields have been introduced by Wegner[6].

e. Scaling[2]

What follows from this intuitive block spin picture for the thermodynami properties in the critical region?
As a cell of size L contains $(\frac{L}{a})^D$ spins it follows in view of (2.10-12) that as we increase the cell size from a to L:

$$f \to f(\tilde{t},\tilde{h}) = (L/a)^{D} f(t,h)$$

Hence

$$\langle\sigma\rangle \to \langle\tilde{\sigma}\rangle = \langle\sigma\rangle(\tilde{t},\tilde{h}) = \frac{\partial f}{\partial \tilde{h}} = (L/a)^{D-x}\frac{\partial f}{\partial h} = (L/a)^{D-x}\langle\sigma\rangle(t,h)$$

(2.20)

$$\langle E\rangle \to \langle\tilde{E}\rangle = \langle E\rangle(\tilde{t},\tilde{h}) = \frac{\partial f}{\partial \tilde{k}} = (L/a)^{D-y}\frac{\partial f}{\partial k} = (L/a)^{D-y}\langle E\rangle(t,h)$$

Thus we are able to express the functions $\langle\sigma\rangle$, $\langle E\rangle$ etc. we are interested in through functions $\langle\tilde{\sigma}\rangle$, $\langle\tilde{E}\rangle$ etc. referring to a system with a reduced number of degrees of freedom.
With $\varkappa = L/a$

$$\langle\sigma\rangle(t,h) = \varkappa^{x-D}\langle\sigma\rangle(\varkappa^{y}t, \varkappa^{x}h) \quad .$$

In order the cell size L to cancel the function on the r.h.s. can only depend on the invariant product : $h|t|^{-x/y}$ hence (set $\varkappa^{y}|t|=1$ or $\varkappa^{x}|h|=1$)

$$\begin{aligned} \langle\sigma\rangle(t,h) &= \operatorname{sign} h\,|t|^{(D-x)/y}\,\varphi_\sigma(h|t|^{-x/y}) \\ &= \operatorname{sign} h\,|h|^{(D-x)/x}\,\chi_\sigma(t|h|^{-y/x}) \end{aligned} \qquad (2.21)$$

(with $\varphi_\sigma(0)$, $\chi_\sigma(0)$ finite) where we used in addition the symmetry properties of the system.
Similarly

$$\begin{aligned} \langle\sigma\sigma\rangle(|\vec{x}|,t,h) &= \varkappa^{2(x-D)}\langle\sigma\sigma\rangle(\varkappa^{-1}|\vec{x}|, \varkappa^{y}t, \varkappa^{x}h) \\ &= |t|^{2\frac{D-x}{y}}\,\varphi_{\sigma\sigma}(|\vec{x}||t|^{1/y}; h|t|^{-x/y}) \\ &= |\vec{x}|^{2(x-D)}\,\psi_{\sigma\sigma}(t|\vec{x}|^{-y}; h|\vec{x}|^{-x}) \end{aligned} \qquad (2.22)$$

$$\begin{aligned} \langle EE\rangle(|\vec{x}|,t,h) &= \varkappa^{2(y-D)}\langle EE\rangle(\varkappa^{-1}|\vec{x}|, \varkappa^{y}t, \varkappa^{x}h) \\ &= |t|^{2\frac{D-y}{y}}\,\varphi_{EE}(|\vec{x}||t|^{1/y}, h|t|^{-x/y}) \\ &= |\vec{x}|^{2(y-D)}\,\psi_{EE}(t|\vec{x}|^{-y}, h|\vec{x}|^{-x}) \end{aligned}$$

The functions $\varphi_{..}$ and $\psi_{..}$ are expected to be finite at zero* and integrable (summable) on $\mathbb{R}^D$ such that at the critical point the behaviour (2.8) for the spin and energy correlation can be read off i.e.

$$x = D - d_\sigma \quad ; \quad y = D - d_E \tag{2.23}$$

Similarly we get for the thermodynamical quantities:

$$\alpha = 2 - D/y \; ; \; \beta = \frac{D-x}{y} \; ; \; \gamma = \frac{2x-D}{y} \; ; \; \delta = \frac{x}{D-x} \tag{2.24}$$

From the comparison of the Ornstein-Zernicke form (2.6) and (2.22) we have

$$\nu = 1/y \tag{2.25}$$

As all critical indices are related to x and y we have the following scaling relations among them:

$$\begin{gathered} \alpha = \alpha' \; ; \; \gamma = \gamma' \; ; \; \nu = \nu' \\ \beta(\delta+1) = 2\beta + \gamma = 2 - \alpha \end{gathered} \tag{2.26}$$

To summarize the Kadanoff scaling picture leads to the following results

(1) Second order phase transitions are described by homogeneous functions. More refined arguments show that at the critical point physics is governed by a scale invariant theory (powerlaws, in exceptional cases also logarithms).

(2) The scaling assumption relates all critical indices to two independent coefficients determined from the knowledge of the spin and the energy twopoint functions.

f. Kadanoff Universality[2)7)]

Our discussion makes plausible, and it is supported by experiment and from model calculations, that the critical behaviour of systems with short range forces is independent on the

---

* (as they refer to a system with a finite number of degrees of freedom by the elimination process)

a) lattice structure and the discretness
b) details of interactions.

Accordingly one expects universality classes of critical theories with identical critical properties. Within an universality class one can perform transformations on the physical parameters such that different systems are described by the same functions. It is well known that e.g. the ferromagnetic transitions and the liquid-gas transitions have the same critical indices.
On the other hand it is found that critical behaviour is differentiated by

a) the dimension of the system;
as Wilson pointed out critical indices seem to depend analytically on D as $D \leq 4$ (This suggested the Wilson-Fisher ε-expansion)
b) Symmetry of the system;
e.g. Lenz-Ising, Heisenberg, Spherical model.
c) ev. other unknown parameters.

This closes our phenomenological discussion of critical phenomena. What has to be done is to make Kadanoff's ideas quantitative. In particular one has to explain the scaling and universality properties and to calculate the critical indices. It was the main benefit from Wilson's RG approach relating Kadanoff's picture to field theory and the ε-expansion and the $1/n$ - expansion that one has approximate solutions for a considerable range of universality classes which also cover many systems realized in nature[2]. In the next section we will discuss how field theory is related to statistical mechanics.

3. The Lenz-Ising System and Euclidean Field Theory

a. The Lenz-Ising system

Migdal, Polyakov and in most powerful manner Wilson used the Kadanoff ideas in order to relate the lattice systems of classical statistical mechanics to euclidean field theory by disgarding details (short range fluctuations) such that one stays in the original universality class, i. e. not changing the critical behaviour. We briefly discuss Wilsons instructive argumentation to manufacture a field theory which is in the universality class of the L. I. model[3].

Let us consider a lattice system of classical spins $\sigma_{\vec{n}}$, with spin distribution $\rho(\sigma_{\vec{n}}^2)$. The generating functional for the spin correlation functions is

$$Z\{J\} = C \int \prod_{\vec{n}} d\sigma_{\vec{n}}\, \rho(\sigma_{\vec{n}})\, e^{\frac{1}{2}\sigma K \sigma + J\sigma} = C \int \prod_{\vec{n}} d\sigma_{\vec{n}}\, e^{-\mathcal{H}[\sigma] + J\sigma} \tag{3.1}$$

$$\left.\frac{\delta^m Z\{J\}}{\delta J_{\vec{n}_1} \cdots \delta J_{\vec{n}_m}}\right|_{J=0} = \langle \sigma_{\vec{n}_1} \cdots \sigma_{\vec{n}_m} \rangle \tag{3.2}$$

with $\sigma K \sigma = \sum_{\vec{n}\vec{m}} K_{\vec{n}\vec{m}}\, \sigma_{\vec{n}}\, \sigma_{\vec{m}}$ and $J\sigma = \sum_{\vec{n}} J_{\vec{n}}\, \sigma_{\vec{n}}$

C serves to normalize Z to $Z\{o\} = 1$. For the ferromagnetic Lenz-Ising model $K_{\vec{n}\vec{m}} \leq o$; $K_{\vec{n}\vec{n}} = o$; $K_{\vec{n}\vec{m}} = K_{\vec{n}-\vec{m}}$

$$\text{with} \quad K_{\vec{n}} = \begin{cases} k < 0 & \text{for} \quad |\vec{n}| = 1 \\ 0 & \text{otherwise} \end{cases} \tag{3.3}$$

i.e. in Fourierspace

$$\tilde{K}(\vec{q}) = (2\pi)^{-D/2} \sum_{\vec{n}} K_{\vec{n}}\, e^{-i\vec{q}\vec{n}} = 2k \sum_{i=1}^{D} \cos q_i \tag{3.4}$$

The spin values are fixed to $\sigma = \pm 1$ with

$$\rho(\sigma_{\vec{n}}^2) = \delta(\sigma_{\vec{n}}^2 - 1) = \lim_{u_o \to \infty} \frac{u_o}{\pi}\, e^{-u_o(\sigma_{\vec{n}}^2 - 1)} \tag{3.5}$$

The approximate Lenz-Ising model we are interested in we obtain for finite u_o ($u_o \gg 1$); the L.I. system will be recovered as $u_o \to \infty$

The bilinear part of $\mathcal{H}[\sigma]$ then reads

$$\mathcal{H}_o = \frac{1}{2} \int_{-\pi}^{+\pi} d^D q\, |\tilde{\sigma}(\vec{q})|^2\, G_o^{-1}(\vec{q}) \tag{3.6}$$

with "propagator"

$$G_o(\vec{q}) = \frac{1}{-2u_o - 2k \sum_{1}^{D} \cos q_i} \tag{3.7}$$

The "interaction" part is:

$$\mathcal{H}_I = \mathcal{H} - \mathcal{H}_o = u_o \sum_{\vec{n}} (\sigma_{\vec{n}}^2)^2 \tag{3.8}$$

The generating functional (3.1) may then be written in the form

$$\begin{aligned} Z\{J\} &= C \int \prod_{\vec{n}} d\sigma_{\vec{n}}\, e^{-\mathcal{H}_I[\sigma]}\, e^{-\mathcal{H}_o[\sigma] + J\sigma} \\ &= \hat{C}\, e^{-\mathcal{H}_I\left[\frac{\delta}{\delta J}\right]} Z_o\{J\} \end{aligned} \tag{3.1'}$$

with

$$Z_o\{J\} = e^{\frac{1}{2} J G_o J}$$

the free generating functional.
A formal power expansion in u_o gives rise to a Feynman graph expansion for the correlation functions (3.2):

$$\begin{aligned} \langle \sigma_{\vec{m}_1} \cdots \sigma_{\vec{m}_n} \rangle = \sum_j \frac{(-u_o)^j}{j!} \sum_{\vec{m}_1,\ldots,\vec{m}_j} \int_{-\infty}^{+\infty} \prod_{\vec{n}} d\sigma_{\vec{n}} \; \times \\ \times\, \sigma_{\vec{m}_1} \cdots \sigma_{\vec{m}_n}\, \sigma^4_{\vec{m}_1} \cdots \sigma^4_{\vec{m}_j}\, e^{-\mathcal{H}_o[\sigma]} \end{aligned} \tag{3.2'}$$

This expression equals by (3.1') and (3.2) to the sum over all total contractions of pairs of σ 's in $\sigma_{\vec{m}_1} \ldots \sigma^4_{\vec{m}_j}$
The Feynman-rules are:

Contractions: $\sigma_{\vec{n}}\, \sigma_{\vec{m}} = G_o(\vec{n} - \vec{m})$: $\vec{n}$ o———o $\vec{m}$

vertices: $\sum_{\vec{n}} \sigma^4_{\vec{n}}$: $\sum_{\vec{n}}$ ✕ $\vec{n}$

As we will see under the RG-transformation u_o transforms in analogy to (2.12) to small effective couplings and perturbation theory becomes applicable near criticality. (In renormalized field theory the renormalized coupling will turn out to be small whereas the bare coupling $\mu_o \to \infty$ as $\Lambda \to \infty$).

For small momenta $p_i = \frac{q_i}{a}$ we observe that

$$G_0(\vec{p}) \simeq a^{-2} k^{-1} \frac{1}{\vec{p}^{\,2} + m^2} \quad \text{with} \quad m^2 = -2a^{-2}(u_0 k^{-1} + D) \tag{3.9}$$

is apart from a factor, which is eliminated by rescaling the field $\sigma \to \hat{\sigma} = a\sqrt{k}\,\sigma$ an euclidean scalar propagator i.e. in the long range region

$$\sigma_{\vec{n}} \longrightarrow \hat{\sigma}(\vec{x}) \tag{3.9a}$$

behaves as a continuous euclidean scalar field. The energy density (2.4) in a similar way

$$E_{\vec{n}} = \sum_{\substack{\vec{m} \\ m.n.}} \sigma_{\vec{n}}\,\sigma_{\vec{m}} \longrightarrow \hat{E}(\vec{x}) = 1/2 \left\{ \partial_i \hat{\sigma}(\vec{x})\, \partial^i \hat{\sigma}(\vec{x}) - m_0^2\, \hat{\sigma}^2(\vec{x}) \right\} \tag{3.9b}$$

behaves as a field.

If one change according to Wilson (3.6) to

$$\mathcal{H}_0 = 1/2 \int_{|\vec{p}|<\Lambda} d^D p \, |\tilde{\hat{\sigma}}(\vec{p})|^2 (\vec{p}^{\,2} + m^2) \tag{3.10}$$

one expects not to change the critical behaviour as the small momentum behaviour (long range) is kept exactly. The rotational invariant cut-off here represents a substitute for the lattice cut-off a^{-1}. The difficulty is that the classical functional (3.1) with the replaced $\mathcal{H}_0$ is illdefined and $\hat{\sigma}(\vec{x})$ has to be replaced by a box field

$$\varphi^{(L)}(\vec{x}) = \frac{1}{L^{D/2}} \sum_{|\vec{p}|<\Lambda} e^{i\vec{p}\vec{x}}\, \tilde{\varphi}(\vec{p}) \,; \quad \vec{p} = \frac{2\pi}{L}\vec{n}$$

For the correlation functions the "thermodynamic limit" $L \to \infty$ may then be carried out:

$$< \varphi(\vec{x}_1) \ldots \varphi(\vec{x}_n) > = \lim_{L\to\infty} < \varphi^{(L)}(\vec{x}_1) \ldots \varphi^{(L)}(\vec{x}_n) >$$

We prefer however to construct directly an euclidean cut-off field theory with $\mathcal{H}_0$ of the form (3.10) avoiding the difficulty of the functional formulation.

b. Euclidean Field Theory [8)]

An euclidean cut-off theory may be constructed as follows: Let A(k) and $A^+(k)$ be annihilation and creation operators subject to the commutation relations

$$[A(k), A(k')] = [A^+(k), A^+(k')] = 0 \tag{3.11}$$

$$[A(k), A^+(k')] = \delta^{(D)}(k-k')$$

From the cyclic euclidean free vacuum $|\Phi_o\rangle_E$

$$A(k)\,|\Phi_o\rangle_E = 0 \tag{3.12}$$

we generate the euclidean Fock-space

$$\mathcal{H}_E = \overline{\{\mathcal{P}\{A^+\}\,|\Phi_o\rangle_E\}} \tag{3.13}$$

Then the free field

$$A_o(x) = (2\pi)^{-D/2} \int_{|k|<\Lambda} \frac{d^Dk}{\sqrt{k^2+m^2}} \{e^{-ikx} A^+(k) + h.c.\} \tag{3.14}$$

leads to the propagator (3.10)

$$G_o(x-y) = {}_E\langle\Phi_o|\,A_o(x)\,A_o(y)\,|\Phi_o\rangle_E \tag{3.15}$$

The commuting fields $A_o(x)$ generate from the euclidean vacuum a cut-off Hilbert-space $\mathcal{H}_{E,\Lambda} \subset \mathcal{H}_E$. In order to obtain a complete set of operators one introduces the canonical conjugate field

$$\Pi_o(x) = \frac{i}{2} \int_{|k|<\Lambda} d^Dk\,\sqrt{k^2+m^2}\,\{e^{-ikx} A^+(k) - h.c.\} \tag{3.16}$$

with $$[\Pi_o(x), A_o(y)] = -i\delta(x-y)\,;\; [\Pi_o(x), \Pi_o(y)] = 0 \tag{3.17}$$

Contrary to the relativistic case (non commuting fields) the euclidean generators of symmetry transformations cannot be represented in terms of the now commuting A's e. g. the euclidean Hamiltonian, generating time translations in $\mathcal{H}_{E,\Lambda}$ is:

$$H = -\frac{i}{2}\int d^D x \; :\Pi_o(x) \overleftrightarrow{\partial_t} A_o(x): \tag{3.18}$$

In the interacting case, with $\mathcal{H}_I[A_o]$ an integral over a local polynomial in A_o, the euclidean Green functions

$$_E\langle o | A(x_1) \ldots A(x_n) | o \rangle_E = {_E\langle} \Phi_o | A_o(x_1) \ldots A_o(x_n) e^{-\mathcal{H}_I[A_o]} | \Phi_o \rangle_E \tag{3.19}$$

are identical with the probabilistic correlation functions (3.2) for

$$\mathcal{H}_I[A_o] = u_o \int d^D x : A_o^4(x): \tag{3.20}$$

There are some peculiar features to euclidean fields: Due to

$$i\int d^D k \, e^{-ikx} A_o(k) \sqrt{k^2+m^2} = i\,\Pi_o(x) + \tfrac{1}{2}(-\Delta + m^2) A_o(x) \tag{3.21}$$

there exist short range fields as

$$\psi_o(x) = (-\Delta + m^2) A_o(x) \tag{3.22}$$

with

$$_E\langle \Phi_o | \psi_o(x) A_o(y) | \Phi_o \rangle_E = \delta_\Lambda(x-y) \tag{3.23}$$

In the relativistic case of course $\psi_o(x) \equiv o$. The set of euclidean local fields therefore consists of the usual Wick ordered fields

$$: A_o^n(x):$$

and short range composite fields like

$$: A_o^n(x)\,\psi_o(x):$$

This situation of course persists in the interacting case. We believe that the so called "redundant" operators introduced by Wegner are related to the short range fields discussed here.

A further serious difference is the following: After renormalization relativistic composite fields remain in the class of operator-valued distribution as $\Lambda \to \infty$. This is not true for euclidean composite fields as e. g. in D = 4

$$\| A^2_{o,\Lambda}(f) | \Phi_o \rangle_E \| \to \infty, \; (\Lambda \to \infty) \tag{3.24}$$

i.e. $:A_o^2(x):$ can only have a meaning as a bilinear form and does not exist as an operator. For non-overlapping test-functions $f_1,\dots,f_n$ composite correlation functions however exist in the limit $\Lambda \to \infty$:

$$_E\langle \Phi_o | A^2_{o,\Lambda}(f_1) \dots A^2_{o,\Lambda}(f_n) | \Phi_o \rangle_E \longrightarrow \text{finite} \qquad (3.25)$$

in D = 2 dimensions: $A^n_{o,\Lambda}$: exists for $\Lambda \to \infty$, for D = 3 only $:A_o^2:$ and A_o in D = 4 only A_o exists as an operator in the limit .
This situation is a handicap for the Kadanoff-Wilson operator product expansion. Either one has to consider it as a statement on correlation functions only or one has to go to the relativistic theory.
The only thing we should learn from the above discussion is that the L. I. model and the A^4-field theory are likely to belong to the same universality class.

4. Construction of Critical Theories

In the construction of critical theories there are two different possibilities. The more ambitious one is to study critical behaviour and deviations from it directly within the global physical theory (e.g. for certain physical systems the Lenz-Ising model in D = 3 dimensions). In this case also non universal properties of the system may be calculated. Recent progress in this approach has been made by Nauenberg and Nienhuis[9] for the LI system.

The other attempt in the spirit of Kadanoff, is to take care only of the universal properties i.e. to construct critical theories lying in a particular universality class where one hopes to find a single scale invariant (and hence conformal invariant)[10] theory. The most reasonable approach in constructing critical theories therefore seems to be the direct construction of conformal invariant theories and to determine the spectrum of e. g. anomalous dimensions of conformal theories (classification of critical theories). This actually was the first attempt in the construction of critical theories by Polyakov[1] in his bootstrap approach as developed further by Parisi-Peliti[11], Mack[12] and others. This ambitious program unfortunately did not yet succeed but we believe that this is the way to construct non trivial critical theories beyond the present approximation schemes.

It was Wilson who succeeded first in the construction of nontrivial critical theories by his RG approach. A quantitative realization of Kadanoff's idea of eliminating irrelevant degrees of freedom (i.e. the short range fluctuations) for the A^4 cut-off theory in functional form (3.1) led him to a study of the RG transformation

$$e^{-\mathcal{H}' + \text{const.}} = \int \prod_{\frac{\Lambda}{s} < |p| < \Lambda} d[\varphi^{(L)}]\, e^{-\mathcal{H}[\varphi^{(L)}]} \Big|_{\varphi^{(L)} \to \alpha_s \varphi^{(sL)}}$$

with $\alpha_s = s^{1-n}/2$. In this procedure all internal lines in Feynman diagrams are integrated out over the short range part $s^{-1}\Lambda < |p| < \Lambda$ giving rise to new effective mass and coupling (renormalized) and new (nonrenormalizable) vertices which however should be irrelevant in the critical region. The external lines have momenta restricted to $|p| < \Lambda s^{-1}$. In an approximate form Wilson was able to determine fixed points of the transformation from computer calculations. Under further approximation using perturbation theory in $\varepsilon = 4-D$ dimensions analytic calculations for nontrivial fixed points have been done by Wilson and Fisher[4] and Wegner and Houghton[4] and others.

According to our philosophy only the universal scaling properties of models (which differ from the global physical model) can be taken seriously. These models have to be choosen within one universality class from the point of view of simplicity and computability. Concerning the universality class of the L.I. system we presented the arguments which suggest that A^4-field theory models are in the same class and we are faced simply with the problem of constructing scale invariant A^4 models. As the direct conformal construction was not yet successfull the next step would be to use renormalized perturbation theory. This is what we will do in the following. We will eliminate the cut-off Λ in the euclidean A^4-model from the beginning and use renormalized field theory[13] for the study of scale invariant theories by looking at the fixed point properties of the dilatation Ward-identity (Callan-Symanzik equation).

We will go one step further and consider the relativistic local A^4-theory avoiding thereby the peculiarities of euclidean theories mentioned in the last section. In doing so we refer to the equivalence statement of Osterwalder and Schrader[8]. By the spectrum condition the time ordered relativistic Green functions in D space-time dimensions $x = (x^0, x^1, \ldots, x^{D-1})$ are analytic in x^0 and are identical with the euclidean Green functions for

$$x_i^0 = i\, x_i^D \, ; \qquad x_i^D \quad \text{real}$$

Criteria on the validity of the Landau mean field approximation as confirmed by model calculations show that the critical theory is a free field theory in $D \geqslant 4$ dimensions and critical indices are likely to depend analytically on D below $D = 4$. This suggested Wilson and Fisher[4] to compute critical theories starting from $D = 4$ by analytic continuation in $1 \gg \varepsilon = D-4 \geqslant 0$ (perturbation around free theory: ε-expansion). Low order ε calculations are in remarkable agreement with LI-calculations and experiments for $\varepsilon = 1$ and even for $\varepsilon = 2$ (see Tab. 2). In our approach we will use the ε - expansion for the construction of critical theories. A direct approach to critical theories in $D = 3$ and $D = 2$ was given by Parisi[14]; see also the investigations of Symanzik[14] and Schroer[29].

To summarize: What we will do in our further discussion within the framework of renormalized perturbation theory, is to

a) construct a long range scale invariant theory
b) calculate critical indices and prove the relations among them
c) formulate field theoretical Kadanoff scale transformations
d) calculate corrections to scaling

5. Renormalized Perturbation Theory and ε-Expansion

a. Parametrizations of Green Functions

We briefly discuss renormalized quantum field theory as used in our further considerations. We start from a Lagrangian cut-off Λ theory with ($\mathcal{L}_{int} = - \mathcal{H}_I$)

$$\mathcal{L} = \mathcal{L}_0 + \mathcal{L}_{int} = \frac{1}{2}(\partial A)^2 - \frac{m_0^2}{2} A^2 - \frac{u_0}{4!} A^4 \qquad (5.1)$$

The correlation functions (time ordered Green functions) are obtained as a formal power series expansion (Feynman graph expansion) in u_0 from the Gell-Mann-Low formula[16]

$$\langle T \prod_{k=1}^{N} A(x_k) \rangle = \langle \Phi_0 | T \prod_{k=1}^{N} A^{(0)}(x_k)\, e^{i\int \mathcal{L}_{int}^{(0)} dx} | \Phi_0 \rangle_{\otimes} \qquad (5.2)$$

$|\Phi_0\rangle$ denotes the free Fock vacuum; $A^{(0)}(x)$ is the free scalar field of mass m_0 and $\mathcal{L}^{(0)}_{int} = \mathcal{L}_{int}(A^{(0)})$. $\otimes$ denotes the omission of vacuum diagrams i.e. the division by $\langle\Phi_0| e^{i\int \mathcal{L}^{(0)}_{int} dx} |\Phi_0\rangle$.
The generating functional for the disconnected Green functions (partition functional) is

$$Z\{J\} = \langle \Phi_0 | e^{i\int(\mathcal{L}^{(0)}_{int} + J(x)A^{(0)}(x))dx} |\Phi_0\rangle_\otimes \tag{5.3}$$

The generating functional of the connected Green functions (Gibbs potential, free entalpy functional) is given by $G\{J\} = \ln Z\{J\}$

$$G^{(N)}(x_1,\ldots,x_N) = \langle T \prod_{k=1}^{N} A(x_k)\rangle^{conn} = (-i)^N \frac{\delta^N G\{J\}}{\delta J(x_1)\ldots\delta J(x_N)}\bigg|_{J=0} = \sum_{conn} \tag{5.4}$$

The parametrization in terms of the bare parameters u_0 and m_0 is not convenient for the purpose of statistical mechanics. At criticality not the bare mass m_0 but the renormalized mass m($\xi = m^{-1}$ correlation length defined by the momentum space location of the propagator pole has to vanish (see (2.6) respectively (2.8)).
Like in particle physics it is therefore much more convenient to use a parametrization in terms of renormalized quantities. To this end a multiplicative renormalization of fields $A \to \hat{A} = Z^{-1/2} A$ and subtractions (by adding appropriate counterterms to the bare Lagrangian (5.1)) are performed to the correlation functions in such a way that certain normalization conditions (defining the physical interpretation of the new parameters) are satisfied.

The re-normalization (re-parametrization) is most conveniently done for the vertex functions, the Legendre-transforms (with respect to the source J(x)) of the connected Green functions. The generating functional (Helmholz potential, free energy functional)[17] reads:

$$\Gamma\{K\} = G\{J\} - i\int K(x)J(x)dx \tag{5.5}$$

with $$K(x) = -i\frac{\delta G\{J\}}{\delta J(x)}$$

The vertex functions

$$\Gamma^{(N)}(x_1,\ldots,x_N) = \frac{\delta^N \Gamma\{K\}}{\delta K(x_1)\ldots\delta K(x_N)}\bigg|_{K=0} = \sum_{prop} \tag{5.6}$$

are represented by a sum over the proper i.e. the connected one-particle irreducible (i.e. connected after cutting a single line) amputated (i.e. no external legs) diagrams and

$$\Gamma^{(2)} = -\{ G^{(2)} \}^{-1} \tag{5.7}$$

The $G^{(N)}$'s which are then trees in the $\Gamma^{(N)}$'s (no extra loops!) are given by the inverse Legendre transformation. The renormalization problem is completly solved by the renormalization of the $\Gamma^{(N)}$'s. The Fourier transforms of $\Gamma^{(N)}$ may be written as:

$$\Gamma^{(N)}(p_1, \ldots, p_N) = \langle T A(0) \tilde{A}(p_2) \ldots \tilde{A}(p_N) \rangle^{prop} \tag{5.8}$$

We now consider the different parametrizations of the correlation functions. The parametrization standard in particle physics (mass shell normalization) is defined through the normalization conditions (see e.g. [18) 19)]):

$$\begin{aligned} &\Gamma^{(2)}\Big|_{p^2=m^2} = 0 \\ &\frac{\partial \Gamma^{(2)}}{\partial p^2}\Big|_{p^2=m^2} = i \\ &\Gamma^{(4)}\Big|_{s.p.\,m^2} = -ig\, m^{\varepsilon} \qquad \text{with } s.p.\, m^2 : \; p_i p_j = \tfrac{1}{4}(3\delta_{ij} - 1)\, m^2 \end{aligned} \tag{5.9}$$

This parametrization is not suitable to our aim of constructing a critical theory $\xi^{-1} = m = 0$ (zero mass theory) as the Green functions are not continuous at $m = 0$ (diverging residue of the propagator pole). A parametrization with a continuous zero mass limit was given by Gell-Mann and Low[16) 20)]

$$\begin{aligned} &\tilde{\Gamma}^{(2)}\Big|_{p^2=\tilde{m}^2} = 0 \\ &\tilde{\Gamma}^{(2)}\Big|_{p^2=-\mu^2} = -i\mu^2 \quad \left(\text{or } \frac{\partial \tilde{\Gamma}^{(2)}}{\partial p^2}\Big|_{p^2=-\mu^2} = i \right) \\ &\tilde{\Gamma}^{(4)}\Big|_{s.p.\,-\mu^2} = -i\tilde{g}\,\mu^{\varepsilon} \end{aligned} \tag{5.10}$$

Here the critical theory is obtained for $\xi^{-1} = \tilde{m} = 0$ where the $\hat{\Gamma}$'s exist (finite residue of the propagator pole).

There is an other parametrization (<u>soft</u> or <u>pre-scaling</u> parametrization) which will be most adequate for our purpose. It is defined by [21)]:

$$\hat{\Gamma}^{(2)}\Big|_{\substack{p^2=0\\ \hat{m}^2=0}} = 0$$

$$\hat{\Gamma}^{(2)}\Big|_{\substack{p^2=-\mu^2\\ \hat{m}^2=0}} = -i\mu^2 \quad \left(\text{or} \quad \frac{\partial \hat{\Gamma}^{(2)}}{\partial p^2}\Big|_{\substack{p^2=-\mu^2\\ \hat{m}^2=0}} = i\right) \tag{5.11}$$

$$\hat{\Gamma}^{(4)}\Big|_{\substack{s.p.-\mu^2\\ \hat{m}^2=0}} = -i\hat{g}\mu^{\varepsilon}$$

and

$$\frac{\partial \hat{\Gamma}^{(2)}}{\partial \hat{m}^2}\Big|_{\substack{p=0\\ \hat{m}^2=\mu^2}} = i$$

This parametrization will be used in the following. The properties we will discuss in section 8. We only mention here that the critical theory again is obtained for $\hat{m} = o$ (however now $\hat{m} \neq \xi^{-1}$) where the $\hat{\Gamma}$'s exist. Equivalent parametrizations have been discussed in Ref. 22) and 2[illegible] in a different context in Ref. 24). We will see that $\hat{m}^2 = t$ is a parameter proportional to the reduced temperatur $(T-T_c)/T_c$ in the critical region. t will simply be called temperature in the following. All the parametrizations mentioned above have a limit $\Lambda \to \infty$ and we are dealing hence with a renormalized local quantum field theory.

Also our model is superrenormalizable in D = 4-ε ($\varepsilon > o$) dimensions we will keep the normalization conditions as for D = 4 in order to have a continuous transition $\varepsilon \to o$ where the critical theory will turn out to be mean field (free theory).

b. Composite fields

For the study of energy fluctuations (2.8), (3.9b) we will also need correlation functions involving composite fields $\mathcal{O}_i(x)$ = local monomial in A and derivatives of A. Composite correlation functions are defined from a corresponding Gell-Mann-Low formula

$$< T \prod_{j=1}^{K} \mathcal{O}_i(y_j) \prod_{k=1}^{N} A(x_k) > = <\Phi_o | T \prod_{j=1}^{K} \mathcal{O}_i^{(0)}(y_j) \prod_{k=1}^{N} A^{(0)}(x_k) e^{i\int \mathcal{L}_{int}^{(0)} dx} | \Phi_o >_{\otimes} \tag{5.12}$$

with $\mathcal{O}_i^{(0)}(x)$ the monomical $\mathcal{O}_i$ in terms of free fields.
The generating functional is

$$Z\{J, h_i\} = <\Phi_o | e^{i\int (\mathcal{L}_{int}^{(0)} + J(x) A^{(0)}(x) + h_i(x) \mathcal{O}_i^{(0)}(x)) dx} | \Phi_o >_{\otimes} \tag{5.13}$$

The connected correlation functions are generated by $G\{J,h_i\} = \ln Z\{J,h_i\}$

$$\langle T \prod_j O_i(y_j) \prod_k A(x_k)\rangle^{conn} = (-i)^{N+K} \frac{\delta^K \delta^N G\{J,h_i\}}{\delta h_i(y_1)\ldots\delta h_i(y_K)\,\delta J(x_1)\ldots\delta J(x_N)}\bigg|_{J=h_i=0} \tag{5.14}$$

$$= \sum_{conn}$$

The Legendre transform (5.5) of $G\{J, h_i\}$ with respect to J generates the composite vertex functions

$$\Gamma^{(N,K)}(x_1,\ldots,x_N; y_1,\ldots,y_K) = \frac{\delta^K \delta^N \Gamma\{K,h_i\}}{\delta h_i(y_1)\ldots\delta h_i(y_K)\,\delta K(x_1)\ldots\delta K(x_N)}\bigg|_{K=h_i=0} \tag{5.15}$$

$$= \sum_{prop}$$

where proper (prop) graphs are connected and one-particle irreducible with respect to all cuts not separating y-vertices. The Fourier transforms may be written as

$$\Gamma^{(N,K)}(p_1,\ldots,p_N; q_1,\ldots,q_K) = \langle T\, A(0)\, \tilde{A}(p_2)\ldots\tilde{A}(p_N)\, \tilde{O}_i(q_1)\ldots\tilde{O}_i(q_K)\rangle^{prop} \tag{5.16}$$

$$= \langle T\, \tilde{A}(p_1)\ldots\tilde{A}(p_N)\, O_i(0)\, \tilde{O}_i(q_2)\ldots\tilde{O}_i(q_K)\rangle^{prop}$$

The composite fields have to be normalized according to the assigned physical interpretation. Composite fields which have an interpretation directly in the critical theory must be renormalized such that the limit $\Lambda \to \infty$ as well as $\hat{m}$ (or $\tilde{m}$ or m) $\to$ o is finite for $D \leq 4$. Corresponding composite fields (normal products) are denoted by

$$\hat{N}[O_i](y)$$

Note that composite fields need apart from multiplicative renormalization

$$O_i(y) \longrightarrow \hat{N}[O_i](y) = Z_i\, O_i(y) \tag{5.17}$$

also additive renormalizations (depending on K and N)

$$\hat{N}[O_i](y_1)\ldots\hat{N}[O_i](y_K)\,A(x_1)\ldots A(x_N) \quad - \quad \text{local distribution} \tag{5.18}$$

Examples will be given in section 6.

For statistical mechanics the euclidean correlation functions are obtained by analytic continuation in x^o to euclidean points:

$$\langle T \prod O_i(y_j) \prod A(x_k) \rangle \Big|_{\substack{x^0 = ix^D \\ x^D \text{ real}}} = \langle \prod O_i(\vec{y}_j) \prod A(\vec{x}_k) \rangle_E \tag{5.19}$$

For structural investigations and proofs of the existence of various limits to all orders of perturbation theory one most conveniently uses the Bogoliubov-Parasiuk-Hepp-Zimmermann (BPHZ)-renormalization scheme[26) 27)]. In this approach the correlation functions ((5.2),(5.12)) are defined directly by a finite part prescription to the Feynman integrands avoiding a cut-off or other regularizations. For technical details we refer to Ref. 21).

c. ε-Expansion

The continuation of a scalar field theory from D integer to non-integer dimensions is possible only via the continuation of Feynman integrals.

Let

$$I(p_j, m, \varepsilon) = \int \prod_{a=1}^{\ell} d^D k_a \left\{ \prod_1^{\ell} \frac{i}{k_a^2 - m^2 + i0} \prod_{j=1}^{n} \delta^{(D)}(p_j - \varepsilon_{ja} k_a) - \text{subtr.} \right\} \tag{5.20}$$

be a Feynman integrand in momentum space to a connected Feynman-diagram with ℓ internal lines and n vertices. D denotes the number of space-time (with metric (+,(D-1)-) dimensions. p_j is the external momentum at vertex j and

$$\varepsilon_{ja} = \begin{cases} +1 & \text{for a line ending at vertex j} \\ -1 & \text{for a line originating at vertex j} \\ 0 & \text{otherwise} \end{cases}$$

The Schwinger-parametric representation of (5.20) (which is defined for D = integer) is obtained with

$$\frac{i}{k^2 - m^2 + i0} = \int_0^\infty d\alpha \, e^{i\alpha(k^2 - m^2 + i0)}$$

$$\delta^{(D)}(k) = \frac{1}{(2\pi)^D} \int_{-\infty}^{+\infty} d^D x \, e^{ikx} \tag{5.21}$$

The four momentum integrals are then all of the Gaussian type

$$\int d^D k \, e^{ia(k^2 + 2bpk)} = \left(\frac{\pi}{ia}\right)^{D/2} e^{-iab^2p^2} \tag{5.22}$$

and lead to (see e. g. 25))

$$I(p_j, m, \varepsilon) = i^L (-i\pi)^{L\frac{D}{2}} \delta^{(D)}(\Sigma p_j) \times$$

$$\times \int_0^\infty d\alpha_1 \dots d\alpha_\ell \, e^{-i\sum_1^\ell \alpha_a (m_a^2 - i0)} \left\{ \frac{e^{ip_i d_{ij}^{-1} p_j}}{P^{D/2}} - \text{subtr. terms} \right\} \tag{5.23}$$

Here L is the number of loops of the graph,
P is a homogeneous polynomial in the α's of degree L
$d_{ij}^{-1} = N_{ij}/P$ with N_{ij} a homogeneous polynomial in the α's
of degree L + 1.

The representation (5.23) may now be analytically continued to complex D. $I(p_j, m, \varepsilon)$ is for $m > 0$ a meromorphic function in D with poles at some negative rational $\varepsilon = 4-D$. Hence $I(p_j, m, \varepsilon)$ has a power expansion in ε for $\varepsilon \geqslant 0$. In this way the correlation functions are obtained as double (formal) power series in g and ε .

For a treatment of field theory in $D = 4-\varepsilon$ dimensions not using the ε-expansion see Parisi and Symanzik[14].

Footnote:

For m = o there are infrared poles at some positive rational values of ε in the region $\varepsilon > \frac{2}{n}$, n the perturbation theoretic order of $I(p_j, 0, \varepsilon)$. Due to these IR divergences the Green functions to all orders in g do not exist at m = o in $4-\varepsilon$ ($\varepsilon > 0$) dimensions in an usual perturbation theory. Symanzik[14] has given a new expansion exhibiting terms non analytic in g which is free of the IR singularities.

6. Critical Theory (Preasymptotic Zero-Mass Theory)

We will first construct the critical theory in order to understand and calculate the behaviour (2.8) field theoretically

$$\langle A(\vec{x})\, A(0)\rangle_E \simeq \frac{\text{const.}}{|\vec{x}|^{2d_A}} \tag{6.1a}$$

$$\langle A^2(\vec{x})\, A^2(0)\rangle_E \simeq \frac{\text{const.}}{|\vec{x}|^{2d_{A^2}}} \tag{6.1b}$$

To this end we have to look for a scale invariant (for long distances) A^4-theory. The only candidate for a Lagrangian that can lead to a scale invariant field theory is (no dimensional parameters!):

$$\text{"}\mathcal{L}^{(0)} = \tfrac{1}{2}(\partial A)^2 - \frac{g\mu^\varepsilon}{4!} A^4\text{"} \tag{6.2}$$

The Lagrangian (6.2) however only makes sense either in a UV-cut-off (Λ) theory where Λ destroys scale invariance or (as UV-subtractions at zero momenta cause infrared divergencies) after performing UV-subtractions at some spacelike normalization spot μ where μ destroys scale invariance. Hence in perturbation theory there is no scale invariance (nonexistence of a zero theory without scale parameter!).

We consider in the following the preasymptotic zero mass theory normaliz by

$$\Gamma_0^{(2)}\big|_{p=0} = 0\; ; \quad \Gamma_0^{(2)}\big|_{p^2=-\mu^2} = -i\mu^2\; ; \quad \Gamma_0^{(4)}\big|_{s.p.\,-\mu^2} = -ig\mu^{4-D} \tag{6.3}$$

The μ-dependence is governed by the Gell-Mann-Low renormalization grour (RG) equation

$$\left\{\mu\frac{\partial}{\partial\mu} + \sigma(g)\frac{\partial}{\partial g} - N\tau(g)\right\}\Gamma_0^{(N)} = 0 \tag{6.4}$$

Here $\mu\,\partial_\mu$ acts as the dilatation operator in the parameter space and (6.4) represents the dilatation Ward-identity.

If we assume (6.4) to be true beyond perturbation theory the vertex functions Γ_0 scale (i. e. are homogeneous functions) provided $\sigma(g) = 0$ for some value $g = g^*$. Hence scale invariance is found in the (by the differential equation) summed up perturbation theory. Expanding the scale invariant solution in g leads back to the leading perturbation terms being individually non scale invariant.

When g^* is small we can use the perturbation theory to calculate $\sigma(g)$ and $\tau(g)$. In this case scaling is in an approximate sense <u>computable</u>.

The global solution of (6.4) is

$$\Gamma_0^{(N)}(\{\varkappa p_i\}; \mu, g) = \varkappa^{D - N d_A}\, r_\tau^{-N}\, \Gamma_0^{(N)}(\{p_i\}; \mu, g(\varkappa)) \tag{6.5}$$

with $$g(\varkappa): \quad \ln \varkappa = \int_g^{g(\varkappa)} \sigma^{-1}(g')\, dg' \tag{6.6}$$

$$r_\tau(\varkappa, g) = \exp \int_g^{g(\varkappa)} dg' \, \frac{\tau(g') - \tau(g^*)}{\sigma(g')} \tag{6.7}$$

with $d_A = d + \gamma_A$ the dynamical dimension of A;
$d = \frac{D-2}{2}$ the canonical dimension of A, and $\gamma_A = \tau(g^*)$ the anomalous dimension of A relative to g^*.
At $g = g^*$ we have

$$\hat{\Gamma}_0^{(N)}(\{\varkappa p_i\}; \mu, g^*) = \varkappa^{D - N d_A}\, \hat{\Gamma}_0^{(N)}(\{p_i\}; \mu, g^*) \tag{6.8}$$

i.e. the $\hat{\Gamma}_0$'s are homogeneous functions in the momenta, μ as a completely passive scale can be eliminated.

There is no reason that in nature $g = g^*$, hence we have to study the case $g \neq g^*$.
If $g \neq g^*$ we distinguish two cases

1) As $\varkappa \to 0$, $g(\varkappa)$ has to go to a value (if any) g_0 where $\sigma(g_0) = 0$ and σ increasing at g_0 (Fig. 2).

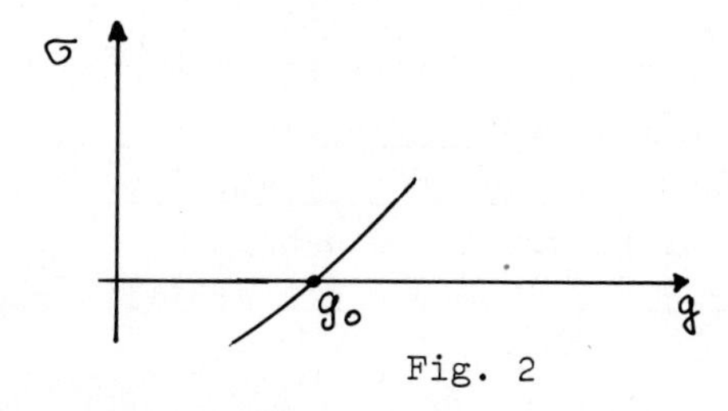

Fig. 2

If $\sigma'(g_o) > o$ then $r_\tau(\varkappa) \to r_o$ finite and

$$\Gamma_o^{(N)}(\{\varkappa p_i\}; \mu, g) \underset{\varkappa \ll 1}{\simeq} \varkappa^{D-Nd_A^o} r_o^{-N} \Gamma_o^{(N)}(\{p_i\}; \mu, g_o) \tag{6.9}$$

Hence if there is a zero g_o of σ with $\sigma'(g_o) > 0$ the long range part of the preasymptotic zero mass theory approaches a scale invariant limit (long range scaling). g_o is called an <u>infrared stable scaling fixed point</u>. This limit is the one relevant for <u>statistical mechanics</u> (i. e. agrees with the critical regime of the L. I. System) where scaling is expected too only in the long range region. The relation is a special case of (6.9) for N = 2. The Lagrangian (6.2) $\mathcal{L}^{(0)}$ can be identified as a critical one.

2) As $\varkappa \to \infty$ by (6.6) $g(\varkappa)$ has to approach a value (if any) g_∞ with $\sigma(g_\infty) = o$ and σ decreasing at g_∞ (Fig. 3).

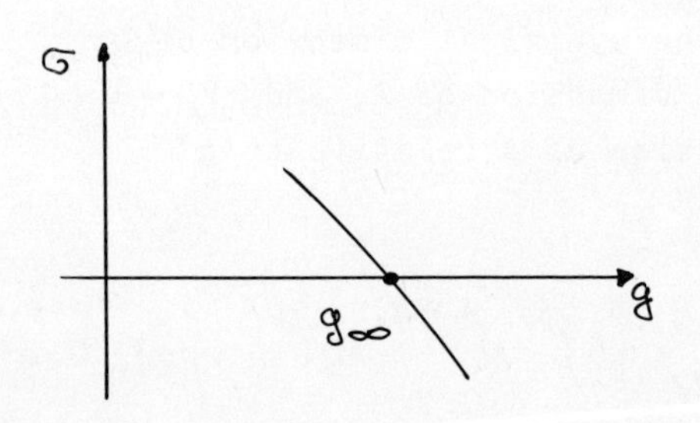

Fig. 3

When $\sigma'(g_\infty) < 0$ then $r_\tau(\varkappa) \to r_\infty$ finite and

$$\Gamma_o^{(N)}(\{\varkappa p_i\}; \mu, g) \underset{\varkappa \gg 1}{\simeq} \varkappa^{D-Nd_A^\infty} r_\infty^{-N} \Gamma_o^{(N)}(\{p_i\}; \mu, g_\infty) \tag{6.10}$$

Thus if there is a zero g_∞ of σ with $\sigma'(g_\infty) < 0$ the short range part of the preasymptotic zero mass theory shows scale invariance (short distance scaling). g_∞ is an <u>ultraviolet stable fixed point.</u> This limit might be relevant for <u>high energy physics.</u> This scaling limit is present only in our renormalized embbeding theory not in the L. I. model or the cut-off field theory which exhibit smooth ultraviolet behaviour.

Note that if $\sigma' = o$, $\tau' \neq o$ at g^* then

$$r_\tau(\varkappa) \longrightarrow r_\tau^{(as)}(\varkappa) = |\ln \varkappa|^{-2\tau'/\sigma''} + const \longrightarrow \begin{cases} o \\ \infty \end{cases} \qquad (6.11)$$

for $\sigma'' \neq o$ i. e. in this case one has logarithmic modifications and no scaling in the strict sense (see also section 8).

From our consideration we see that the preasymptotic theory contains all information about the scaling structure of A^4-theory whether long range or short range. The question of computable scaling we will discuss below.

We turn now to the consideration of composite fields in the preasymptotic theory, in order to derive (6.1b).
The energy density by (3.9b) is of the form $E(x) \propto \frac{1}{2}(\partial A)^2 + \frac{1}{2} m_o^2 A^2$
In the long range region (relevant for statistical mechanics) however the term of lowest dimension is dominant and hence

$$E(x) \propto A^2$$

We thus consider the field

$$\frac{1}{2} \hat{N}[A^2](x)$$

The composite vertex-functions are

$$\Gamma_o^{(N,K)}(p_i, q_j; \mu, g) = z^{-K} \langle T A(o) \tilde{A}(p_2) \ldots \hat{\tilde{N}}[A^2](q_1) \ldots \rangle^{prop} \qquad (6.12)$$

normalized by (6.3) and

$$\Gamma_o^{(2,1)}\left(\frac{p}{2}, \frac{p}{2}; -p\right)\Big|_{p^2=-\mu^2} = 1 \quad ; \quad \Gamma_o^{(0,2)}(; q, -q)\Big|_{q^2=-\mu^2} = 0 \qquad (6.13)$$

They obey the RG-equation

$$\left\{ \mu \frac{\partial}{\partial \mu} + \sigma(g) \frac{\partial}{\partial g} - N\tau(g) + K\delta(g) \right\} \Gamma_o^{(N,K)} = -i \mu^{D-4} \omega(g) \delta_{N0} \delta_{K2} \qquad (6.14)$$

The term $\delta(g)$ is due to multiplicative renormalization of A^2 whereas the inhomogeneous term occurs from the additive renormalization of the "energy fluctuation" $\langle T \hat{N}[A^2](x) \hat{N}[A^2](o) \rangle$ which is already present in the free field case.

In A^4-theory there are no other dynamically independent composite fields with $d_\mathcal{O} < D$; the A^3 field is connected by the equation of motion to A and A^2. Non renormalizable fields as

$$\frac{1}{6!}\hat{N}[A^6](x)$$

can be included in a similar manner[21].

From the normalization conditions (6.2) and (6.13) the coefficients in (6.14) are given by:

$$\begin{aligned}
\tau(g,\varepsilon) &= i\left.\frac{\partial\Gamma_0^{(2)}}{\partial\mu^2}\right|_{p^2=-\mu^2}\\
\sigma(g,\varepsilon) &= -2i\mu^{2-\varepsilon}\left.\frac{\partial\Gamma_0^{(4)}}{\partial\mu^2}\right|_{s.p.-\mu^2}+4g\tau\\
\delta(g,\varepsilon) &= -2i\mu^2\left.\frac{\partial\Gamma_0^{(2,1)}}{\partial\mu^2}\right|_{p^2=-\mu^2}+2\tau\\
\omega(g,\varepsilon) &= i\mu^{1-\varepsilon}\left.\frac{\partial\Gamma_0^{(0,2)}}{\partial\mu}\right|_{q^2=-\mu^2}
\end{aligned}\tag{6.15}$$

and may be calculated in perturbation theory (see Appendix A). In perturbation theory to n-th order these functions are holomorphic in ε for $\frac{2}{n} > \mathrm{Re}\,\varepsilon \geqslant 0$ (see however Appendix C). In $D = 4-\varepsilon$ dimensions the leading terms in the ε-expansion are[20]

$$\begin{aligned}
\sigma &= -\varepsilon g+\frac{3}{(4\pi)^2}g^2+O(\varepsilon g^2, g^3)\\
\tau &= \frac{1}{12}\frac{1}{(4\pi)^4}g^2+O(\varepsilon g^2, g^3)\\
\delta &= \frac{1}{(4\pi)^2}g+O(\varepsilon g, g^2)\\
\omega &= \frac{1}{(4\pi)^2}+O(\varepsilon, g)
\end{aligned}\tag{6.16}$$

The fixed point condition $\sigma(g^*,\varepsilon) = 0$ can now be solved explicitly for g^* (= power series in ε) in an approximate sense (computable scaling) (Fig. 4).

For $\varepsilon > 0$ there is a long range scaling fixed point ($\sigma' > 0$)

$$g_o = \frac{(4\pi)^2}{3}\varepsilon + O(\varepsilon^2) \tag{6.17}$$

and a Gaussian (free field) short range fixed point ($\sigma' < o$) $g_\infty = o$

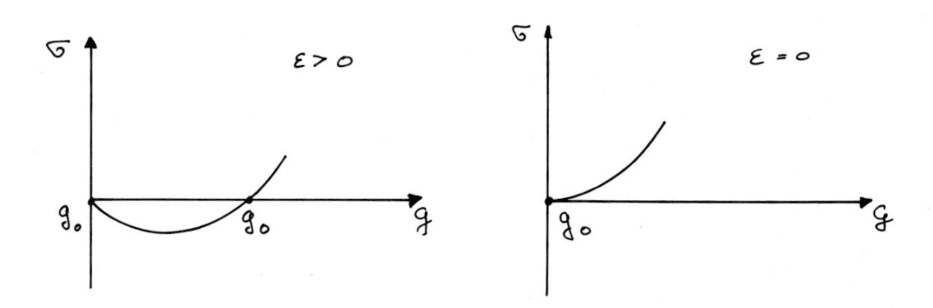

Fig.4

Hence in D = 4-ε (ε > o) dimensions we have a non-trivial critical scaling theory with anomalous dimensions

$$\tau(g_o) = \gamma_A = \frac{\varepsilon^2}{12 g} + O(\varepsilon^3)$$

$$\delta(g_o) = \gamma_{A^2} = \frac{\varepsilon}{3} + \frac{19}{2g^2}\varepsilon^3 + O(\varepsilon^3) \tag{6.18}$$

For calculations to order $O(\varepsilon^4)$ see Ref. 19).
We have actually calculated the critical indices

$$d_\sigma = d_A = d + \gamma_A \tag{6.19}$$

and

$$d_E = d_{A^2} = 2d + \gamma_{A^2}$$

appearing in formula (2.8) and by (2.23) we have calculated the two independent Kadanoff coefficients.
The short range (high energy) asymptote is a canonical theory.

In D = 4 dimensions there is (in perturbation theory) only a second order infrared scaling fixed point at $g_o = o$. We thus have reproduced the well known result that in D = 4 dimensions the critical theory (associated with a ferromagnetic Lenz-Ising system) is a mean field theory (mean field critical indices). The ε -expansion appears hence as a perturbation expansion around a free field theory.

As $\tau' = o$ in view of (6.11) the $\Gamma_0^{(N)}$'s scale. However as $\delta' \neq o$ (an analogue of (6.11) holds for Γ_δ) the $\Gamma_0^{(N,K)}$'s (K ≠ o) show up logarithms in the leading terms[26].

In order to study the singular behaviour (2.9) we have now to consider the non critical (t,H) ≠ (o,o) theory.

7. Non Critical Theory (Linearly broken massive A^4-theory)

a) Scaling parametrization[20) - 24)]

We will now perturb the preasymptotic (critical) theory by the relevant fields (the temperature and the magnetization) in the sense of Kadanoff-Wegner (2.16) (remember $E(x) \propto A^2(x)$):

$$\mathcal{L} = \mathcal{L}^{(0)} + \delta\mathcal{L} = \mathcal{L}^{(0)}(x) - \frac{t}{2} N[A^2](x) + H A(x) \tag{7.1}$$

in order to study the singular behaviour (2.9). By a translation of the field A

$$A \to \bar{A} = A - M \; ; \quad M = \langle A \rangle \; ; \quad \langle \bar{A} \rangle = 0 \tag{7.2}$$

our Lagrangian takes the form $\mathcal{L} = \mathcal{L}_0 + \mathcal{L}_{int}$

$$\mathcal{L}_0 = \tfrac{1}{2}(\partial \bar{A})^2 - \tfrac{1}{2}\left(t + \frac{\bar{g}M^2}{2}\right)\bar{A}^2$$

$$; \; \bar{g} = g\mu^{\varepsilon} \tag{7.1'}$$

$$\mathcal{L}_{int} = -\frac{\bar{g}}{4!}\bar{A}^4 - \frac{\bar{g}M}{3!}\bar{A}^3 + C\bar{A}$$

with $C = H - M\left(t + \frac{\bar{g}M^2}{3!}\right)$ determined by $\langle \bar{A} \rangle = o$

As independent parameters we choose

t, M, g and μ .

In the perturbation expansion g M^2 is (as a mass term) treated as $O(1)$. The equation of state reads

$$H = H(\mu, t, M, g) = M\left(t + \frac{\bar{g}M^2}{3!}\right) + C \tag{7.3}$$

All technical details are given in Ref. 21).

In o.th order we see that the phase diagram is of the correct form (Fig. 1).

The theory is normalized by (5.11)

a) the preasymptotic normalizations

$$\Gamma^{(2)}\Big|_{\substack{p=0\\ t,M=0}} = 0 \; ; \quad \Gamma^{(2)}\Big|_{\substack{p^2=-\mu^2\\ t,M=0}} = -i\mu^2 \; ; \quad \Gamma^{(4)}\Big|_{\substack{s.p.\,-\mu^2\\ t,M=0}} = -ig\mu^{4-D} \tag{7.4}$$

i. e. at (t,M) = (o,o) μ and g are the parameters of the preasymptotic theory.

b) The normalizations of the "perturbation" terms

$$\frac{\partial\Gamma^{(2)}}{\partial t}\Big|_{\substack{p=0\\ t=\mu^2\\ M=0}} = -i \quad ; \quad \Gamma^{(1)} \equiv 0 \tag{7.5}$$

these conditions define the "temperature" t and the "magnetization" M.

There are three <u>independent</u> (linear) parametric differential equations (PDE's)

$$\left\{\mu\frac{\partial}{\partial\mu} + G(g)\frac{\partial}{\partial g} - \tau(g)\left(N + M\frac{\partial}{\partial M}\right) + \delta(g)K + \hat{\delta}(g)\,t\frac{\partial}{\partial t}\right\}\Gamma^{(N,K)} =$$

$$= -i\mu^{D-4}\omega(g)\,\delta_{N0}\,\delta_{K2} \tag{7.6}$$

$$\partial_t\,\Gamma^{(N,K)} = -\tilde{\Delta}_t\,\Gamma^{(N,K)}$$

$$\tilde{\Delta}_t = \frac{i}{2}\int dx\,\tilde{N}[A^2](x) - \frac{\partial C}{\partial t}\,i\int dx\,A(x) \tag{7.7}$$

$$\partial_M\,\Gamma^{(N,K)} = -\tilde{\Delta}_M\,\Gamma^{(N,K)}$$

$$\tilde{\Delta}_M = g\mu^{\varepsilon}\frac{i}{3!}\int dx\,\bar{N}[A^3](x) + g\mu^{\varepsilon}M\,\frac{i}{2}\int dx\,\bar{N}[A^2](x) - \frac{\partial C}{\partial M}\,i\int dx\,A(x) \tag{7.8}$$

$\tilde{\Delta}_t$ and $\tilde{\Delta}_M$ are <u>soft insertions</u> (in the high energy sense) i.e. $\tilde{\Delta}_{t,M}\Gamma^{(N,K)}$ falls off relative to $\Gamma^{(N,K)}$ for large nonexceptional euclidean momenta by powers (up to logarithms) to all orders of perturbation theory. This implies that the t and M dependence of Green functions drops out for large nonexceptional momenta. We therefore call this parametrization soft.

For $M = o$ $C \equiv o$ and (7.7) tells us that t is actually the parameter conjugate to $N[A^2]$ i. e. the temperature. That M is the magnetization is guaranted by $M = \langle A \rangle$. For comparison the PDE's for the parametrizations (5.9) and (5.10) are given in Appendix B.

We will see below that the (pre)-scaling equation (7.6) (replacing the usual RG equation) is nothing but a differential form of Kadanoff scaling (scaling substitution law). Actually our parametrization is a global (pre)-scaling parametrization in the sense of Wegner. We observe that the hard (in the high energy sense) dilatation symmetry breaking terms are exactly those already present in the preasymptotic theory.

The dilatation-Ward-identity (Callan-Symanzik) (CS) equation) follows from (7.6) and (7.7,8):

$$\{ D + \sigma(g)\frac{\partial}{\partial g} - \tau(g) N + \delta(g) K \} \Gamma^{(N,K)} = \\ = \{ (1+\tau)\hat{\Delta}_M + (2-\hat{\delta})\hat{\Delta}_t \} \Gamma^{(N,K)} \tag{7.9}$$

where $D = \mu\frac{\partial}{\partial\mu} + 2t\frac{\partial}{\partial t} + M\frac{\partial}{\partial M}$ is the dilatation operator in the parameter space. Our parametrization has the particular property that the two limits:

(i) large nonexceptional momenta

(ii) preasymptotic $(t,M) \to (o,o)$

are identical to $\Gamma_0^{(N,K)}$. In both limits the RG equation (7.6) and the CS-equation (7.9) coincide.

In the soft parametrization t only appears in the propagators not however in (the symmetric) counterterms; this explains our observations of $\Gamma_{as}^{(N)} = \Gamma_0^{(N)}$ (Γ_{as} the nonexceptional large momentum asymptote

The main feature of the pre-scaling parametrization is that the hard dilatation symmetry breaking is completely controlled by a <u>globally solvable</u> pre-scaling equation. At the same time it is the appropriate parametrization (as we will see) for the study of statistical mechanics aspects of the model.

From the normalization condition (7.5) we have

$$\hat{\delta}(g) = 2\tau(g)f(g) - \sigma(g)\frac{\partial f}{\partial g} - 2i\left.\frac{\partial \Gamma^{(2)}}{\partial \mu^2}\right|_{\substack{p=o \\ t=\mu^2 \\ M=0}} \tag{7.10}$$

with

$$f(g) = i\mu^{-2}\,\Gamma^{(2)}\Big|_{\substack{p=0\\ t=\mu^2\\ M=0}}$$

In $D = 4-\varepsilon$ dimensions

$$\hat{\delta}(g) = \frac{g}{(4\pi)^2} + O(\varepsilon g, g^2) \qquad (= \delta(g) \text{ to this order}) \tag{7.11}$$

Generally $\hat{\delta}(g) \neq \delta(g)$, at the fixed point g^* ($\sigma(g^*) = 0$) however $\hat{\delta}(g^*) = \delta(g^*)$
This is shown in Appendix C (see (C.7)).

b) Global solution of the pre-scaling equation

The global solution of (7.6) reads

$$\Gamma^{(N,K)}(\{\varkappa p_i\}; \mu, t, M, g) =$$

$$= \varkappa^{D-Nd_A+K(d_{A^2}-D)}\, r_\tau^{-N} r_\delta^{K}\, \Gamma^{(N,K)}(\{p_i\}; \mu, \varkappa^{d_{A^2}-D} F_\delta t, \varkappa^{-d_A} r_\tau^{-1} M, g^*+h) \tag{7.12}$$

$$+\, i\mu^{D-4} E_\omega\, \delta_{N0}\, \delta_{K2}$$

where $g(\varkappa)$ and r_τ are defined in (6.6,7) and

$$\overset{(\wedge)}{r}_\delta = \exp \int_g^{g(\varkappa)} dg' \,\frac{\overset{(\wedge)}{\delta}(g') - \delta(g^*)}{\sigma(g')} \tag{7.13}$$

$$E_\omega = \int_g^{g(\varkappa)} dg'\, \omega(g')\, \sigma^{-1}(g') \exp \int_g^{g'} dg'' \,\frac{\delta(g'') - \varepsilon/2}{\sigma(g'')} \tag{7.14}$$

Apart from the E_ω term (7.12) represents a global substitution law (analogue of (2.11,12) and (2.17) under momentum dilatations.

$$\Gamma \to f_\Gamma(g,\varkappa)\Gamma\,;\; t \to f_t(g,\varkappa)\,t\,;\; M \to f_M(g,\varkappa)\,M\,;\; g \to g(g,\varkappa) \tag{7.15}$$

this is a generalization of Kadanoff's scaling.

Now if there is a scaling fixed point g^* : $\sigma(g^*) = o$ and if τ, $\hat{\delta}'$ and ω are continuous (at least one side) then for $g = g^*$ (where no hard breaking of dilatation symmetry is present) (7.12) takes a homogeneous substitution form (Kadanoff in narrow sense). We have then strict global Kadanoff scaling as

$$\hat{\Gamma}^{(N,K)} = \lim_{g \to g^*} \Gamma^{(N,K)} \tag{7.16}$$

satisfies

$$\begin{aligned} &\hat{\Gamma}^{(N,K)}(\{\varkappa p_i\}; \mu, t, M, g^*) = \\ &= \varkappa^{D-Nd_A+K(d_{A^2}-D)} \hat{\Gamma}^{(N,K)}(\{p_i\}; \mu, \varkappa^{d_{A^2}-D} t, \varkappa^{-d_A} M, g^*) \\ &+ i\,\delta_{N0}\,\delta_{K2}\,\omega(g^*)\,\mu^{D-4} \times \begin{cases} (2\gamma_{A^2}-\varepsilon)^{-1}[\varkappa^{2\gamma_{A^2}-\varepsilon}-1] \; ; & 2\gamma_{A^2} \neq \varepsilon \\ \ln \varkappa \; ; & 2\gamma_{A^2} = \varepsilon \end{cases} \end{aligned} \tag{7.17}$$

μ may be eliminated completely by introducing quantities of canonical dimension zero:

$$\bar{\hat{\Gamma}} = \frac{\hat{\Gamma}}{\mu^{D-Nd+2Kd}} \; ; \quad \bar{p}_i = \frac{p_i}{\mu} \; ; \quad \bar{t} = \frac{t}{\mu^2} \; ; \quad \bar{M} = \frac{M}{\mu^d} \tag{7.18}$$

From positivity the dynamical dimensions of the fields A and A^2 are larger than d.

We <u>assume</u> (always true in the region where perturbation theory applies i.e. for small anomalous dimensions) d_A and d_{A^2} to be smaller than D. Hence

$$0 \leq d \leq d_A \; ; \; d_{A^2} < D \tag{7.19}$$

In view of (2.17,18) we may then classify the fields. In order to have also an example of an irrelevant field (in the long range region) we add to the Lagrangian (7.1) a non renormalizable perturbation term

$$\mathcal{L} \longrightarrow \mathcal{L} - \frac{u}{6!} N[A^6](x)$$

and assume $d_{A^6} > D$[21]. We than have

(i) As $\varkappa \to \infty$

$$t' = \varkappa^{d_{A^2} - D}\, t \to 0$$

$$M' = \varkappa^{-d_A} M \to 0$$

short range irrelevant

$$g' - g_\infty = g(g,\varkappa) - g_\infty \simeq \begin{cases} (\ln \varkappa)^{-1}\,; & D = 4\,, \text{ marginal} \\ \varkappa^{\sigma'(g_\infty)}\,; & D < 4\,, \text{ irrelevant} \end{cases}$$

$$u' = \varkappa^{d_{A^6} - D}\, u \to \infty$$

short range relevant

(ii) As $\varkappa \to 0$

$$t' \to \infty$$

$$M' \to \infty$$

long range relevant

$$g' - g_0 = g(g,\varkappa) - g_0 \simeq \begin{cases} (\ln \varkappa)^{-1}\,; & D = 4 \quad \text{marginal} \\ \varkappa^{\sigma'(g_0)}\,; & D < 4 \quad \text{irrelevant} \end{cases}$$

$$u' \to 0$$

long range irrelevant

The marginal variables lying at the boundary of UV and IR-criticality are those determining the fixed point structure of the theory.
As we will see below for $g \neq g^*$ the power laws appearing in (7.17) are (under certain conditions) at most modified by logarithms and they do not change the character of the fields.

The critical surfaces and trajectories under momentum dilatations for a A^4-theory with A^2 and A^6 perturbations, normalized such that we have a global scaling parametrization

$$\left\{ \mu \frac{\partial}{\partial \mu} + \sigma(g) \frac{\partial}{\partial g} - N\,\tau(g) + \delta(g)\, t \frac{\partial}{\partial t} + 6\,\tau(g)\, u \frac{\partial}{\partial u} \right\} \Gamma^{(N)} = 0$$

are depicted in Fig. 5

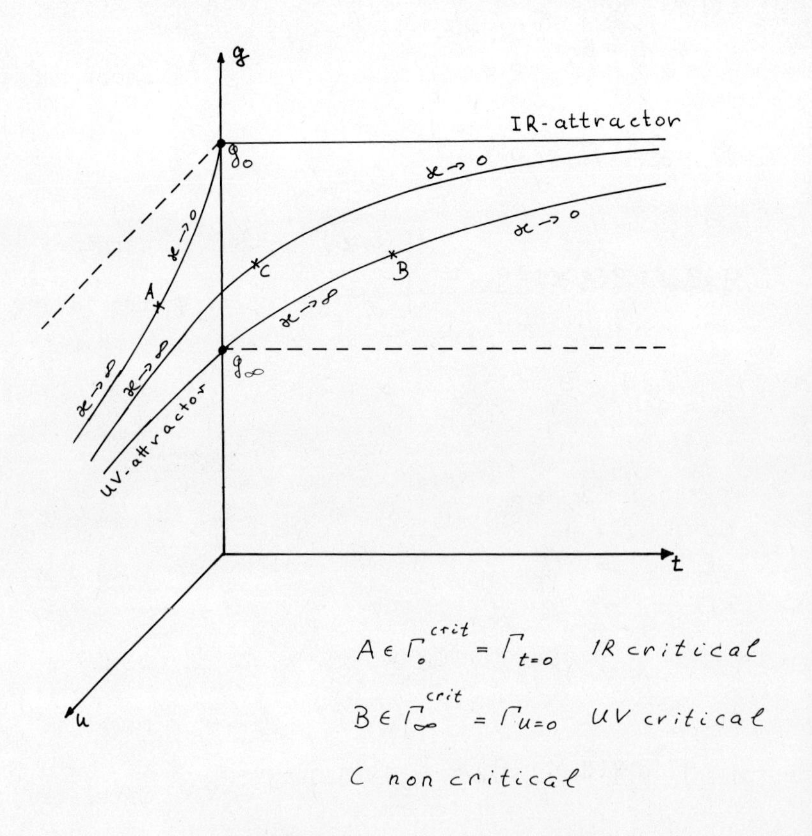

Fig. 5

8. Marginal Corrections to Kadanoff Scaling

As shown in the last section the homogeneous substitution laws (7.17) are violated by the non vanishing marginal variable $\Delta = g - g^*$. If we assume (ev. beyond perturbation theory)

1. the existence of a fixed point g^*
2. $\sigma, \tau, \overset{(\wedge)}{\delta}$ and ω have asymptotic expansion in g at g^*

$$\alpha(g) = \sum_{n=0}^{N} \alpha_n(g^*) \frac{\Delta^n}{n!} + R_\alpha^N \;; \quad \alpha = \sigma, \tau, \delta, \hat{\delta}, \omega$$
$$R_\alpha^N = O(\Delta^{N+1}) \;; \quad \Delta \to 0 \tag{8.1}$$

we may expand the "correction" terms appearing in formula (7.12) in the region

$$|\varepsilon_0| < |\Delta| \ll 1 \tag{8.2}$$

with ε_0 leading term of $g(\varkappa) - g^*$.

Leading corrections:

a)

$$\sigma_1 \neq 0 ; \quad \varepsilon_0 = \varkappa^{\sigma_1} \Delta ; \quad \sigma_1 \ln \varkappa < 0$$

$$h = \varepsilon_0 \left\{ 1 - \frac{\sigma_2}{2\sigma_1} (\Delta - \varepsilon_0) + O(\Delta^2) \right\}$$

$$r_\alpha = 1 - \frac{\alpha_1}{\sigma_1} (\Delta - \varepsilon_0) + O(\Delta^2) ; \quad \alpha = \tau, \delta, \hat{\delta} \tag{8.3}$$

$$r_\alpha^{(as)} = r_\alpha (\varepsilon_0 = 0, \Delta) \qquad \text{finite.}$$

In this case scaling holds in the strict sense.
We find this situation for the A^4 theory in $D = 4 - \varepsilon$ $(\varepsilon > 0)$ dimensions in the infrared and the ultraviolet region.

b)

$$\sigma_1 = 0, \sigma_2 \neq 0 ; \quad \varepsilon_0 = \frac{\Delta}{1 - \Delta \frac{\sigma_2}{2} \ln \varkappa} ; \quad \sigma_2 \ln \varkappa < 0$$

$$h = \varepsilon_0 \left\{ 1 - \frac{\sigma_3}{3\sigma_2} \varepsilon_0 \ln \frac{\varepsilon_0}{\Delta} + O(\Delta^2) \right\}$$

$$r_\alpha = \left(\frac{\varepsilon_0}{\Delta} \right)^{2 \frac{\alpha_1}{\sigma_2}} \left\{ 1 - \frac{3 a_2 \sigma_2 - 2 \alpha_1 \sigma_3}{3 \sigma_2^2} \Delta + O\left(\varepsilon_0 \ln \frac{\varepsilon_0}{\Delta}\right) \right\} \tag{8.4}$$

$$r_\alpha^{(as)} = | \ln \varkappa |^{-\frac{2\alpha_1}{\sigma_2}} \text{ const.}$$

Hence if $\tau_1 \neq 0 ; \delta_1^{(1)} = 0$ there are logarithmic corrections to scaling

if $\tau_1 = 0$ the $\Gamma^{(N)}$'s scale

if $\delta_1 = 0$ the $\Gamma^{(q,K)}$'s scale

This situation happens in D = 4 dimensions for the A^4 - theory at the infrared fixed point $g_0 = 0$, where $\sigma_1 = 0 ; \tau_1 = 0 ; \sigma_2 \neq 0 ; \delta_1 \neq 0$

c)

$$\sigma_1=\sigma_2=0\,;\ \sigma_3\neq 0\,;\ \varepsilon_0=\frac{\Delta}{(1-\Delta^2\frac{\sigma_3}{3}\ln\kappa)^{1/2}}\,;\qquad \sigma_3\ln\kappa<0$$

$$h=\varepsilon_0\{1-\frac{\sigma_4}{4\sigma_3}\varepsilon_0+\frac{\sigma_4}{4\sigma_3}\varepsilon_0(\frac{\varepsilon_0}{\Delta})+O(\Delta^2\ln\frac{\varepsilon_0}{\Delta}\}$$

$$r_\alpha=\exp\{\frac{6\alpha_1}{\sigma_3}(\frac{1}{\Delta}-\frac{1}{\varepsilon_0})-\frac{3\sigma_4\alpha_1}{2\sigma_3^2}(1-\frac{\varepsilon_0}{\Delta})\}(\frac{\varepsilon_0}{\Delta})^{\frac{3(2\alpha_2\sigma_3-\alpha_1\sigma_4)}{2\sigma_3^2}}\{1+O(\Delta^2)\}$$

$$r_a^{(as)}=\exp\{\frac{6\alpha_1}{\sigma_3}\frac{1}{\Delta}(1-\sqrt{|\ln\kappa|})-\frac{3\alpha_1\sigma_4}{2\sigma_3^2}(1-\frac{1}{\sqrt{|\ln\kappa|}})\}|\ln\kappa|^{-\frac{3(2\alpha_2\sigma_3-\alpha_1\sigma_4)}{4\sigma_3}}\cdot const.$$

This situation holds for the A^3-theory in D = 6 dimensions and for a class of non-abelian gauge theories (with $\tau_1=0,\ \tau_2\neq 0,\ \delta_1\neq 0$). Note that the structure of the marginal corrections are completely determined from the universal preasymptotic theory.

9. Thermodynamical quantities

From the field theoretical analogues of the definitions (2.5) and the Kadanoff relation (7.17) we obtain the singularities of the thermodynamical quantities (2.9 or 2.21,22). Using the expansion (8.1) in the region (8.2) we immediately get the corrections to scaling by expanding the r.h.s. of the Kadanoff relations. The corrections are given below for $\varepsilon>0$; they are by (8.3) powers in $g-g^*$ and κ^ω with $\omega=\sigma_1$ [30].

a) $M=0$: $\kappa=(\frac{t}{\mu^2})^{\frac{1}{D-d_{A^2}}}$

Field susceptibility χ_A : With $\gamma=\frac{D-2d_A}{D-d_{A^2}}$

$$\chi_A^{-1}=\Gamma^{(2)}(0;\mu,t,0,g)=(\frac{t}{\mu^2})^{\gamma}r_\tau^{-2}\,\Gamma^{(2)}(0,\mu,\mu^2\hat{r}_\delta,0,g^*+h)$$

$$=-i\mu^2(\frac{t}{\mu^2})^{\gamma}\{C_0(g^*)+(\frac{t}{\mu^2})^{\nu\omega}(g-g^*)C_1(g^*)+(g-g^*)\hat{C}_1(g^*)+O(\Delta^2)\}\qquad(9.1)$$

The numbers $C_i(g^*)$ are given by

$$C_0(g^*) = i\mu^{-2}\,\Gamma^{(2)}(0;\mu,\mu^2,0,g^*)$$

$$\hat{C}_1(g^*) = 2i\mu^{-2}\left\{\frac{\hat{\delta}_1}{\sigma_1}\mu^2\frac{\partial\Gamma^{(2)}}{\partial t} - \frac{\tau_1}{\sigma_1}\Gamma^{(2)}\right\}(0;\mu,\mu^2,0,g^*) \qquad (9.2)$$

$$C_1(g^*) = i\mu^{-2}\,\frac{\partial\Gamma^{(2)}}{\partial g}(0;\mu,\mu^2,0,g^*) - \hat{C}_1(g^*)$$

For $M \neq 0$ a similar expansion can be obtained from:

$$\chi_A^{-1} = \Gamma^{(2)}(0;\mu,t,M,g) = \left(\frac{t}{\mu^2}\right)^{\gamma} r_\tau^{-2}\,\Gamma^{(2)}(0;\mu,\mu^2\hat{r}_\delta,\mu^d x r_\tau^{-1},g^*+h) \qquad (9.3)$$

where
$$x = \frac{M}{\mu^d}\left(\frac{t}{\mu^2}\right)^{-\beta} \qquad (9.4)$$

Equation (9.3) has a power expansion in x.

<u>Energy susceptibility</u> $\chi_{A^2} = C$ (specific heat): with $\alpha = \dfrac{D-2d_{A^2}}{D-d_{A^2}}$

$$\chi_{A^2} = \Gamma^{(0,2)}(0;\mu,t,0,g) = \left(\frac{t}{\mu^2}\right)^{-\alpha} r_\delta^2\,\Gamma^{(0,2)}(0;\mu,\mu^2\hat{r}_\delta,0,g^*+h) + i\mu^{-\varepsilon}E_\omega$$

$$= -i\mu^{-\varepsilon}\Big[\left(\frac{t}{\mu^2}\right)^{-\alpha}\Big\{B_0(g^*) + \left(\frac{t}{\mu^2}\right)^{\nu\omega}(g-g^*)B_1(g^*) + (g-g^*)\hat{B}_1(g^*) + O(\Delta^2)\Big\} \qquad (9.5)$$

$$- \left\{b_0(g^*) + (g-g^*)b_1(g^*) + O(\Delta^2)\right\}\Big] \quad ; \ \alpha \neq 0$$

where

$$B_0(g^*) = i\mu^{+\varepsilon}\,\Gamma^{(0,2)}(0;\mu,\mu^2,0,g^*) - i\,\frac{\omega_0\nu}{\alpha}$$

$$\hat{B}_1(g^*) = 2i\mu^{+\varepsilon}\left\{\frac{\hat{\delta}_1}{\sigma_1}\mu^2\frac{\partial\Gamma^{(0,2)}}{\partial t} + \frac{\delta_1}{\sigma_1}\Gamma^{(0,2)}\right\}(0;\mu,\mu^2,0,g^*) - 2i\,\frac{\delta_1}{\sigma_1}\,\frac{\omega_0\nu}{\alpha}$$

$$B_1(g^*) = i\mu^{+\varepsilon}\left\{\frac{\partial\Gamma^{(0,2)}}{\partial g} + \frac{2\hat{\delta}_1}{\sigma_1}\mu^2\frac{\partial\Gamma^{(0,2)}}{\partial t} + \frac{2\delta_1}{\sigma_1}\Gamma^{(0,2)}\right\}(0;\mu,\mu^2,0,g^*) - i\,\frac{\omega_1\sigma_1 + 2\omega_0\delta_2}{2\delta_0+\sigma_1-\varepsilon} \qquad (9.6)$$

$$b_0(g^*) = \frac{\omega_0\nu}{\alpha}\ ; \quad b_1(g^*) = -\frac{2\delta_1}{\sigma_1}\,\frac{\omega_0\nu}{\alpha} + \frac{\omega_1\sigma_1 + 2\omega_0\delta_1}{2\delta_0+\sigma_1-\varepsilon}$$

The leading terms for t → o are:

$$\alpha > 0 \quad : \chi_{A^2}(t) \propto t^{-\alpha}$$

$$\alpha = 0 \quad : \chi_{A^2}(t) \propto \ln\left(\frac{t}{\mu^2}\right)^{\nu} \tag{9.7}$$

$$\alpha < 0 \quad : \chi_{A^2}(t) \propto const.$$

As from (7.19) $-(1-\varepsilon/2) < \delta_0 < 2$, the critical index $\alpha = -\nu(2\delta_0 - \varepsilon)$ actually can take positive and negative values.

In D = 4 we have $\alpha = 0$ and C behaves logarithmically due to the additive renormalization term E_ω ! In D = $4-\varepsilon$ ($\varepsilon > 0$) dimensions $\alpha = \nu\frac{\varepsilon}{3} + O(\varepsilon^2) > 0$ the power singularity is present.

For M ≠ o one may again expand the expression

$$\begin{aligned} \chi_{A^2} &= \Gamma^{(0,2)}(0;\mu,t,M,g) \\ &= \left(\frac{t}{\mu^2}\right)^{-\alpha} r_\delta^2\, \Gamma^{(0,2)}(0;\mu,\mu^2\hat{r}_\delta,\mu^d x\, r_\varepsilon^{-1}, g^*+h) + i\mu^{-\varepsilon}E_\omega \end{aligned} \tag{9.8}$$

in Δ and x .

Correlation length ξ : with $\nu = \frac{1}{D - d_{A^2}}$

$$\begin{aligned} \xi^{-2} = m^2 &= -\frac{\Gamma^{(2)}}{\frac{\partial\Gamma^{(2)}}{\partial p^2}}(0;\mu,t,0,g) = \\ &= -\left(\frac{t}{\mu^2}\right)^{2\nu}\frac{\Gamma^{(2)}}{\frac{\partial\Gamma^{(2)}}{\partial p^2}}(0;\mu,\mu^2\hat{r}_\delta,0,g^*+h) \\ &= -\mu^2\left(\frac{t}{\mu^2}\right)^{2\nu}\left\{A_0(g^*) + \left(\frac{t}{\mu^2}\right)^{\nu\omega}(g-g^*)A_1(g^*) + (g-g^*)\hat{A}_1(g^*) + O(\Delta^2)\right\} \end{aligned} \tag{9.9}$$

with $\quad (\dot{\Gamma} \doteq \frac{\partial\Gamma}{\partial p^2})$

$$A_0(g^*) = \mu^{-2}\,\Gamma^{(2)}\dot{\Gamma}^{(2)^{-1}}(0;\mu,\mu^2,0,g^*)$$

$$\hat{A}_1(g^*) = -\mu^{-2}\left\{\frac{\hat{\delta}_1}{\sigma_1}\mu^2\frac{\partial\Gamma^{(2)}}{\partial t} - \frac{\hat{\delta}_1}{\sigma_1}\mu^2\frac{\partial\dot{\Gamma}^{(2)}}{\partial t}\mu^2 A_0(g^*)\right\}\dot{\Gamma}^{(2)^{-1}}(0;\mu,\mu^2,0,g^*) \tag{9.10}$$

$$A_1(g^*) = \mu^{-2}\left\{\frac{\partial\Gamma^{(2)}}{\partial g} - \frac{\partial\dot{\Gamma}^{(2)}}{\partial g}\mu^2 A_0(g^*)\right\}\dot{\Gamma}^{(2)^{-1}}(0;\mu,\mu^2,0,g^*) - \hat{A}_1(g^*)$$

Again for $M \neq o$ we may expand

$$\xi^{-2} = - \frac{\Gamma^{(2)}}{\frac{\partial \Gamma^{(2)}}{\partial p^2}} (o; \mu, t, M, g) \tag{9.11}$$

$$= - \left(\frac{t}{\mu^2}\right)^{2\nu} \frac{\Gamma^{(2)}}{\frac{\partial \Gamma^{(2)}}{\partial p^2}} (o; \mu, \mu^2 \hat{r}_\delta, \mu^d x \xi_\tau^{-1}, g^* + h)$$

in Δ and x .

b) Equation of state: $\quad \varkappa = \left(\frac{M}{\mu^d}\right)^{1/d_A}$

The equation of state (7.3)

$$H = M \hat{m}^2 (\mu, t, M, g)$$

satisfies the scaling equation [21]

$$\left\{ \mu \frac{\partial}{\partial \mu} + \sigma \frac{\partial}{\partial g} - \tau (1 + M \frac{\partial}{\partial M}) + \hat{\delta} t \frac{\partial}{\partial t} \right\} H = 0 \tag{9.12}$$

and hence with $y = \frac{t}{\mu^2} \left(\frac{M}{\mu^d}\right)^{1/\beta}$ and $\delta = \frac{D - d_A}{d_A}$

$$H(\mu, t, M, g) = \left(\frac{M}{\mu^d}\right)^{\delta} r_\tau^{-1} H(\mu, \mu^2 y \hat{r}_\delta, \mu^d r_\tau^{-1}, g^* + h) \tag{9.13}$$

On the critical isotherme y=o:

$$H(\mu, o, M, g) = \left(\frac{M}{\mu^d}\right)^{\delta} r_\tau^{-1} H(\mu, o, \mu^d r_\tau^{-1}, g^* + h)$$

$$= \mu^{D-d} \left(\frac{M}{\mu^d}\right)^{\delta} \left\{ c_o(g^*) + \left(\frac{M}{\mu^d}\right)^{\frac{\nu\omega}{\beta}} (g - g^*) c_1(g^*) + (g - g^*) \hat{c}_1(g^*) + O(\Delta^2) \right\} \tag{9.14}$$

Here the numbers $c_i(g^*)$ are given by

$$c_o(g^*) = \mu^{d-D} H(\mu, o, \mu^d, g^*)$$

$$\hat{c}_1(g^*) = \mu^{d-D} \frac{\bar{c}_1}{\sigma_1} \left\{ H + \mu^d \frac{\partial H}{\partial M} \right\} (\mu, o, \mu^d, g^*) \tag{9.15}$$

$$c_1(g^*) = \mu^{d-D} \frac{\partial H}{\partial g} (\mu, o, \mu^d, g^*) - \hat{c}_1(g^*)$$

On the other hand on the coexistence curve H=o[21]:

$$\left\{ \mu \frac{\partial}{\partial \mu} + \sigma \frac{\partial}{\partial g} - \tau M \frac{\partial}{\partial M} - \hat{\delta} \right\} t(\mu, M, g) = 0 \tag{9.16}$$

With $\beta = \frac{d_A}{D - d_{A^2}}$ the equation of state takes the form:

$$t(\mu, M, g) = \left(\frac{M}{\mu^d}\right)^{1/\beta} \hat{r}_\delta^{-1}\, t(\mu, \mu^d r_\tau^{-1}, g^* + h)$$

$$= \mu^2 \left(\frac{M}{\mu^d}\right)^{1/\beta} \left\{ a_0(g^*) + \left(\frac{M}{\mu^d}\right)^{\frac{\nu\omega}{\beta}} (g-g^*)\, a_1(g^*) + (g-g^*)\, \hat{a}_1(g^*) + O(\Delta^2) \right\} \tag{9.17}$$

with

$$a_0(g^*) = \mu^{-2}\, t(\mu, \mu^d, g^*)$$

$$\hat{a}_1(g^*) = \mu^{-2} \left\{ \frac{\hat{\delta}_1}{\sigma_1} t + \frac{\tau_1}{\sigma_1} \mu^d \frac{\partial t}{\partial M} \right\} (\mu, \mu^d, g^*)$$

$$a_1(g^*) = \mu^{-2} \frac{\partial t}{\partial g}(\mu, \mu^d, g^*) - \hat{a}_1(g^*) \tag{9.18}$$

We have now determined all the critical indices from (2.9) and we may check the scaling relations (2.26) to be satisfied exactly. For at the fixed point (6.17) the corrections to scaling for the thermodynamical quantities are completely governed by the exponent.

$$\omega = \sigma_1 = \sigma'(g^*) = \varepsilon + O(\varepsilon^2) \tag{9.19}$$

These corrections are due to non vanishing marginal variable $\Delta = g - g^*$ In the "relevant" variables x and y of the equations (9.3, 8, 11, 13) the thermodynamical quantities are analytic (to any order of perturbation theory). For the correlation functions in contrast non vanishing relevant fields give rise to corrections of non analytic type.

10. Correlation Functions

At the critical point the long range parts of the correlation functions scale according to (7.12) and (8.3)

$$\Gamma^{(N,K)}(\{\varkappa p_i\};\mu,t,M,g) \simeq \varkappa^{D-Nd_A+K(d_{A^2}-D)}\{\Gamma_0^{(N,K)}(\{p_i\};\mu,0,0,g^*) \tag{10.1}$$

+ correction terms proportional to (g-g*) and

$$(g-g^*)\varkappa^{\omega}+\ldots\} + i\,\delta_{N0}\,\delta_{K2}\,\mu^{-\varepsilon}E_{\omega}$$

In particular with $\varkappa^2\mu^2 = -p^2 = \vec{p}^{\,2}$

$$\Gamma^{(2)}(-\vec{p}^{\,2};\mu,0,0,g) \simeq \left(\frac{\vec{p}^{\,2}}{\mu^2}\right)^{1-\eta/2}\Gamma^{(2)}(\mu^2;0,0,g^*)\;;\quad \eta = 2\gamma_A \tag{10.2}$$

and

$$\Gamma^{(0,2)}(-\vec{p}^{\,2};\mu,0,0,g) \simeq \left(\frac{\vec{p}^{\,2}}{\mu^2}\right)^{-\frac{\alpha}{2\nu}}\Gamma^{(0,2)}(\mu^2;\mu,0,0,g^*) - i\mu^{-\varepsilon}a(g^*)\frac{\nu}{\alpha}\left[\left(\frac{\vec{p}^{\,2}}{\mu^2}\right)^{-\frac{\alpha}{2\nu}} - 1\right] \tag{10.3}$$

For (t,M) ≠ (o,o) relevant corrections to scaling occur. In the region

$$\vec{p}_i^{\,2},\ t,\ M^2 \ll \mu^2 \tag{10.4}$$

we have

$$\Gamma^{(N,K)}(\{\varkappa p_i\};\mu,\varkappa^2 t,\varkappa^d M,g) =$$

$$= \varkappa^{D-Nd_{A^2}+K(d_{A^2}-D)}\Big\{1+\varepsilon_0\frac{\partial}{\partial g}+(\Delta-\varepsilon_0)\Big[\Big(N\frac{\tilde{c}_1}{\sigma_1}-K\frac{\delta_1}{\sigma_1}\Big)-\frac{\hat{\delta}_1}{\sigma_1}\varkappa^{d_{A^2}-D}t\frac{\partial}{\partial t}+\frac{\tilde{c}_1}{\sigma_1}\varkappa^{-d_A}M\frac{\partial}{\partial M}\Big]$$

$$+\,O(\Delta^2)\Big\}\,\Gamma^{(N,K)}(\{p_i\},\mu,\varkappa^{\gamma_{A^2}}t,\varkappa^{-\gamma_A}M,g^*)$$

Now we further may calculate the corrections for

$$t,\ M^2 \ll \vec{p}_i^{\,2}\ (\ll \mu^2) \tag{10.6}$$

using the inhomogeneous PDE's (7.7) and (7.8) for the vertex functions on the r.h.s. of (10.5) together with short distance expansions (SDE), for the r.h.s. of these PDE's.

For simplicity we only consider the leading relevant corrections for vanishing marginal field Δ = o i.e. of

$$\Gamma^{(N,K)}(\{p_i\};\mu,t,M,g^*)$$

Integration of the PDE's (7.7) and (7.8) leads to:

$$\Gamma^{(N,K)}(\{p_i\};\mu,t,M,g) = \Gamma_0^{(N,K)}(\{p_i\};\mu,0,0,g)$$

$$- \int_0^t dt' \hat{\Delta}_t \Gamma^{(N,K)}(\{p_i\};\mu,t',0,g) \tag{10.7}$$

$$- \int_0^M dM' \hat{\Delta}_M \Gamma^{(N,K)}(\{p_i\};\mu,0,M',g)$$

$$+ \int_0^t \int_0^M dt' dM' \hat{\Delta}_t \hat{\Delta}_M \Gamma^{(N,K)}(\{p_i\};\mu,t',M',g) \quad ; \text{ N even.}$$

As $\Gamma^{(N,K)}(\{p_i\};\mu,t,0,g) = 0$ for N odd and

$\partial_M \Gamma^{(N,K)}(\{p_i\};\mu,0,0,g)$ finite (10.7) holds for N odd also with $\Gamma^{(N,K)}$ replaced by $\partial_M \Gamma^{(N,K)}$.

The small t, M behaviour may be obtained (using homogeneity and the fact that the vertex functions depend on μ only logarithmically) from the large momentum expansion (SDE):

N even:

$$\hat{\Delta}_t \Gamma^{(N,K)}(\{p_i\};\mu,t,M,g) = F^{(N+2,K)}(\{p_i,0,0\};\mu,\mu^2,0,g)\, \hat{\Delta}_t^2 \Gamma^{(0,0)}(\mu,t,M,g) + R_t^{(N,K)} \tag{10.8}$$

$$\hat{\Delta}_M \Gamma^{(N,K)}(\{p_i\};\mu,t,M,g) = F^{(N+2,K)}(\{p_i,0,0\};\mu,\mu^2,0,g)\, \hat{\Delta}_M \hat{\Delta}_t \Gamma^{(0,0)}(\mu,t,M,g) + R_M^{(N,K)} \tag{10.9}$$

N odd:

$$\hat{\Delta}_t \hat{\Delta}_M \Gamma^{(N,K)}(\{p_i\};\mu,t,M,g) = \begin{cases} F^{(N+1,K)}(\{p_i,0\};\mu,\mu^2,0,g)\, \hat{\Delta}_t \hat{\Delta}_M G^{(1)}(\mu,t,M,g) + R_1; \ (K \neq 0) \\ F^{(N+3)}(\{p_i,0,0,0\};\mu,\mu^2,0,g)\, \hat{\Delta}_t \hat{\Delta}_M^2 \Gamma^{(0,0)}(\mu,t,M,g) + R_2; \ (K=0) \end{cases} \tag{10.10}$$

$$\hat{\Delta}_M \hat{\Delta}_M \Gamma^{(N,K)}(\{p_i\};\mu,t,M,g) = \begin{cases} F^{(N+1,K)}(\{p_i,0\},\mu,\mu^2,0,g)\, \hat{\Delta}_M^2 G^{(1)}(\mu,t,M,g) + R_3; \ (K \neq 0) \\ F^{(N+3)}(\{p_i,0,0,0\};\mu,\mu^2,0,g)\, \hat{\Delta}_M^3 \Gamma^{(0,0)}(\mu,t,M,g) + R_4; \ (K=0) \end{cases} \tag{10.11}$$

The remainders $R^{(N,K)}_{..}$ drop out to each order of perturbation theory by powers up to logarithms. The coefficient functions $\Gamma^{(N+..,K)}$ are represented by graphs which get one particle irreducible with respect to cuts not separating the K-vertices after connecting the N-external legs to a point. The singular parts on the r.h.s. of (10.8 - 11) obey the scaling equations:

$$\left\{\mu\frac{\partial}{\partial\mu}+\sigma\frac{\partial}{\partial g}-\tau\left(m+M\frac{\partial}{\partial M}\right)+\hat{\delta}\left(n+t\frac{\partial}{\partial t}\right)\right\}\Gamma_{mn}=-i\,Q_{mn} \tag{10.12}$$

with Q_{mn} a polynomial in t and M of degree $\delta = 4-2n-m \geq 0$, the coefficients depending on g, and

$$\left\{\mu\frac{\partial}{\partial\mu}+\sigma\frac{\partial}{\partial g}-\tau\left(m+M\frac{\partial}{\partial M}-1\right)+\hat{\delta}\left(n+t\frac{\partial}{\partial t}\right)\right\}G^{(1)}_{mn}=-i\,P_{mn} \tag{10.13}$$

where P_{mn} is a polynomial of degree $\delta = 3-2n-m \geq 0$.

Here we used the notation

$$\Gamma_{mn}(\mu,t,M,g)=\hat{\Delta}_M^m\,\hat{\Delta}_t^n\,\Gamma^{(0,0)}(\mu,t,M,g)$$

and

$$G^{(1)}_{mn}(\mu,t,M,g)=\hat{\Delta}_M^m\,\hat{\Delta}_t^n\,G^{(1)}(\mu,t,M,g)$$

The inhomogeneous terms are due to the additive renormalizations of Γ_{mn} and $G^{(1)}_{mn}$. In particular:

$$Q_{02}=\mu^{-\varepsilon}\omega_{02}(g)\,;\quad Q_{21}=\omega_{21}(g)\,;\quad Q_{11}=\omega_{21}(g)\,M$$

$$Q_{40}=\mu^{\varepsilon}\omega_{40}(g)\quad\text{and}\quad Q_{30}=\mu^{\varepsilon}\omega_{40}(g)\,M$$

are determined from:

$$\Gamma_{02}\Big|_{\substack{t=\mu^2\\ M=0}}=0\;;\quad \Gamma_{11}\Big|_{\substack{t=\mu^2\\ M=0}}=0\;;\quad \Gamma_{40}\Big|_{\substack{t=\mu^2\\ M=0}}=0 \tag{10.14}$$

as

$$\omega_{02}=i\mu^{\varepsilon}(\mu\partial_\mu+\hat{\delta}t\partial_t)\Gamma_{02}\Big|_{\substack{t=\mu^2\\ M=0}}\,;\quad \omega_{21}=i(\mu\partial_\mu+\hat{\delta}t\partial_t)\Gamma_{11}\Big|_{\substack{t=\mu^2\\ M=0}}$$

$$\omega_{40}=i\mu^{-\varepsilon}(\mu\partial_\mu+\hat{\delta}t\partial_t)\Gamma_{40}\Big|_{\substack{t=\mu^2\\ M=0}}$$

With the solutions of (10.12) at g = g* :

$$\Gamma_{02}(\mu,t,0,g^*) = \left(\frac{t}{\mu^2}\right)^{-\alpha}\Gamma_{02}(\mu,\mu^2,0,g^*) + i\mu^{-\varepsilon}\omega_{02}(g^*)\,\nu\,\alpha^{-1}\left[\left(\frac{t}{\mu^2}\right)^{-\alpha}-1\right]$$

$$\Gamma_{11}(\mu,0,M,g^*) = \left(\frac{M}{\mu^d}\right)^{\beta^{-1}(1-\alpha)-1}\Gamma_{11}(\mu,0,\mu^d,g^*) + i\mu^{d}\omega_{21}(g^*)\,\nu(\gamma-1)^{-1}\left[\left(\frac{M}{\mu^d}\right)^{\beta^{-1}(1-\alpha)-1}-\left(\frac{M}{\mu^d}\right)\right]$$

(10.15)

$$\Gamma_{21}(\mu,t,0,g^*) = \left(\frac{t}{\mu^2}\right)^{\gamma-1}\Gamma_{21}(\mu,\mu^2,0,g^*) + i\,\omega_{21}(g^*)\,\nu(\gamma-1)^{-1}\left[\left(\frac{t}{\mu^2}\right)^{\gamma-1}-1\right]$$

$$\Gamma_{30}(\mu,0,M,g^*) = \left(\frac{M}{\mu^d}\right)^{\delta-2}\Gamma_{30}(\mu,0,\mu^d,g^*) + i\mu^{d+\varepsilon}\omega_{40}(g^*)\,\nu(\gamma-2\beta)^{-1}\left[\left(\frac{M}{\mu^d}\right)^{\delta-2}-\left(\frac{M}{\mu^d}\right)\right]$$

we obtain from (10.5 - 15) the leading corrections to $\Gamma_0^{(N,K)}$ and $\partial_M\Gamma_0^{(N)}$ in the region

$$|t|,\ M^2 \ll p^2 \ll \mu^2$$

With $p_i = pn_i$, $p > 0$, $|n_i| = 1$; n_i euclidean, nonexceptional using the notation (7.18) we have:

For <u>N even</u>:

$$\bar{\Gamma}^{(N,K)}(\{pn_i\};\mu,t,M,g) \simeq \qquad (10.16)$$

$$\simeq \bar{p}^{\beta/\nu[\delta-(K+N-1)]}\left[A_0(g^*) - A_1(g^*)\begin{cases}\left(\bar{t}/\bar{p}^{1/\nu}\right)^{1-\alpha}D_1(g^*) - \left(\bar{t}/\bar{p}^{1/\nu}\right)D_2(g^*); & (M=0)\\ \left(\bar{M}/\bar{p}^{\beta/\nu}\right)^{\frac{1-\alpha}{\beta}}D_3(g^*) - \left(\bar{M}/\bar{p}^{\beta/\nu}\right)^2 D_4(g^*); & (t=0)\end{cases} + \dots\right]$$

with

$$A_0(g^*) = \bar{\Gamma}_0^{(N,K)}(\{\mu n_i\};\mu,0,0,g^*)\ ;\ A_1(g^*) = \bar{\Gamma}^{(N+2,K)}(\{\mu n_i,0,0\};\mu,\mu^2,0,g^*)$$

$$D_1(g^*) = (1-\alpha)^{-1}\{\mu^{\varepsilon}\Gamma_{02}(\mu,\mu^2,0,g^*) + D_2(g^*)\}\ ;\ D_2(g^*) = i\,\omega_{02}(g^*)\,\nu\,\alpha^{-1}$$

$$D_3(g^*) = (1-\alpha)^{-1}\beta\{\mu^{-d}\Gamma_{11}(\mu,0,\mu^d,g^*) + 2D_4(g^*)\}\ ;\ D_4(g^*) = \frac{i}{2}\,\omega_{21}(g^*)\,\nu(\gamma-1)^{-1}$$

For <u>N odd:</u>

$$\partial_M\bar{\Gamma}^{(N)}(\{pn_i\};\mu,t,M,g) \simeq$$

$$\simeq \bar{p}^{\beta/\nu[\delta-(K+N)]}\left[B_0(g^*) - B_1(g^*)\begin{cases}\left(\bar{t}/\bar{p}^{1/\nu}\right)^{\gamma}E_1(g^*) - \left(\bar{t}/\bar{p}^{1/\nu}\right)E_2(g^*); & (M=0)\\ \left(\bar{M}/\bar{p}^{\beta/\nu}\right)^{\gamma/\beta}E_3(g^*) - \left(\bar{M}/\bar{p}^{\beta/\nu}\right)^2 E_4(g^*); & (t=0)\end{cases} + \dots\right]$$

with

$$B_0(g^*) = \partial_M\bar{\Gamma}_0^{(N)}(\{\mu n_i\};\mu,0,0,g^*)\ ;\ B_1(g^*) = \bar{\Gamma}^{(N+3)}(\{\mu n_i,0,0,0\};\mu,\mu^2,0,g^*)$$

with

$$E_1(g^*) = \gamma^{-1}\{\bar{\Gamma}_{21}(\mu,\mu^2;0,g^*) + E_2(g^*)\}; \quad E_2(g^*) = i\,\omega_{21}(g^*)\,\nu(\gamma-1)^{-1}$$

(10.19)

$$E_3(g^*) = \gamma^{-1}/\beta\,\{\mu^{-(d+\epsilon)}\bar{\Gamma}_{30}(\mu,0,\mu^d;g^*) + 2E_4(g^*)\}; \quad E_4(g^*) = \tfrac{i}{2}\,\omega_{40}(g^*)\,\nu(\gamma-2\beta)^{-1}$$

Higher correction terms may be calculated by taking into account further terms in the SDE's (10.8 - 11) and by applying SDE to the non leading terms in (10.5).

In view of (9.14) and (9.17) we easily obtain the relevant corrections on the coexistence curve from (10.16).

11. Conclusions

Within our field theoretical framework we are able to give a precise meaning to many of Kadanoff's considerations and we have a model matching Wegner's phenomenological scheme[6)].

Our discussion shows that the soft parametrization is most transparent for the discussion of scaling behaviour. We want to point out that using the soft renormalization technique[21)] all perturbation calculations can be performed in a usual sense (no loopwise-summation) and that all perturbation theoretical statements have been proved to all orders using PDE's.

What we have shown is that the A^4-model exhibits:

1. Long range scaling for

   $$t,\ M^2 \ll \vec{p}_i^{\,2} \ll \mu^2 \qquad \text{euclidean nonexceptional}$$

2. There are two independent critical indices and the scaling relations among the critical indices are exactly valid (calculable in $4-\varepsilon$ dim. $0 \leq \varepsilon \ll 1$).
3. In $D = 4-\varepsilon$ $(\varepsilon > 0)$ dimensions strict homogeneous Kadanoff substitution laws (Kadanoff scaling) are valid in the long range asymptote.
4. There is a global (pre-)scaling parametrization in the sense of Wegner.
5. The scaling structure (singular behaviour) and the structure of

corrections are universal in the sense that they are intrinsic to the preasymptotic theory (i.e. known from $\sigma(g)$, $\tau(g)$ and $\delta(g)$). Whereas thermodynamical quantities besides the "marginal" corrections are analytic in the relevant variables, the correlation functions exhibit non analytic "relevant" corrections.

In Tab. 1 we have listed the values for the critical exponents obtained from the ε-expansion to order ε^2 and ε^3 in comparison to the experimental values, the mean field values (MFA) and results from Lenz-Ising model calculations for D=3. To order ε^2 the agreement is stricking for D=3 and even for D=2 the results have the right orders of magnitude. There is of course (at present) no explanation why starting from an asymptotic expansion for small ε by setting ε = 1 (or 2) one gets reasonable answers.

Concerning the structural investigations our results immediatly generalize to n-component scalar models:

$$\hat{\varphi} = (\varphi_1, \ldots, \varphi_n)$$

$$\mathcal{L} = \frac{1}{2}(\partial\hat{\varphi})^2 - \frac{t}{2}\hat{\varphi}^2 - \frac{g}{4!}(\hat{\varphi}^2)^2 + H\varphi_n$$

The functions σ, τ, δ and ω now depend on n and so do the corresponding critical indices[19)]. For n > 1 the Goldstone phenomenon takes place in the spontaneous limit otherwise there is no principal structural change.

For similar investigations of other models we refer to the review articles Ref. 33).

Appendix A: Graphical Representation of Green Functions

a) Zero mass

$\Gamma_0^{(2)} = \quad + \quad + \ldots$

$\Gamma_0^{(4)} = \quad + \quad + \quad + \quad +$ 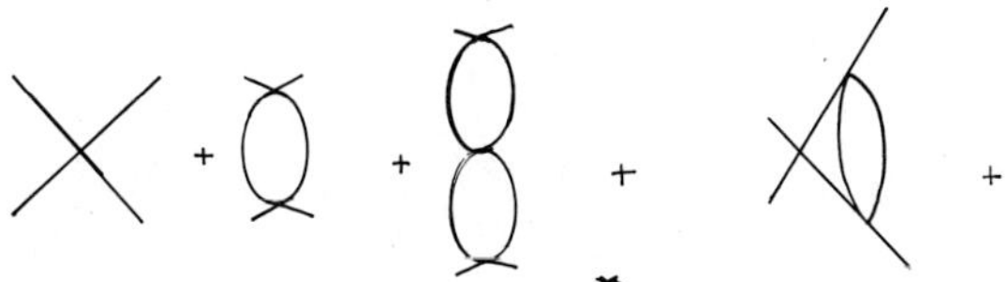 crossed terms + ...

$\Gamma_0^{(2,1)} = \quad + \quad + \quad + \quad + \ldots$

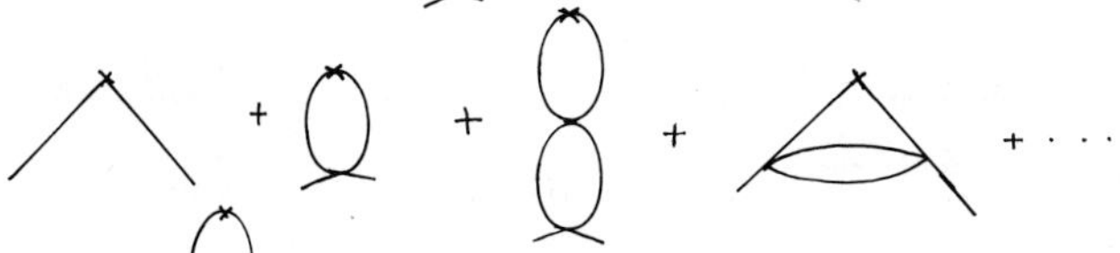

$\Gamma_0^{(0,2)} = \quad + \quad + \ldots$

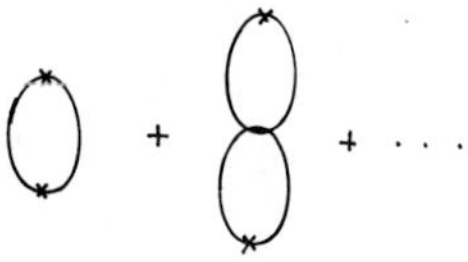

For a evaluation of the integrals see e.g. Ref. 20).

b) $(t, M) \neq (o,o)$

$\Gamma^{(2)} = \quad + \quad + \quad + \quad + \quad + \quad + \quad + \quad + \quad + \quad + \quad + \ldots$

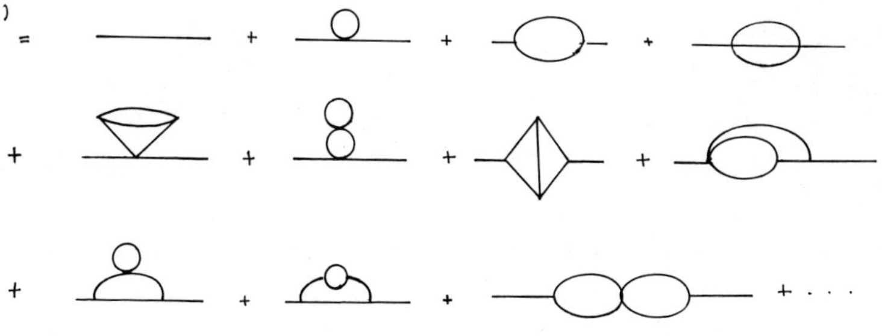

$\Gamma^{(4)} = \quad + \quad + \quad + \quad +$ 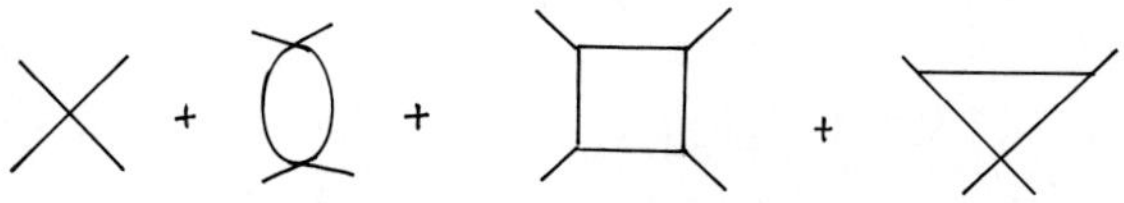 crossed terms + ...

In D=4 to lowest order we have for M = o:

$$\Gamma^{(2)} = i \left\{ -t - \vec{p}^{\,2} - (4\pi)^{-2} \frac{g}{2} \left(t \ln \frac{t}{\mu^2} - t \right) + O(g^2) \right\}$$

$$\Gamma^{(4)} = i \left\{ -g - (4\pi)^{-2} \frac{g^2}{2} \int_0^1 d\alpha \ln \frac{\alpha(1-\alpha)(\vec{p}_1 + \vec{p}_2)^2 + t}{\alpha(1-\alpha)\, 4/3\, \mu^2} + \text{crossed terms} + O(g^3) \right\}$$

Equation of state to lowest order:

$$H = i \sum_{prop} \text{—⊗} + M(t + g\frac{M^2}{3!}) = M(t + g\frac{M^2}{3!}) + i \text{—○} + \cdots$$

$$= M\{t + g\frac{M^2}{3!} - (4\pi)^{-2}\frac{g}{2}[t + g\frac{M^2}{2} + t\ln\mu^2 + g\frac{M^2}{2}\ln 4/3\,\mu^2$$

$$- (t + g\frac{M^2}{2})\ln(t + g\frac{M^2}{2}) + g\frac{M^2}{2}\int_0^1 d\alpha\,\ln\alpha(1-\alpha)] + O(g^{3/2})\}$$

Appendix B: Structure of PDE's for the Parametrizations (5.9) and (5.10)

In the symmetric (M = o) mass shell normalized theory (m = physical mass) there is one PDE, the standard CS-equation (Dilatation Ward-identity)[29]

$$\{m\frac{\partial}{\partial m} + \beta(g)\frac{\partial}{\partial g} - \gamma(g)N\}\bar{\Gamma}^{(N)} = -\bar{\Delta}_m\bar{\Gamma}^{(N)} \tag{B.1}$$

i.e. the physical mass breaks dilatation symmetry necessarily in a hard way as $\beta \neq o$ for generic g and the large momentum (nonexceptional) asymptote $\bar{\Gamma}_{as}^{(N)}$ differs from $\Gamma_o^{(N)}$ by a complicated wave-function - and coupling-constant - renormalization. Also the $\bar{\Gamma}_{as}^{(N)}$ are vertex-functions of a zero mass theory (mass in propagators dropped out) the $\Gamma_{as}^{(N)}$'s still depend on m (through the m-dependent counter terms) The $\Gamma_{as}^{(N)}$ are solutions to the homgeneous CS-equation.

$$\{m\frac{\partial}{\partial m} + \beta(g)\frac{\partial}{\partial g} - \gamma(g)N\}\bar{\Gamma}_{as}^{(N)} = 0 \tag{B.2}$$

Note that the functions β (g) and γ (g) are expressed in terms of massi vertex functions and are hence holomorphic in ε for Re $\varepsilon \geq o$. The widely used μ-normalization of Gell-Mann and Low with mass shell normalization of the propagator pole (m physical mass) has similar properties, however with Green-functions continuous at m = o[16]. The PDE's are (see e. g. 22))

$$\{\mu\frac{\partial}{\partial\mu} + \sigma(g,\frac{m}{\mu})\frac{\partial}{\partial g} - \tau(g,\frac{m}{\mu})N\}\tilde{\Gamma}^{(N)} = 0 \tag{B.3}$$

(RG-equation)

$$\left\{ m\frac{\partial}{\partial m} + \mu\frac{\partial}{\partial\mu} + \beta(g,\tfrac{m}{\mu})\frac{\partial}{\partial g} - \gamma(g,\tfrac{m}{\mu})N \right\} \hat{\Gamma}^{(N)} = -\hat{\Delta}_m \hat{\Gamma}^{(N)} \tag{B.4}$$

(CS-equation)

In this parametrization the RG-equation is not globally integrable for $m \neq o$. There are two regimes:

(i) Large nonexceptional momenta: $\breve{\Delta}_m \hat{\Gamma}$ drops, however: CS $\hat{\Gamma}_{as}$ - RG $\tilde{\Gamma}_{as}$ $\neq$ o still m dependence!

(ii) zero mass: $\tilde{\Delta}_m \hat{\Gamma}$ drops and σ, τ, β and γ simplify such that CS $\hat{\Gamma}_o$ = RG $\hat{\Gamma}_o$.
Again $\tilde{\Gamma}_{os}$ and $\hat{\Gamma}_o$ are related by complicated wave function and coupling constant renormalization.

Appendix C: Universality Properties of Γ_o.

Two zero mass A^4-theories Γ_o and $\hat{\Gamma}_o$ with length scales μ and $\tilde{\mu}$ can differ at $\mu = \tilde{\mu}$ only by a finite wave-function-and coupling constant renormalization:

$$\tilde{\Gamma}_o^{(N)}(p;\mu,V(g)) = Z(g)^{N/2}\,\Gamma_o(p;\mu,g) \tag{C.1}$$

As Γ_o and $\tilde{\Gamma}_o$ obey the PDE's

$$\left\{ \mu\frac{\partial}{\partial\mu} + \sigma(g)\frac{\partial}{\partial g} - \tau(g)N \right\} \Gamma_o(p;\mu,g) = 0$$

and

$$\left\{ \mu\frac{\partial}{\partial\mu} + \beta(V)\frac{\partial}{\partial V} - \gamma(V)N \right\} \tilde{\Gamma}_o(p;\mu,V) = 0 \tag{C.2}$$

we have

$$\beta(V) = \sigma(g)\,\frac{\partial V(g)}{\partial g}$$

$$\gamma(V) = \tau(g) + \tfrac{1}{2}\,\sigma(g)\frac{\partial}{\partial g}\ln Z(g) \tag{C.3}$$

As in perturbation theory $V(g) = g\,P(g)$ and $Z(g) = 1 + g\,Q(g)$ with P and Q polynomials in g we see that

$$\beta(V^*) = 0 \iff \sigma(g^*) = 0\,; \quad V^* = V(g^*)$$

$$\gamma(V^*) = \tau(g^*) \tag{C.4}$$

and $\frac{d\beta}{dV} = \frac{d\sigma}{dg} = \omega$; $\frac{d^2\beta}{dV^2} = \frac{d^2\sigma}{dg^2} \left(\frac{dV}{dg} \right)^{-1}$; etc.

i.e. fixed points and their nature as well as anomalous dimensions are universal.

Note that the composite field $\tilde{N}[A^2]$ in (7.7) normalized by

$$\frac{1}{2} \langle T \tilde{N}[A^2](0) \hat{A}(p_1) \hat{A}(p_2) \rangle^{prop} \Big|_{\substack{p_i = 0 \\ t = \mu^2}} = 1$$

and the field $\hat{N}[A^2]$ in (6.12) normalized by

$$\frac{1}{2} \langle T \hat{N}[A^2](0) \hat{A}(p_1) \hat{A}(p_2) \rangle^{prop} \Big|_{\substack{p_1 = p_2 = \frac{p}{2} \\ p^2 = -\mu^2 \\ t = 0}} = 1$$

are similarly related by

$$\tilde{N}[A^2] = Z_2(g) \hat{N}[A^2]$$

i.e.

$$\tilde{\Gamma}^{(N,1)}(p_1, \ldots, p_N, q) = Z_2(g) \Gamma^{(N,1)}(p_1, \ldots, p_N, q) \tag{C.5}$$

As $\partial_t \Gamma^{(N)} = \tilde{\Gamma}^{(N,1)}(p_1, \ldots, p_N, 0)$ we have from (7.6) for non-exceptiona momenta $q \neq 0$ as $t \to 0$:

$$\{ \mu \frac{\partial}{\partial \mu} + \sigma \frac{\partial}{\partial g} - \tau N + \hat{\delta} \} \tilde{\Gamma}_0^{(N,1)}(p_1, \ldots, p_N, q) = 0 \tag{C.6}$$

whereas

$$\{ \mu \frac{\partial}{\partial \mu} + \sigma \frac{\partial}{\partial g} - \tau N + \delta \} \Gamma_0^{(N,1)}(p_1, \ldots, p_N, q) = 0$$

Hence it follows

$$\delta(g) = \hat{\delta}(g) + \sigma(g) \frac{\partial}{\partial g} \ln Z_2(g) \tag{C.7}$$

and $\delta(g^*) = \hat{\delta}(g^*)$ at any fixed point g^*.

We see that if $\tilde{\Gamma}_o$ above is identified with the vertex functions $\bar{\Gamma}_{as}$ in (B.2) the functions $\beta(g)$ and $\gamma(g)$ as well as $\hat{\delta}(g)$ are expressed in terms of massive vertex functions. They are hence holomorphic in the dimension ε = D-4 for Re $\varepsilon \geqslant$ o. On the other hand the functions V(g) and Z(g) may show up infrared singularities (see Symanzik[14]) and so do the massless functions $\tau(g)$, $\sigma(g)$ and $\delta(g)$. From (C.4) and (C.7) however we see that these infrared sigularities do not give troubles at the fixed points.

Table 1. 32)

Critical Exponents

D = 3 (i.e. ε = 1)

Exp.	Exp	MFA	L. - I.	ε^2	ε^3
α	small	0	0.125 ±.015	0.077	0.196
β	0.3÷0.4	1/2	0.312 ±.003	0.340	0.304
γ	1.2÷1.4	1	1.250 ±.003 (1.250 HTE)	1.244	1.195
δ	-	3	5.15 ± .02	4.463	-
ν	0.6÷0.7	1/2	0.642 ±.003	0.627	-
η	small	0	0.056±.01 (0.041 HTE)	0.037	0.029

D = 2 (i.e. ε = 2)

	α	β	γ	δ	ν	η
L.I.	ln	0.125	1.75	15.0	1	0.25
ε^2	-0.025	0.191	1.642	6.852	0.840	0.235

References

1) A.A. Migdal, ZhETF 55, 1964 (1969), 59, 1015 (1970)
A.M. Polyakov, ZhETF 55, 1026 (1969)

2) L. P. Kadanoff, Physics 2, (1966) 263
L. P. Kadanoff et. al., Rev. Mod. Phys. 39, 359 (1967)
see also: L.P. Kadanoff, Cargese Lect. 1973

3) K. Wilson, Phys. Rev. B4, 3174, 3184 (1971)
K. Wilson, J. Kogut, Cornell Preprint (1973)

4) K. Wilson, E. Fisher: Phys. Rev. Lett 28, 240 (1972)
F. J. Wegner, A. Houghton, Phys. Rev. A8, 401 (1973)

5) H. E. Stanley: "Introduction to Phase Transitions and Critical Phenomena", Oxford Univ. (1971)

6) F. J. Wegner, Phys. Rev. B5, 4529 (1972)

7) G. Jona-Lasinio: Proc. Nobel Symp. 24, Aspensäsgarden
F. J. Wegner: J. Phys. C7, 2096 (1974)

8) J. Schwinger, Proc. Nat. Acad. Sci. 44, 956 (1958)
K. Symanzik in Proc. Int. School "Enrico Fermi" R. Jost Ed. Acad. Press (1969)
K. Osterwalder, R. Schrader, Comm. Math. Phys. 31, 83 (1973)
E. Nelson, J. Funct. Anal. 12, 97 (1973)

9) M. Nauenberg: Renormalization Group Solution of One Dimensional Ising Model
M. Nauenberg, B. Nienhuis: - Critical Suface for Square Ising Spin Lattice - RG Approach to Solution of General Ising Models, Preprints Univ. of Utrecht (1974)

10) B. Schroer: Lett. Nuovo Cimento 2, 867 (1971)

11) G. Parisi, L. Peliti: Lett. Nuovo Cimento 2, 627 (1971)

12) G. Mack, I. T. Todorov: Phys. Rev. D8, 1764 (1973)
G. Mack: in "Renormalization and Invariance in Quantum Field Theory" (Ed.) E. R. Caianello, New York Plenum Press and Ref. cited therein.

13) C. di Castro: Lett.Nuovo Cimento 5, 69 (1972)

14) G. Parisi: "Field Theory Approach to Second Order Phase Transitions in Three and Two Dimensional Systems" Cargese Summe School 1973
K. Symanzik: "Mass less ϕ^4-Theory in $4-\varepsilon$ Dimensions" DESY 73/58

15) B. Schroer: "Recent Developement in the Theory of Critical Phenomena" FU-Berlin Preprint (1974)

16) M. Gell-Mann, F.E. Low: Phys. Rev. 95, 1300 (1954)

17) G. Jona-Lasinio: Nuovo Cimento 34, 1790 (1964)

18) K. Symanzik: Comm. Math. Phys. 18, 227 (1970)

19) E. Brezin, J. C. Le Guillou, J. Zinn-Justin: Phys. Rev. D8, 434, 2418 (1973)
J. Zinn-Justin: Cargèse Lect. (1973)
E. Brezin et. al.: Phys. Rev. D9, 1121 (1974)

20) B. Schroer: Phys. Rev. B8, 4200 (1973)
F. Jegerlehner, B. Schroer: Acta Physica Austriaca Suppl. XI 389 (1973)

21) F. Jegerlehner: FU-Berlin Preprint (1974)

22) M. Gomes, B. Schroer: Univ. of Sao Paulo Preprint (1974)

23) C. di Castro, G. Jona-Lasinio, L. Peliti: Univ. Rome Preprint (1973)

24) G. t'Hooft: Nucl. Phys. B61, 455 (1973)
S. Weinberg: Phys. Rev. D8, 3497 (1973)

25) M. Bergère, Y. M. Lam: "Asymptotic Expansion of Feynman Amplitudes" FU-Berlin Preprint (1973)

26) W. Zimmermann: Comm. Math. Phys. 15, 208 (1969)
Annals of Physics 77, 536 (1973)
J. Lowenstein: Phys. Rev. D4, 2281 (1971) Comm. Math, Phys. 24, 1 (1971)

27) M. Gomes, J. Lowenstein, W. Zimmermann: to be published
see also: F. Jegerlehner, B. Schroer: Nucl. Phys. B68, 461 (1974)
J. Lowenstein, A. Rouet, R. Stora, W. Zimmermann: C.N.R.S. Marseille Preprint 1973

28) A. I. Larkin- D.E. Khmel'nitzky: JETP 29, 1123 (1969)

29) C. G. Callan: Phys. Rev. D2, 1541 (1970)
K. Symanzik: Comm. Math. Phys. 18, 227 (1970)

30) G. Mack: in Lecture Notes in Physics, Vol. 17, Springer (1972)
F. J. Wegner: Phys. Rev. B5, 4529 (1972)
E. Brezin, J.C. Le Guillou, J. Zinn-Justin: Phys. Rev. D8,2418(1973)

31) K. Symanzik: Comm. Math. Phys. 23, 49 (1971) and Capri Lectures (1973)
E. Brezin, Le Guillou, J. Zinn-Justin: Phys. Rev. Lett. 32, 473 (1974)
F. Wegner: "Correlation Functions near the Critical Point" Univ. of Heidelberg Preprint 1974

32) see e.g. M. Wortis: in "Renormalization Group in Critical Phenomena and Quantum Field Theory" Ed. J. D. Gunton, M.S. Green, Temple University 1973

33) S. Ma: Rev. Mod. Phys. 45, 589 (1973)
R. Brout: Physics Reports 10C, 1 (1974)

CRITICAL PHENOMENA AND THE RENORMALIZATION GROUP

Franz J. Wegner

Institut für Theoretische Physik, Universität Heidelberg

Abstract

The recent theory of critical phenomena and the renormalization group as promoted by Wilson is considered on an introductory level. The main emphasis is on the idea of the fixed point Hamiltonian (asymptotic invariance of the critical Hamiltonian under change of the length scale) and the resulting homogeneity laws.

1. CRITICAL BEHAVIOR

A. Critical Points

The transition[1] from one phase to another like melting or boiling changes the properties of a system discontinuously. Such a phase transition is called a first order transition or discontinuous transition. By varying one or several thermodynamic variables like the temperature, it is frequently possible to follow the coexistence curve so that the two distinct phases become more and more similar until both phases become equal at a certain point. If beyond this point only one homogeneous phase exists and all changes are smooth and continuous, then this point is called a critical point.[2]

Examples of critical points are (a) the termination point of the coexistence curve of a liquid and its vapor (or two phases of different density of a lattice gas like hydrogen in metals) at the critical temperature T_c and pressure p_c, (b) the critical point of separation of mixtures and alloys above (or below) which the components mix without a miscibility gap, (c) the ordering temperature of a homogeneous binary crystal below which one sublattice is primarily occupied by one species.

A second class of systems exhibits domains of magnetic or electric moments of different orientation which vanish at the critical temperature. Examples are (d) ferromagnets with ferromagnetic domains of different orientation whose spontaneous magnetization vanish continuously at the Curie temperature, (e) ferroelectrics with ferroelectric domains whose spontaneous polarization go to zero at the critical temperature, and (f) NH_4 compounds whose electric octupole moments order primilary in one or the other direction below T_c. (g) The alternating spin order of antiferromagnets goes to zero at the Neel point so that two counterphase domains become indistinguishable. Analogously one observes (h) alternating ordering of electric dipole moments in antiferroelectrics and (i) alternating ordering of electric octupole moments in NH_4 compounds.

Thirdly (j) superfluid helium and (k) superconductors are characterized by a condensate associated with a phase below T_c. This condensate vanishes continuously approaching T_c from below so that domains with different phase cannot be distinguished above T_c. This list does not exhaust the types of critical points observed. But it gives an impression of the variety of phenomena which can be described by the theory of critical phenomena.

To unify the description of critical phenomena one has introduced the concept of the order parameter. For the liquid-vapor transition and other transitions characterized by a difference of densities in both phases (sublattices) the order parameter is the difference between the expectation value of the density of the phase (sublattice) from its value at criticality. For orientational transitions the expectation value of the electric (magnetic) moment (or the difference on the sublattices) serves as the order parameter. In superfluid helium and superconductors the expectation of the condensate wave function is the order parameter. The amount of the order parameter is (approximately) the same in all phases but it differs in sign, direction and phase, respectively, in different phases (domains).

The field conjugate to the order parameter is often called the symmetry breaking field, since it breaks the symmetry of the Hamiltonian in the case of orientational phase transitions and the transitions to the superfluid and superconducting state. Without this field the Hamiltonian is invariant under certain rotations of the order parameter or invariant under the change of the phase (gauge transformation). This symmetry breaking field is the magnetic field for ferromagnets, the electric field for ferroelectrics, the chemical potential for the liquid vapor transition, the difference of chemical potentials for mixtures. In several cases the symmetry breaking field is not experimentally accessible as in superfluids and in superconductors. But it is often introduced in theoretical physics for conceptual reasons like the staggered field in antiferromagnets and antiferroelectrics.

This unified description allows us to restrict to one class of system in explaining the main features of critical phenomena, provided we neglect a number of peculiar features of certain systems. Two features we will often neglect are (i) the quantum mechanic (or discrete) nature of the microscopic origin of many phase transitions (superfluid He, superconductors, spin and exchange interaction in magnets, etc). Since critical phenomena become apparent on a macroscopic scale, it is assumed that the commutators can be neglected and the order parameter can be handled like a continuous classical variable. (ii) In many cases we will neglect long range interactions or the long range part of these interactions. Therefore we will neglect dipolar interactions and the coupling of the interaction to lattice distortions which induce long range interactions.

We will mainly use the magnetic language. Thus we will discuss the critical behaviour of a ferromagnet consisting of classical spins on a rigid lattice for which the exchange interaction dominates so that the dipolar interaction can be neglected.

B. Critical Exponents - the Homogeneity Assumption

The first theory to explain the critical behavior of ferromagnets was the molecular field theory by P. Weiss.[3] According to this theory the spontaneous magnetization m is zero above T_c and goes to zero from below like $\sqrt{T_c - T}$. The susceptibility diverges like $|T-T_c|^{-1}$ from below and above T_c and the specific heat shows a finite jump at T_c. Experimentally however one finds $m \propto (T_c-T)^{\beta}$ with $\beta \approx 1/3$ for the spontaneous magnetization, $\chi \propto |T_c-T|^{-\gamma}$ with $\gamma \approx 4/3$ and a singular contribution to the specific heat like $c_{sing} \propto |T_c-T|^{-\alpha}$ with α close to zero. For negative α the specific heat shows a cusp, for positive α it diverges. The exponents α,β and γ are called critical exponents. The deviation of the molecular field exponents from the experimental critical exponents has led to the search of soluble models. Unfortunately most models (approximations) lead back to the molecular field behavior. Two models which give different sets of critical exponents are the spherical model and the two dimensional Ising model. The exponents are listed in table 1. None of these models give exponents which are close to the experimentally observed exponents. The reason is that the molecular field theory completely neglects the critical fluctuations (apart from the homogeneous component) which leads to a γ which is too small; the spherical model overestimates the critical fluctuations which leads to a γ which is too large. The two dimensional Ising model describes the

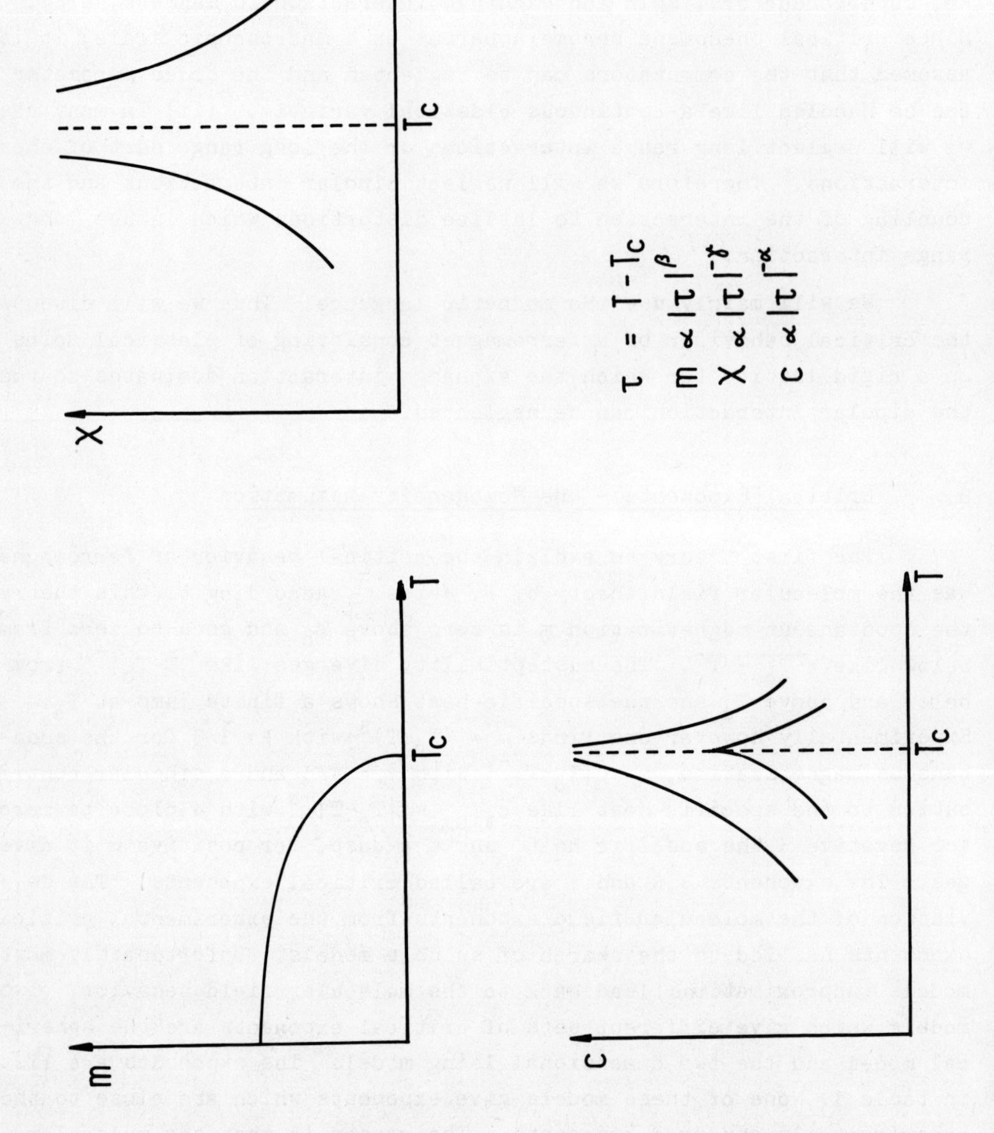

Fig. 1. The schematic behavior of the spontaneous magnetization, susceptibility and specific heat near T_c.

Power law	Exponent	Molecular Field Approxim.	Spherical Model d = 3	Ising Model d = 2	Experi-ments d = 3	High temperature Expansions d = 3		
						n=1	n=2	n=3
$m \propto \|\tau\|^{\beta}$	β	1/2	1/2	1/8	$\approx 1/3$	.31		
$\chi \propto \|\tau\|^{-\gamma}$	γ	1	2	7/4	$\approx 4/3$	1.25	1.32	1.38
$c_{sing} \propto \|\tau\|^{-\alpha}$	α	0 (disc.)	-1 (kink)	0 (log)	≈ 0	.13	.00	-.10

Table 1. Critical Exponents of various models.

fluctuations properly. However the dimensionality of the system plays an important role in critical phenomena so that the two dimensional Ising model does not yield a reasonable approximation for the three dimensional Ising model.

Apart from some other two dimensional models (F-model, KDP-model, eight-vertex-model), there are no exactly soluble models available. Therefore one has tried a different approach to determine critical exponents by means of series expansions. One expands for example the susceptibility or the specific heat of a model like the Ising model in powers of the inverse[4] temperature $\beta = (k_B T)^{-1}$, assumes that the quantity considered shows a power law behavior close to T_c and analyzes the series accordingly. This yields estimates for the critical exponents listed in the last columns of table 1 for three models: The Ising model (a model of spins S with two states $S = \pm 1$), the XY-model (a model of planar spins S, that is spins with two components $S_x = \cos\varphi$, $S_y = \sin\varphi$) and the classical Heisenberg model (a model of three dimensional (classical) vectors S with $S^2 = 1$). The spins are located at the sites of a lattice and interact via an (isotropic) short range (in most cases nearest neighbor) interaction . Low temperature expansions are only available for the Ising model. Therefore β is quoted only for the Ising model. One can estimate the low temperature exponents for the specific heat and the susceptibility of the Ising model. They are slightly different from the high temperature exponents. Since it is hard to estimate the accuracy of the exponents determined, it is hard to decide whether high and low temperature exponents are equal within the error bars. One finds that the exponents determined from the expansions are quite close to the experimentally observed ones. But unfortunately one does not learn from these expansions why the systems exhibit these broken power laws near T_c. It is the aim of this paper to review on an introductory level the ideas which provide an understanding of the critical behavior.

A first step to link different aspects was the homogeneity assumption by Widom[5]. We bring a modified version of it. (Widom's assumption included the possibility of logarithmic singularities which will not be considered in this section). Widom assumes that the free energy can be separated as a function of the magnetic field h and the temperature difference $\tau = T - T_c$ into a regular and a singular part

$$F = F_{reg}(\tau) + F_{sing}(\tau, h) \quad , \qquad (1.1)$$

where the singular part is responsible for the critical behavior. He assumes that the singular part is a homogeneous function of the variables τ and h, that is

$$F_{sing}(\tau,h) = |\tau|^{2-\alpha} f_{\pm}\left(\frac{h}{|\tau|^{\Delta}}\right) \tag{1.2}$$

where the $\pm$ denotes that the function is different for positive and negative τ. Δ is called the gap exponent. Homogeneity means that multiplying τ by a factor c and h by a factor c^{Δ} multiplies the function by a factor $c^{2-\alpha}$.

Let us discuss some consequences. We obtain the specific heat by differentiating[6] F twice with respect to α. This gives the singular part of the specific heat at constant vanishing field h

$$c_{sing} \propto |\tau|^{-\alpha} f_{\pm}(0) \tag{1.3}$$

which was the reason for calling the exponent in eq. (1.2) $2-\alpha$. The magnetization is obtained from eq. (1.2) by differentiating with respect to h

$$m = -|\tau|^{2-\alpha-\Delta} f'_{\pm}\left(\frac{h}{|\tau|^{\Delta}}\right) \quad . \tag{1.4}$$

At h = 0 this leads to

$$m = -|\tau|^{\beta} f'_{\pm}(0) \tag{1.5}$$

with

$$\beta = 2-\alpha-\Delta \quad . \tag{1.6}$$

Differentiating twice with respect to h we obtain

$$\chi = -|\tau|^{2-\alpha-\Delta} f''_{\pm}\left(\frac{h}{|\tau|^{\Delta}}\right) \tag{1.7}$$

which yields

$$\gamma = \alpha + 2\Delta - 2 \quad . \tag{1.8}$$

From eqs. (1.6) and (1.8) we find a relation between the exponents α, β, γ

$$\alpha + 2\beta + \gamma = 2 \quad . \tag{1.9}$$

A look at table 1 shows that this relation is fulfilled for all listed sets of exponents. We note that from eqs. (1.6) and (1.8) we obtain

$$\Delta = \beta + \gamma \quad . \tag{1.10}$$

Then eq. (1.4) can be easily cast in the form

$$\frac{m}{|\tau|^{\beta}} = -f'_{\pm}\left(\frac{h}{|\tau|^{\Delta}}\right) \quad . \tag{1.11}$$

Solving with respect to $h/|\tau|^{\Delta}$ yields

$$\frac{h}{|\tau|^{\Delta}} = g_{\pm}\left(\frac{m}{|\tau|^{\Delta}}\right) \tag{1.12}$$

with some function g, which can be written

$$\frac{h}{m|\tau|^{\gamma}} = w_{\pm}\left(\frac{m}{|\tau|^{\beta}}\right) \tag{1.13}$$

with $w(x) = g(x)/x$. Thus $h/|\tau|^{\Delta}$ should be a function of $m/|\tau|^{\beta}$ only. In Fig.2 data[7] of the magnetization m of $CrBr_3$ as a function of the two variables, temperature and field, are plotted in the variables $m/|\tau|^{\beta}$ and $h/(m|\tau|^{\gamma})$. If the homogeneity assumption would not hold, the data points would be scattered in the whole plot. Since the data follow the homogeneity assumption, they lie on two lines corresponding on the behaviour above and below T_c.

In the following section we will show how the homogeneity relation can be derived.

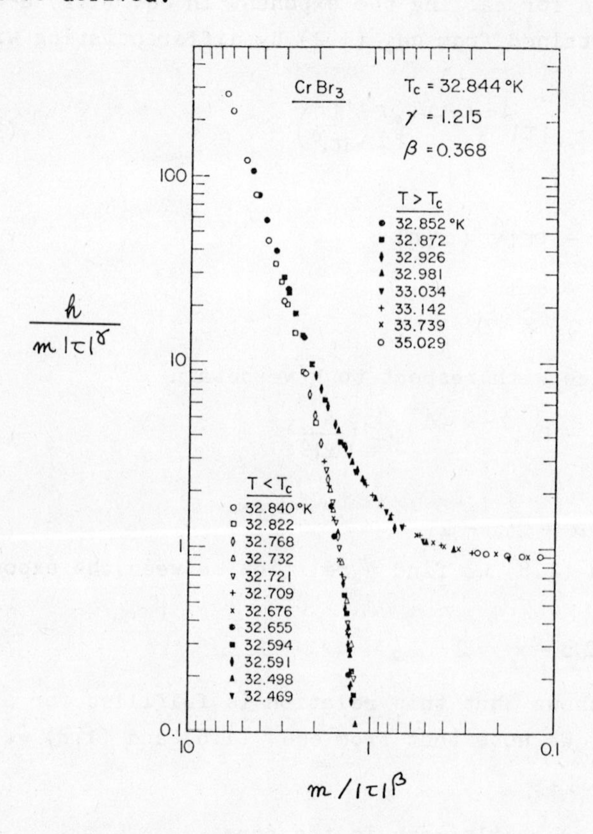

Fig. 2. This plot of $h/(m|\tau|^{\gamma})$ against $m/|\tau|^{\beta}$ confirms the scaling hypothesis for $CrBr_3$. The two branches are for $T>T_c$ and $T<T_c$. After J.T. Ho and J.D. Litster, J. Appl. Phys. 40, 1270 (1969)

2. RENORMALIZATION GROUP EQUATION

A. Motivation

A hint on how the critical state can be characterized can be obtained from the correlation functions. Let us consider the auto correlation function of the spins $S_o(r)$. From what one knows from exactly solvable systems this correlation function decays at criticality with a power law for large distances

$$\langle S_o(o)\, S_o(r)\rangle_{crit} = \frac{c}{r^{d-2+\eta}} \qquad (2.1)$$

where η is a new critical exponent and d the dimensionality of the system. η describes the deviation from the Ornstein-Zernicke-behavior of the correlation function. Let us now consider the same ferromagnet under a different length scale. To accomplish this we divide the sample into cubic cells of length b lattice spacings in each direction. Then the magnetization of a cell

$$s = \sum_{cell} S_o(r') \qquad (2.2)$$

obeys asymptotically

$$\langle s(o)\, s(r)\rangle = \frac{c\, b^{2d}}{r^{d-2+\eta}} \qquad (2.3)$$

since each cell contains b^d spins. Now we change the length scale by a factor b and the scale for the magnetization by a factor $b^{(d+2-\eta)/2}$

$$r = bR \;, \quad s(r) = b^{(d+2-\eta)/2}\, S_1(r) \qquad (2.4)$$

Then we obtain the asymptotic behavior of our new spin variables

$$\langle S_1(o)\, S_1(R)\rangle = \frac{c}{R^{d-2+\eta}} \quad . \qquad (2.5)$$

Therefore the correlation function is invariant under the change of the scale (2.4). This invariance of the correlation function suggests that the effective interaction at criticality is invariant with respect to the change of the length scale. We call the procedure which changes the scale of the hamiltonian (effective interaction) renormalization group (RG) procedure and the corresponding transformation is called RG transformation. In the remainder of this section we outline some requirements and properties of RG transformation and derive one of the RG equations.

B. Properties of the RG transformation[8-12]

We denote the hamiltonian function $\mathcal{H}$ and the free energy $\mathcal{F}$.

We introduce

$$H = \beta \mathcal{H} \quad , \quad F = \frac{\beta \mathcal{F}}{V} \quad , \quad \beta = \frac{1}{k_B T} \tag{2.6}$$

where V is the volume of the system. For simplicity's sake we call H and F hamiltonian and free energy, resp.

$$-F = \frac{1}{V} \ln \operatorname{trace} \exp(-H) \quad . \tag{2.7}$$

The RG transformation consists of
(i) a change of the length scale by a factor $b=e^{\ell}$ in all linear dimensions (we leave the partition function Z=trace exp (-H) invariant). Since the volume shrinks by a factor $e^{-d\ell}$ we obtain

$$F_0 = e^{-d\ell} F_\ell \tag{2.8}$$

(ii) a transformation and/or elimination of the spin variables S which leaves the free energy invariant. The transformation shall not generate long-range interactions. The new hamiltonian H_ℓ has to be comparable with the original hamiltonian H_0 (same Hilbert or function space). This demands an extension of the system to the original volume for finite systems. The RG transformation transforms H_0 into H_ℓ

$$H_\ell = R_\ell(H_0) \tag{2.9}$$

$$F(H_0) = e^{-d\ell} F(H_\ell) \quad . \tag{2.10}$$

C. RG equation with smooth momentum cut-off

There are various ways to construct RG equations which transform hamiltonians:
(i) Wilson's recurrence relation (approximation)[13]. Numerical solution for d=3 see Refs[14]. Expansion in ε=4-d see Refs. 15.
(ii) Wilson's differential RG equation with smooth momentum cut-off[10], generalization Ref.[16].
(iii) Differential RG equation with sharp momentum cut-off (generates long-range interactions): Wegner and Houghton[17]
(iv) Aharony's method[18]
(v) Two-dimensional Ising models: Niemeijer and van Leeuwen [19], Nauenberg and Nienhuis[20].

We do not discuss the other varieties of the RG, which transform the correlation functions. (Compare the review by Zinn-Justin[21]).

Now we derive the RG equation with smooth momentum cut-off. We represent the hamiltonian as a functional of the Fourier components.

$$S_q = \int d^d r \; e^{-iqr} \; s(r) \tag{2.11}$$

of the variable S(r)

$$\begin{aligned} H = H\{S_q\} &= V u_0 + u_1 S_0 + \frac{1}{2}\int u_2(q) S_q S_{-q} \, d^d q \\ &+ \frac{1}{3!}\int u_3(q_1,q_2) S_{q_1} S_{q_2} S_{-q_1-q_2} \, d^d q_1 \, d^d q_2 \\ &+ \frac{1}{4!}\int u_4(q_1,q_2,q_3) S_{q_1} S_{q_2} S_{q_3} S_{-q_1-q_2-q_3} \, d^d q_1 \, d^d q_2 \, d^d q_3 \; + \ldots \end{aligned} \tag{2.12}$$

We perform an infinitesimal change of the length scale

$$r \to r(1-\delta) \qquad \delta \text{ infinitesimal} \tag{2.13}$$

$$q \to q(1+\delta) \tag{2.14}$$

$$S_q^0 = S_{q+\delta q}^\delta = S_q^\delta + \delta q \, \nabla_q S_q^\delta \tag{2.15}$$

$$V^0 = V^\delta(1 + d\cdot\delta) = V^\delta + \delta\cdot d\cdot V^\delta \; . \tag{2.16}$$

Therefore the hamiltonian H transforms into

$$H\{S_q\} \to H\{S_q\} + \delta\left(\int d^d q \left(q \nabla S_q \frac{\delta H}{\delta S_q} + \frac{d}{2}\right) + d\cdot V\cdot\frac{\partial H}{\partial V}\right). \tag{2.17}$$

The transformation (2.11) is unitary apart from a volume dependent constant. This volume dependence produces the term d/2 in eq. (2.17). From this eq. we obtain the generator G_{dil} of the dilatation

$$G_{dil} H = \int d^d q \left(q \nabla S_q \frac{\delta H}{\delta S_q} + \frac{d}{2}\right) + d\cdot V\cdot\frac{\partial H}{\partial V} \; . \tag{2.18}$$

Secondly we allow a transformation of the variables

$$S_q \to S_q + \delta\cdot\varphi_q\{S\} \tag{2.19}$$

which transforms the hamiltonian according to

$$H \to H + \delta \sum \varphi_q \frac{\partial H}{\partial S_q} \tag{2.20}$$

where φ_q is a functional of the spin variables. The volume element in phase space transforms according to

$$\prod dS_q \to \prod d(S_q + \delta \cdot \psi_q) = \prod dS_q \cdot (1 + \delta \sum \frac{\partial \psi_q}{\partial S_q}) \tag{2.21}$$

which with

$$\sum \frac{\partial \psi_q}{\partial S_q} = \int d^d q \frac{\delta \psi_q}{\delta S_q} \tag{2.22}$$

yields

$$\int \prod dS_q \exp(-H) = \int \prod dS_q \exp(-H - \delta \int (\psi_q \frac{\delta H}{\delta S_q} - \frac{\delta \psi_q}{\delta S_q}) d^d q) \tag{2.23a}$$

so that the hamiltonian transforms into

$$H \to H + \delta \int (\psi_q \frac{\delta H}{\delta S_q} - \frac{\delta \psi_q}{\delta S_q}) d^d q \quad . \tag{2.23b}$$

The generator G_{tra} (ψ) of this transformation is

$$G_{tra}(\psi) H = \int (\psi_q \frac{\delta H}{\delta S_q} - \frac{\delta \psi_q}{\delta S_q}) d^d q \quad . \tag{2.24}$$

The transformation is performed so that the free energy is invariant (eq. (2.23a)).

From eqs. (2.18) and (2.24) we obtain the renormalization group equation

$$\frac{\partial H}{\partial \ell} = G(\psi) H = (G_{dil} + G_{tra}(\psi)) H \tag{2.25}$$

in differential form. Wilson[10] has choosen a special dependence for ψ

$$\psi_q = \frac{d}{2} S_q + (c + 2q^2)(S_q - \frac{\delta H}{\delta S_q}) \tag{2.26}$$

where the constant c has to be adjusted properly. This choice guarantees that the Fourier components S_q with large q are eliminated and survive only in $u_2(q)$ which for large q approaches unity.

3. SCALING AND THE LINEARIZED RG EQUATION

A. Fixed point, classification of operators

In the Wilson theory of critical phenomena the following two assumptions are made:

(i) It is assumed that a fixed point hamiltonian H^* exists

$$G H^* = 0 \tag{3.1}$$

This is a hamiltonian which maps into itself.

(ii) It is assumed that for a critical hamiltonian

$$\lim_{l\to\infty} H_l = H^* \quad . \tag{3.2}$$

The RG equation

$$\frac{\partial H}{\partial l} = G_{\text{Wilson}} H = \int d^d q \left(q \nabla_q S_q + \frac{d}{2} S_q + (c+2q^2) S_q\right) \frac{\delta H}{\delta S_q} + \int d^d q (c+2q^2)\left(-\frac{\delta H}{\delta S_q}\frac{\delta H}{\delta S_{-q}} + \frac{\delta^2 H}{\delta S_q \delta S_{-q}}\right) \tag{3.3}$$

yields in linear order in ΔH

$$G_{lin}(H^* + \Delta H) = \int d^d q \left(q \nabla_q S_q + \frac{d}{2} S_q + (c+2q^2) S_q\right) \frac{\delta \Delta H}{\delta S_q} + \int d^d q (c+2q^2)\left(-2\frac{\delta H^*}{\delta S_{-q}} + \frac{\delta}{\delta S_{-q}}\right)\frac{\delta \Delta H}{\delta S_q} \equiv L\Delta H \quad . \tag{3.4}$$

We define eigenoperators O_i by the eigenvalue equation

$$L O_i = y_i O_i \quad . \tag{3.5}$$

We assume in the following that the eigenoperators form a complete set of operators so that any hamiltonian H_o can be expanded

$$H_o = H^* + \sum \mu_i O_i \quad . \tag{3.6}$$

Then we obtain in linear order in μ

$$H_l = H^* + \sum \mu_i e^{y_i l} O_i \quad . \tag{3.7}$$

Corresponding to the eigenvalues y one distinguishes

$y > 0$ <u>relevant</u> operator,

$y = 0$ <u>marginal</u> operator,

$y < 0$ <u>irrelevant</u> operator.   (3.8)

From equation (3.7) we find immediately that at the critical point the fields (in high energy physics sources) μ_i of all relevant operators have to vanish.

Depending on the nonlinear terms marginal operators may act as relevant, irrelevant, and substantially marginal operators (as in the eight-vertex-model), resp.

There is a <u>special</u> operator, the constant $V(u_o - u_o^*)$, eq. (2.12) which formally has

$$O_o = 1, \qquad y_o = d \tag{3.9}$$

However the addition of a constant to the hamiltonian does not change its critical behavior. Therefore $\mu_o = 0$ is not necessary for criticality. This is the origin of the regular part of the free energy.

The type of the critical behavior depends on the number of symmetry conserving relevant operators. (Symmetry conserving means that the symmetry of the hamiltonian is conserved, it does not exclude a spontaneously broken symmetry of the system). Let us expand

$$\mathcal{H} = \sum \mu_i^o O_i \tag{3.10}$$

$$H^* = \sum \mu_i^* O_i \tag{3.11}$$

then we obtain

$$\mu_i = \beta \mu_i^o - \mu_i^* \quad . \tag{3.12}$$

For a normal critical point one has one relevant symmetry conserving operator (apart from O_o) O_E which determines the critical temperature

$$\mu_E \equiv \tau = \beta \mu_E^o - \mu_E^* = (\beta - \beta_c)\mu_E^o \; ; \; \beta_c = \frac{\mu_E^*}{\mu_E^o} \quad . \tag{3.13}$$

Crudely speaking O_E is proportional to the hamiltonian minus its expectation value at the critical point. At a tricritical point one has two relevant symmetry conserving operators (apart from O_o) and consequently two conditions for criticality.

Redundant operators:

We state a few results on redundant operators (see ref. 16). The hamiltonian H^* is not uniquely defined, it depends on the functional ψ. Varying this functional one can show that any hamiltonian $H^* + \delta\, G_{tra}(\varphi)\, H^*$ can be a fixed point. (δ infinitesimal). We call these hamiltonians equivalent to H^* and the operators $G_{tra}(\varphi)\, H^*$ redundant operators. One shows that L applied to a redundant operator yields again a redundant operator. Since equivalent hamiltonians can be obtained from H^* by means of the transformation (2.23) both hamiltonians have equal free energy. Therefore the redundant operators O_i do not contribute to the critical behavior. Therefore $\mu_i = 0$ need not be fulfilled for redundant operators at the critical point.

The eigenvalues y for redundant operators are not uniquely defined. They depend on the choice of ψ for the RG equation. The eigenvalues of the other operators are uniquely defined. An example of a redundant operator (see Hubbard and Schofield[22]) which corresponds to a shift of S(r) by a constant: $\varphi_q = \delta(q)$

$$G_{tra}(\varphi)\, H = \frac{\delta H}{\delta S_0} \tag{3.14}$$

For the hamiltonian

$$H = \frac{1}{2}\int (r_0+q^2) S_q S_{-q}\, d^dq + \frac{1}{4!} u_0 \int S_{q_1} S_{q_2} S_{q_3} S_{-q_1-q_2-q_3}\, d^{3d}q \tag{3.15}$$

one obtains

$$G_{tra}(\varphi) H = r_0 S_0 + \frac{u_0}{3!}\int S_{q_1} S_{q_2} S_{-q_1-q_2}\, d^{2d}q = \int \left(r_0 S(r) + \frac{u_0}{3!} S^3(r)\right) d^dr . \tag{3.16}$$

B. Scaling of the free energy

Within a simplified picture (Kadanoff's cell model[23]) we consider only two operators O_E and the magnetization O_h

$$H_0 = H^* + \tau O_E + h O_h \tag{3.17}$$

which yields

$$H_\ell = H^* + \tau e^{y_E \ell} O_E + h e^{y_h \ell} O_h \tag{3.18}$$

$$F(\tau, h) = e^{-d\ell} F(\tau e^{y_E \ell}, h e^{y_h \ell}) . \tag{3.19}$$

We choose τ by

$$|\tau| e^{y_E \ell} = 1 \tag{3.20}$$

and obtain Widom's scaling law (1.2)

$$F(\tau, h) = |\tau|^{d/y_E} F\left(\pm 1, \frac{h}{|\tau|^{y_h/y_E}}\right) \tag{3.21}$$

with

$$d/y_E = 2-\alpha \qquad y_h/y_E = \Delta . \tag{3.22}$$

Normally one has an infinite number of perturbations O_i in equation (3.17). To study their effect on the scaling law we add at least one further operator pars pro toto

$$\mathcal{H}_0 = \mathcal{H}^* + \tau O_E + h O_h + \mu_i O_i \tag{3.23}$$

and obtain

$$\mathcal{H}_\ell = \mathcal{H}^* + \tau e^{y_E \ell} O_E + h e^{y_h \ell} O_h + \mu_i e^{y_i \ell} O_i \tag{3.24}$$

$$F(\tau, h, \mu_i) = |\tau|^{d/y_E} F\left(\pm 1, \frac{h}{|\tau|^{y_h/y_E}}, \frac{\mu_i}{|\tau|^{y_i/y_E}}\right) . \tag{3.25}$$

We are interested in the critical behavior that is in the limit $\tau \to 0$

$$\lim_{\tau \to 0} \frac{\mu_i}{|\tau|^{y_i/y_E}} \to \begin{cases} 0 & y_i < 0 \text{ or } \mu_i = 0 \\ \pm\infty & y_i > 0 \text{ and } \mu_i \neq 0 \end{cases} . \tag{3.26}$$

If O_i is relevant ($y_i > 0$) then μ_i has explicitly to be taken into account. For irrelevant operators the term $\mu_i/|\tau|^{y_i/y_E}$ can be neglected if F can be expanded in powers of μ_i. Note that the right hand side of eq. (3.26) contains the free energy well apart from the critical point. The irrelevant operator yields a correction to scaling

$$F = |\tau|^{d/y_E} F\left(\pm 1, \frac{h}{|\tau|^{y_h/y_E}}, 0\right) + |\tau|^{(d-y_i)/y_E} F'\left(\pm 1, \frac{h}{|\tau|^{y_h/y_E}}, 0\right) + \cdots \tag{3.27}$$

as observed in superfluid He (Ahlers[24]). If F cannot be expanded in powers of μ_i, then Fisher's idea of the anomalous dimension of the vacuum might apply[25].

C. Correlations

Until now we considered only translational invariant perturbations. Let us consider the eigenvalue equation for localized operators $\tilde{O}_i$ (Wilson and Kogut[10])

$$L \tilde{O}_i = - x_i \tilde{O}_i . \tag{3.28}$$

From the representation

$$\tilde{O}_i = \tilde{O}_i \{ S_q \} \tag{3.29}$$

we define

$$\tilde{O}_i(r) = \tilde{O}_i \{ S_q e^{iqr} \} . \tag{3.30}$$

It follows that

$$L\tilde{O}_i(r) = -x_i \tilde{O}_i(r) - r\nabla_r \tilde{O}_i(r) \quad . \tag{3.31}$$

In linear approximation we obtain from

$$H_0 = H^* + \lambda_i \tilde{O}_i(r) \tag{3.32}$$

$$H_\ell = H^* + \lambda_i e^{-x_i \ell} \tilde{O}_i(re^{-\ell}) \quad . \tag{3.33}$$

Let us define

$$O_i(q) = \int \tilde{O}_i(r) e^{-iqr} d^d r \quad . \tag{3.34}$$

Then

$$H_0 = H^* + \lambda_i O_i(q) \tag{3.35}$$

yields

$$H_\ell = H^* + \lambda_i e^{(d-x_i)\ell} O_i(qe^\ell) \quad . \tag{3.36}$$

By comparison for q=0 with (3.5) we see

$$y_i = d - x_i \tag{3.37}$$

From

$$H_0 = H^* + \sum \mu_i O_i + \lambda_1 O_1(q) + \lambda_2 O_2(-q) \tag{3.38}$$

we obtain in linear approximation

$$H_\ell = H^* + \sum \mu_i e^{y_i \ell} O_i + \lambda_1 e^{y_1 \ell} O_1(qe^\ell) + \lambda_2 e^{y_2 \ell} O_2(-qe^\ell) . \tag{3.39}$$

Differentiating the free energy of the Hamiltonian (3.38) yields the correlation function

$$G(q, H_0) = \langle O_1(q) O_2(-q) \rangle_0 = \frac{\partial^2}{\partial\lambda_1 \partial\lambda_2} F(H_0) \tag{3.40}$$

Therefore we obtain

$$\langle O_1(q) O_2(-q) \rangle_0 = e^{(y_1+y_2-d)\ell} \langle O_1(qe^\ell) O_2(-qe^\ell) \rangle_\ell \tag{3.41}$$

$$G(q, \tau, h) = e^{(y_1+y_2-d)\ell} G(qe^\ell, \tau e^{y_E \ell}, h e^{y_H \ell}) . \tag{3.42}$$

As an example we consider the spin-spin-correlation ($y_1=y_2=y_h$)

(i) $\tau=h=0$

$$G_c(q) = e^{(2y_h-d)\ell}\, G_c(qe^{\ell}) \quad . \tag{3.43}$$

With $qe^{\ell}=1$ we obtain

$$G_c(q) = q^{d-2y_h}\, G_c(1) \tag{3.44}$$

and identify the exponent

$$d-2y_h = -2+\eta \tag{3.45}$$

(ii) $\tau \neq 0$, $h=0$

$$G(q,\tau) = e^{(2y_h-d)\ell}\, G(qe^{\ell}, \tau e^{y_E \ell}) \quad . \tag{3.46}$$

With $|\tau| e^{y_E \ell} = 1$ we find

$$G(q,\tau) = |\tau|^{-\frac{2y_h-d}{y_E}}\, G(q|\tau|^{-1/y_E}, \pm 1) = |\tau|^{-\gamma} g_0(q\xi) . \tag{3.47}$$

Apart from a constant factor ξ is called the correlation length and scales with an exponent ν . We obtain

$$\gamma = \frac{2y_h-d}{y_E} \quad , \quad \xi \propto |\tau|^{-\nu} \quad , \quad \nu = \frac{1}{y_E} \quad , \quad \gamma = \nu(2-\eta) \; . \tag{3.48}$$

4. NONLINEAR CONTRIBUTIONS

A. Scaling fields[9]

Apart from certain exceptions which will be discussed below the nonlinearities of the RG equation can be absorbed in scaling fields g_i which depend nonlinearly on the fields μ_j, so that g can be formally expanded in powers of μ and

$$F\{g_i\} = e^{-d\ell}\, F\{g_i e^{y_i \ell}\} \quad . \tag{4.1}$$

From equation (3.3) we find

$$\frac{d\mu_i}{d\ell} = y_i \mu_i + \frac{1}{2} \sum_{jk} a'_{ijk} \mu_j \mu_k \tag{4.2}$$

with

$$-2\int d^dq\,(c+2q^2)\,\frac{\delta O_j}{\delta S_q}\,\frac{\delta O_k}{\delta S_{-q}} = \sum a'_{ijk}\,O_i \quad . \tag{4.3}$$

To obtain equation (4.1) we require

$$\frac{dg_i}{d\ell} = y_i\,g_i \tag{4.4}$$

and expand

$$\mu_i = g_i + \frac{1}{2}\sum b_{ijk}\,g_j\,g_k + O(g^3) \tag{4.5}$$

which yields

$$\frac{d\mu_i}{d\ell} = y_i g_i + \frac{1}{2}\sum (y_i b_{ijk} + a'_{ijk}) g_j g_k = y_i g_i + \frac{1}{2}\sum (y_j + y_k) b_{ijk} g_j g_k + O(g^3) \tag{4.6}$$

$$(y_j + y_k - y_i)\,b_{ijk} = a'_{ijk} \tag{4.7}$$

which can be solved provided $y_i \neq y_j + y_k$. Similarly the terms of nth order in g can be calculated if y_i differs from any sum of n exponents y.

B. Logarithmic Corrections[9]

If $y_i = y_j + y_k$, then logarithmic factors arise. We give an example in which we neglect all terms in the equations (4.2) which do not contribute to the logarithm. Suppose

$$2y_E = y_0 = d \tag{4.8}$$

$$\frac{d\mu_0}{d\ell} = d\cdot\mu_0 + \frac{1}{2}\,a'_{0EE}\,\mu_E^2 \tag{4.9}$$

$$\frac{d\mu_E}{d\ell} = y_E\,\mu_E \tag{4.10}$$

then we obtain ($\tau = \mu_E$)

$$\mu_E(\ell) = \mu_E(0)\,e^{d/2\,\ell} \tag{4.11}$$

$$\frac{d\mu_0}{d\ell} = d\cdot\mu_0 + \frac{1}{2}\,a'_{0EE}\,\mu_E^2(0)\,e^{d\ell} \tag{4.12}$$

$$\mu_0(\ell) = \mu_0(0)\,e^{d\ell} + \frac{1}{2}\,\ell\,a'_{0EE}\,\mu_E^2(0)\,e^{d\ell} \tag{4.13}$$

$$F(0,\mu_E) = e^{-dl}\mu_0(l) + e^{-dl} F(0,\mu_E e^{d/2\, l})$$
$$= \frac{1}{2}\, l\, a'_{0EE}\mu_E^2 + \mu_E^2 F(0,\pm 1) \quad . \tag{4.14}$$

Again we choose $|\mu_E| e^{d/2\, l} = 1$ and obtain

$$F(0,\mu_E) = -\frac{a'_{0EE}}{d}\mu_E^2 \ln|\mu_E| + \mu_E^2 F(0,\pm 1) \quad . \tag{4.15}$$

C. Broken powers of Logarithms[31]

If $y_u=0$, then logarithms to some broken powers arise. Again we start from simplified equations to demonstrate the singularity

$$\frac{d\mu_u}{dl} = \frac{1}{2} a'_{uuu}\mu_u^2 \tag{4.16}$$

$$\frac{d\mu_i}{dl} = y_i\mu_i + a'_{iui}\mu_i\mu_u \tag{4.17}$$

and obtain

$$\mu_u = s\,(l+l_0)^{-1} \qquad s = -\frac{2}{a'_{uuu}} \tag{4.18}$$

$$\mu_i(l) = \left(\frac{l+l_0}{l_0}\right)^{p_i} e^{y_i l}\mu_i(0) \quad , \quad p_i = s\, a'_{iui} \quad . \tag{4.19}$$

Such singularities appear in four dimensions at a critical point and in three dimensions at a tricritical point[31-33].

D. Correlation Functions

The operators $O_i(q)$ become extremely small, if $q \gg q_0$ (q_0 momentum cut-off). Therefore the correlations G become extremely small as soon as $q \gg q_0$ and equation (3.42) will not apply for $qe^l \gg q_0$. The perturbations $\lambda_1 O_1(q) + \lambda_2 O_2(-q)$ will generate contributions $O_0, O_E, \ldots$ because of the nonlinear terms of the RG equation. Similarly homogeneous perturbations and nonhomogeneous perturbations generate contributions nonlinear in μ and γ. To discuss these effects we make the following simplifying assumption:

(i) We assume that the linear approximation is good for $q \lesssim q_0$.

(ii) In a narrow region around q_0 the nonlinear contributions dominate.

(iii) For $q \gtrsim q_0$ we neglect the inhomogeneous perturbations.

Then from

$$\mathcal{H}_0 = \mathcal{H}^* + \tau O_E + \lambda_1 O_h(q) + \lambda_2 O_h(-q) \qquad (4.20)$$

we obtain according to (i) with $e^{\ell} = q_0/q$

$$\mathcal{H}_\ell = \mathcal{H}^* + \tau\left(\frac{q_0}{q}\right)^{y_E} O_E + \lambda_1\left(\frac{q_0}{q}\right)^{y_h} O_h(q_0) + \lambda_2\left(\frac{q_0}{q}\right)^{y_h} O_h(-q_0) \qquad (4.21)$$

where actually $q\ e^{\ell}$ should be slightly smaller than q_o. According to our assumption (ii), we obtain with a slight change of ℓ (which we do not indicate explicitly in the next equation)

$$\tilde{\mathcal{H}}_\ell = \mathcal{H}^* + \tau\left(\frac{q_0}{q}\right)^{y_E} O_E + \lambda_1\lambda_2\left(\frac{q_0}{q}\right)^{2y_h}\left(A + B O_E + C\tau\left(\frac{q_0}{q}\right)^{y_E} + \cdots\right). \qquad (4.22)$$

We have already neglected the inhomogeneous perturbations in this equation according to (iii). Now we have a homogeneous interaction and we can apply the inverse RG transformation

$$\tilde{\mathcal{H}}_0 = \mathcal{H}^* + \tau O_E + \lambda_1\lambda_2\left(\frac{q_0}{q}\right)^{2y_h-d}\left(A + B\left(\frac{q_0}{q}\right)^{d-y_E} O_E + C\tau\left(\frac{q_0}{q}\right)^{y_E} + \cdots\right). \qquad (4.23)$$

Since the free energy is conserved under the total of these transformations we obtain

$$\langle O_h(q)\, O_h(-q)\rangle = \frac{\partial^2 F(\tilde{\mathcal{H}}_0)}{\partial\lambda_1\,\partial\lambda_2} = \left(\frac{q_0}{q}\right)^{2y_h-d}\left(A + B\left(\frac{q_0}{q}\right)^{d-y_E}\langle O_E\rangle + C\tau\left(\frac{q_0}{q}\right)^{y_E} + \cdots\right) \qquad (4.24)$$

a result suggested by Fisher and Langer[26] and which has also been derived by means of the Callan-Symanzik-equation[27]. We emphasize that the conditions (i) and (ii) are not necessary to derive eq. (4.23). It is only necessary[30] that the operators $O_i(q)$ can be neglected for $q \gg q_o$. Then, however, the derivation of (4.23) becomes more complicated. We note that for a linear RG equation (3.42) holds exactly which means that $O_i(q)$ does not become negligible for $q \gg q_o$. Therefore a linear RG does not eliminate the Fouriers components for large q. We see that the elimination of short wave length fluctuations and the linearity of a RG equation exclude each other.

5. FINAL REMARKS

We have outlined the basic ideas of the RG procedure as initiated by Wilson. These ideas can be applied to actual calculations.

We display in table 2 the critical exponents α and γ for three-dimensional systems as obtained by various methods. In section 2C we mentioned already a number of them. We can distinguish three types of calculations:

(i) Approximate calculations. Wilson's recurrence relation[13] can be used to calculate numerically the critical exponents. They are shown in the table (r.r. numerical).

(ii) The critical exponents can be expanded around dimensionality 4 (ε-expansion, $\varepsilon = 4-d$). Unfortunately, however, the series seem to be asymptotically. As a thumb-rule one finds that the exponents in order ε^2 yield a good approximation. In 4 dimensions one obtains molecular-field behaviour (with logarithmic corrections) which can be described by a free fixed-point. It is possible to expand around this fixed point since the coupling constant g for the four-spin interaction (which is marginal for $d = 4$) can be expanded in powers of ε .

(iii) The critical exponents can be expanded in powers of $1/n$. For $n = \infty$ one obtains the critical exponents for the spherical model[34]

$$\gamma = \frac{2}{d-2} , \quad \alpha = \frac{d-4}{d-2} , \quad 2 \leq d < 4 . \tag{5.1}$$

One can perform a systematic expansion[35] around this limit which yields for $d = 3$

$$\gamma = 2 - \frac{24}{\pi^2 n} + O\left(\frac{1}{n^2}\right), \quad \alpha = -1 + \frac{32}{\pi^2 n} + O\left(\frac{1}{n^2}\right). \tag{5.2}$$

These numbers are not yet good approximations although they tend into the correct direction. One has to wait for terms in order $1/n^2$.

Table 2. Critical exponents as obtained by various methods for d=3

	n=1	n=2	n=3	Ref.
α high temp.exp.	.13	.00	-.10	28, compare 1
α r. r. numerical	.17	.07	-.04	13,14
α in $O(\varepsilon)$	.17	.10	.05	15
α in $O(\varepsilon^2)$	.08	-.02	-.10	15
α in $O(\varepsilon^3)$	.20	+.08	.01	15
α experiment	.16	-.02	-.14	24. 29
γ high temp. exp.	1.25	1.32	1.38	28, compare 1
γ r. r. numerical	1.22	1.29	1.36	13, 14
γ in $O(\varepsilon)$	1.17	1.20	1.23	15
γ in $O(\varepsilon^2)$	1.24	1.30	1.35	15
γ in $O(\varepsilon^3)$	1.19	1.26	1.32	15

REFERENCES AND FOOTNOTES

1. In the beginning we follow widely the introduction of the review article by M.E. Fisher, Rept. Prog. Phys. 30, 615 (1967). Compare also the review article by L.P. Kadanoff et al, Rev. Mod. Phys. 39, 395 (1967)

2. Depending on the number of experimentally available variables one may have a critical point, a critical line (λ-line) or even a critical surface, etc.

   There exist other special points, for example the common end point of a triple point line and three critical lines, called a tricritical point. These points can be described in the framework of this theory, too. They differ from normal critical points by the number of conditions necessary to reach such a point.

3. P. Weiss, J. Physique 6, 661 (1907); Arch. Sc. phys. et nat. 31, 401 (1911). The first theory which described a phase transition (for a liquid-vapor system) is due to J.D. van der Waals, Doctoral dissertation, Leiden (1873); Die Kontinuität des gasförmigen und flüssigen Zustands I und II. Verlag J.A. Barth (Leipzig, 1899 und 1900).

4. We hope that the reader does not get confused since as usual β is used for the critical exponent of the order parameter and the inverse temperature.

5. B. Widom, J. Chem. Phys. 43, 3898 (1965).

6. A number of factors $k_B T$ is missing which however does not affect the critical exponents.

7. J.T. Ho and J.D. Litster, J. Appl. Phys. 40, 1270 (1969)

8. K.G. Wilson, Phys. Rev. B4, 3174 (1971)

9. F.J. Wegner, Phys. Rev. B5, 4529 (1972)

10. K.G. Wilson and J. Kogut, Cornell preprint 1972, to be published in Phys. Reports.

11. S. Ma, Rev. Mod. Phys. 45, 589 (1973).

12. F.J. Wegner, Critical Phenomena and the Renormalization Group. An Introduction (VIIIth Finnish Summer School in Theoretical Solid State Physics, August 1973, Siikajärvi, Finnland).

13. K.G. Wilson, Phys. Rev. B4, 3184 (1971).

14. M.K. Grover, L.P. Kadanoff and F.J. Wegner, Phys. Rev. B6, 311 (1972), M.K. Grover, Phys. Rev. B6, 3546 (1972), J. Swift and M.K. Grover, Phys. Rev. A9, 2579 (1974).

15. For example:
K.G. Wilson and M.E. Fisher, Phys. Rev. Lett. 28, 240 (1972)
M.E. Fisher and P. Pfeuty, Phys. Rev. B6, 1889 (1972)
F.J. Wegner, Phys. Rev. B6, 1891 (1972)
K.G. Wilson, Phys. Rev. Lett. 28, 548 (1972)
E. Brezin, J.C. Le Guillou and J. Zinn-Justin, Phys. Rev. B8, 5330 (1973)
A. Houghton and F.J. Wegner, Phys. Rev. A10, 435 (1974).

16. F.J. Wegner, J. Physics C7, 2098 (1974)

17. F.J. Wegner and A. Houghton, Phys. Rev. A8, 401 (1973).

18. A. Aharony and M.E. Fisher, Phys. Rev. B8, 3323 (1973), A. Aharony, Phys. Rev. B8, 3342, 3349, 3358, 3363 (1973), A.D. Bruce and A. Aharony, preprint.

19. Th. Niemeijer and J.M.J. van Leeuwen, Phys. Rev. Lett. 31, 1411 (1973), Physica, to appear.

20. M. Nauenberg and B. Nienhuis, preprint.

21. J. Zinn-Justin, Lectures given at the 1973 Cargese Summer School.

22. J. Hubbard and P. Schofield, Phys. Lett. 40A, 245 (1972).

23. L.P. Kadanoff, Physics 2, 263 (1966).

24. G. Ahlers, Phys. Rev. A8, 530 (1973).

25. M.E. Fisher, Nobel Symposium 24, 16 (1973)

26. M.E. Fisher and J.S. Langer, Phys. Rev. Lett. 20, 665 (1968), compare M.E. Fisher, Phil. Mag. 7, 1731 (1962).

27. E. Brezin, J.C. Le Guillou, and J. Zinn-Justin, Phys. Rev. Lett. 32, 473 (1974).

28. D. Jasnow and M. Wortis, Phys. Rev. 176, 739 (1968)

29. G. Ahlers, A. Kornblit, and M.B. Salamon, Phys. Rev. B9, 3932 (1974).

30. F.J. Wegner, to be published.

31. F.J. Wegner and E.K. Riedel, Phys. Rev. B7, 248 (1973).

32. E.K. Riedel and F.J. Wegner, Phys. Rev. Lett. 29, 349 (1972).

33. A.I. Larkin and D.E. Khmel'nitskii, Zh. Eksperim. i. Teor. Fiz. 56, 2087 (1969) [Sov. Phys. JETP 29, 1123 (1969)] .

34. H.E. Stanley, Phys. Rev. 176, 718 (1968).

35. R.A. Ferrell and D.J. Scalapino, Phys. Rev. Lett. 29, 413 (1972)
Phys. Lett. 41A, 371 (1972)
R. Abe, Prog. Theor. Phys. 48, 1414 (1972); 49, 113 (1973)
R. Abe and S. Hikami, Prog. Theor. Phys. 49, 442 (1973)
S. Hikami, Prog. Theor. Phys. 49, 1096 (1973)
S. Ma, Phys. Rev. Lett. 42A, 5 (1972), Phys. Rev. A7, 2172 (1973
M.E. Fisher, S. Ma and B.G. Nickel, Phys. Rev. Lett. 29, 917 (1972)
M. Suzuki, Phys. Lett. 42A, 5 (1972); Prog. Theor. Physik 49, 424 (1973).

RENORMALIZATION GROUP SOLUTION OF ISING SPIN MODELS

Michael Nauenberg

University of California, Santa Cruz, California

An excellent series of lectures on the renormalization group theory for critical phenomena have been given at this school by Professor Wegner and I will assume in my discussion that the basic ideas of this theory are known to you. I would like to discuss some new developments based on recent work done in collaboration with a graduate student, Bernard Nienhuis, at the University of Utrecht. We have extended the renormalization group approach to evaluate the complete free energy for general Ising spin models, which give a concrete realization of the scaling operators introduced by Professor Wegner. In particular we can evaluate not only the critical exponents and critical temperature but also the coefficients of the singular terms which had not been determined previously except for the special case of a logarithmic singularity. Two basic assumptions of renormalization group theory, the existence of a fixed point Hamiltonian and the analyticity of the renormalization group transformations have been verified for planar Ising models in a cell cluster approximation of Niemeijer and van Leeuwen, but the third assumption introduced by Professor Wegner, the continuity of the renormalization transformations as functions of the dimension of the Kadanoff cells, cannot be justified in this model.

To start, I will discuss a new method of solution of the basic equations of renormalization group theory and later on I will illustrate the derivation of these equations in the simplest case, the original one dimensional Ising model. This model consists of spins with only components $s = \pm 1$ arranged on an infinite one dimensional lattice with nearest neighbor interactions. It was invented by Lenz in 1920 to explain ferromagnetism and later solved by his student Ising who found that in fact it did not give rise to such a phase transition. Not until 1944 did Onsager solve the corresponding model in two dimensions where a phase transition does occur, but up to now no one has been able to extend his methods to three dimensions or even to include an external magnetic field in two dimensions. At the end of my lecture I will show you the results we have obtained for a square Ising lattice in a four cell cluster approximation and compare them with Onsager's and with approximate numerical results. Our approximate renormalization group method can be extended straightforwardly to include a magnetic field[7] and to three dimensions.

After summing over specific degrees of freedom of the Kadanoff cells,[1] we find a scaling equation for the free energy f(K) as a function of the spin interaction coupling constants K_α of the form[2]

$$f(K') = L\,\{f(K) - g(K)\} \qquad 1.$$

where L is the number of spins in a Kadanoff cell, Lg(K) is the self energy of this cell, and K' are the effective spin cell interaction coupling constants determined by the renormalization transformations

$$K'_\alpha = F_\alpha(K) \qquad 2.$$

The coupling constants K_α and K'_α correspond to the fields associated with the scaling operators introduced by Professor Wegner except for one important difference: we do not include a field associated with the unit operator or what Niemeijer and van Leeuwen have referred to as the "empty set".[2] This accounts for the explicit appearance of the self energy function g(K) in the scaling equation. This field has been previously identified incorrectly as the regular part of the free energy and consequently discarded in considering the singularities associated with the phase transition, but as we shall see it plays an essential role in our approach.

The mathematical problem we face now is to solve the scaling equation, eq. 1, for f(K) given the self energy g(K) and the corresponding renormalization transformation eq. 2, subject to an appropriate physical boundary condition, e.g. in the case that all spin interactions vanish, i.e. K = 0, f(0) = ln 2. It is then straightforward to prove uniqueness of the solution of the scaling equation. I will show you a practical method of solution via an infinite series expansion based on the semi-group property of the renormalization transformations. First re-write the scaling equation in the form

$$f(K) = \frac{1}{L}\, f(K') + g(K) \qquad 3.$$

and apply the renormalization transformation on the argument of both sides of eq. 3 to obtain

$$f(K') = \frac{1}{L}\, f(K'') + g(K') \qquad 4.$$

where $K''_\alpha = F_\alpha(K')$. Substituting eq. 4 in eq. 3 we have

$$f(K) = \frac{f(K'')}{L^2} + g(K) + \frac{1}{L} g(K') \qquad 5.$$

and repeating this procedure n times we obtain

$$f(K) = \frac{f(K^{(n)})}{L^n} + \sum_{m=0}^{m=n-1} \frac{g(K^{(m)})}{L^m} \qquad 6.$$

where $K_\alpha^{(n)}$ is given by the recurrence relation $K_\alpha^{(n)} = F_\alpha(K^{(n-1)})$ obtained from eq. 2.

Although the series for f(K) eq. 6 is valid for every integer n it is still not very useful because it depends on the unknown function $f(K^{(n)})$. However if we take the limit $n \to \infty$ we obtain

$$f(K) = h(K) + \sum_{n=0}^{\infty} \frac{g(K^{(n)})}{L^n} \qquad 7.$$

where the function $h(K) = \lim_{n\to\infty} \frac{f(K^{(n)})}{L^{(n)}}$ can then be calculated. It satisfies the homogeneous scaling equation

$$h(K') = L\, h(K) \qquad 8.$$

and therefore it is singular at the critical point, unless, of course, the coefficient of all singular terms vanishes. In fact it turns out that h(K) = 0 although the derivative with respect to the magnetic field H, $\frac{\partial h(K)}{\partial H} = \lim_{n\to\infty} \frac{1}{L^n} \frac{\partial f(K^n)}{\partial H}$ is finite below the critical temperature and gives the spontaneous magnetization.

We will now show that the infinite series in eq. 7 must be singular on the critical surface defined by an unstable fixed point K^*. Recall that this critical surface is defined by the domain of points K which map into K^* in the limit of an infinite number of consecutive renormalization mappings, i.e. $\lim_{n\to\infty} K^{(n)} = K^*$. Now suppose that a given point K is arbitrarily close to one side of this critical surface. Then repeated renormalization transformations map K along points which remain close to the critical surface until for some integer n_o, $K^{(n_o)}$ approaches closest to the fixed point. Then further transformations must map $K^{(n)}$ away from the fixed point without crossing the critical surface. In fact, it turns out that two points K_1 and K_2 arbitrarily close but on opposite sides of the critical surface map in the limit $n\to\infty$ into two different fixed points, one at $K^* = 0$ and $K^* = \infty$.

It is clear therefore that the infinite series for $f(K_1)$ and $f(K_2)$ differ for all terms with integer $n > n_o$ and hence $f(K)$ is singular on the critical surface.

In order to obtain the singular part of $f(K)$ we introduce a new set of variables ξ_i by a non-linear transformation[3] [4]

$$K_\alpha = G_\alpha(\xi) \qquad 9.$$

which is defined by the condition that the renormalization transformation, eq. 2, in the ξ-space is

$$\xi_i' = \lambda_i \xi_i \qquad 10.$$

The constants λ_i are the eigenvalues of the matrix $\partial F_\alpha / \partial K_\beta$ at the fixed point K_α^* corresponding to $\xi_i = 0$, and define relevant, marginal and irrelevant variables ξ_i according to whether λ_i is greater, equal or less than one respectively. The critical surface is determined by the condition that the relevant variables vanish. In terms of the ξ_i variables, the expansion for the free energy, eq. 7, takes the form

$$f(\xi_1,\xi_2,\ldots) = h(\xi_1,\xi_2,\ldots) + \sum_{n=0}^{\infty} \frac{1}{L^n} g(\lambda_1^n\xi_1,\lambda_2^n\xi_2,\ldots) \qquad 11.$$

Let us now assume for simplicity that $\lambda_1 > 1$ and $\lambda_j < 1$ for all $j \neq 1$. Then $\lambda_1^m > L$ for some smallest integer m and the m-th partial derivation $f^{(m)}(\xi_1,\xi_2\ldots) \equiv \partial^m f(\xi_1\ldots)/\partial\xi_1^n$ becomes

$$f^{(m)}(\xi_1, \xi_2\ldots) = h^{(m)}(\xi_1,\xi_2\ldots) + \sum_{n=0}^{\infty} \left(\frac{\lambda_1^m}{L}\right)^n g^{(m)}(\lambda_1^n\xi_1,\lambda_2^n\xi_2\ldots) \qquad 12.$$

Hence in the limit $\xi_1 \to 0$ the infinite series in eq. 12 diverges. To obtain an explicit representation for the singular part f_s of $f(\xi)$, neglecting the irrelevant variables ξ_j, $j \neq 1$, we must first find the regular solution $f_r^{(m)}$ of the m-th derivative with respect to ξ_1 of the scaling equation, eq. 1. Expanding each side of eq. 1 in a power series in ξ_1 and re-summing we obtain

$$f_r^{(m)}(\xi_1) = \sum_{n=1}^{\infty} \left(\frac{\lambda_1^m}{L}\right)^{-n} g^{(m)}(\lambda_1^{-n}\xi_1) \qquad 13.$$

as can also be verified by direct substitution in eq. 1. Subtracting eq. 13 from eq. 12, we then obtain the singular part

$$f_s^{(m)}(\xi_1) = h^{(m)}(\xi_1) + \sum_{n=-\infty}^{\infty} \left(\frac{\lambda_1^m}{L}\right)^n g^{(m)}(\lambda_1^n \xi_1) \qquad 14.$$

which is a solution of the homogeneous scaling equation

$$f_s^{(m)}(\lambda_1 \xi_1) = \left(\frac{L}{\lambda_1^m}\right) f_s^{(m)}(\xi_1) \qquad 15.$$

Setting $f_s^{(m)}(\xi_1) = C_\pm\ (\xi_1)\ |\xi_1|^{-\alpha}$ for $\xi_1 \gtrless 1$, where $\alpha =$ $m - \ln L/\ln\lambda_1$, it follows from eq. 15 that

$$C_\pm\ (\lambda_1 \xi_1) = C_\pm\ (\xi_1) \qquad 16.$$

Expanding $C_\pm(\xi_1)$ in a Fourier series in $\ln|\xi_1|/\ln \lambda_1$

$$C_\pm\ (\xi_1) = \sum_{n=-\infty}^{\infty} C_n^{(\pm)}\ e^{2\ i\pi n\ \ln|\xi_1|/\ln\lambda_1} \qquad 17.$$

we obtain the Fourier coefficients $C_n^{(\pm)}$ from eq. 14,

$$C_n^{(\pm)} = \frac{1}{\ln\lambda_1} \int_0^\infty d\xi\ \ \xi^{\alpha-1}\ g^{(m)}(\pm\xi)\ e^{-2\ i\pi n \ln\xi\ /\ln\lambda_1}$$

$$+ \frac{1}{\ln\lambda_1} \int_1^{\lambda_1} d\xi\ \xi^{\alpha-1}\ h^{(m)}(\pm\xi)\ e^{-2\ i\pi n \ln\xi/\ln\lambda_1} \qquad 18.$$

In particular for n=0, we obtain from eq. 18 the coefficient of the well known power singularities discussed in previous lectures. However, we find also apparently additional oscillating terms in $\ln\xi_1$. Up to this point we have not made any assuptions concerning the dependence of the renormalization transformations and the free energy on the number of spins L in a Kadanoff cell. Recall that Professor Wegner assumed in his derivation of the power law singularity that these transformations were continuous functions in L, but this cannot be applied to the Ising model where L could only have integer values. However, since the period $\ln\lambda_1$ of the oscillatory terms in the variable $\ln\xi$, does depend on L while the exact free energy is independent of L, we should expect $C_n = 0$ for $n \neq 0$.

Now I would like to illustrate the derivation of the basic equations of renormalization group theory in a simple example, namely the one dimensional Ising model.[5]

We start with the familiar hamiltonian $H_N(K)$ for the one dimensional Ising spin model for N spins, $S_i = \pm 1$, $i = 1,2...N$, with nearest neighbour interaction coupling constant K,

$$H_N(K) = K \sum_{i=1}^{N} S_i S_{i+1} \qquad 19.$$

where $S_{N+1} = S_1$. Following Kramers and Wannier[6], we introduce the 2 x 2 transfer matrix $\mathbb{P}_{S_1 S_2} = e^{KS_1S_2}$ which enables us to write the Boltzmann probability function $e^{-H_N(K)}$,(kT = 1), in the form

$$e^{-H_N(K)} = \mathbb{P}_{S_1S_2} \mathbb{P}_{S_2S_3} \cdots \mathbb{P}_{S_NS_1} \qquad 20.$$

Instead of computing the usual partition sum, $\sum_{\{S\}} e^{-H_N(K)}$ = trace $\mathbb{P}^N$, we consider here only the partial sum of $e^{H_N(K)}$ over all possible values of the even spins, $S_i = \pm 1$, i=2,4... and obtain for N even

$$\sum_{\{S_2S_4...S_N\}} e^{-H_N(K)} = \mathbb{P}^2_{S_1S_3} \mathbb{P}^2_{S_3S_5} \cdots \mathbb{P}^2_{S_{N-1}S_1} \qquad 21.$$

The idea behind this partial summation is to find a renormalization transformation $K \to K'$ such that

$$\mathbb{P}^2(K) = e^{2g(K)}\mathbb{P}(K') \qquad 22.$$

where g(K) is a scalar of function K. Then K' can be interpreted as an effective Ising coupling constant for the remaining odd spins S_i, i=1,3,5... N-1 and eq. 21 takes the form

$$\sum_{\{S_2S_4...S_N\}} e^{-H_N(K)} = e^{-H_{\frac{N}{2}}(K) + N g(K)} \qquad 23.$$

This is the basic equation of the renormalization group approach. The matrix condition, eq. 22, is readily satisfied by

$$K' = \frac{1}{2} \ln \cosh 2K \qquad 24.$$

and

$$g(K) = \frac{1}{2} K' + \frac{1}{2} \ln 2 \qquad 25.$$

The non-linear renormalization transformation, eq. 24, has fixed points at $K^* = 0$ and $K^* = \infty$ with associated eigenvalues $\lambda = 0$ and $\lambda = 1$ respectively, where $\lambda = dK'/dK$ evaluated at $K = K^*$. Since a necessary condition for a critical transition is the existence of an eigenvalue $\lambda > 1$, this establishes the well known result that there is no phase transition for the one dimensional Ising model. After applying the renormalization n times, the mapping $K \to K^{(n)}$ can be obtained from the recurrence relation

$$K^{(n)} = \frac{1}{2} \ln \{\cosh 2K^{(n-1)}\} \qquad 26.$$

where $K^{(0)} = K$. It can be readily shown that $\lim_{n\to\infty} K^{(n)} = 0$, i.e. every finite point K is mapped towards the fixed point at the origin $K^* = 0$. In order to solve eq. 26 we introduce a new variable ζ related to K by a nonlinear transformation in such a way that the renormalization transformation in the ζ variable becomes simpler. For $\lambda \neq 0,1$ this transformation is defined by the condition [3,4] $\zeta' = \lambda\zeta$, but this is not possible in the present case. Instead we require

$$\zeta' = \zeta^2 \qquad 27.$$

and find the solution

$$\zeta = \tanh K \qquad 28.$$

and

$$K^{(n)} = \frac{1}{2} \ln \left(\frac{1 + \zeta^{2^n}}{1 - \zeta^{2^n}}\right), \quad -1 < \zeta < 1 \qquad 29.$$

Introducing the free energy per spin for N spins

$$f_N(K) = \frac{1}{N} \ln \sum_{\{S\}} e^{-H_N(K)} \qquad 30.$$

we obtain from eq.23 the functional relation

$$f_{\frac{N}{2}}(K') = 2 \{f_N(K) - g(K)\} \qquad 31.$$

In the thermodynamic limit, eq. 31 then leads to the scaling equation for $f(K) = \lim_{N\to\infty} f_N(K)$, eq. 1 with L = 2

$$f(K') = 2 \{f(K) - g(K)\} \qquad 32.$$

To obtain a unique solution of eq. 32, we must impose a boundary condition on f(K), e.g. for K = 0, absence of spin interactions, $f(0) = \ln 2$.

To prove uniqueness suppose there are two solutions $f_1(K)$ and $f_2(K)$ of eq. 32 which satisfy this boundary condition. Then the difference $f_-(K) = f_1(K) - f_2(K)$ satisfies the homogeneous scaling equation

$$f_-(K') = 2\, f_-(K) \qquad 33.$$

and applying the renormalization mapping n-times leads to the relation

$$f_-(K) = \frac{1}{2^n}\, f_-(K^{(n)}) \qquad 34.$$

Since $\lim_{n\to\infty} K^{(n)} = 0$, and $f_-(0) = 0$, eq. 34 implies $f_-(K) = 0$, Q.E.D. Actually, this proof shows that we need to demand only the weaker boundary condition that $f(K)$ be finite at $K = 0$, because the solution of eq. 32 determines the value of $f(0)$.

We obtain the solution of the scaling equation by substituting eq. 25 for $g(K)$ in eq. 7 noting that $h(K) = 0$ because $f(0) = \ln 2$,

$$f(K) = \ln 2 + \sum_{n=1}^{\infty} \frac{K^{(n)}}{2^n} \qquad 35.$$

This series converges very rapidly and can be readily used to evaluate $f(K)$. For example, for $K = 1$, the sum of the first four terms of this series gives an accuracy of 10^{-4}. We can also sum this series by substituting eq. 29 in eq. 35 to abtain

$$f(K) = \ln 2 + \ln \prod_{n=1}^{\infty} \left(\frac{1+\zeta^{2^n}}{1-\zeta^{2^n}}\right)^{\frac{1}{2^{n+1}}} \qquad 36.$$

Applying the easily proven identity

$$\frac{1}{1-\chi} = \prod_{n=o}^{\infty} \left(\frac{1+\chi^{2^n}}{1-\chi^{2^n}}\right)^{\frac{1}{2^{n+1}}}, \quad -1 < \chi < 1 \qquad 37.$$

we find

$$f(K) = \ln\left(2/\sqrt{1-\zeta^2}\right) \qquad 38.$$

and from eq. 27, we obtain

$$f(K) = \ln(2 \cosh K) \qquad 39.$$

which is the well known solution of the one dimensional Ising model. We can verify that this solution satisfies the scaling equation by substituting eq. 39 together with eq. 24 into eq. 32. A second solution of the scaling equation is $\tilde{f}(K) = \ln(2\sinh K)$ for which $h(K) = \ln$

(tanh K). This solution is also of physical interest because the spin correlation function $C_{|i-j|}(K) = \langle s_i s_j \rangle$ is given by

$$C_{|i-j|}(K) = e^{-|i-j|/\ell(K)} \qquad 40.$$

where the correlation length $\ell(K) = |f(K) - \tilde{f}(K)|^{-1}$. Hence $\ell(K)$ satisfies the excepted homogeneous scaling relation

$$\ell(K') = \frac{1}{2}\,\ell(K) \qquad 41.$$

These results can be readily extended to include a magnetic field, and to higher spins, eq. $s = \pm 1$, and 0.

For two dimensional models the renormalization transformations are much more complicated and an exact analytic treatment has not been given. However, an excellent approximation obtained by keeping only a finite number of spins has been developed by Niemeijer and van Leeuwen[2] for triangular lattices and by us for square lattices.[4,6] Time does not permit me to discuss these interesting developments in this lecture and I refer you for details to our papers.[4,5,6,7] The results of a numerical calculation for 16 spins are shown in Fig. 1 which shows the free energy, energy and specific heat compared with Onsager's exact result, and Fig. 2 which shows the critical surface in the subspace of nearest neighbor coupling constant K_1, next to nearest neighbor constant K_2, and four spin interactions K_4. Note in particular the ridge in the upper half of the critical surface; it corresponds precisely to Baxter's critical curve for the solution of the eight vertex model. The intersection of the critical surface with $K_3 = 0$ plane agrees with the approximate calculation by Dalton and Wood[8] using high temperature expansions, but their method fails to converge for $K_1 > 0$ and $K_2 < 0$, while the renormalization group approach works also quite well for this case.

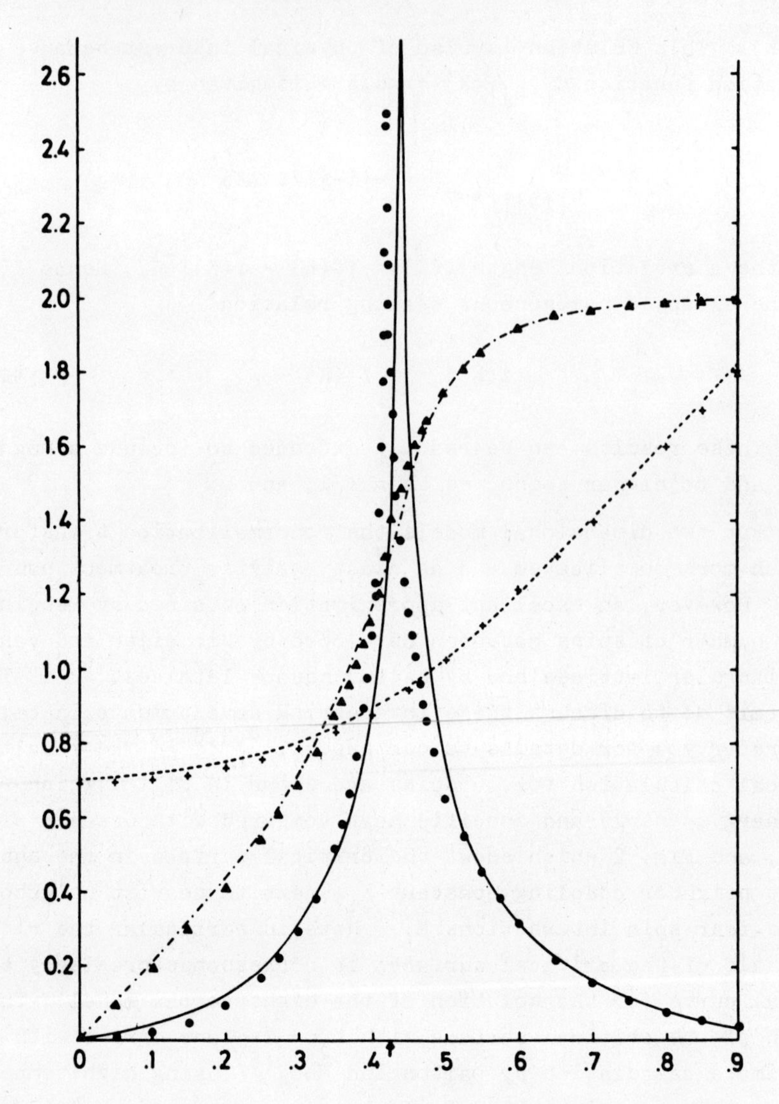

Fig. 1

dashed curve ----		Onsager's free energy
crosses +	$f(K_1)$	free energy, eq. 7
dashed-dot curve -·-·-		Onsager's energy
triangles ▲	$\frac{\partial f(K_1)}{\partial K_1}$	energy from first derivative eq.7
solid curve ——		Onsager's specific heat
dots ●	$K_1^2 \frac{\partial^2 f(K_1)}{\partial K_1^2}$	specific heat from second derivative eq. 7

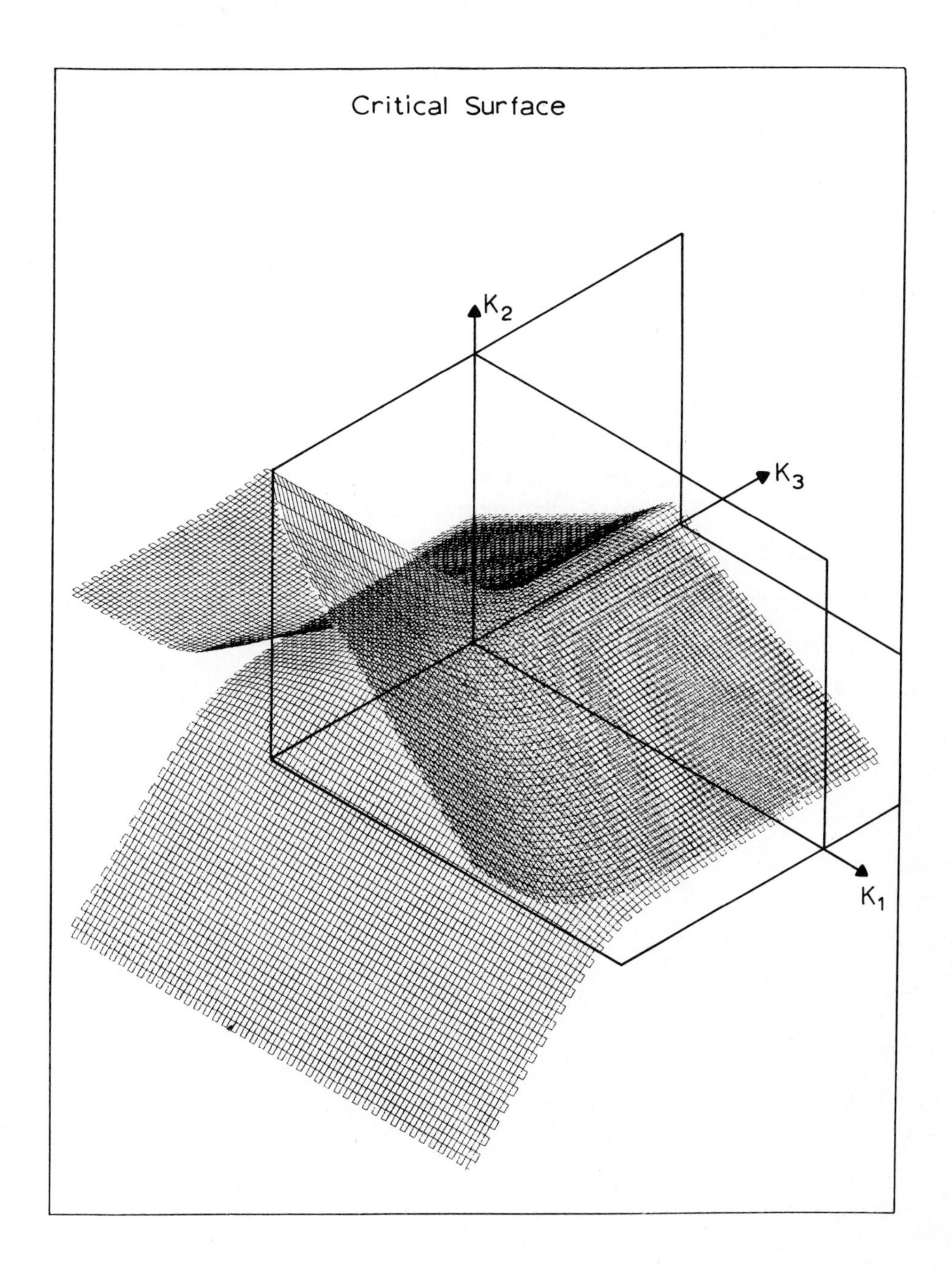

Fig. 2

Critical surface in the range $-2 \leq K_1, K_3 \leq +2$ seen along the direction $K_1 = 1$, $K_2 = 1$ and $K_3 = -1$.

REFERENCES

1. L.P. Kadanoff, Physics 2, 263 (1966).
2. Th. Niemeijer and J.M.J. van Leeuwen, Phys. Rev. Letters 31, 1412 (1973) and Physics 71, 17 (1974).
3. F.J. Wegner, Phys. Rev. B5, 4529 (1972).
4. M. Nauenberg and B. Nienhuis, Phys. Rev. Letters, Dec. 1974. Non linear transformations of this type were introduced by Poincaré and play an important role in stability studies in mechanics. See for example C.L. Siegel and C.K. Moser, Lectures on Celestial Mechanics, Springer Verlag (1971).
5. M. Nauenberg, J. Math. Phys. (to be published).
6. M. Nauenberg and B. Nienhuis, Phys. Rev. Letters 33, 944 (1974).
7. M. Nauenberg and B. Nienhuis (to be published).
8. N.W. Dalton and D.W. Wood, J. Math. Phys. 10, 1271 (1969).

PARTONS

COVARIANT PARTON MODEL

J. C. Polkinghorne, University of Cambridge, England

I. INTRODUCTION

Experiments at NAL and the ISR have revealed a new regime when particles with transverse momentum greater than about 2 - 3 GeV are observed. This regime is characterised by:

(i) A power law rather than exponential decrease with p_T;
(ii) A marked energy dependence at fixed p_T;
(iii) Particle ratios are different; in particular, kaons and protons are relatively copiously produced.

In these lectures, I shall seek to show that a natural explanation of these phenomena is obtained if one supposes them to be another manifestation of the granular nature of matter; that is, of partons. I shall assume that the partons are quarks, though I must confess that I do not know how to solve the crucial problem of the observability of quarks in the final state, which such an assumption inevitably poses. But first let us see how parton models work by applying them in the context where they were first successful: deep inelastic electroproduction.

II. DEEP INELASTIC SCATTERING[1]

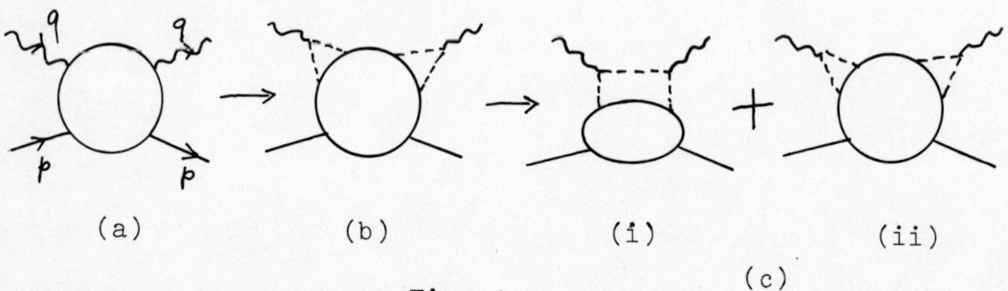

Fig. 1

The current-hadron process of Fig.1a is pictured as being due to the current coupling in a point-like fashion to the constituent partons denoted by broken lines in Fig.1b. This term decomposes into the sum of a "handbag" diagram, Fig. 1c(i), and the "cat's ears", Fig. 1c(ii). We wish to calculate the structure functions by taking the imaginary part of the process Fig. 1 in the forward direction. We are interested in the Bjorken limit

$$\nu = p \cdot q \to \infty \quad , \quad |q^2| \to \infty \quad ,$$

$$\omega = \frac{2p \cdot q}{-q^2} \quad \text{fixed} . \tag{1}$$

I shall now show that in general, Fig. 1c(i) dominates in that limit. To calculate its contribution, we use Sudakov parameters, writing momenta in the form

$$k = xp + yq + \underset{\sim}{\kappa} \quad ,$$
$$\underset{\sim}{\kappa} \cdot p = \underset{\sim}{\kappa} \cdot q = 0 . \tag{2}$$

This is technically convenient because it acknowledges the different roles of longitudinal momenta, parallel to p and q, and transverse momenta, like κ. The lower blob of (i) is forward parton-hadron amplitude with energy

$$s' = (p-k)^2 = (1-x)^2 M^2 + 2\nu(x-1-\frac{y}{\omega})y - \underset{\sim}{\kappa}^2 \tag{3}$$

and the parton has virtual mass

$$\mu^2 = k^2 = x^2M^2 + 2\nu(x - \frac{y}{\omega})\, y - \kappa^2 . \tag{4}$$

The parton line connecting the two current vertices in (i) carries momentum

$$\sigma^2 = (k+q)^2 = x^2M^2 + 2\nu\, (x - \frac{1+y}{\omega})(y+1) - \underset{\sim}{\kappa}^2 \tag{5}$$

The Jacobian of the transformation (2) is such that

$$\int d^4k \to \nu \int dx\, dy\, d^2 \underset{\sim}{\kappa} \quad . \tag{6}$$

The basic idea of the covariant parton model is that parton-hadron amplitudes decrease off-shell sufficiently rapidly with virtual parton mass μ^2 that the region of integration in which μ^2 is finite dominates. Thus we must transform to variables which make (4) finite. This achieved by writing

$$y = \bar{y}/2\nu \quad , \tag{7}$$

so that

$$\mu^2 = x^2 M^2 + \bar{y}\, x - \underset{\sim}{\kappa}^2 \tag{8}$$

$$s' = (1-x)^2 M^2 + \bar{y}(x-1) - \underset{\sim}{\kappa}^2 \tag{9}$$

$$\sigma^2 = x^2 M^2 + 2\nu(x - \frac{1}{\omega}) - \underset{\sim}{\kappa}^2 \tag{10}$$

and

$$\int d^4k \rightarrow \int dx\, d\bar{y}\, d^2 \underset{\sim}{\kappa} \quad . \tag{11}$$

The parton propagator at the top of (i) behaves like

$$\frac{1}{\sigma^2} \sim \frac{1}{2\nu} \cdot \frac{1}{x-\omega^{-1}} \quad , \qquad \nu \to \infty \tag{12}$$

(or the equivalent for spin $\frac{1}{2}$ partons[1]). On taking the imaginary part (12) will become

$$\frac{1}{2\nu}\, \delta(x-\omega^{-1}) \quad , \tag{13}$$

identifying ω^{-1} with the fraction of momentum p carried by the struck parton.

We must now consider the $\bar{y}$ integration. It avoids by $+i\varepsilon$ presriptions the cuts in μ^2, s', and the left hand cut variable of the parton-hadron amplitude

$$u' = (1+x)^2 M^2 + \bar{y}(x+1) - \underset{\sim}{\kappa}^2 \tag{14}$$

If all these cuts lie on the same side of the $\bar{y}$ contour, the latter can be completed in the opposite half plane to give zero, so that in fact a non-vanishing term arises only if x, (x-1), (x+1) (the coefficients of $\bar{y}$ in (8), (9) and (14)) are not of the same sign. This requirement imposes the (physically clear) constraint

$$\omega > 1 \ . \tag{15}$$

The cuts then lie with respect to the $\bar{y}$ integration as in Fig. 2 and the contour can be distorted so as to pick up the s' discontinuity only.

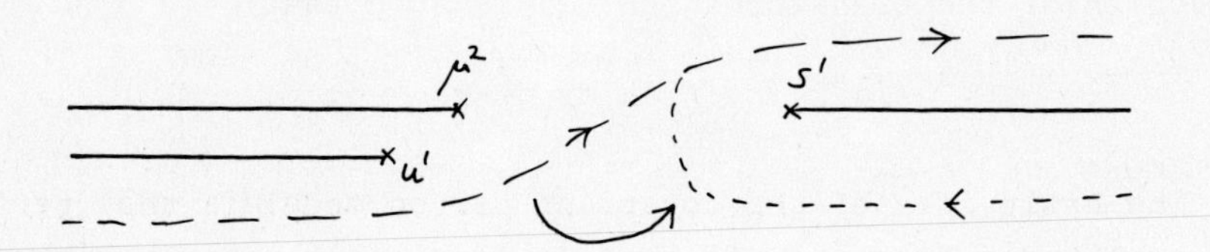

Fig. 2

Collecting all factors together one gets a non-vanishing contribution to νW_2 in the Bjorken limit (1), of the form

$$\nu W_2 = \frac{2}{(2\pi)^3} \frac{1}{\omega(\omega-1)} \int ds' d^2\underset{\sim}{K} \, \mathrm{Disc}_{s'} \, T(s',\mu^2) \, , \tag{16}$$

$$\mu^2 = \frac{s'}{1-\omega} + \frac{M^2}{\omega} - \underset{\sim}{K}^2 \frac{\omega}{\omega-1} \, . \tag{17}$$

with T the parton-hadron amplitude. The variable of integration has been changed from $\bar{y}$ to s'.

Many interesting consequences[1] follow from (16) and (17) which we shall not have the opportunity to discuss now. The calculation has served to illustrate Sudakov techniques and their utility. They may also be used to show that in general, Fig.1c(ii) is not significant in the scaling limit. Each loop of (ii) contributes a factor ν^{-1} so that to contribute to νW_2 a compensating factor of ν must be obtained from the connected amplitude C. One might think this possible because of the mechanism of Fig.3, where the zig-zag line is a spin 1 object, like the Pomeron. However, the relevant integral can be shown to be zero by completing contours.[1] This only fails if

(a) there is a point-like coupling (vector gluon field theory[2]);

(b) β is an exponential coupling with an essential singularity at infinity in the complex plane (as in the massive quark model of Preparata[3]).

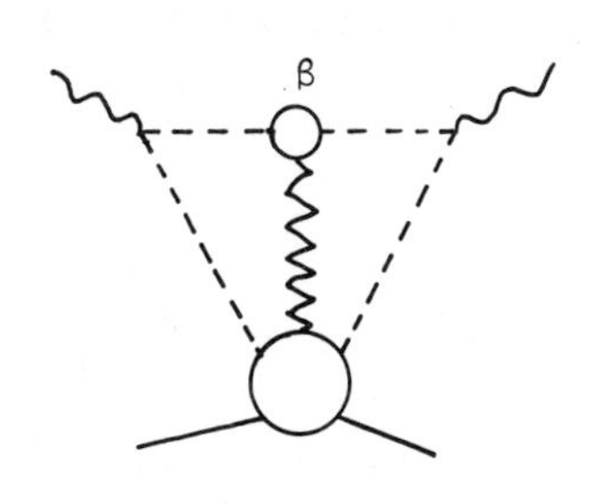

Fig. 3

Except for these two extreme possibilities, Fig. 1c(ii) does not contribute in the Bjorken limit.

III. QUARK FUSION[4]

Let us now try to apply these ideas to large p_T processes. Because we are dealing with hadronic processes, partons are no longer expected to have a point-like coupling. Their essential character is simply that of propagating fields which will decrease less slowly off mass shell than do fields associated with composite particles.

It will be easiest to consider first inclusive processes. They are less model dependent than exclusive ones, though we shall find (in contrast to deep inelastic current scattering where Fig. 1c(i) is the dominant diagram) that there are a number of possible mechanisms whose relative roles can only be evaluated by appeal to experiment. A very simple mechanism for producing a large p_T final state hadron is that shown in Fig. 4, in which partons from each hadron fuse to form the detected final state hadron. If one of the partons has large p_T, so will the detected hadron. The cross-section will be small because there is small probability of finding a large p_T parton within a hadron. If the partons are quarks or antiquarks, only mesons can be formed this way.

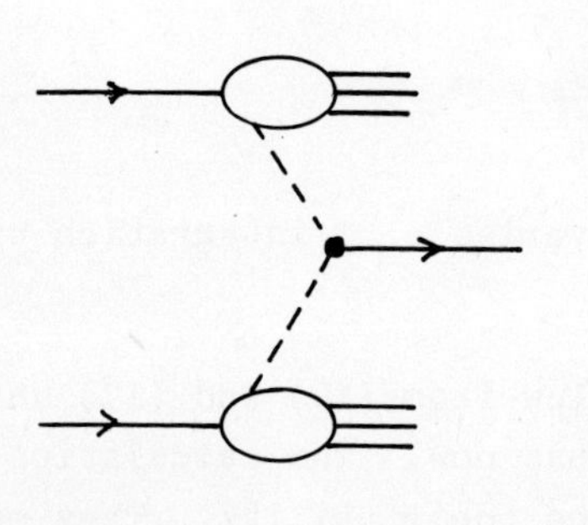

Fig. 4

If one tries to calculate the cross-section for the process Fig. 4, it is best to apply Mueller-Regge ideas to the diagram of Fig. 5, taking the discontinuity indicated to calculate the cross-section. I shall now explain how to modify Sudakov type calculations in order to do so.

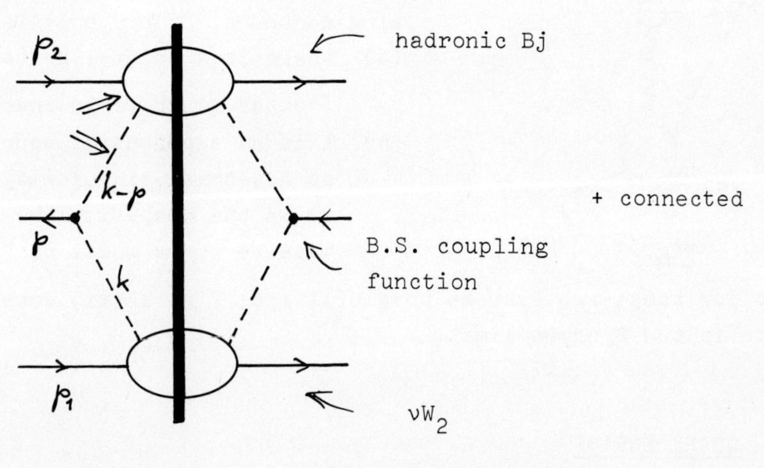

Fig. 5

The process is

$$p_1 + p_2 \rightarrow p + X \quad , \tag{18}$$

with

$$\nu = p_1 \cdot p_2 \rightarrow \infty \quad ,$$

$$p \cdot p_1 = x_2 \nu \,, \qquad p \cdot p_2 = x_1 \nu \,, \qquad x_1, x_2 \text{ fixed,} \tag{19}$$

so that the large transverse momentum of the detected hadron is

$$p_T{}^2 = 2x_1 x_2 \nu \,, \tag{20}$$

its centre of mass angle

$$\cot^2 \tfrac{1}{2}\theta = x_1/x_2 \tag{21}$$

Because there are now three important momenta in the problem, in place of (2) we write

$$k = up_1 + vp_2 + wp + \lambda \,, \tag{22}$$

where λ is a one-dimensional space-like vector satisfying

$$\lambda \cdot p_1 = \lambda \cdot p_2 = \lambda \cdot p = 0 \,. \tag{23}$$

In the limit (19)

$$\int d^4k \to (x_1 x_2)^{\frac{1}{2}} (2\nu)^{\frac{3}{2}} \int du\, dv\, dw\, d\lambda \,, \tag{24}$$

and

$$k^2 = 2\nu(uv + x_1 vw + x_2 wu) + M^2(u^2+v^2)+\mu^2 w^2 - \lambda^2 \,, \tag{25}$$

where M is the nucleon mass and μ the meson mass.

According to the ideas of the covariant parton model, one is tempted to look for the region of integration where both parton masses k^2 and $(k-p)^2$ are finite. However, a detailed discussion shows that while there is such a region in the amplitude of Fig. 5, it does not contribute to the discontinuity we wish to take. It is only possible to make one mass, k^2 say, finite and the other, $(k-p)^2$, is then allowed to become large. This can be exhibited by the change of variables

$$v = \left(\frac{x_2}{2x_1\nu}\right)^{1/2} \chi + \bar{y}/2\nu \,, \qquad w = -\frac{\chi}{(2x_1x_2\nu)^{1/2}} \tag{26}$$

In terms of the new variables u, $\bar{y}$, χ and λ one finds

$$k^2 = u\bar{y} + u^2 M^2 - (\chi^2 + \lambda^2) ,$$
$$(k - p_1)^2 = (u-1)\bar{y} + (u-1)^2 M^2 - (\chi^2 + \lambda^2) , \tag{27}$$

and

$$(k - p)^2 \sim - 2x_2 u\nu ,$$
$$(k - p + p_2)^2 \sim (1-x_2)(u-\omega^{-1})\, 2\nu , \tag{28}$$

where

$$\omega = \frac{1-x_2}{x_1} , \tag{29}$$

and

$$\int d^4k \to \int du\, d\bar{y}\, d\chi\, d\lambda . \tag{30}$$

Thus the lower bubble in Fig. 5 is evaluated at finite energy and parton mass. A comparison of (27) with (8), (9) and (13) shows that the contribution of this bubble to the integral we are evaluating is identical with the corresponding parton contribution F(u) to νW_2. Our knowledge of current processes will thus enable us to evaluate this part of the integrand.

The upper bubble in Fig. 5 (which is the one from which the large p_T parton emerges) is quite different. Its s' and μ^2 (indicated by arrows in the figure) are both becoming large with ν in a ratio fixed by u and x_1 and x_2. In other words, this bubble is evaluated in what one might term a hadronic Bjorken limit (in analogy with (1)). We shall return later to the question of how to evaluate this. We shall now simply suppose that in the limit

$$s' = (k - p + p_2)^2 \to \infty , \quad \mu^2 = (k - p)^2 \to \infty ,$$
$$- s'/\mu^2 \text{ fixed} , \tag{31}$$

the imaginary part of the parton-hadron amplitude behaves like

$$\mathrm{Im}\, T(s',\mu^2) \sim (-\mu^2)^{-\gamma_2} (s')^{\delta-1}\, \Phi(-s'/\mu^2) , \tag{32}$$

with $\Phi(0) \neq 0$.

Meanwhile, consider the middle of Fig. 5. This involves the coupling of two partons, one far off shell, to make a meson. The study of Bethe-Salpeter models suggests that such a coupling $\Gamma(k_1 = k, k_2 = k - p)$, behaves like

$$\Gamma \sim c(-k_2^2)^{-\gamma_1} , \qquad k_2^2 \to \infty , \quad k_1^2 \text{ fixed} , \tag{33}$$

where c and γ_1 are constants, independent of k_1^2. They are parameters of the model.

Putting all this together, one obtains an expansion for the inclusive differential cross-section of the form

$$E \frac{d\sigma}{d^3\underset{\sim}{p}} \sim s^{-2\gamma_1-\gamma_2+\delta-2} \, x_1^{\delta-1} \, x_2^{-2\gamma_1-\gamma_2} \int_{\omega^{-1}}^{1} du \, u^{-2\gamma_1-\gamma_2-2} (u\omega-1)^{\delta-1} F(u) \, \Phi \left(\frac{x_1(u\omega-1)}{x_2 u}\right) , \quad + x_1 \longleftrightarrow x_2 \tag{34}$$

where the second term arises from interchanging the roles of the bubbles in Fig. 5.

Finally, one may note that if some exchange (of a Pomeron or otherwise) is allowed between the two bubbles of Fig. 5, the general nature of the energy dependence is not altered. Thus, as indicated in the figure, one should also consider connected contributions. An explicit example of this will be given later.

IV. GENERAL FEATURES

Some aspects of Equation (34) are much more general in parton models than the specific case considered. These features are:

(i) Transverse Scaling

Eq. (34) can be rewritten in the form

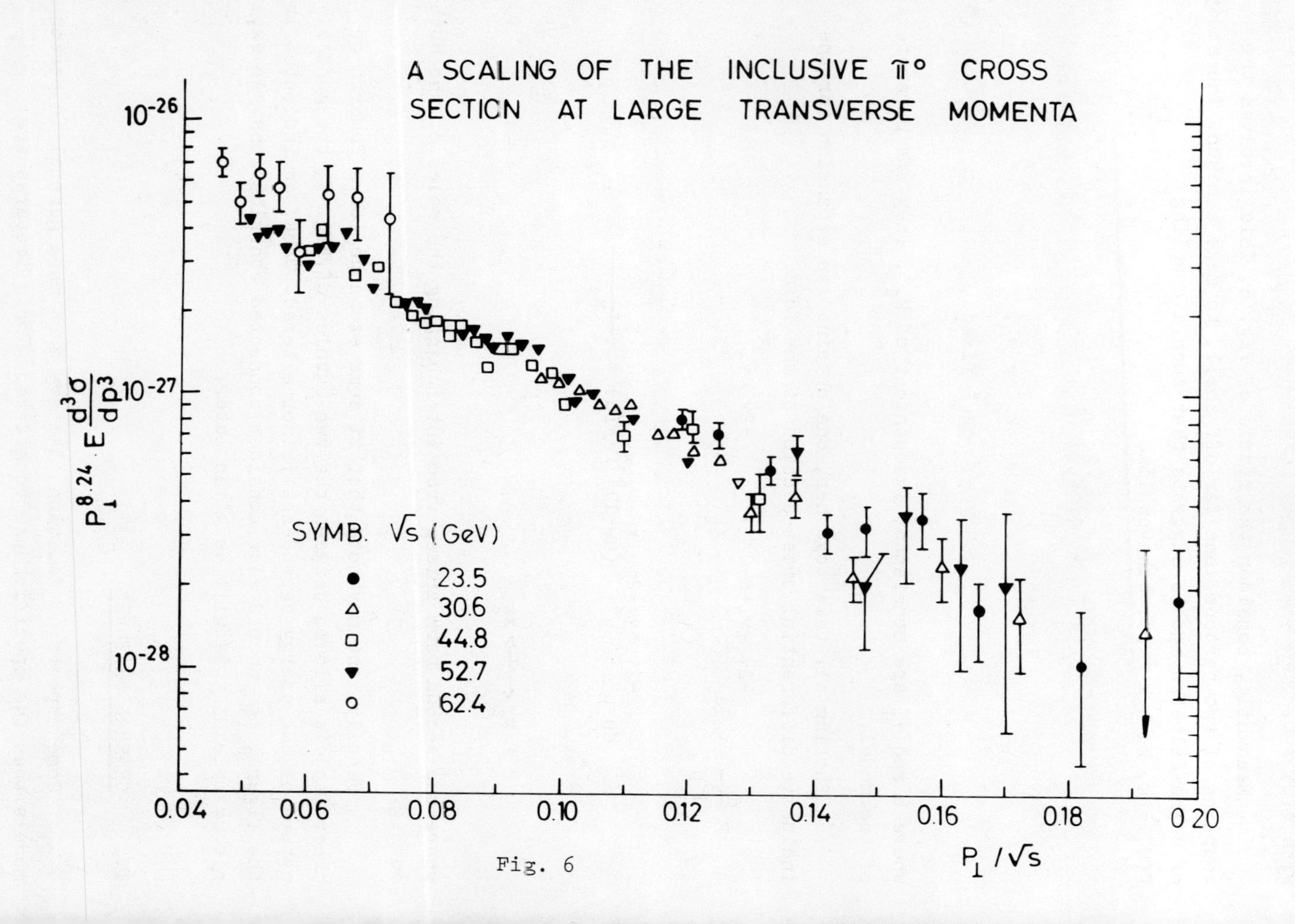

A SCALING OF THE INCLUSIVE π° CROSS SECTION AT LARGE TRANSVERSE MOMENTA
SYMB. √s (GeV)
23.5
30.6
44.8
52.7
62.4
$p_\perp^{8.24}\, E \frac{d^3\sigma}{dp^3}$
10-26
10-27
10-28
0.04
0.06
0.08
0.10
0.12
0.14
0.16
0.18
0.20
P⊥ /√s

Fig. 6

$$E \frac{d\sigma}{d\underset{\sim}{p}^3} \sim (p_T^2)^{-n} F(p_T/\sqrt{s}, \theta) \tag{35}$$

Results of this type hold in many parton mechanisms, as we shall see, though with various exponents n and different functions F. It is one of the great encouragements to this point of view that relations of this sort seem to hold experimentally also. The CCR data[5] at 90°, shown in Fig. 6, seems to satisfy a relation of this sort with $n \sim 4$. Lower energy experiments at NAL[6] seem to prefer a rather larger value of $n \gtrsim 5$, at least at larger values of $x_T = p_T/\sqrt{s}$. The relation of this to disentangling competing parton mechanisms will be discussed later.

(ii) Smooth Regge Connection.

If parton hadron amplitudes are assumed to behave in a Regge fashion in the appropriate regime (large energy and finite parton virtual mass) then one can show that relations of the form (34) will lead to a smooth transition into the Regge region.
That is, $F(0,\theta) \neq 0$, so that eventually

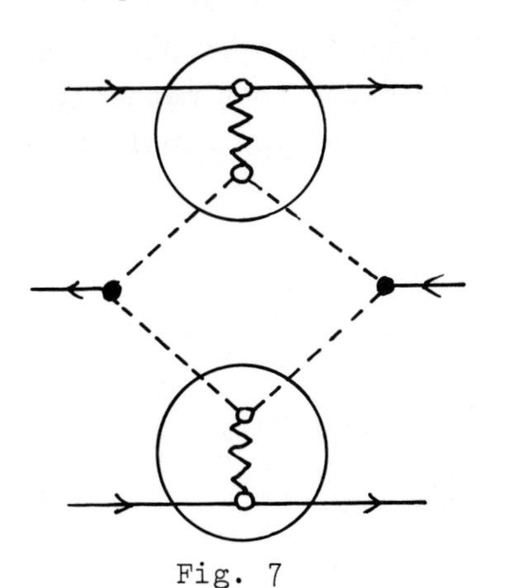

Fig. 7

$$E \frac{d\sigma}{d\underset{\sim}{p}^3} \sim (p_T^2)^{-n} , \quad s \to \infty \tag{36}$$

p_T large and fixed.

I do not give the calculation. It is simply the formal reflection of the diagrammatic fact that putting Pomerons into one or other or both of the parton amplitudes of Fig. 5, as indicated in Fig. 7 generates the terms of Fig. 8 corresponding to fragmentation and pionization regimes à la Mueller-Regge. (The energy dependence seen at fixed large p_T at NAL and ISR is, of course, due to the fact that the corresponding values of x_T are far from zero and $F(x_T,\theta)$ is varying significantly with x_T.)

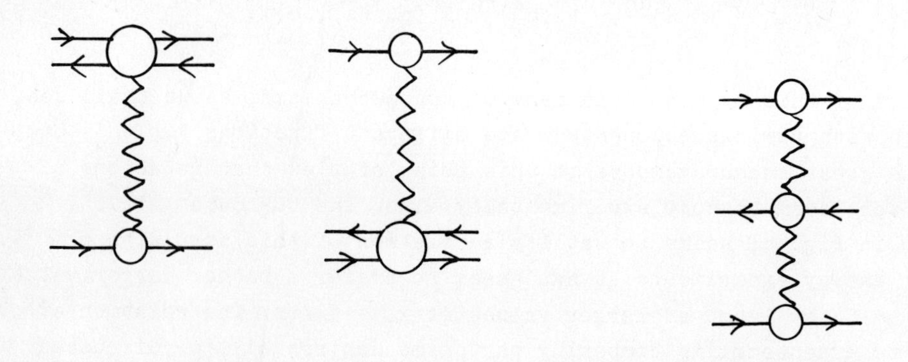

Fig. 8

V. A SIMPLE MODEL[7]

To make further progress with the quark fusion model, one must be able to say more about the hadronic Bjorken limit of parton amplitudes. An attractive assumption is that it is due to a mechanism as closely analoguous to that of Fig. 1 as possible. The simplest suggestion appears to be that of Fig. 9,

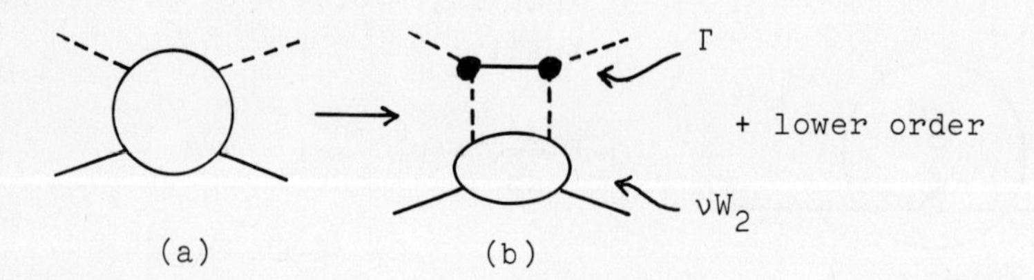

Fig. 9

where the solid internal line is an on-shell meson. It is straightforward to use Sudakov techniques to work out Fig. 9 in this limit (try it!) The bottom bubble is again a parton-hadron amplitude at finite s' and μ^2; that is, it can be identified with a contribution to νW_2. The contribution of the top part of Fig. 9b depends on the behaviour of the parton-meson coupling function Γ in the limit (33). In other words, this model only contains two parameters, c and γ_1! (This assumes one can determine the different quark contributions to νW_2. For this, we use our dual model which gives good agreement with deep inelastic current processes[8]). One finds that the exponent in (35) is given by

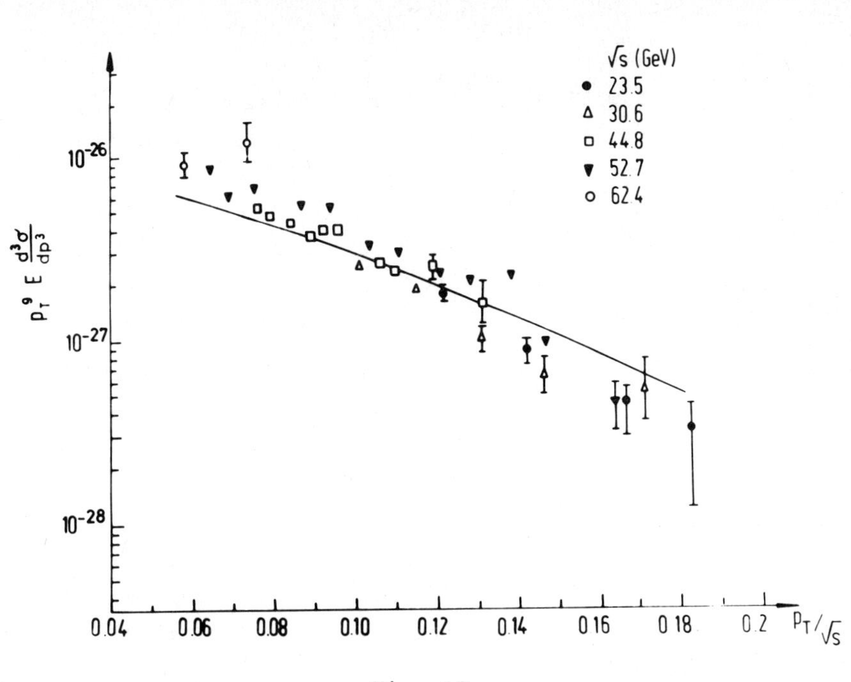
√s (GeV)
23.5
30.6
44.8
52.7
62.4
10^{-26}
10^{-27}
10^{-28}
$p_T^9\, E\, \frac{d^3\sigma}{dp^3}$
0.04
0.06
0.08
0.10
0.12
0.14
0.16
0.18
0.2
$p_T/\sqrt{s}$

Fig. 10

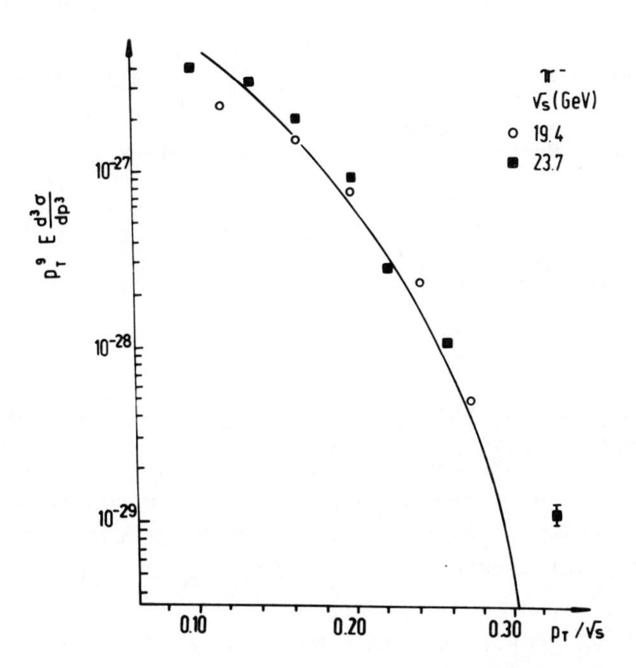
π^-
√s (GeV)
19.4
23.7
10^{-27}
10^{-28}
10^{-29}
$p_T^9\, E\, \frac{d^3\sigma}{dp^3}$
0.10
0.20
0.30
$p_T/\sqrt{s}$

Fig. 11

$$n = 2 + 4\gamma_1 . \tag{37}$$

For reasons I will discuss later, this simple picture cannot be the whole story, but it may well contain important aspects of the truth. Figs. 10 and 11 show the comparison with ISR and NAL data of the predictions of the model with γ_1 chosen equal to 5/8 (to give $n = 4\frac{1}{2}$ as compromise number).* Remember that F is calculated as a complicated convolution of deep inelastic structure functions. Our paper[7] gives many more detailed predictions. Among features one may note are:

(i) The model (with SU3 invariance for the vertices Γ) gives $d\sigma(\pi^+) = d\sigma(K^+)$. This is not true to better than a factor 2 experimentally.

(ii) F is strikingly slowly varying with centre of mass angle for $45^\circ \lesssim \theta < 135^\circ$.

The calculated curves correspond to a slightly more complicated model than I have so far revealed. Putting Fig. 5 and Fig. 9b together gives a process of the form of Fig. 12a. It is natural also to associate

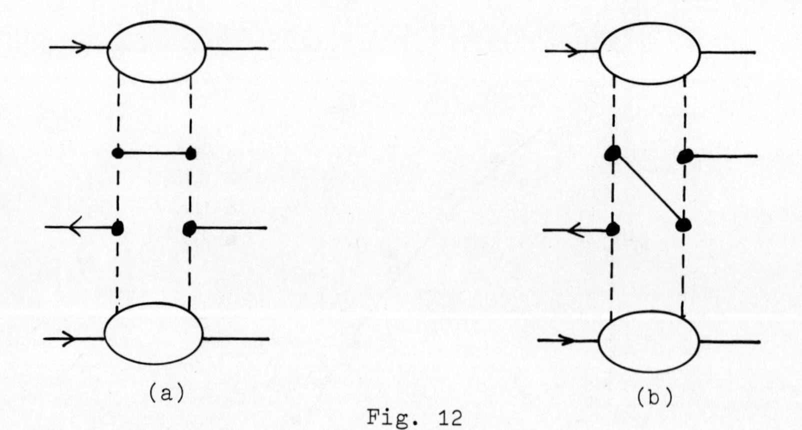

Fig. 12

with this the process of the type of Fig. 12b. These correspond to the connected terms in Fig. 5 referred to earlier.

In Fig. 12, the large p_T of the detected meson is balanced by the large p_T of the internal meson shown. The partons which emerge from the top and bottom bubbles have small p_T. Thus Fig. 12 can be thought of in

*Two slightly different values of C were chosen for the two figures. They are compatible with the uncertainty of the relative normalizations of the two sets of data.

a different way from that by which we have reached it. Namely, we can picture its mechanism for producing large p_T final state particles as being due to a wide angle scattering of the small p_T constituents of the colliding hadrons, the scattering process being

$$q + \bar{q} \rightarrow M + \bar{M} \quad . \tag{38}$$

This leads us on to consider a wider class of such scattering mechanisms. But note that quark fusion only becomes a subset of this wider class if Fig. 9 is the right picture for hadronic Bjorken limits.

VI. SCATTERING MODELS

A wide class of models can be constructed on the basis of the scattering picture given at the end of the last section. The earliest discussion[9] of large p_T processes was framed in terms of just such a model. The corresponding Mueller-Regge diagram is shown in Fig. 13.

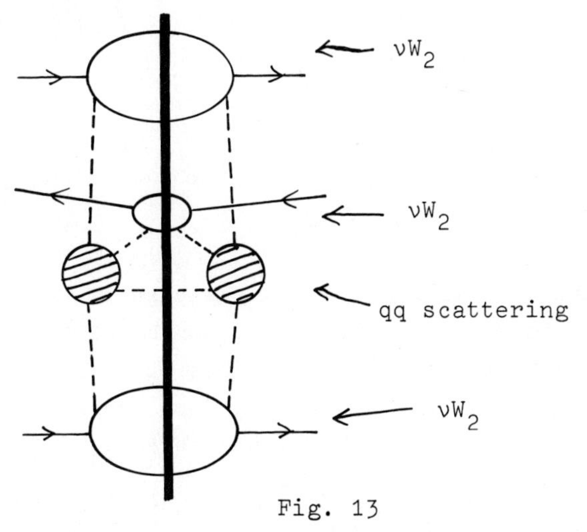

Fig. 13

Its analysis in terms of Sudakov variables is straightforward[10] and we need not go into it in detail now. It turns out that all the masses associated with the internal (broken) parton lines can be held finite. Thus the top and bottom bubbles in Fig. 13 can be identified with contributions to νW_2 . In the middle bubble one has a parton of large p_T fragmenting to produce the detected final state hadron of large p_T. Such a term can be identified with a contribution to $\nu \bar{W}_2$, the structure function associated with e^+e^- annihilation. The unknown terms correspond to the shaded bubbles, which are high energy wide angle parton-parton scattering amplitudes.

If the broken lines in Fig. 13 represent quarks, these bubbles will be the amplitudes for the process

$$q + q \rightarrow q + q \,. \tag{39}$$

Later in these lectures, I will discuss an idea of Brodsky and Farrar[11] which suggests that the cross section for such a process should be scale free at high energies. (A particular model which gives such a scale free result is the exchange of a single vector gluon, which was the original B.B.K. suggestion.) This then leads to a scale free result for the inclusive cross section (35), that is

$$n = 2 \,. \tag{40}$$

This is not in agreement with experiment. However, one can obtain the value $n = 4$ by using the B.F. rules but reinterpreting the broken lines in Fig. 13 so that (39) is replaced by one or more of the processes

$$M + q \rightarrow M + q \,, \tag{41a}$$

$$q + \bar{q} \rightarrow M + \bar{M} \,, \tag{41b}$$

$$q + q \rightarrow B + \bar{q} \,, \tag{41c}$$

where M is a meson and B a baryon. Only in the case of (41b) or (41c) can one continue to interpret the top and bottom bubbles of Fig. 13 as contributions to νW_2. The process (41b) is just our simple model of the previous section which took, in Fig. 12, Born approximations for the high energy wide angle amplitudes. If one takes $\gamma_1 = \frac{1}{2}$, then the B.F. rules are satisfied.

VII. LEADING PARTICLE MODEL

The final type of mechanism we shall consider is that in which a parton from one hadron scatters coherently off the other hadron (Ref. 10 and B.B.G.). The Mueller-Regge diagram is given in Fig. 14 . Again all parton masses may be made finite and the bottom bubble corresponds to a contribution to νW_2 . The shaded bubbles are parton-

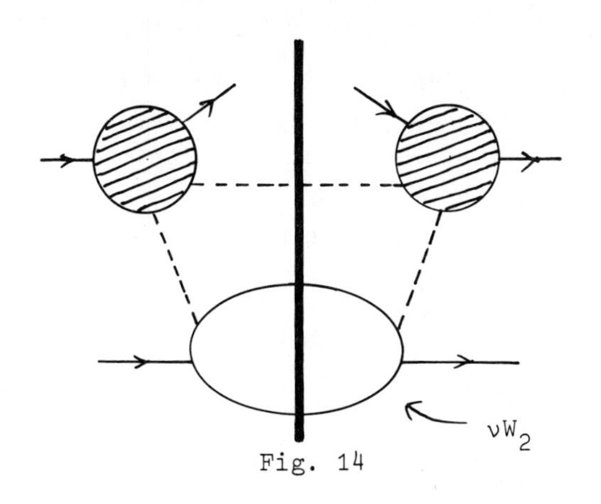

Fig. 14

hadron scattering amplitudes at high energy and fixed angle. Their detailed form is unknown but if we use the B.F. rules, then this process gives for

$$p + p \to p + X$$

a cross section of the form (35) with

$$n = 6 . \qquad (42)$$

Thus this mechanism will not dominate over (41) at sufficiently high energies. However, we shall find that it is of importance in a certain extreme region of phase space. Its contribution is also very simple in form and can be written without a convolution integral[10].

An interesting feature of this mechanism is that it has $F(0,\theta)=0$, $\theta \neq 0$. This is because by substituting Pomerons into the bubbles of Fig. 14 one can only produce diagrams of the type of Fig. 8 corresponding to the fragmentation region of the upper hadron. Thus Fig. 14 must give a vanishing contribution in the pionization region.

VIII. COMPARISON OF MECHANISMS AND EXPERIMENTAL RESULTS

We now discuss the relation of these mechanisms to observations at large p_T. At the present stage of both experiment and theory, the comparison can at best be semiquantitative. I shall try to give a survey which draws attention to the salient features.

(i) The Value of n.

If the B.F. rules are accepted then quark-quark scattering must either be numerically small or absent. We will discuss this again later. The ISR and NAL results are broadly compatible with a mixture of scattering mechanisms of the type (41) together with a leading particle mechanism (42). NAL now report that they see the larger values of n at the large values of x_T. This is readily understood in terms of the next point.

(ii) Edge of Phase Space.

As the missing mass becomes small near the edge of phase space, the inclusive cross section must vanish. This reflects itself in the formalism by the fact that the unshaded blobs in Figs. 5, 13, and 14, which represent contributions to νW_2 or $\nu\bar{W}_2$ are evaluated nearer and nearer their thresholds $\omega = 1$. Thus the rate of vanishing of the inclusive cross section as

$$M^2/s = \varepsilon = (1-x_1-x_2) \to 0 \tag{43}$$

will depend on how many such blobs there are. The leading particle mechanism of Fig. 14 has the only one such blob and so it must dominate in the limit $\varepsilon \to 0$. This has the remarkable consequence that in p-p scattering near the edge of phase space one of the large p_T particles should be a baryon!

Scott[12] has shown that the dominance of Fig. 14 as $\varepsilon \to 0$ is consistent with the correspondence notion[13] of a smooth relation between inclusive and exclusive processes.

(iii) Particle Ratios.

These will provide in due course one of the best ways of hoping to make quantitative discrimination between the various processes. The process of Fig. 5 can only account for mesons. One would not expect the process of Fig. 13 to produce baryons copiously except through the scattering process (41c). This is because otherwise the baryon must come from a fragmenting quark, but this would also give baryons in e^+e^- annihilation where they do not seem to be produced in large numbers so that the probability must be small. Of course, the process of Fig. 14 will certainly give protons of large p_T; in fact, it was constructed for this purpose[10].

At NAL, it appears that many large p_T protons are produced (comparable to π^+), whilst at the ISR there are rather less ($\lesssim$ 30 % of π^+). Perhaps this is because the leading particle mechanism is largely responsible but with its large value of n decreases in importance relative to pion production mechanism at the higher energy.

Of course, the simple model[7] of Fig. 12 discussed above gives many detailed predictions of meson ratios, for which reference may be made to the paper. It cannot, however, be the whole story because of baryon production at large p_T.

(iv) Correlations.

All the mechanisms discussed will tend to produce balancing jets of large p_T particles. This is seen in its simplest form in the model of Fig. 12 where the detected large p_T meson is balanced by just one meson of opposite p_T (though in general different longitudinal momentum so that the two mesons are not back to back in the centre of mass). This could, of course, be modified by adapting the model to produce resonances decaying in cascades. For all the other mechanisms detailed predictions are uncertain because of the lack of understanding of the mechanism for quark fragmentation (even in electroproduction phenomena this is far from detailed understanding).

The experimental situation is also rather unclear at present. Some jet-like effects are seen but it is hard to know how much is simply a reflection of momentum conservation.

(v) Meson Effects.

If one considers πp scattering rather than pp scattering, some quantitative features change because the pion provides a ready source of antiquarks as well as quarks, whilst the proton appears to act as if it is predominantly a 3q system. Combridge[14] has shown, using a model based on a combination of the mechanisms of Figs. 12 and 14, that this might lead to substantially larger cross sections for large p_T production by mesons than by protons. This observation may prove of particular importance in assessing experiments (like those at NAL) which use heavy nuclear targets exposed to proton beams since the production of secondary pion beams within the nucleus may in consequence have significant effects.

IX. EXCLUSIVE PROCESSES

The discussion of exclusive processes at large p_T, such as high energy wide angle elastic scattering, is much more model dependent. The reason is easy to see.

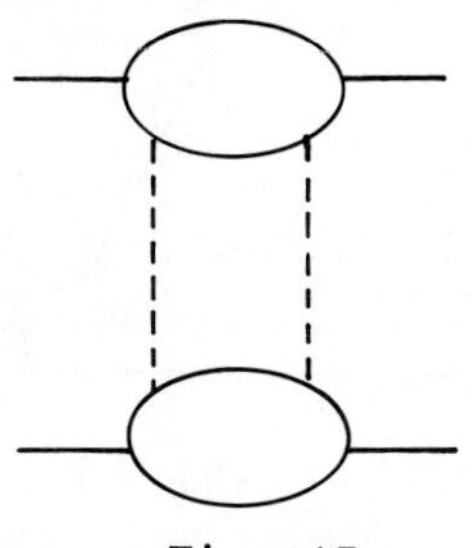

Fig. 15

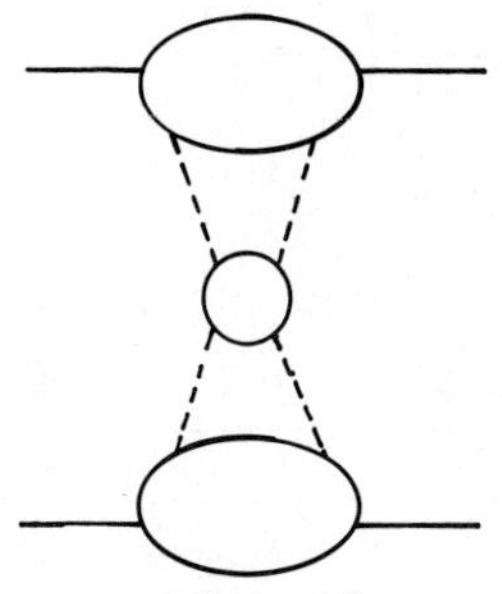

Fig. 16

Because of the large momentum transfer, it is natural to try to picture the process as due to a single basic interaction of some sort. Popular assumptions have been that this basic interaction is either constituent interchange[15, 4] or parton scattering[16], as shown in Figs. 15 and 16 respectively. These figures are necessarily different in character to the diagrams we draw for inclusive processes. In the latter case, the blobs represented complete parton-hadron amplitudes, etc. This cannot be so for Figs. 15, 16 since in that case Fig. 15 would give a double pole in the t-channel and Fig. 16 a triple pole! Thus the blobs in these figures must be reduced amplitudes in some sense, and what this sense is must be specified by a much more specific dynamical scheme of parton-hadron interactions than we have needed to use so far. I shall now describe one such scheme which has attractive features.

X. DIMENSIONAL COUNTING[11]

The scheme is the B.F. dimensional counting ansatz. It pictures baryons as being 3q systems and mesons as $q\bar{q}$ systems (or more strictly that their composite wave functions contain components of this type, since more complicated components, such as $qqqq\bar{q}$ in B, would give non-leading contributions). The quarks are to interact through the exchange of point-coupled scalar or vector gluons.

The nature of the conclusions of the model can be seen by considering a simplified picture in which the constituent quarks are treated as free,

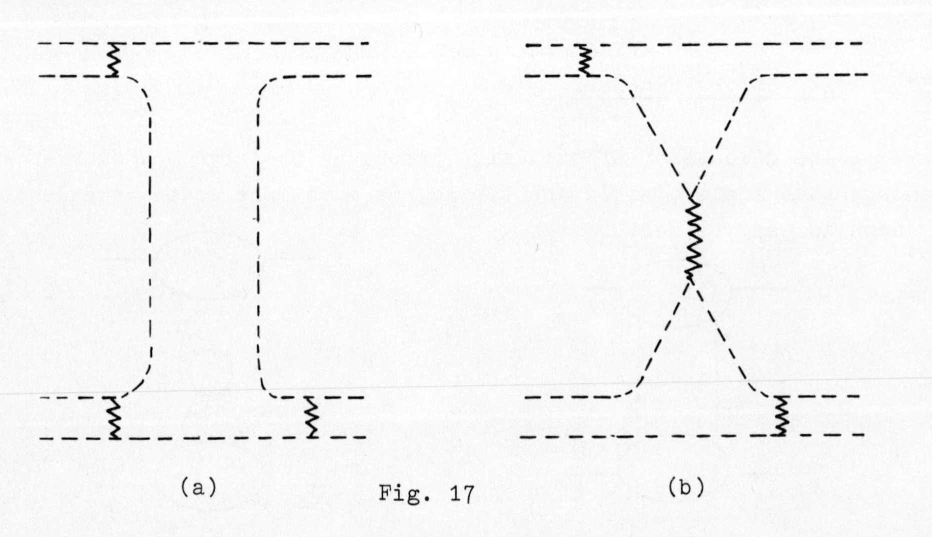

Fig. 17

each carrying a fraction of the momentum of the parent hadron. Fig. 17 gives two examples of the minimum connected diagrams which would correspond to π-π scattering in such a picture. The zig-zag lines represent gluons. Fig. 17a is an interchange process and Fig. 17b is a scattering process. It is straightforward to calculate that diagrams of this type all give contributions to the differential cross section in the high energy fixed angle θ limit which behave like

$$\frac{d\sigma}{dt} \sim s^{-6} F(\theta) \quad . \tag{44}$$

More generally, one finds

$$\frac{d\sigma}{dt} \sim (s)^{2-\Sigma n_i} F(\theta) \ , \tag{45}$$

where n_i is the number of constituents in the ith participating hadron in the process. The function $F(\theta)$ can also be calculated by considering the different terms arising from diagrams of the type of Fig. 17.

Of course, in actual fact the contributions of Fig. 17 should be convoluted with hadronic wave functions to give the true scattering amplitude. If the point-like gluon picture is taken literally, this leads to unbounded powers of logarithms modifying (45) in an unknown way. This is characteristic of renormalizable field theory, and to get the B.F. scheme, one must assume some mechanism softens the theory to remove these logarithms. Ezawa[17] has shown how to construct a softened theory which can be made arbitrarily close in its behaviour to (45).

Equation (45) has many interesting consequences. We shall concentrate on its prediction of the exponent m, where we write

$$\frac{d\sigma}{dt} \sim s^{-m} F(\theta) \tag{46}$$

It implies the following predictions:

$$p + p \to p + p \ : \quad m = 10, \tag{47a}$$

$$\pi + p \to \pi + p \ : \quad m = 8, \tag{47b}$$

$$\gamma + p \to \pi + p \ : \quad m = 7 \ . \tag{47c}$$

The best measured process is (47a) and Fig. 18 shows our analysis[18] of machine energy pp data which accords with (46) with m = 9.7. It is interesting to note that this regime only appears to set in for events with $|t| > 2.6$. The situation is less clear at ISR energies, where some largely energy independent structure seems to be revealed in the range $1.5 < |t| < 4$.

The predictions (47b) and (47c) appear in accord with the scantie experimental data, within the errors. Also (45) applied to electron scattering implies that the nucleon form factors decrease like $(q^2)^{-2}$ and the pion form factor like $(q^2)^{-1}$, in accord with popular belief.

Thus the B.F. scheme has many attractive features. There is, however, one substantial difficulty, to which I now turn.

XI. LANDSHOFF MECHANISM[19]

Landshoff has shown that in fact the diagrams of Fig. 17 do not give the asymptotically dominant term! Instead, this comes from Fig. 19 for π-π scattering, and in similar diagrams invo ving three quark-quark interactions for pp scattering. In Fig. 19 all the parton lines have finite masses. The values of m given by the Landshoff mechanism are:

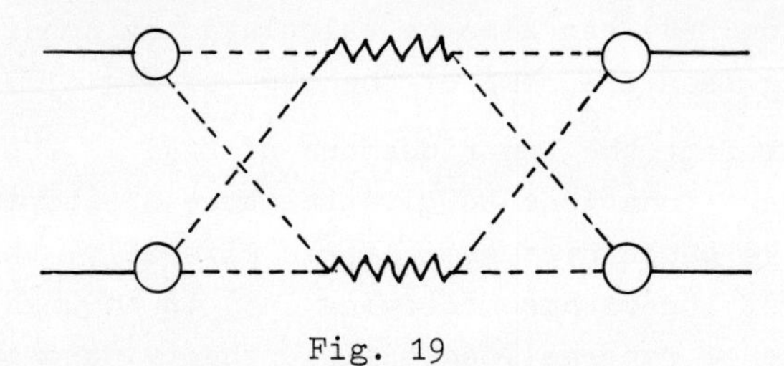

Fig. 19

$$\pi + \pi \rightarrow \pi + \pi \quad : m = 5 \text{ (B.F., } m = 6\text{)}; \qquad (48a)$$

$$p + p \rightarrow p + p \quad : m = 8 \text{ (B.F., } m = 10\text{)} . \qquad (48b)$$

It is interesting to note that the B.F. terms correspond to what, in the terminology of the asymptotic behaviour of Feynman integrals[20], are called "end point" contributions, whilst the Landshoff mechanism is a "pinch" contribution.

We must now address ourselves to possible explanations of why the prediction (48b) does not appear to agree with the existing data. There appear to be three possible types of explanation:

(i) The term is present but its numerical coefficient is small compared with that of the B.F. terms so that at moderate energies it do not manifest itself. This view receives some support from the fact that formally the term is multiplied by the eighth power of the "quark mass"

(ii) There is some dynamical mechanism (presumably related to

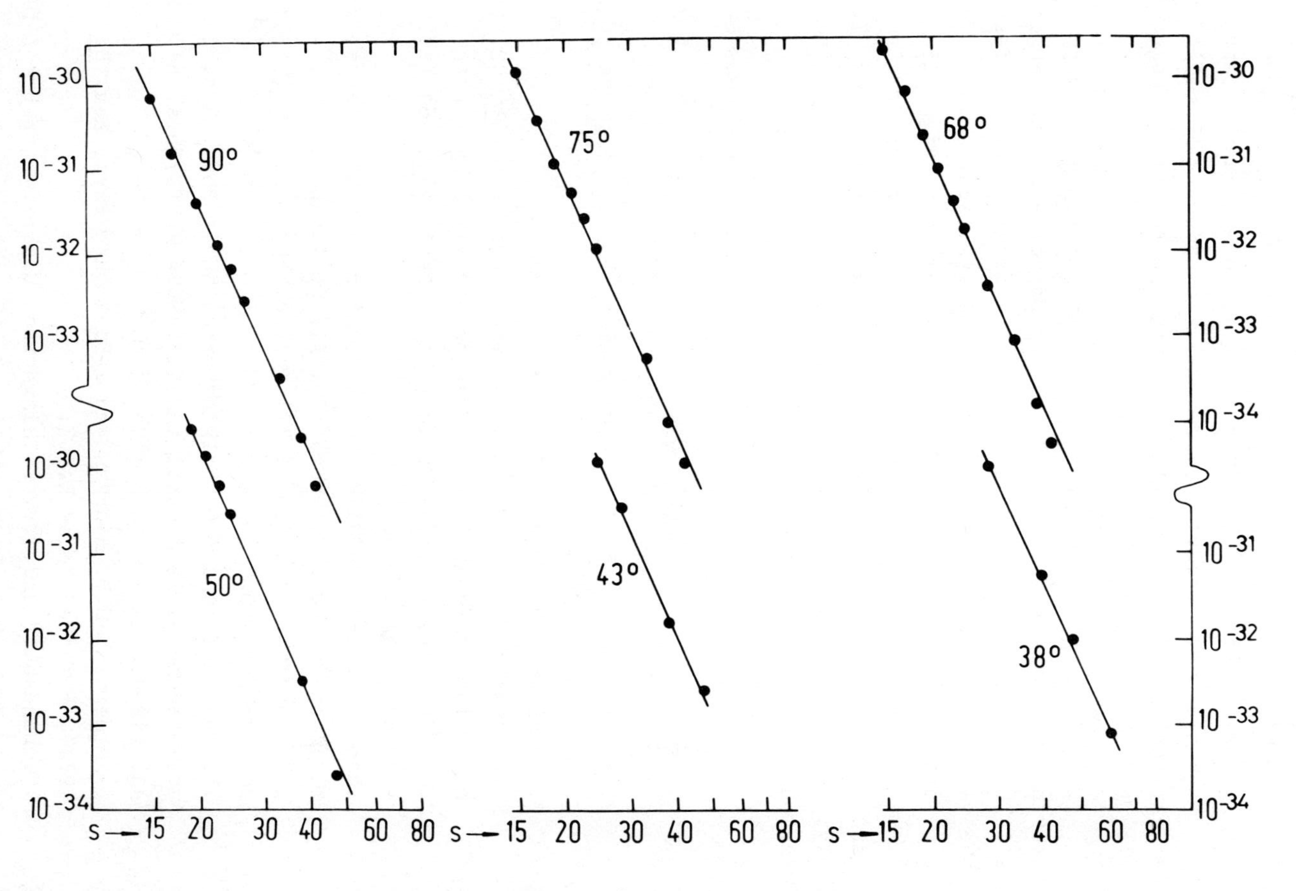

Fig. 18

whatever confines the quarks within hadrons) which does not permit direct interactions between quarks in different hadrons. This would then exclude Fig. 19 and also Fig. 17b, but not the interchange process, Fig. 17a. The same prohibition would remove the quark-quark scattering contribution in Fig. 13. Since (45) gives m = 2 for quark-quark scattering, that is a scale free result, this removal avoids n = 2 in (35). Thus this "explanation" is an attractive one, though really it replaces a puzzle by a deeper mystery.

(iii) It is possible[21] that (45) does not apply to quark-quark scattering unless some or all of the quark masses are also large. That this is consistent with relativistic quantum mechanics is in fact shown by the vector gluon exchange model since multiple scattering effects seem to produce just this sort of behaviour. Since in Figs. 13 and 19 the quark masses are all finite these processes would no longer give the (unwanted) results corresponding to scale free q-q scattering. It appears likely, however, that the interchange processes could still give B. F. dimensional counting results.

REFERENCES

1. P. V. Landshoff and J.C. Polkinghorne, Physics Reports 5C,1 (1972
2. C. Nash, Nucl. Phys. B61, 351 (1973).
3. G. Preparata, Phys. Rev. D7, 2973 (1973).
4. P.V. Landshoff and J.C. Polkinghorne, Phys. Rev. D8, 927 (1973). An earlier non-covariant discussion was given by R.Blankenbecler, S. J. Brodsky, and J.F. Gunion, Phys. Lett. 42B, 461 (1972). Referred to as B.B.G.
5. CERN-Columbia-Rockefeller Collaboration. Aix Conference, 1973.
6. Princeton-Chicago Collaboration. Aix Conference, 1973.
7. P. V. Landshoff and J.C. Polkinghorne, Phys. Rev. D8, 4157 (1973) See also Ref. 10.
8. P.V. Landshoff and J.C. Polkinghorne, Nucl. Phys. B28, 240 (1971) A related model is J. Kuti and V.F. Weisskopf, Phys. Rev. D4, 3418 (1971).
9. S. M. Berman, J.D. Bjorken and J.B. Kogut, Phys. Rev. D4, 3388 (1971). Referred to as B.B.K.
10. P.V. Landshoff and J.C. Polkinghorne, Cambridge preprint DAMTP 73/31, Phys. Rev. to be published. See also S.D. Ellis and M. B. Kislinger, Phys. Rev. D9, 2027 (1974).

11. S.J. Brodsky and G. R. Farrar, Phys. Rev. Lett. 31, 1153 (1973). Referred to as B.F.

12. D.M. Scott, Cambridge preprint DAMTP 73/37, Nucl. Phys. to be published.

13. J. D. Bjorken and J. B. Kogut, Phys. Rev. D8, 1341 (1973).

14. B. L. Combridge, Cambridge preprint DAMTP 74/8.

15. R. Blankenbecler, S.J. Brodsky and J.F. Gunion, Phys. Lett. 39B, 649 (1972) and Phys. Rev. D8, 287 (1973). Also referred to as B.B.G.

16. D. Horne and M. Moshe, Nucl. Phys. B57, 139 (1973).

17. Z.F. Ezawa, Cambridge preprint DAMTP 74/5.

18. P.V. Landshoff and J.C. Polkinghorne, Phys. Lett. 44B, 293 (1973).

19. P.V. Landshoff, Cambridge preprint DAMTP 73/36.

20. R.J. Eden, P.V. Landshoff, D.I. Olive and J.C. Polkinghorne, *The Analytic S-Matrix* (C.U.P., 1966). Chapter 3.

21. J.C. Polkinghorne, Phys. Lett. B49, 277 (1974), and in preparation.

QUARK CONFINEMENT IN GAUGE THEORIES OF STRONG INTERACTIONS

Leonard Susskind†

Belfer Graduate School of Science, New York, N.Y., USA
and
Tel Aviv University, Tel Aviv, Israel

I. ABELIAN MODELS

a) Introduction

I will not begin by telling you all the reasons why you have to believe in quarks as hadron constituents[1]. Lets just suppose that I had and that we all believe hadrons are loosely bound collections of quarks*. Secondly I want you to suppose that nobody will ever discover a free isolated quark. We are then faced with the puzzling problem of explaining how finite forces conspire to confine quarks to the interior of hadrons.

To begin with we must realize that most of our intuitions, even our idea that a puzzle exists, come from our experience with weakly coupled quantum electrodynamics and its perturbative solution. We are often led astray into asking questions which make the phenomenon sound much more complicated then it really is. For example: What class of graphs is important to confine quarks? Or: Do the catastrophic infrared divergences of Yang Mills theory combine to screen quarks? We ought to understand that these questions do not really refer to the behaviour of the system but rather to the method of solution - perturbation theory about free fields. The reason that quark confinement seems so odd to us is because we start with all the wrong ideas about how (the correct strong interaction) field theory behaves and then attempt to perturb our way to an infinitely distant behaviour.

---

† Most of the work described in these lectures was carried out in collaboration with J. Kogut while the author was a visitor at Cornell University.

* Loosely bound in the sense that they behave almost freely at short distances.

In these lectures I will show you three examples of theories with confinement. In each case it is easy to see that quarks are confined although perturbation theory buries the obvious in a jungle of complicated graphs. The three examples share a key element, namely local gauge invariance. The importance of local gauge invariance is that it connects additive conserved charges to long range fields through Gauss's theorem. The most familiar case is the long range Coulomb field accompanying every isolated charge in electrodynamics. Similarly in the non-abelian color-gauge theory of quarks every state with a non-zero color must have a long range color-electric field in order to be gauge invariant. This includes all states with non-vanishing triality.

The quark confining mechanism does not directly deal with the quarks but rather with their long range color-electric fields. If the color-electric fields are confined so that the electric flux lines are prevented from radiating to infinity then the finite energy states must be color-neutral.

The three examples are Schwinger's one dimensional QED[2,3,4]; a semiclassical model based on unusual dielectric properties of the vacuum[5,6]; and a hamiltonian formulation of Wilson's lattice gauge theory[7,8,9].

b) The Schwinger Model

I will now make two approximations on the real problem. First I will replace the three colors of the 3-triplet model by a single abelian color called charge. Instead of three kinds of quarks (red, yellow, blue) I now have only one. The confinement mechanism will operate to eliminate all objects except neutral bosons.

Having agreed to approximate three by one I will apply the approximation again, this time on the number of space dimensions. The result of these approximations is the Schwinger model or QED in one dimension[2,3,4].

I am not going to derive the formal solution[3] to the model or give a rigorous demonstration of confinement. This is partly because you can find these things in the literature, but more importantly, I want to avoid the special features of one dimension which make the model solvable. In fact I will work in a gauge which is particularly inconvenient for exact solutions. The gauge is defined by setting the time component of the vector potential to zero

$$A_t = 0 \tag{I.1}$$

The space component of the vector potential will be called A. The gauge invariant field tensor has only one independent component, the electric field, which is given by

$$E = \frac{d}{dt} A \tag{I.2}$$

The hamiltonian is given by

$$H = \int dz \left\{ \psi^\dagger \alpha (\partial_z + igA) \psi + \frac{1}{2} \dot{A}^2 \right\} \tag{I.3}$$

Properly speaking eq. (I.3) defines a class of gauges related by time independent gauge transformations

$$\begin{aligned} \psi &\longrightarrow e^{ig\Lambda(z)} \psi \\ A &\longrightarrow A + \partial_z \Lambda \end{aligned} \tag{I.4}$$

for time independent Λ. The hamiltonian in eq. (I.3) is of course gauge invariant for this class of gauge transformations. Furthermore, and this is important, all physical states must be invariant under (I.4). The reason I have chosen this special class of gauges and restricted the gauge invariance to time independent Λ is because the restricted gauge transformations can be represented using unitary operators $U(\Lambda)$.

$$\begin{aligned} U \psi U^{-1} &= e^{ig\Lambda} \psi \\ U A U^{-1} &= A + \partial_z \Lambda \end{aligned} \tag{I.5}$$

Infinitesimal generators $G(\Lambda)$ can also be introduced for infinitesimal Λ

$$\begin{aligned} [G, \psi] &= ig\Lambda(z) \psi(z) \\ [G, A] &= \partial_z \Lambda \end{aligned} \tag{I.6}$$

When t-dependent gauge transformations are considered operators like U and G no longer exist and I don't know how to express the gauge invariance of the physical states. Gauge invariance under the restricted gauge group simply requires every physical state $|\rangle$ to satisfy

$$U(\Lambda)\,|\,\rangle = |\,\rangle$$

$$G(\Lambda)\,|\,\rangle = 0 \tag{I.7}$$

for all Λ.

Fortunately it is very easy to find $G(\Lambda)$. Using the canonical commutation relations

$$[A(z), \dot{A}(z')] = i\,\delta(z-z') = [A(z), E(z')]$$

$$[\psi(z), \rho(z')] = g\,\psi(z)\,\delta(z-z') \tag{I.8}$$

where $\rho = g\psi^{+}\psi$, you can easily verify that

$$G = \int \Lambda(z)\left[\rho(z) - \frac{\partial E}{\partial z}\right] dz \tag{I.9}$$

We see that gauge invariance requires

$$\left\{\rho(z) - \frac{\partial E}{\partial z}\right\}|\,\rangle = 0 \tag{I.10}$$

This equation expresses the familiar fact that the charge density is the source of electric field. It is not an equation of motion but a constraint on the physical states. In the physical subspace it implies

$$E(z) = E(-\infty) + \int_{-\infty}^{z} \rho(z')\,dz' \tag{I.11}$$

$$= E(+\infty) - \int_{z}^{\infty} \rho(z')\,dz' \tag{I.12}$$

Lets suppose $E(\infty)$ or $E(-\infty)$ is non-zero. Then since H contains the term $\int E^2 dz$ it is evident that the energy will be infinite. To remain in the space of finite energy $E(\pm\infty)$ must be zero. But then the two expressions (I.11) and (I.12) will not be equal unless

$$\int_{-\infty}^{\infty} \rho(z')\, dz' = 0 \qquad \text{(I.13)}$$

What I have proved is that the finite energy, gauge invariant states have zero total charge. But I would be cheating if I told you that this proves charged particles don't exist. What I really want to show is that the finite energy, gauge invariant states do not contain well separated quarks and antiquarks.

For definiteness I will use eq. (I.11) for the electric field. We can picture a charge at z_0 as being the source of an electric field which vanishes for $z < z_0$. This is shown in Fig. 1

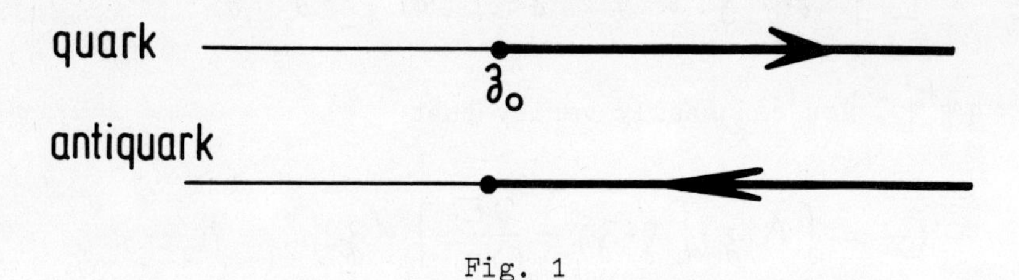

Fig. 1

I want to digress briefly to describe the objects in Fig. 1 by operators. Since the field ψ is not gauge invariant the state $\psi(z_0)|0\rangle$ is not a good description of a physical quark. To make a state which satisfies gauge invariance we have to do something to create the line of electric flux which must accompany the charge. Consider the operator

$$\sqcup(f) = \exp\left[i \int_{-\infty}^{\infty} A(z)\, f(z)\, dz \right] \qquad \text{(I.14)}$$

where f is a c-number function of position. Using the fact that A and E are canonical conjugates we find

$$\sqcup(f)\, E(z)\, \sqcup^{-1}(f) = E(z) + f(z) \qquad \text{(I.15)}$$

This means that $\sqcup$ acts on a state to shift the electric field by amount f. This is useful because we need to shift E by amount $g\,\theta(z - z_0)$ when a quark is created at z_0. Therefore it makes sense to multiply $\psi(z_0)$ by the factor exp $ig \int_{z_0}^{\infty} A(z)\, dz$. The resulting gauge invarian operator can be used to create the physical states shown in Fig. 1. We define

$$\Psi(z) = \exp\left[ig\int_z^\infty A(z')\,dz'\right]\psi(z)$$

or (I.16)

$$\Psi(z) = U(z)\,\psi(z)$$

Exercise: Prove Ψ is gauge invariant. Show that the expectation value of E in the state $\Psi|0\rangle$ is $g\,\Theta(z-z_0)$.

Now the reason why quarks are confined in this model is not because it costs an infinite energy to apply $\psi(z)$ to a state but rather because the factor $U(z)$ costs an infinite energy. This of course is due to the uniform electric field which fills space from z_0 to ∞.

Next lets consider a high energy $q\bar{q}$ pair which is produced, perhaps by a lepton annihilation,at the origin. My discussion is going to be at the impressionistic level so I suggest you look at Ref. 3 for formal arguments. The initial state is something like

$$|\,initial\,\rangle = \bar{\psi}(0)\,\psi(0)\,|0\rangle \tag{I.17}$$

Since the two operators ψ and $\bar{\psi}$ are evaluated at the same point the initial state is gauge invariant.

As the system evolves the quark pair will separate. If you forgot gauge invariance you might guess that the state at a later time is something like

$$|\,later\,\rangle = \bar{\psi}(-z)\,\psi(z)\,|0\rangle \tag{I.18}$$

But this is impossible because the state (I.18) is not gauge invariant and could not have been obtained if the system's evolution is governed by a gauge invariant hamiltonian.

A more correct guess is

$$\begin{aligned}|\,later\,\rangle &= \bar{\Psi}(-z)\,\Psi(z)\,|0\rangle \\ &= \bar{\psi}(-z)\,e^{ig\int_{-z}^{z} A\,dz'}\,\psi(z)\,|0\rangle \end{aligned} \tag{I.19}$$

which describes a $q\bar{q}$ pair at the ends of a line of electric flux. (See Fig. 2)

Fig. 2

Since the electric field is uniform between the quarks, the energy stored in the field is $\sim g^2 |x|$ where $|x|$ is the distance separating the pair. Thus the quarks can separate to a distance proportional to their initial energy.

Of course the real state $|\text{later}\rangle$ is not really as simple as (I.19). Since ψ is a relativistic field the interaction between the quark field and the electric field can create pairs in the region between the original pair. For example $|\text{later}\rangle$ will have a piece like

$$\overline{\Psi}(-x)\ \bar{\Psi}(x_1)\ \overline{\Psi}(x_2)\ \bar{\Psi}(x)\ |0\rangle \tag{I.20}$$

which looks like Fig. 3.

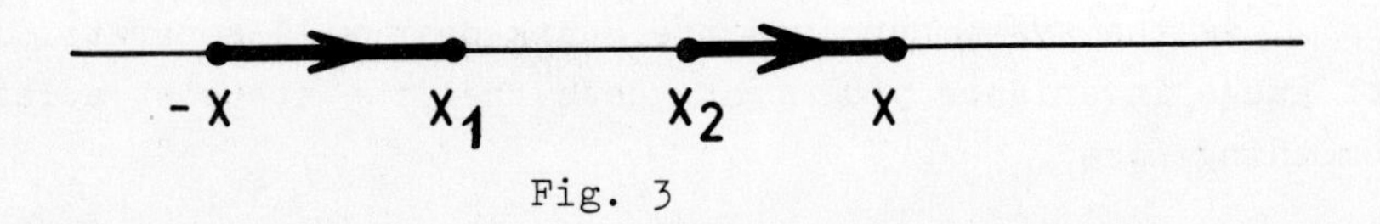

Fig. 3

However it is obvious that the evolution of the system can never lead to an isolated quark which is separated from compensating charges by more than a distance proportional to the initial energy. The exact solution of the Schwinger model shows that the real final state consists of a number of $q\bar{q}$ pairs, each with its connecting flux line and that the probability to find a quark at a distance $> \frac{1}{g}$ falls exponentially.

I have dwelled at length on this trivial model so that you would get a clear picture of the connection between gauge invariance, continui[ty] of electric flux and confinement. Just in case it was not clear I will say it again: Gauge invariance requires every quark to be the end of a flux line with uniform energy density and every end to be a quark. Since every flux line has two ends (unless it is infinite and therefore infinitely heavy) quarks must occur in pairs. This idea will be repeated throughout the rest of these lectures.

c) Semiclassical Model

At first sight the situation in 3 space dimensions looks very unfavorable for confinement. Gauge invariance still requires the charges to be the sources of electric flux

$$\nabla \cdot E = \rho \tag{I.21}$$

but this time the flux lines have two more dimensions to spread out into. (See Fig. 4)

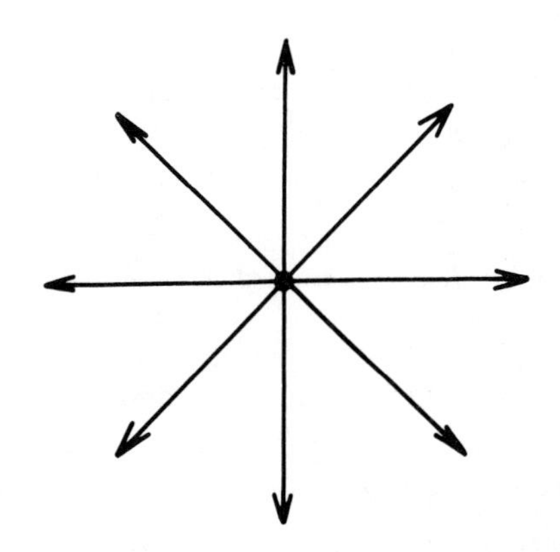

Fig. 4

If the field spreads with spherical symmetry then continuity of flux insures that it falls like

$$|E| \sim \frac{1}{r^2} \tag{I.22}$$

and the total energy is finite except for ultraviolet $(r \to 0)$ effects which are removed by renormalization.

I am going to describe a model, cooked up by Kogut and myself[5] and independently by 't Hooft[6] which forces the electric field to behave very differently from Fig. 4. The model is very unrealistic but it will help prepare you for the more ambitious model of lecture 3.

The model assumes that the vacuum is a dielectric medium with some unusual properties. I will begin by reminding you of the electrostatics of dielectrics.

The free charges (quarks) are sources of the Maxwell D field

$$\vec{\nabla} \cdot \vec{D} = \rho \tag{I.23}$$

The electric field E is curl free and is related to D by the dielectric permeability $\varepsilon(x)$

$$\vec{D}(x) = \varepsilon(x) \vec{E}(x) \tag{I.24}$$

In this model $\varepsilon(x)$ can take one of two values, namely zero and one at any point. The regions where $\varepsilon = 0$ will be called forbidden because the D field is excluded from such regions. Wherever $\varepsilon = 1$ the material is normal.

The energy consists of two terms, the first being electrostatic energy and the second being the internal energy stored in the dielectric The electrostatic energy is

$$W_{e.s.} = \int d^3x \, \vec{D} \cdot \vec{E} = \int d^3x \, \frac{D^2}{\varepsilon} \tag{I.25}$$

From (I.25) it is evident that D is excluded from regions where $\varepsilon = 0$. The internal energy of the dielectric will be chosen so that the forbidden regions have less energy than the normal. Thus the ground state or vacuum is forbidden. We will write the internal energy as

$$W_{\varepsilon} = c \int \varepsilon(x) \, d^3x \tag{I.26}$$

remembering that ε has only 2 values. The total energy is

$$W = \int \frac{D^2}{\varepsilon} d^3x + c \int \varepsilon(x) \, d^3x \tag{I.27}$$

The model was invented so that the long range component of the D field would cost an infinite amount of energy. To see how the model works let's suppose the dielectric material fills a sphere of radius R and outside the sphere $\varepsilon = 0$. Suppose a charge g is placed at the origin. The D field then satisfies

$$\nabla \cdot \vec{D} = g \, \delta^3(x) \tag{I.28}$$

The first type of solution to try is a spherically symmetric distribution of flux

$$\vec{D} = g\,\frac{\hat{r}}{r^2} \tag{I.29}$$

Since any forbidden region with non-vanishing D costs infinite energy, the entire dielectric must be normal. The resulting energy is

$$W = \frac{4\pi g}{a} + \frac{4}{3}\pi R^3 c \tag{I.30}$$

In this formula a represents the size over which the charge is smeared. The first term is the electrostatic energy and the second term is the internal energy of the dielectric when the whole sphere is normal. The second term diverges as R^3 when the volume of the dielectric goes to infinity.

The energy can be lowered by allowing the electric flux to be distributed non-symmetrically. For example, suppose all of the flux is distributed over a solid angle Ω within which the dielectric is normal. The D field is given by

$$\vec{D} = \frac{4\pi}{\Omega}\,g\,\frac{\hat{r}}{r^2} \tag{I.31}$$

within the solid angle Ω and is zero outside. This time the total energy is

$$W = g\,\frac{16\pi^2}{\Omega a} + \frac{\Omega}{3}R^3 c \tag{I.32}$$

The electrostatic energy has increased because the field lines are squeezed but the internal energy is lowered. Since when $R \to \infty$ the internal energy dominates it always pays to decrease Ω .

The limiting form of field which lowers the energy to its absolute minimum is to allow all the flux to go through a long thin tube of normal material until it reaches the surface of the dielectric. (See Fig. 5)

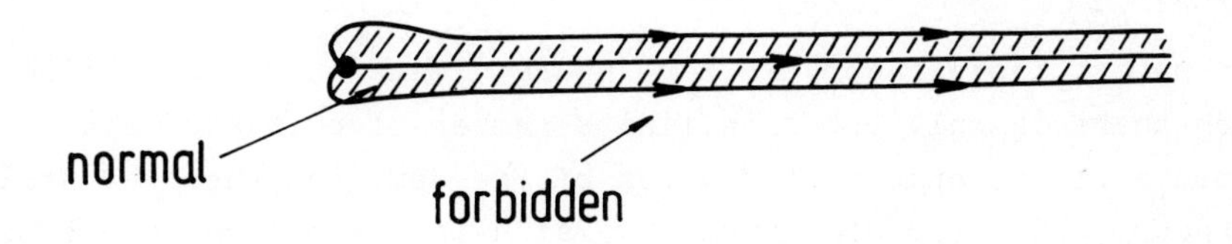

Fig. 5

The thickness of the tube is obtained by varying the energy per unit length with respect to the radius. If the radius is r the D field (which is parallel to the tube) is

$$D = \frac{4g}{r^2} \tag{I.32}$$

and the electrostatic energy per unit length is

$$\frac{W_{es}}{L} = \frac{16\pi g^2}{r^2} \tag{I.33}$$

and the internal energy/L is $\pi r^2 c$. Thus

$$\frac{W}{L} = \pi r^2 c + \frac{16\pi g^2}{r^2} \tag{I.34}$$

and the minimum occurs at

$$r^2 = \frac{4g}{\sqrt{c}} \tag{I.35}$$

Eq. (I.35) represents the thickness of the tube far from the charge. Near the charge the situation is more complicated. What is clear is that the minimum energy of an isolated charge grows linearly with R since, far from the charge, the energy per unit length is constant.

The remaining arguments now parallel the one dimensional case. The separation of quarks can only take place until the available energy is used up or until the tube breaks by pair production.

In the next two lectures I will show you how the nonlinearities of quantized Yang Mills theory can squeeze the electric flux into one dimensional tubes.

II. YANG MILLS IN ZERO DIMENSIONS

a) Gauge Invariance in Zero Space Dimensions

What is field theory in zero space dimensions? It is field theory in which there is only one or a finite number of points of space and therefore a finite number of degrees of freedom. For the free scalar field theory, the zero dimensional version is a single harmonic oscillator or a finite number of coupled oscillators. The first step in understanding a field theory is to understand its zero dimensional analog.

The second step involves a lattice of elementary zero dimensional systems with some form of coupling between the neighboring systems. If the lattice spacing is not too large a qualitative understanding of the large scale behaviour of the field theory is usually possible at this level. Of course the short distance behaviour is absent.

The final and most difficult step is allowing the lattice spacing to go to zero. Typically this involves renormalization of the parameters so that the low energy (long wave length) behaviour is prevented from varying as the spacing tends to zero.

In this lecture I am going to show you how to do step one. We will formulate Yang Mills theory for two spatial points (one point is too trivial). Then in lecture III we will do step 2 and show how quarks may be confined in the strongly coupled theory. Unfortunately the third step will have to wait until someone figures out how to do it.

We begin with a universe consisting of a pair of points 1 and 2 and a continuum of time. The presence of colored quarks on site 1 and 2 is described by fields $\psi_j(i)$ and $\psi_j^\dagger(i)$. Here i labels the 2 points 1 and 2 and j is the color index*. The field ψ may be represented in terms of fermion creation and annihilation operators for each site

$$\psi^\dagger(i) = a^+(i) + b^-(i) \tag{II.1}$$

where $a^+(i)$ $\left(b^-(i)\right)$ creates (annihilates) a quark (antiquark) at site i.

We will begin with a very simple hamiltonian which just assigns an energy μ to a quark

$$H = \mu \sum_i :\psi^\dagger(i)\,\psi(i): \tag{II.2}$$

In addition to global color rotations

$$\psi(i) \longrightarrow V\psi(i) \tag{II.3}$$

H is invariant under separate color rotations at sites 1 and 2

$$\psi(i) \longrightarrow V(i)\,\psi(i) \tag{II.4}$$

In equs. (II.3) and (II.4) the quantities V, $V(1)$ and $V(2)$ are any special unitary 2x2 matrices.

---

* For illustrative purposes the color group will be SU_2 instead of SU_3.

Transformations like (II.4) in which different color rotations may act at 1 and 2 are called local gauge transformations. They are symmetries of the hamiltonian in (II.2) since the degrees of freedom at 1 and 2 are completely uncoupled. But the lack of coupling is not necessary for local gauge invariance. For example the term

$$\psi^\dagger(1)\,\psi(1)\,\psi^\dagger(2)\,\psi(2) \tag{II.5}$$

couples sites 1 and 2 and is gauge invariant. The important feature of hamiltonians like (II.2) and (II.5) is that they do not transport quarks from one site to another.

To make H a little more interesting we can introduce terms which do transport quarks from 1 to 2. For example

$$i\left[\psi^\dagger(1)\,\psi(2) - \psi^\dagger(2)\,\psi(1)\right] \tag{II.6}$$

annihilates a quark at 2 and creates one at 1. This term is still globally color invariant but local gauge invariance is lost. This implies an absolute standard of comparison between color directions at 1 and 2.

I don't know of any mathematical principle which forbids such an absolute standard but it does seem to me to endow space with some extra machinery to keep track of the relative phases between 1 and 2. Let me make this machinery more explicit in the form of a matrix U which relates the two color reference frames. If the two frames are parallel then $U = 1$ and H is given by (II.6). Now let's imagine that the color frame at 2 was secretly rotated relative to 1. The relative rotation would be detected because the dynamics would now involve a nontrivial matrix U in the form

$$i\left[\psi^\dagger(1)\,U\,\psi(2) - \psi^\dagger(2)\,U^{-1}\,\psi(1)\right] \tag{II.7}$$

In the Yang Mills theory the relative rotation would be undetectable. The gauge invariance is restored by making the connecting matri U a dynamical variable with time dependence, an equation of motion and quantum fluctuations. The new degree of freedom U belongs to neither site but jointly to the two sites, or better yet, to the space between the sites.

Since U is an SU_2 matrix connecting the color frames at sites 1 and 2 it can be written in the form

$$\mathcal{U} = \exp \frac{i}{2} \tau \cdot \mathcal{B} \qquad \text{(II.8)}$$

where τ are the three Pauli matrices. The two indices of $\mathcal{U}$ are associated with the two sites. Under a local gauge transformation $\mathcal{U}$ transforms as

$$\mathcal{U} \rightarrow V(1)\, \mathcal{U}\, V^{-1}(2) \qquad \text{(II.9)}$$

in order to keep the hamiltonian in (II.7) unchanged.

We will soon introduce gauge invariant terms into H which do not commute* with $\mathcal{U}$. When this is done $\mathcal{U}$ will no longer be a static set of numbers but will become a full fledged quantum dynamical variable. We will no longer be able to transform $\mathcal{U}$ away by a color rotation at one site. And finally, although H permits processes in which quarks hop from site to site, the dynamics remains invariant under local gauge transformation.

b) Kinematics and Dynamics of $\mathcal{U}$ [9]

The real heart of non-abelian gauge mechanics is in the properties of the operators $\mathcal{U}$. On what space of states do components of $\mathcal{U}$ act? What are the variable conjugate to $\mathcal{U}$ and what are the commutation relations? The answer to these questions in the simplified zero dimensional model will determine the principles of quantization of the infinitely richer lattice model of lecture III [9].

The system described by $\mathcal{U}$ has as its configuration space the set of all possible rotations in 3-dimensional color space (More exactly elements of the universal covering group SU_2). The elements of $\mathcal{U}$ are a particular set of coordinates in this space. There are many other possible ways to coordinitize this space. For example the Euler angles can be used to parametrize rotations. Or the matrices $\mathcal{U}$ may be written in terms of a vector potential $\mathcal{B}$ as in (II.8). A particularly useful family of coordinates is defined by the representation matrices for color spin j. These $(2j+1) \times (2j+1)$ matrices may be written

---

* At present $\mathcal{U}$ is a matrix in the 2x2 color space. The individual components of $\mathcal{U}$ will become operators in the quantum space of states. We are using the term commute in the latter sense.

$$U_j = \exp \frac{i}{2} [T_j \cdot B] \tag{II.10}$$

where T_j are the Pauli matrices for spin j. The U of eq. (II.7) is the special case $U_{1/2}$. Whenever U occurs without a subscript j it will be understood as $U_{1/2}$.

The symmetry group associated with local gauge invariance is $SU_2 \times SU_2$. The two SU_2 groups are the local gauge transformations at sites 1 and 2 and each has its own generators. The 3 generators at site (i) are called $E_\alpha(i)$ and have the commutation relations

$$[E_\alpha(i), E_\beta(j)] = i\, \varepsilon_{\alpha\beta\gamma}\, \delta_{ij}\, E_\gamma(i) \tag{II.11}$$

From the transformation laws (II.9) it follows that the E's and U's satisfy the commutation relations

$$[E_\alpha(1), U_j] = \frac{1}{2} (T_j)_\alpha U_j \qquad \text{(no sum on } j\text{)}$$

$$[E_\alpha(2), U_j] = -U_j \frac{T_j}{2} \tag{II.12}$$

Since $U_{1/2}$ completely determines an element of the rotation group, all the U_j are functions of $U_{1/2}$. Therefore the quantum conditions are completely specified by the relations (II.12) for $j = 1/2$. Furthermore the three sets of variables $E(1)$, $E(2)$ and U are not independent. In fact $E(2)$ is given in terms of $E(1)$ and U_1 by

$$E(2) = -U_1\, E(1) \tag{II.13}$$

This can be shown by substituting

$$(U_1)_{\alpha\beta} = \frac{1}{2} \mathrm{Tr}\, U \tau_\alpha U^{-1} \tau_\beta$$

Then the second of eq. (II.12) follows from the first and (II.13). Eq. (II.13) says that the color vectors $E(1)$ and $E(2)$ are related by the rotation described by U. This observation will play a central role in our understanding of electric flux in Y. M. theory.

From (II.13) it follows that

$$E(2)^2 = E(1)^2 \tag{II.14}$$

In general states classified under the group $SU_2 \times SU_2$ are labelled by two total angular momenta $j(1)$ and $j(2)$ and two magnetic quantum numbers $m(1)$ and $m(2)$ such that

$$-j(i) \leq m(i) \leq j(i)$$

In the present case eq. (II.14) requires

$$j(1) = j(2) \equiv j \tag{II.15}$$

so that the states form $(2j+1)^2$ degenerate multiplets.

The conditions (II.11) - (II.15) can be realized on a space of states generated as follows. We begin with a "base" state $|0\rangle$ which is invariant under $SU_2 \times SU_2$. We then construct a $(2j+1)^2$ dimensional multiplet by acting with the $(2j+1)^2$ elements of U_j on $|0\rangle$. Thus we define a unique $|0\rangle$ such that

$$E(1)|0\rangle = E(2)|0\rangle = 0 \tag{II.16}$$

The $(2j+1)^2$ states forming the (j,j) representation of $SU_2 \times SU_2$ are given by

$$U_j|0\rangle \tag{II.17}$$

It is easy to prove that the states in (II.17) are eigenvectors of $E(1)^2 = E(2)^2$.

$$E(1)^2 U_j|0\rangle = E_\alpha(1) E_\alpha(1) U_j|0\rangle$$

$$= E_\alpha(1)[E_\alpha(1), U_j]|0\rangle \quad \text{(see eq. (II.16))}$$

$$= E_\alpha(1) \frac{(T_j)_\alpha}{2} U_j|0\rangle \quad \text{(see eq. (II.12))}$$

$$= \frac{1}{2}[E_\alpha(1), (T_j)_\alpha U_j]|0\rangle$$

$$= \frac{1}{4}(T_j)_\alpha (T_j)_\alpha U_j|0\rangle$$

$$= j(j+1) U_j|0\rangle \tag{II.18}$$

It is also possible to generate the space of states using only the matrices U of the $1/2$ color representation. This is done by expressing U_j as a homogeneous polynomial of order $2j$ in the components of U and U^{-1}. This corresponds to the fact that any angular momentum can be built from spin $1/2$ systems. As an example we express U_1 in terms of $U_{1/2}$

$$(\mathcal{U}_1)_{\alpha\beta} = \frac{1}{2} \, Tr \left[\mathcal{U} \, \tau_\alpha \, \mathcal{U}^{-1} \tau_\beta \right] \tag{II.19}$$

The matrices $\mathcal{U}$ are the zero dimensional analogs of

$$\exp \; i \, a \, g \, \frac{\tau}{2} \cdot A \tag{II.20}$$

where a is the spatial distance between sites 1 and 2, g is the coupling constant and A is the vector potential. Similarly the generators $E(1)$ and $E(2)$ have analogs in the conventional Y. M. theory. The generators E are the non-abelian analogs of electric field. More precisely $E(1)$ ($E(2)$) is the electric field at site 1 (2) pointing toward 2 (1).

You should notice a certain formal similarity between the abelian and non abelian theories. In the abelian theory the operator $\exp[igA\cdot\varepsilon]$ acts to create an electric field along the direction $\vec{\varepsilon}$. In the non abelian theory $\mathcal{U}_j = \exp \frac{1}{2} i a g \, T_j A$ creates a non abelian electric field with magnitude $E^2 = j(j+1)$. However in the abelian theory the electric flux adds linearly, in the Y. M. theory it adds like angular momentum. The interpretation of E as electric field will become clearer when it is shown that $\nabla \cdot E = \rho$ in the next lecture.

The total color carried by the system consists of the color carrie by fermions plus the color carried by the gauge field $\mathcal{U}$. The color carried by $\mathcal{U}$ is defined as the quantity which generates global color rotations of $\mathcal{U}$. A global rotation rotates both frames equally

$$\mathcal{U} \rightarrow V \, \mathcal{U} \, V^{-1} \tag{II.21}$$

and under an infinitesimal rotation about the color axis α

$$\delta \mathcal{U} = \left[c(\alpha), \mathcal{U} \right] = \left[\tau_\alpha \mathcal{U} - \mathcal{U} \tau_\alpha \right] / 2 \tag{II.22}$$

where $c(\alpha)$ is the total α component of color. From (II.12) it is evident that the color carried by the gauge field is $E_\alpha(1) + E_\alpha(2)$. The total color is then

$$\sum_{i=1}^{2} : \psi^\dagger(i) \frac{\tau_\alpha}{2} \psi(i) : + E_\alpha(1) + E_\alpha(2) \tag{II.23}$$

The color carried by the gauge field $E(1) + E(2)$ may be thought of as the zero dimensional analog of $\nabla \cdot E$.

Not all the states of the system are physical. As in 1-dimensional QED the constraint of local gauge invariance must be applied to the physical states. To derive these conditions we note that the local color rotation at site i is generated by

$$G^{\alpha}(i) = \frac{1}{2}\psi^{+}(i)\,\tau_{\alpha}\psi(i) + E_{\alpha}(i) \tag{II.24}$$

The terms $\frac{1}{2}\psi^{+}(i)\tau\psi(i)$ rotate the quark fields while the E's rotate U. As in abelian theory the gauge constraints state that $G(i)$ annihilate any physical state

$$\left\{E(i) + \psi^{+}(i)\frac{\tau}{2}\psi(i)\right\}|\,\rangle = 0 \tag{II.25}$$

When eq. (II.25) and (II.13) are combined an interesting physical picture emerges. We can visualize sites 1 and 2 as sources and sinks of electric field. This is shown in Fig. 6.

Fig. 6

Eq. (II.25) tells us that the total flux leaving site i is equal to the charge at that point. However in going from site 1 to 2 the electric flux undergoes a color rotation as indicated by eq. (II.13). This change in electric flux can be viewed as a source if we recall that the total color carried by the gauge field between 1 and 2 is

$$C_{\alpha}(\mathit{field}) = E_{\alpha}(1) + E_{\alpha}(2) \tag{II.26}$$

The point which I will reemphasize in the Lecture III is that the color in the gauge field does not originate new lines of flux but rather it twists them in color space.

The construction of the physical space of states begins by defining a gauge invariant product state $|0\rangle$ by means of the relations

$$\begin{aligned} a^{-}(i)|0\rangle &= b^{-}(i)|0\rangle = 0 \\ E(i)|0\rangle &= 0 \end{aligned} \tag{II.27}$$

where a^{-} (b^{-}) annihilates quarks (antiquarks) at site i. The next step is to define enough gauge invariant operators to generate the whole space when acting on $|0\rangle$. We will do this by considering products of $\psi(i)$, $\psi^{+}(i)$ and $U_{1/2}$. Let's first formulate a rule which will allow us to easily recognize gauge invariant operators. The rule is that if we focus attention on the indices associated with one site (or the other) the operator should form a scalar. This can only happen if all the indices at a given site are contracted among themselves. I'll give some examples. First the operator $\psi_i^\dagger(1)\,\psi_i(1)$ (here i is a color index) is gauge invariant because the contracted indices (i) belong to the same site. However $\psi^\dagger(1)\,\psi(2)$ is not gauge invariant. The list of gauge invariant operators which are necessary to create the full space of states is given by

$$\begin{aligned} &\psi^\dagger(1)\,\psi(1) \\ &\psi^\dagger(2)\,\psi(2) \\ &\psi^\dagger(1)\,U\,\psi(2) \\ &\psi^\dagger(2)\,U^{-1}\psi(1) \end{aligned} \qquad \text{(II.28)}$$

The idea of local contraction of indices is rather trivial for the zero dimensional case but it will be very useful in constructing the gauge invariant operators in the more complex lattice theory.

Now lets examine the states that can be made by repeated application of the operations (II.28).

1) $\psi^\dagger(i)\,\psi(i)\,|0\rangle$. This is a $q\bar{q}$ pair in the color singlet state. Both particles are at site i .
2) $\psi^\dagger(1)\psi(1)\,\psi^\dagger(2)\psi(2)|0\rangle$. This is a color singlet pair at each site.
3) $\psi^\dagger(1)\,U\,\psi(2)$. This is a quark at (1) and an antiquark at (2). The operator U creates an electric flux satisfying (II.25).
4) $\{\psi^\dagger(1)U\psi(2)\}\{\psi^\dagger(2)\,U^{-1}\psi(1)\}\,|0\rangle$.
   This state is more complex than the others since it contains two superimposed electric fluxes.

I will illustrate the technique for adding flux by using a simple identity whose proof you can supply.

$$U_{ij}\,(U^{-1})_{kl} = \frac{1}{2}\delta_{il}\,\delta_{jk} + \frac{1}{4}\left(\mathrm{Tr}\,U^{-1}\tau_\beta U\tau_\alpha\right)(\tau_\beta)_{il}\,(\tau_\alpha)_{jk}$$

$$= \frac{1}{2}\delta_{il}\,\delta_{jk} + \frac{1}{2}\,(U_1)_{\beta\alpha}\,(\tau_\beta)_{il}\,(\tau_\alpha)_{jk} \qquad \text{(II.29)}$$

When (II.29) is substituted into the state (4) we get a superposition of states

$$\frac{1}{2}\,\psi^{\dagger}(1)\,\psi(1)\;\psi^{\dagger}(2)\,\psi(2)$$

$$+\frac{1}{2}\left[\psi^{\dagger}(1)\,\tau_{\alpha}\,\psi(1)\right]\left(U_{1}\right)_{\alpha\beta}\left[\psi^{\dagger}(2)\,\tau_{\beta}\,\psi(2)\right] \qquad \text{(II.30)}$$

The first term we have already talked about. The second represents a new object composed of a color-spin 1 pair at each site. The colored pairs are accompanied by a color-1 flux line created by U_1. This example illustrates how you must combine flux in a non-abelian theory.

In general the states 1-4 are not energy eigenvectors. If H is gauge invariant it will not lead out of this subspace but it may have transition elements within the subspace.

I will choose H to be as close as possible to a real covariant Y. M. hamiltonian. For this purpose we can write

$$U_{1/2} = \exp\frac{ig}{2}\,\tau\cdot A \;=\; 1+\frac{ig}{2}\,\tau\cdot A+\ldots \qquad \text{(II.31)}$$

First consider $\psi^{\dagger}(1)\,U\,\psi(2)$. Applying (II.31) gives

$$\psi^{\dagger}(1)\,U\,\psi(2) = \psi^{\dagger}(1)\,\psi(2) + ig\,\psi^{\dagger}(1)\,\frac{\tau}{2}\,\psi(2)\cdot A$$

adding the h. c. gives

$$i\Big[\psi^{\dagger}(1)\,\psi(2)-\psi^{\dagger}(2)\,\psi(1)\;+$$

$$+\,ig\,A\cdot\Big(\psi^{\dagger}(1)\,\frac{\tau}{2}\,\psi(2)+\psi^{\dagger}(2)\,\frac{\tau}{2}\,\psi(1)\Big)\Big] \qquad \text{(II.32)}$$

These terms are analogs of the kinetic and interaction terms in a conventional gauge theory.

The next term represents the energy stored in the electric field. It is given by the gauge invariant operator

$$\frac{E(1)^{2}}{2} = \frac{E(2)^{2}}{2} \qquad \text{(II.33)}$$

You should compare these terms with eq. (I.3) to see how they are similar to ordinary terms in a gauge theory. The only terms which are not present in the zero dimensional model (and one dimensional models) are the magnetic energy. Magnetic fields do not occur for spatial dimensions < 2. In the lattice theory in 3 dimensions they will be included.

Exercise: Construct a Yang Mills theory for 4 points arranged in a square. Each corner has a ψ and each side a U. What is the significance of the operator

$$\mathrm{Tr}\ U(1)\,U(2)\,U(3)\,U(4) \qquad ?$$

What is the effect of adding this operator into the hamiltonian?

III. LATTICE YANG MILLS THEORY

a) Degrees of Freedom of a Lattice

The usual continuous coordinates (x, y, z) of space are replaced by a triplet of integers $(r_x, r_y, r_z) = (\vec{r})$. The points $(\vec{r})$ are called sites. At each site there are six lattice vectors $\hat{n}_x$, $\hat{n}_y$, $\hat{n}_z$, $\hat{n}_{-x}$, $\hat{n}_{-y}$, $\hat{n}_{-z}$ shown in Fig. 7 .

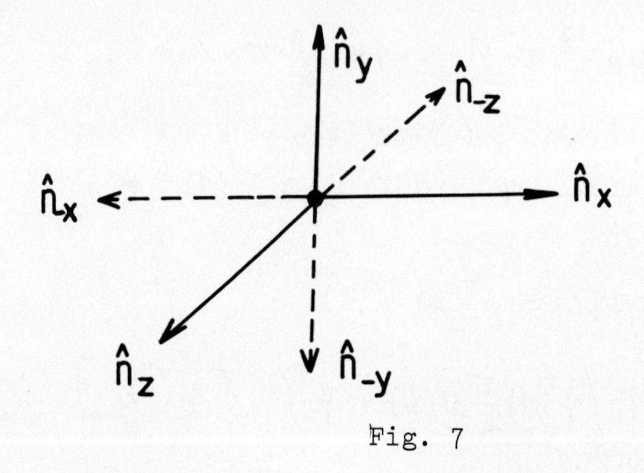

Fig. 7

In general sums over the lattice vectors will include all 6 directions.

The spaces between sites will be called links. The links will usually be considered to be directed and will be labelled by a site and a lattice vector. For example $(r, \hat{n})$ and $(r+\hat{n}, -\hat{n})$ describe the two directed links associated with the space between r and $r+\hat{n}$.

A 4 component fermion field $\psi(r)$ can be represented in terms of creation and annihilation operators for quarks and antiquarks at site (r)

$$\psi(r) = a_i^{+}(r)\,\xi_i + b_i^{-}(r)\,\overline{\xi}_i \tag{III.1}$$

where $\xi_1 = \begin{pmatrix}1\\0\\0\\0\end{pmatrix}$, $\xi_2 = \begin{pmatrix}0\\1\\0\\0\end{pmatrix}$, $\xi_3 = \begin{pmatrix}0\\0\\1\\0\end{pmatrix}$, $\xi_4 = \begin{pmatrix}0\\0\\0\\1\end{pmatrix}$

in a representation in which

$$\gamma_0 = \begin{pmatrix} 1 & 0 & 0 & 0 \\ 0 & 1 & 0 & 0 \\ 0 & 0 & -1 & 0 \\ 0 & 0 & 0 & -1 \end{pmatrix}$$

Each link will carry a degree of freedom $U(r,\hat{n})$ to describe how color information is transported between neighboring sites. The two directed links associated with the same lattice space do not have independent degrees of freedom. In fact the two U's are inverses of one another

$$U(r,\hat{n}) = U^{-1}(r+\hat{n},-\hat{n}) \qquad \text{(III.2)}$$

Each link has two generators analogous to $E(1)$ and $E(2)$ in the last lecture. I will use a labelling scheme for the E's defined as follows. If we consider the link $(r,\hat{n})$ it has two ends, one at (r) and one at $(r+\hat{n})$. The two generators for the degree of freedom $U(r,\hat{n})$ at r and $r+\hat{n}$ will be called

$$E(r)\cdot\hat{n} \quad \text{and} \quad E(r+\hat{n})\cdot(-\hat{n}) \qquad \text{(III.3)}$$

The notation indicates that the two generators represent electric flux in opposing directions as shown in Fig. 8.

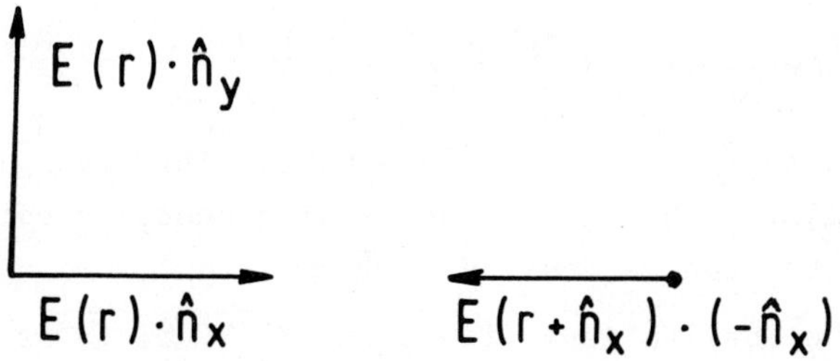

$E(r)\cdot\hat{n}_y$

$E(r)\cdot\hat{n}_x$

$E(r+\hat{n}_x)\cdot(-\hat{n}_x)$

Fig. 8

Thus the generators $E\cdot\hat{n}$ are electric fluxes flowing outward from r in the direction $\hat{n}$. The commutation relations between the components of $E\cdot\hat{n}$ are the usual SU_2 commutation relations. Eq. (II.13) is replaced by

$$E(r+\hat{n})\cdot(-\hat{n}) = -U(r,\hat{n})\,E(r)\cdot\hat{n} \qquad \text{(III.4)}$$

and (II.14) by

$$\left(E(r)\cdot\hat{n}\right)^2 = \left(E(r+\hat{n})\cdot(-\hat{n})\right)^2 \tag{III.5}$$

Evidently the source of electric flux on link $(r,\hat{n})$ is

$$c(r,\hat{n}) = E(r)\cdot\hat{n} + E(r+\hat{n})\cdot(-\hat{n}) \tag{III.6}$$

b) The Gauge Invariant Subspace

The physical constraint of gauge invariance again requires every quark to be a source of electric flux. The way to show this is to follow the same logic we used in one dimensional QED and zero dimension Y. M. theory - construct the local generator of gauge transformations and then set it to zero. The gauge transformation at site r acts on $\psi(r)$ and on the six gauge fields $U(r,\hat{n})$. Accordingly the generator is the sum of seven terms

$$G(r) = \psi^\dagger(r)\frac{\tau}{2}\psi(r) + \sum_{\hat{n}} E(r)\cdot\hat{n} \tag{III.7}$$

The physical subspace is then defined by

$$\left\{\psi^\dagger(r)\frac{\tau}{2}\psi(r) + \sum_{\hat{n}} E(r)\cdot\hat{n}\right\}|\ \rangle = 0 \tag{III.8}$$

The quantity $\sum_{\hat{n}} E(r)\cdot\hat{n}$ is the total flux diverging from the point r. Eq. (III.8) then gives the usual connection between the divergence of E and the charge density $\psi^\dagger\frac{\tau}{2}\psi$.

Unlike the abelian gauge field the Y. M. field is also a source. This can be seen from eq's. (III.4) and (III.6) which say that the field varies along a link by an amount equal to the color carried by that link It is evident that the flux passing through a closed surface is the sum of the colors carried by the sites (quarks) and links (gauge field) enclosed.

We can now construct the physical space of states beginning with a vector $|0\rangle$ satisfying

$$a^-(r)|0\rangle = b^-(r)|0\rangle = 0$$

$$E(r)\cdot\hat{n}|0\rangle = 0 \qquad (\text{all } r \text{ and } \hat{n}) \tag{III.9}$$

Let's ignore the quarks and concentrate on the gauge invariant operators which can be built from the U's. The principle for forming gauge invariants is again the local contraction of indices. To form the general class of gauge invariant operators we first specify a closed oriented path of links Γ. The path may cover any link one or more times (see Fig. 9)

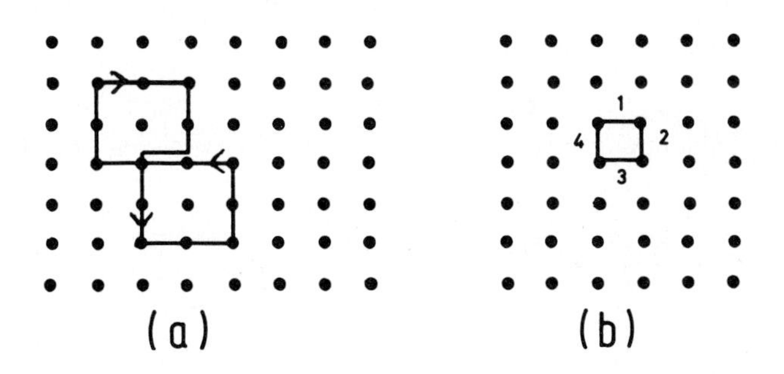

Fig. 9

Now beginning with an arbitrary link on the path, multiply (2x2 matrix multiplication) the U's in the order indicated by Γ. For example for the path shown in Fig. 9b the required product is

$$U(1)\,U(2)\,U(3)\,U(4) \tag{III.10}$$

There are still two open indices which are contracted by taking the trace. The resulting object is called $U(\Gamma)$. Since the indices in $U(\Gamma)$ are all locally contracted $U(\Gamma)$ is gauge invariant. It can be shown that the entire gauge invariant space is generated by repeated application of the $U(\Gamma)$ operators applied to $|0\rangle$.

The physical properties of $U(\Gamma)|0\rangle$ are very simple and interesting. First consider any link not on the path Γ. Since no U has acted to create electric field these links have no electric flux through them. The links which appear in Γ (suppose no link appears more than once) have an electric flux satisfying

$$E^2 = \frac{1}{2}\left(\frac{1}{2}+1\right) = \frac{3}{4} \tag{III.11}$$

Accordingly the closed curve Γ can be described as a closed line of electric flux. The fact that electric flux lines must form closed lines in the absence of quarks is of course the familiar idea of continuity of electric flux originally envisioned by Faraday. However there are two differences between lines of flux in ordinary electrodynamics and Y. M. theory on a lattice.

The first difference is due to the fact that the Y. M. field is its own source. However it is a particularly simple kind of source which according to (III.4) causes the electric field to undergo a color rotation between the two ends of a link. The important observation is that the color on a link is not a source or origin of a new flux line but rather it color-twists the flux lines.

The second difference is due to the fact that the color group is compact. This means that the generators $E \cdot \hat{n}$ are quantized in the sense that

$$(E \cdot \hat{n})^2 = \frac{n}{2}\left(\frac{n}{2}+1\right) \quad , \quad n = 0, 1, 2, \ldots \qquad \text{(III.12)}$$

The flux through a link can not be arbitrarily small. To see what this means we can compare the situation with conventional electrodynamics formulated on a spatial lattice. In this case the flux can be arbitrarily subdivided. The flux emanating from a charge can spread out so that the flux through a distant link goes as $1/r^2$. This contrasts sharply with the flux in a non abelian theory which comes in quantized units.

When quark fields are included the electric flux lines can begin and end on sites occupied by quarks. This is because the open indices of an expression like $U(1)\,U(2)\,U(3)\,\ldots\,U(6)$ (see Fig. 10) can be contracted with quark field indices.

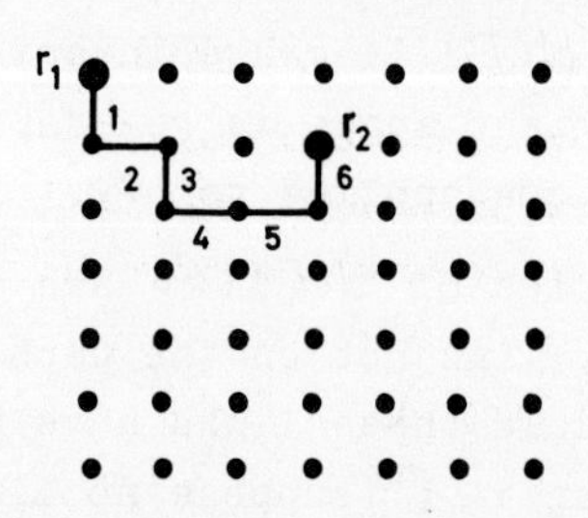

Fig. 10

For example we can form

$$\psi^{\dagger}(r_1)\,\mathcal{O}\,U(1)\,\ldots\,U(6)\,\psi(r_2) \qquad \text{(III.13)}$$

where σ is an arbitrary Dirac matrix. In general the full set of gauge invariant functions of ψ, ψ^{+} and U will depend on the group describing color. If the group is SU_2 then we have operators like (III.13) as well as operators

$$(\psi^{c}(r))^{+} \sigma\, U(1) \ldots U(n)\, \psi(r_2) \qquad \text{(III.14)}$$

where ψ^{c} is the charge conjugate to ψ. These operators describe quark pairs as opposed to quark antiquark pairs. If the color group is SU_3 the diquark operators like (III.14) are replaced by qqq operators. These are formed by considering a connected collection of links with the topology of a Y as in Fig. 11.

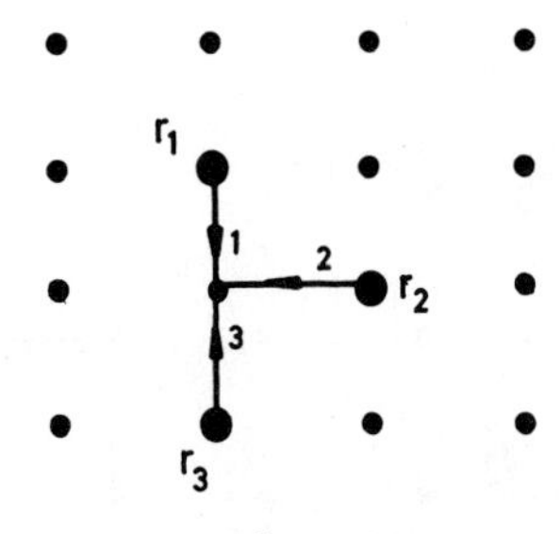

Fig. 11

The operator for Fig. 11 is

$$\psi_i^{+}(r_1)\, U_{ij}(1)\, \psi_k^{+}(r_2)\, U_{kl}(2)\, \psi_m^{+}(r_3)\, U_{mn}(3)\, \varepsilon_{jln}$$

The entire space of states can be represented in terms of arbitrary products of closed flux-loops and open flux-lines with quark ends. However a more useful representation exists for cases in which a given link is covered more than once. In this case it is useful to combine the flux according to the rules of angular momentum addition. I will illustrate this for the example in Fig. 12.

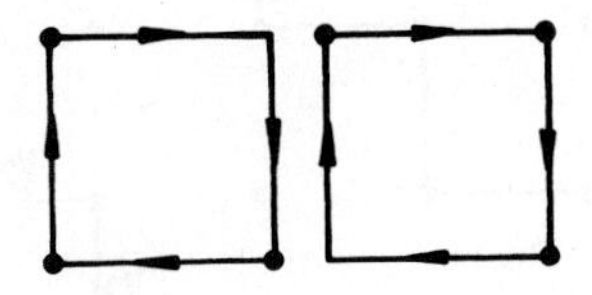

Fig. 12

The doubly covered link can be treated according to the method used in eq. (II.29). The resulting state is a linear superposition of two states in which that particular link carries electric flux of 0 and 1. This is shown in Fig. 13.

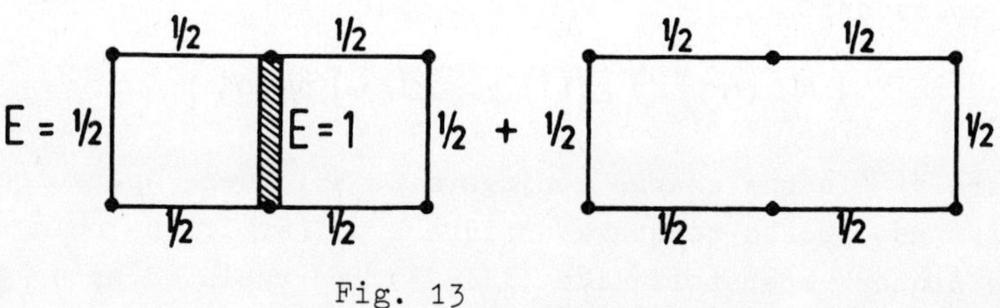

Fig. 13

This method may be generalized in order to introduce a representation of states in which flux may branch as in Fig. 14 as long as E_1 is

Fig. 14

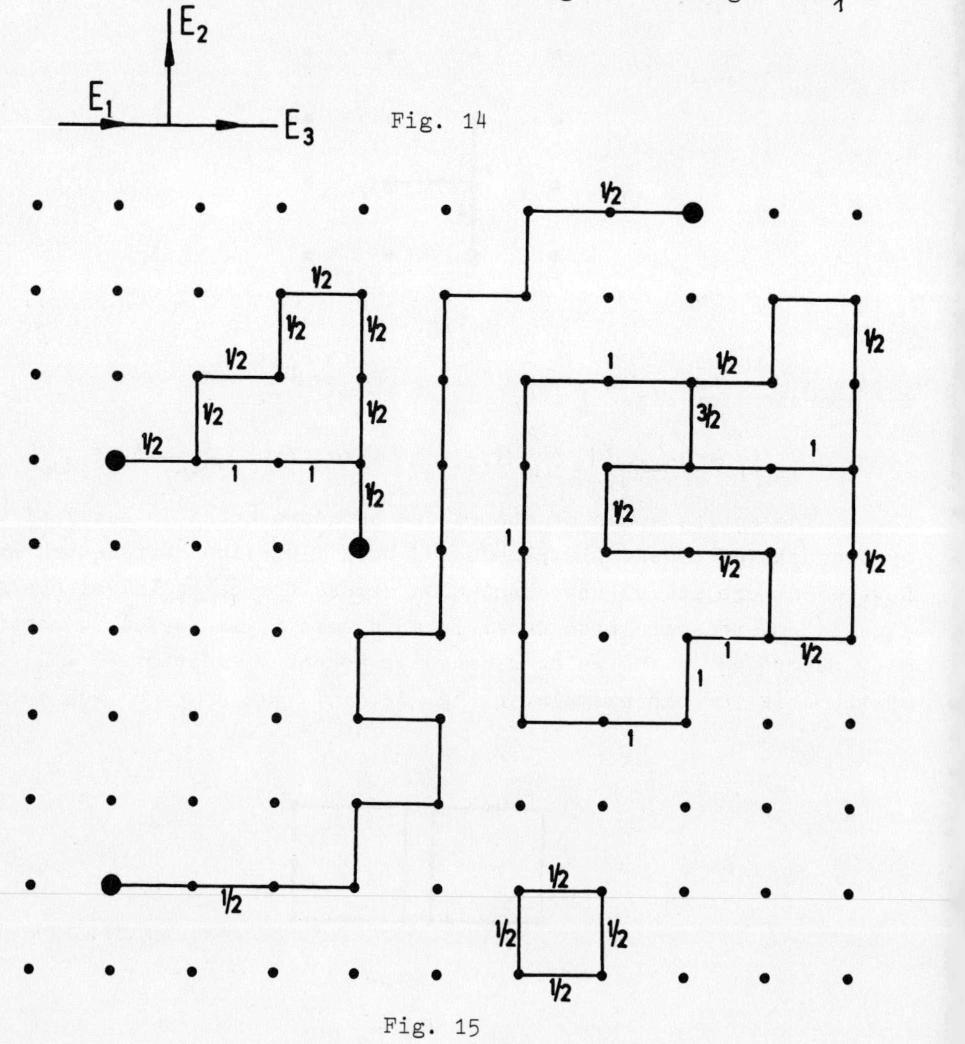

Fig. 15

included in the addition of E_2 and E_3 (added as angular momenta). A typical state is shown in Fig. 15.

Evidently we can characterize the space of states in terms of arbitrary branching strings or electric flux lines subject only to the constraint of flux continuity. All string ends must be quarks and all quarks must be string ends. This defines the kinematics of lattice Y. M. theory.

c) The Hamiltonian

The dynamics is defined by a gauge invariant hamiltonian whose matrix elements do not lead out of the gauge invariant subspace. In choosing H two principles, in addition to gauge invariance will guide us. The first is that in the limit of zero lattice spacing $(a \to 0)$, conventional Y. M. theory should be recovered. The second condition is that H shall be as local as possible. We will restrict our choice so that no links or sites are coupled if they are more than a single lattice space apart.

I will not do the algebra involved in taking the continuum limit but I will tell you how to do it yourself. You first define new variables $\chi(x)$, $\vec{A}(x)$, $\vec{\mathcal{E}}(x)$ by the equations

$$\psi(x) = a^{3/2} \chi(x)$$

$$E(r) \cdot \hat{n} = \frac{a^2}{g} \mathcal{E}(x) \cdot \hat{n}$$

$$U(r, \hat{n}) = \exp\left[iga A \cdot \hat{n} \frac{\tau}{2}\right] = 1 + iga \frac{\tau}{2} A \cdot \hat{n} + \dots \tag{III.15}$$

where a is the lattice spacing. Finite differences are replaced by derivatives

$$a^{-1}\left[f(r) - f(r - \hat{n})\right] = \hat{n} \cdot \partial f$$

and sums by integrals

$$\sum a^3 f(r) = \int d^3x \, f(x)$$

The hamiltonian will contain the following terms :

1) $$\frac{g^2}{2a}\sum_{r,\hat{n}}\left\{E(r)\cdot\hat{n}\right\}^2 \tag{III.16}$$

In the limit $a\to 0$ this becomes the usual electrostatic energy $\frac{1}{2}\int d^3x\ \mathcal{E}^2(x)$.

2) $$a^{-1}\sum_{r,\hat{n}}\left\{\psi^\dagger(r)\frac{\alpha\cdot\hat{n}}{i}U(r,\hat{n})\psi(r+\hat{n})+\mu\bar{\psi}(r)\psi(r)\right\} \tag{III.17}$$

when the continuum limit is taken this becomes the free quark hamiltonian plus the interaction energy (in the gauge $A_0=0$)

$$\int d^3x\left\{\bar{\chi}\,\gamma_i\,\partial_i\chi+ig\,\bar{\chi}\gamma_i A_i\frac{\tau}{2}\chi+\mu\bar{\chi}\chi\right\}$$

3) A term which has been absent in our simplified zero and one dimensional models is the magnetic energy. These terms are associated with elementary boxes or squares on the lattice. For each oriented unit square we include

$$\frac{4}{ag^2}\,\mathrm{Tr}\,UUUU \tag{III.18}$$

In the limit $a\to 0$ this term becomes the usual magnetic energy $\frac{1}{2}\int d^3x\ \mathcal{H}(x)^2$

where

$$\mathcal{H}_\alpha(x)=\nabla\times A_\alpha+g\,\varepsilon_{\alpha\beta\gamma}A_\beta\times A_\gamma$$

Each term in H has a particular significance for the string-like flux lines. We shall study these terms in the order of their importance when the coupling g is large.
The most important term for $g\gg 1$ is the electric energy $\frac{g^2}{2a}\sum E^2$. This term gives an energy

$$\frac{g^2}{2a}\left(\frac{n}{2}\right)\left(\frac{n}{2}+1\right) \tag{III.19}$$

to every link carrying flux $E^2=\frac{n}{2}\left(\frac{n}{2}+1\right)$. The vacuum of the strongly coupled limit is the state which minimizes this term. Therefore it is evident that the vacuum is the state $|0\rangle$ in which no flux lines are excited.

If we consider states in which the electric flux lines cover no link more than once then the electric energy gives each state an energy proportional to the total length of flux lines. This is the source of quark confinement in the strongly coupled Y. M. theory. For example if we consider a quark pair located at sites r_1 and r_2 then the minimum energy configuration of the gauge field will involve an electric flux line of minimal number of links. Therefore the energy will be stored on a straight line between the quarks and will grow linearly with their separation. Evidently the strongly coupled Y. M. theory is behaving exactly like the dielectric model of lecture I. If the lattice spacing is a then the minimum energy for a $q\bar{q}$ pair separated by distance D is

$$\frac{g^2}{2a}\left(\frac{1}{2}\right)\left(\frac{1}{2}+1\right)\frac{D}{a} = \frac{3g^2}{8a^2}D \tag{III.20}$$

Quarks will be confined if the electric energy dominates.

The electric energy is also instrumental in giving the pure gauge field excitations mass. For example consider the state

$$\mathrm{Tr}\, UUUU\, |0\rangle \tag{III.21}$$

in which a single box of electric flux is excited. The electric energy of this state is

$$H(\mathrm{Box}) = 4\,\frac{g^2}{2a}\cdot\frac{3}{4} = \frac{3g^2}{2a}\,.$$

Since each electric flux configuration is an eigenvector of the electric energy, no propagation of signals through the lattice will take place until the other terms are included.

The next term in importance is

$$a^{-1}\sum \psi^{\dagger}(r)\,\frac{\alpha\cdot\hat{n}}{i}\,U(r,\hat{n})\,\psi(r+\hat{n}) + \mu\bar{\psi}\psi\,.$$

This term allows the fermions to propagate through the lattice. It describes a process in which a quark and antiquark are created or annihilated at two neighboring points. The factor U creates or cancels the flux between them. The sequence of events in Fig. 16 shows how fermions may move through the lattice.

Processes induced by this term allow an electric flux line to break as quarks separate. Of course this only occurs when a pair is produced as in our earlier examples. Finally this term causes the physical vacuum to have a fluctuating number of $q\bar{q}$ pairs.

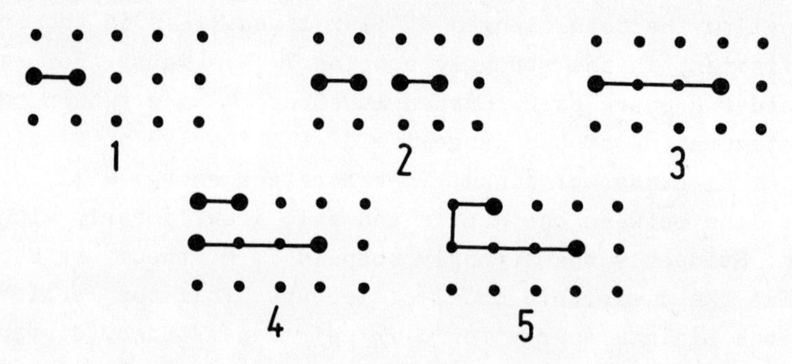

Fig. 16

The last term is the magnetic energy

$$\frac{4}{ag^2}\sum_{Boxes} \mathrm{Tr}\, UUUU$$

This term causes fluctuations in the position and structure of the string like flux lines. For example consider a static $q\bar{q}$ pair with a straight electric flux line as in Fig. 17

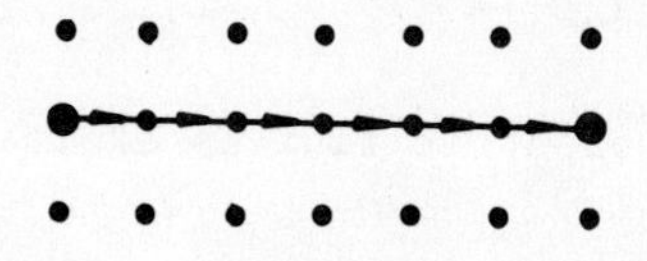

Fig. 17

Suppose we consider a box with a side in common with the flux line. If we apply $\mathrm{Tr}\, UUUU$ we create the superposition shown in Fig. 18 .

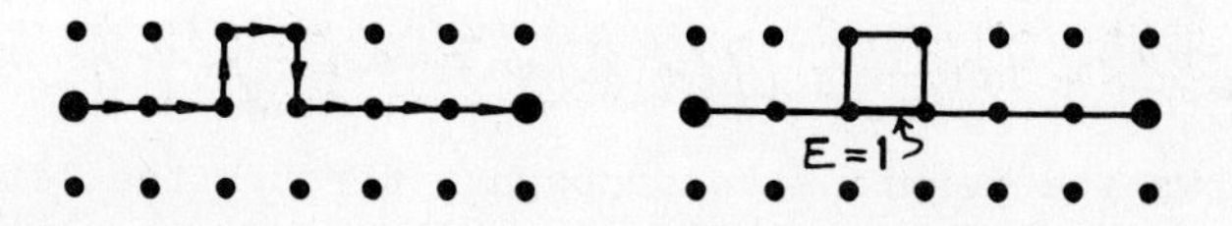

Fig. 18

In addition this term allows the vacuum to contain a fluctuating sea of closed flux lines.

If the fluctuations due to the magnetic energy become too large the quark confining mechanism can become undone. To get a rough idea of how this can happen we suppose the vacuum contains a dense sea of closed flux lines as in Fig. 19.

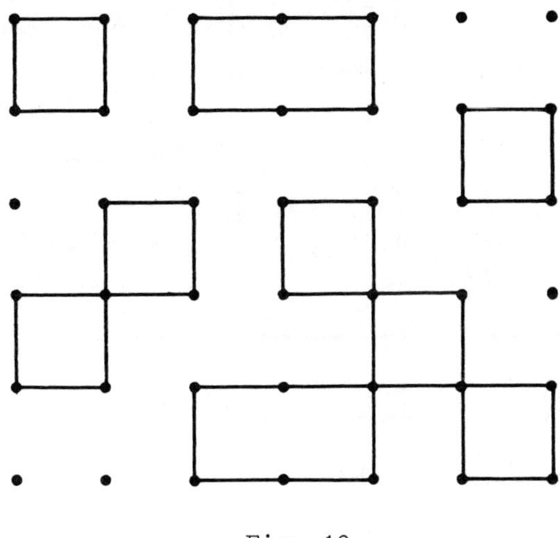

Fig. 19

Now suppose a quark is placed in the lattice as in Fig. 20. The flux due to the quark (which must go to ∞) is represented by the dark line.

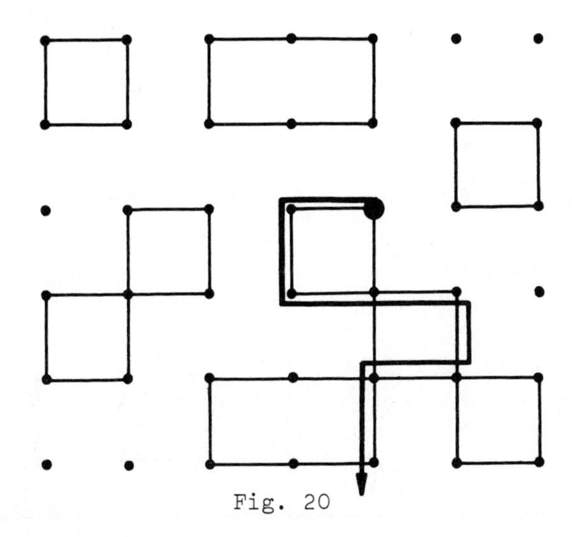

Fig. 20

As usual the doubly occupied links can be resolved into a coherent superposition with flux zero and flux one. Let us consider the particular contribution in which all these links carry $E = 0$. It is shown in Fig. 21.

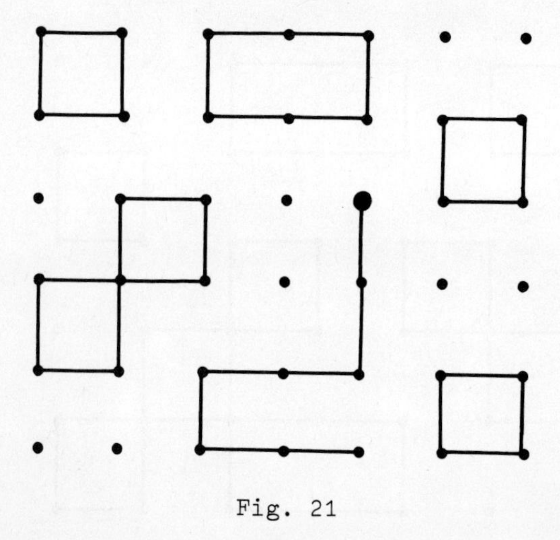

Fig. 21

Fig. 19 has 40 excited links but Fig. 21 has only 33. You see we have actually lowered the electric energy by adding a quark. However this obviously can not happen if the flux density of the vacuum is very low, i.e. if $g \gg 1$. The term

$$\sum \frac{4}{ag^2} \mathrm{Tr}\, UUUU$$

must have sufficient strength to fill the vacuum with a high density of flux loops.

The term $\mathrm{Tr}\, UUUU$ also causes motion through the lattice. The sequence in Fig. 22 shows how the gauge field excitations are caused to move through the lattice:

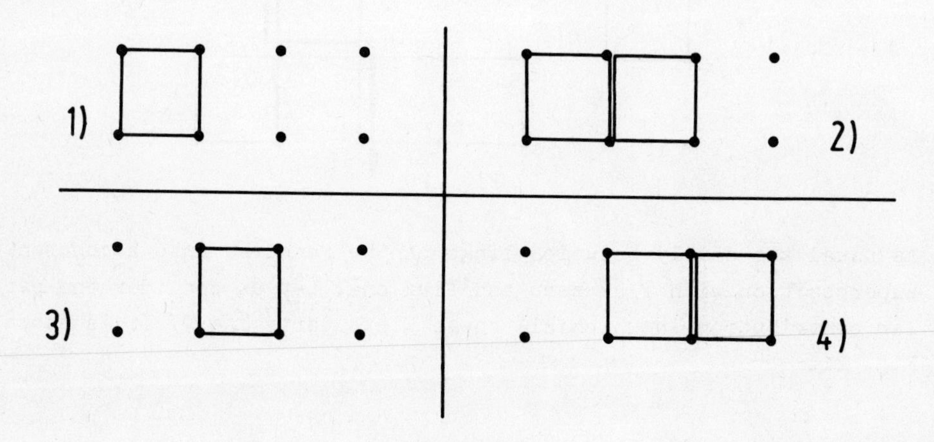

Fig. 23

d) Removing the Lattice and Infrared Slavery

The most difficult unanswered question posed by lattice Yang Mills theory concerns the removal of the lattice from the theory[10]. A proper discussion of this point is well beyond the range of these lectures and also this lecturer. Nevertheless I will try to give you a vague idea of how I think things should go. First of all we must realize that taking a to zero is far more delicate in quantum field theory than in classical theory. This is because the large scale behaviour of the theory is sensitive to a unless the bare parameters of the theory are continuously readjusted as $a \to 0$. This is the process of renormalization.

For example suppose that with lattice spacing a we use a coupling $g \gg 1$. The energy stored in a $q\bar{q}$ pair separated by distance D is

$$\frac{3g^2}{8a^2} D$$

Now suppose we wish to represent the same physics (for large D) by a new model in which the lattice spacing is $a/2$. In order to keep the energy unchanged we must use a new coupling constant which satisfies

$$(g')^2 = \frac{1}{2} g^2$$

Thus as the lattice spacing decreases the squared coupling constant must also decrease in order to keep the large scale physics unchanged.

The right theory probably requires

$$g^2 \sim \frac{1}{\log a}$$

as $a \to 0$. This was discovered by 't Hooft, Politzer and by Gross and Wilczek. This means that an accurate representation of continuum Yang Mills theory on a very fine lattice would require a very small coupling. However renormalization effects cause the effective coupling to increase with a until we (hopefully) reach a point where a is comparable to the hadron radius and $g > 1$. We can then apply the strong coupling methods outlined in this lecture.

If this view is correct then there is no "phase transition" between large g and small g so that no discontinuous effects occur as a and g^2 become small. Under these conditions quark confinement can be decided by examination of the large g limit.

In this regard I should mention the relation between the mechanism described here and the idea of infrared slavery[11]. The quark confining mechanism I've described begins with the idea that the "running" coupling constant is $\gg 1$ for large a and then provides a picture of how quarks are trapped by the electric field. It does not tell us why the coupling is large. On the other hand the infrared slavery ideas tell us why the coupling increases with a but fail to explain how a strong coupling confines quarks. The quark confining mechanism of lattice Y. M. theory and infrared slavery are not different mechanisms but are complementary aspects of the same thing.

ACKNOWLEDGMENTS

The ideas presented in these lectures are the combined efforts of John Kogut and myself. The lattice theory was inspired by Ken Wilson. I am very grateful to him for explaining many things about lattices and gauge theories to me.

REFERENCES

1. H. J. Lipkin, Physics Reports, Vol. 8c, Number 3, Aug. 1973
2. J. Schwinger, Theoretical Physics (International Atomic Energy Agency, Vienna, 1963) p. 89
3. Lowenstein and Swieca, Annals of Physics 68, 172 (1971)
4. A. Casher, J.Kogut and L.Susskind,Phys. Rev. Lett.31, 792 (1973)
5. J. Kogut and L. Susskind, Vacuum Polarization and the Absence of Free Quarks in 4 Dimensions, Phys. Rev. D (to appear)
6. G. 't Hooft, private communication
7. K. G. Wilson, C.L.N.S. 262 (Feb. 1974) to appear Phys. Rev. D
8. Lattice Gauge Theories have also been studied by Polyakov. Private communication from J. Bjorken
9. J. Kogut and L. Susskind, Hamiltonian Formulation of Wilson's Lattice Gauge Theories (to appear Phys. Rev. D)
10. The process of letting $a \to 0$ in a quantum field theory is called renormalization group. The most complete approach to the renormalization group as a computational tool is due to K. G. Wilson. See for example K. G. Wilson and J. Kogut, The Renormalization Group and the ε Expansion, perhaps to appear in Physics Reports C
11. It is known that in Y. M. theory the running coupling constant can increase as the cutoff distance becomes large. G.'t Hooft, Marseille Conference on Gauge Theories, June 1972,H.D. Politzer, Phys. Rev. Lett. 30, 1346 (1973), D. J. Gross and F. Wilczek, Phys. Rev. Lett. 30, 1343 (1973). Speculations that this effect can account for quark confinement in some way have been made by 't Hooft, Weinberg, Georgi and Glashow and probably many more.

PARTON MODELS FOR WEAK AND ELECTROMAGNETIC INTERACTIONS

M. Gourdin

Laboratoire de Physique Théorique et Hautes Energies,
Université Paris VI, France

INTRODUCTION

The aim of these lectures is to give a review of the situation concerning inclusive reactions induced by charged and neutral leptons in the light of the quark parton model. They will be divided into three parts.

PART A is devoted to theoretical generalities. We first describe the kinematics in order to introduce the notation. The consequences of a scaling à la Bjorken are presented for differential and total cross sections. A general formulation of the parton model is presented and structure functions are computed when the interacting partons are identified with the basic quarks of a symmetry group of strong interactions. As a byproduct the Adler and Gross-Llewellyn Smith sum rules are written in the quark parton language.

PART B is a study of three inclusive processes
- electroproduction
- neutrino and antineutrino induced reactions with charge changing current
- neutrino and antineutrino induced reactions with charge conserving current.

We begin with a review of the main experimental facts. The scaling of electroproduction structure functions and of neutrino and antineutrino total cross sections is exhibited. Then the particular quark parton model based on SU(3) symmetry is described and various experimental data are analysed in this framework. It is shown how the model is consistent with experiments. In particular, the naive Weinberg model for hadronic neutral current fits nicely the Gargamelle bubble chamber data and a value of the mixing angle is computed from experiment and turns out to be compatible with the range of values permitted by purely leptonic processes.

PART C is more speculative in the sense that a comparison with experiments is not yet possible and belongs to the future. Always in the framework of the SU(3) quark parton model we study two applications of the previous techniques

(i) The polarization effects in electroproduction, the target namely a nucleon, and the incident beam being polarized.

(ii) The weak effects in electroproduction due to the possible exchange of a neutral vector boson interfering with the usual one-photon exchange contribution.

Only simple cases have been considered and straightforward extensions can be made, in case (i) to weak reactions, in case (ii) to a polarized nucleon target.

Finally the last section of this part is concerned with a schematic description of a modern approach of scaling, using the renormalization group techniques.

PART A THEORETICAL GENERALITIES

I CURRENTS

1) In unified gauge theories of electromagnetic and weak processes the interaction between leptons and hadrons is mediated by vector bosons, one of them, the photon, is massless, the other ones being expected very heavy.

To each physical vector boson field corresponds a current with one leptonic part and one hadronic part. In models based on the $SU(2) \otimes U(1)$ gauge group there are four vector boson fields

i) the electromagnetic field
ii) the two charged boson fields
iii) the neutral boson field

The corresponding Lagrangians involving only known leptons have the following structure,

a - electromagnetic Lagrangian associated to the photon field

$$\mathcal{L}_Q = e\left[i\sum_l \bar{l}\gamma_\mu l + J_\mu^Q\right]A^\mu \qquad (1)$$

b - weak charged Lagrangian associated to the charged boson fields

$$\mathcal{L}_c = g_c\left[i\sum \bar{l}\gamma_\mu(1+\gamma_5)\nu_l + J_\mu^W\right]W^\mu + \text{Hermitian conjugate} \qquad (2)$$

c - weak neutral Lagrangian associated to the neutral boson field

$$\mathcal{L}_N = g_N \left[i \sum_\ell \left(\bar{\nu}_\ell \gamma_\mu (1+\gamma_5) \nu_\ell + \bar{\ell} \gamma_\mu (a - b\gamma_5) \ell \right) + c\, J_\mu^Z \right] Z^\mu \tag{3}$$

where all the coupling constants - but e - are model dependent.

2) Let us call q_μ the energy momentum four vector of the intermediate boson. At actual available energies, $\sqrt{q^2}$ remains small as compared to the vector boson masses expected to be as large as 40 GeV or more. Therefore the local Fermi effective interaction becomes a good approximation.
The Fermi constant G measured with the μ life time is defined by

$$\frac{G}{\sqrt{2}} = \frac{g_C^2}{m_W^2} \tag{4}$$

The normalization C of the neutral hadronic current is defined by the convention

$$\frac{G}{\sqrt{2}} = \frac{g_N^2}{m_Z^2} \tag{5}$$

II KINEMATICS

1) Inclusive reactions induced by leptons are described to lowest order by the diagram of Fig. 1 where the kinematical notations are indicated.

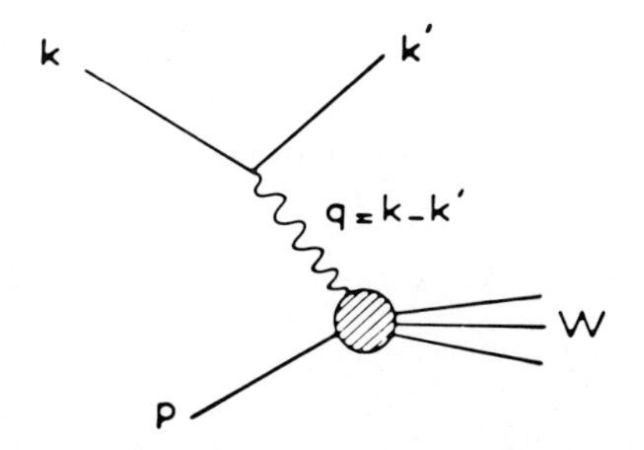

Fig. 1 One vector boson exchange in inelastic lepton scattering.

We introduce, as usual, the scalar variables q^2, W^2 and ν defined by

$$q^2 = (k-k')^2 \qquad W^2 = -(p+q)^2 \qquad \nu M = -\, p\cdot q$$

with the relation $W^2 = M^2 + 2M\nu - q^2$, M being the nucleon mass.
In the laboratory frame the lepton variables are

E incident lepton energy
E' final lepton energy
θ scattering angle between leptons.

They can expressed in terms of the invariant q^2 and ν by

$$q^2 = 4EE' \sin^2 \frac{\theta}{2} \qquad \nu = E - E'$$

Let us recall that the physical region is restricted by the inequality

$$o \leq q^2 \leq 2M\nu$$

The elastic case $W^2 = M^2$ corresponds to $q^2 = 2M\nu$.

2) As a consequence of the one vector boson exchange B the transition matrix element is factorized into the product of two matrix elements of the B current, one for leptons and one for hadrons

$$T^B = \frac{C^B}{q^2+m_B^2} \left[\bar{u}(k')\gamma^\mu(a-b\gamma_5)u(k)\right] \langle\Gamma|J_\mu^B(o)|p\rangle \qquad (6)$$

where u(k) and u(k') are lepton free Dirac spinors and C^B is a product of coupling constants.
The cross-section is therefore written as the product of a leptonic tensor $m^{\mu\nu}$ by a hadronic tensor $M_{\mu\nu}$. We shall work, in what follows, in the zero mass limit for charged leptons and only a longitudinal polarization survives. We call η the helicity of the incident lepton

$\eta = +1$ Right-hand R

$\eta = -1$ Left-hand L

The leptonic tensor is easily computed in this approximation and the result is

$$m^{\mu\nu} = (a^2+b^2)(t^{\mu\nu} + \eta s^{\mu\nu}) + 2ab(s^{\mu\nu} + \eta t^{\mu\nu}) \qquad (7)$$

where

$$t^{\mu\nu} = \frac{1}{2}\left[k^\mu k'^\nu + k'^\mu k^\nu + \frac{1}{2} q^2 g^{\mu\nu}\right] \qquad (8)$$

$$s^{\mu\nu} = \frac{i}{2} \varepsilon^{\mu\nu\rho\sigma} k_\rho k'_\sigma \qquad (9)$$

Let us specify the leptonic tensor in the various cases considered a - electromagnetic current: incident charged lepton a = 1 b = 0

$$m^{\mu\nu} = t^{\mu\nu} + \eta\ s^{\mu\nu} \qquad (10)$$

b - weak currents: incident left-hand neutrinos and right-hand antineutrinos a = 1 b = -1

$$m^{\mu\nu} = 4\ (\ t^{\mu\nu} \mp\ s^{\mu\nu}\ ) \qquad (11)$$

c - interference between electromagnetic and weak neutral currents: incident charged leptons

$$m^{\mu\nu} = a(t^{\mu\nu} + \eta\ s^{\mu\nu}) + b(s^{\mu\nu} + \eta\ t^{\mu\nu}) \qquad (12)$$

3) The hadronic tensor $M_{\mu\nu}$ is generally defined by

$$M^{\alpha\beta}_{\mu\nu} = M \sum_{\Gamma} \int_{\Gamma} (2\pi)^3 \delta_4(p+q-p_\Gamma) \langle \Gamma | J^{\alpha}_{\mu}(0) | p \rangle \langle \Gamma | J^{\beta}_{\nu}(0) | p \rangle^* \qquad (13)$$

where $\int_\Gamma$ means a phase space integration and a summation over polarization for all the particles belonging to Γ.

We shall use four hadronic tensors in the various processes studied

$M^{QQ}_{\mu\nu}$ for electroproduction with one photon exchange

$M^{WW}_{\mu\nu}, M^{ZZ}_{\mu\nu}$ for neutrino and antineutrino scattering

$\frac{1}{2}(M^{QZ}_{\mu\nu} + M^{ZQ}_{\mu\nu})$ for one photon, one neutral boson interference in electroproduction.

The hadronic tensor $M^{\alpha\beta}_{\mu\nu}$ can also be written as the Fourier transform of the one particle matrix element of the product of two current operators. From equation (13) we get

$$M^{\alpha\beta}_{\mu\nu} = \frac{M}{2\pi} \int e^{-iq\cdot x} \langle p | J^{\beta}_{\nu}(x)\ J^{\alpha}_{\mu}(0) | p \rangle\ d_4x$$

This equality shows that the hadronic tensor is proportional to the imaginary part of a forward Compton scattering amplitude as appears in Fig. 2.

In the polarization space the hadronic tensor is repesented by a 8 x 8 matrix which is Hermitian by construction when $\alpha = \beta$. The total helicity is a conserved quantity for the forward Compton amplitude and

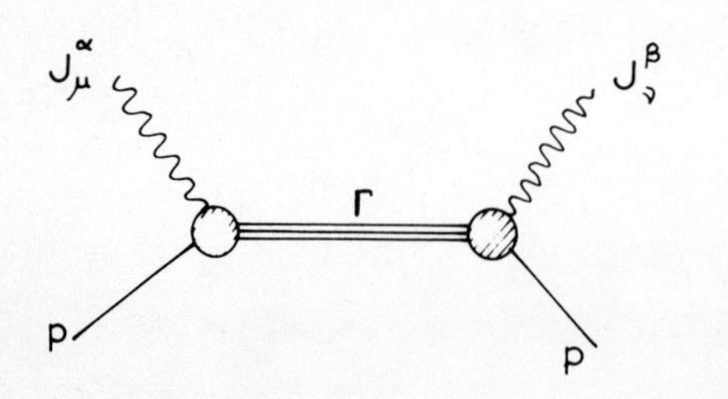

Fig. 2 Compton like amplitude for the hadronic tensor.

the 8 x 8 matrix is reducible into:

two 1 x 1 matrices for total helicity $\pm \frac{3}{2}$

two 3 x 3 matrices for total helicity $\pm \frac{1}{2}$

It follows that the total number of structure functions is 20 and it reduces to 6 when the target is unpolarized.

Furthermore the structure functions associated to a scalar polarization of the current give contributions to the cross section proportional to the lepton mass. In the zero lepton mass limit it is then sufficient to restrict to vector polarizations for the current and the 3 x 3 matrices associated to the total helicity $\pm \frac{1}{2}$ reduce to 2 x 2 matrices. In this approximation the total number of structure functions is 10 for a polarized spin $\frac{1}{2}$ target and 3 for an unpolarized one. With respect to space and time reflections these ten structure functions are classified as shown in Table 1.

	Parity conserving	Parity violating
Time Reversal invariant	2 + 2	1 + 3
Time Reversal violating	0 + 1	0 + 1

Table 1

In this table the first number refers to the unpolarized part $T_{\mu\nu}$ of the hadronic tensor and the second number to the polarized part $S_{\mu\nu}$

$$M_{\mu\nu} = T_{\mu\nu} + S_{\mu\nu}$$

The diagonal elements of the hadronic matrix for $\alpha = \beta = B$ can be interpreted as the total cross sections $\sigma^B_{\lambda s}(q^2, W^2)$ for the process

$$B + p \rightarrow \text{HADRONS}$$

where λ is the helicity of the virtual boson B ($\lambda = \pm 1, 0$) and s the helicity of the spin $\frac{1}{2}$ target ($s = \pm \frac{1}{2}$) .

This is the case for the three unpolarized structure functions and three polarized structure functions the other ones being transverse longitudinal correlations for total helicity $\pm \frac{1}{2}$. Because of the semi definite positive character of the hadronic tensor for $\alpha = \beta = B$ the structure functions $\sigma^B_{\lambda s}(q^2,W^2)$ are positive in the physical region $q^2 \geq 0$ $W^2 \geq M^2$ and the correlations are bounded by total cross sections using the Schwartz inequality in the 2 x 2 matrices.

4) We now consider the case of an unpolarized target. Three structure functions $\sigma^B_\lambda(q^2, W^2)$ will describe the hadronic tensor and the double differential cross section for inelastic scattering of polarized leptons off unpolarized targets has the general structure

$$\frac{d^2\sigma^B_\eta}{dq^2 dW^2} = \frac{A^B(q^2)}{8\pi^2} \frac{q^2}{M\sqrt{\nu^2+q^2}} \frac{E'}{E} \Big\{ \cos^2\frac{\theta}{2}\left[\sigma^B_T(q^2,W^2) + \sigma^B_L(q^2,W^2)\right] +$$

$$+ 2\sin^2\frac{\theta}{2} \frac{\nu^2+q^2}{q^2} \sigma^B_T(q^2,W^2) -$$

$$- \eta \sin^2\frac{\theta}{2} \frac{(E+E')\sqrt{\nu^2+q^2}}{q^2} \left[\sigma^B_{-1}(q^2,W^2) - \sigma^B_{+1}(q^2,W^2)\right] \quad (14)$$

where

$$A^Q(q^2) = \frac{e^4}{q^4}$$

$$A^W(q^2) = \frac{4g_c^2}{(q^2+m_W^2)^2} = \frac{2G^2}{(1+q^2/m_W^2)^2}$$

$$A^Z(q^2) = \frac{4c^2 g_N^2}{(q^2+m_Z^2)^2} = \frac{2G^2}{(1+q^2/m_Z^2)^2}$$

In the particular case of electromagnetic interactions, parity is conserved and $\sigma^Q_{-1} = \sigma^Q_{+1}$. The last term in equation (14) disappears and the cross section is independent of the lepton helicity η. Analogous expressions can be written with a polarized target. The particular case of electroproduction will be discussed in part C.

III SCALING

1) Let us define new dimensionless variables

$$\xi = \frac{q^2}{2M\nu} \qquad \rho = \frac{p \cdot k'}{p \cdot k} = \frac{E'}{E}$$

and new structure functions

$$\sigma_\lambda^B(q^2, W^2) = \frac{\pi}{M\sqrt{\nu^2+q^2}} F_\lambda^B(q^2, \xi)$$

$$F_\lambda(q^2, \xi) = \left| \text{Forward Compton Amplitude} \right|^2$$

The double differential cross sections (14) are equivalently written as

$$\frac{d^2\sigma_\eta^B}{d\rho\, d\xi} = \frac{A^B(q^2)}{2\pi} M E \xi \Big\{ (1-\rho^2) F_T^B(q^2,\xi) + \tag{16}$$
$$+ \frac{\nu^2}{\nu^2+q^2}\left(2\rho - \frac{q^2}{2E^2}\right)\left[F_T^B(q^2,\xi) + F_L^B(q^2,\xi)\right] - \eta \frac{1-\rho^2}{2} \frac{\nu}{\sqrt{\nu^2+q^2}} \left[F_{-1}^B(q^2,\xi) + F_{+1}^B(q^2,\xi)\right]\Big\}$$

2) When the variables ρ and ξ are fixed the high energy limit E → ∞ implies the Bjorken limit LIM for the structure functions where LIM means q^2, W^2 → ∞ with ξ fixed

$$\text{LIM}\; F_\lambda^B(q^2, \xi) = F_\lambda^B(\xi)$$

For the differential cross sections (16) we simply obtain

$$\lim_{\substack{E\to\infty \\ \rho,\xi \text{ fixed}}} \frac{d^2\sigma_\eta^B}{d\rho\, d\xi} = \frac{A^B(q^2)}{2\pi} M E \xi \Big\{ (1+\rho^2) F_T^B(\xi) +$$
$$+ 2\rho F_L^B(\xi) - \eta \frac{1-\rho^2}{2}\left[F_{-1}^B(\xi) - F_{+1}^B(\xi)\right] \tag{17}$$

For electroproduction

$$A^Q(q^2) = \frac{e^4}{4M^2E^2(1-\rho)^2\xi^2}$$

and the fixed ρ, ξ differential cross section tends to zero like 1/E at high energy.

For neutrino and antineutrino induced processes we assume that there exists in the q^2, W^2 plane a region where scaling takes place and where the Fermi theory is still valid. Then

$$A^W = A^Z = 2\, G^2$$

and the differential cross sections (17) increase linearly with the incident energy E and their ρ dependence becomes simply a second order polynomial

$$\lim_{\substack{E\to\infty \\ \rho,\xi \text{ fixed}}} \frac{d^2\sigma^\nu}{d\rho\, d\xi} = \frac{G^2 M E}{\pi}\, \xi \left[\rho^2 F_+^\nu(\xi) + F_-^\nu(\xi) + 2\rho\, F_0^\nu(\xi)\right]$$

$$\lim_{\substack{E\to\infty \\ \rho,\xi \text{ fixed}}} \frac{d^2\sigma^{\bar\nu}}{d\rho\, d\xi} = \frac{G^2 M E}{\pi}\, \xi \left[\rho^2 F_-^{\bar\nu}(\xi) + F_+^{\bar\nu}(\xi) + 2\rho\, F_0^{\bar\nu}(\xi)\right] \tag{18}$$

3) Because of the zero mass of the photon the electroproduction total cross section is infrared divergent. The situation is different for weak processes the bosons W and Z being massive.

In order to compute total cross sections we integrate over the variables ρ and ξ in the square 0 - 1. The assumption generally made is that in such an integration we can use everywhere the form of $d^2\sigma/d\rho\, d\xi$ obtained in the scaling region. The result is simply a linear rising with energy of the total neutrino and antineutrino cross sections

$$\lim_{E\to\infty} \sigma_{TOT} = \frac{G^2 M E}{\pi} A \tag{19}$$

where the coefficients A are given by

$$A^\nu = \frac{1}{3} I_+^\nu + I_-^\nu + I_0^\nu$$

$$A^{\bar\nu} = \frac{1}{3} I_-^{\bar\nu} + I_+^{\bar\nu} + I_0^{\bar\nu} \tag{20}$$

and the integrals I_λ by

$$I_\lambda = \int_0^1 \xi\, F_\lambda(\xi)\, d\xi$$

4) The final lepton energy distributions can be computed with the same assumptions and the result is

$$\lim_{E\to\infty} \frac{d\sigma^\nu}{d\rho} = \frac{G^2 M E}{\pi}\left[\rho^2 I_+^\nu + I_-^\nu + 2\rho\, I_0^\nu\right]$$

$$\lim_{E\to\infty} \frac{d\sigma^{\bar\nu}}{d\rho} = \frac{G^2 M E}{\pi}\left[\rho^2 I_-^{\bar\nu} + I_+^{\bar\nu} + 2\rho\, I_0^{\bar\nu}\right] \tag{21}$$

5) The structure functions $\sigma_\lambda(q^2,W^2)$ being total cross sections, they are positive functions of q^2 and W^2 in the physical region $q^2 \geq 0$,

$W^2 \geq M^2$. Therefore the scaling function $F_\lambda(\xi)$ and the first moment integrals I_λ are also positive

$$F_\lambda(\xi) \geq 0 \qquad \text{for} \qquad 0 \leq \xi \leq 1$$

$$I_\lambda \geq 0$$

IV PARTON MODELS

1) The hadrons are assumed to be composite systems of elementary constituents called partons. The structure functions are Lorentz invariant quantities so that they can be computed in any frame of reference.

A simple description of the hadron occurs in the $P \to \infty$ system where the hadron momentum P becomes very large as compared to the hadron mass. Then the partons appear to be quasi-free particles and the impulse approximation can be used for the interaction of the current with the hadron. The partons have an instantaneous interaction with the current which is point-like and only the parton charge associated to the current is seen. After interaction the partons gain a transverse momentum q^2 and they remain quasi-free on mass shell.

The main condition for the impulse approximation to be valid is that the time of interaction of the current with the parton must be small as compared with the typical life time of metastable states in the hadron. In other words the effective mass W of the final hadronic system must be large as compared with a typical resonance energy W_R so that the scattering must be deeply inelastic.

2) As pointed out in section II the structure functions $\sigma_\lambda^B(q^2,W^2)$ are directly proportional to the imaginary part, in the forward direction, of a Compton type amplitude $B + p \to B + p$.
In the parton model we make an incoherent summation of the various parton contributions as shown in Fig. 3, and the partons are assumed to interact in a point like manner with the current J^B.

Let us call $D_j(\xi)$ the distribution function of the parton of type j in the hadron, its momentum being $\xi\vec{P}$. The normalization integral of these distributions

$$\int_0^1 D_j(\xi)\, d\xi = \langle N_j \rangle \qquad (22)$$

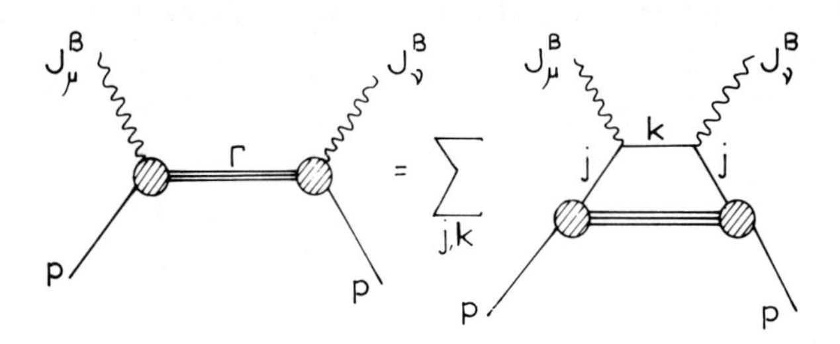

Fig. 3
Parton models for the hadronic tensor.

gives the average value of the number of type j parton in the hadron. One important property of the D_j's which are essentially probability distributions is their positivity.

3) The point-like matrix element of the current J^B of helicity λ between the partons j and k is called charge $a_{jk}^{B\lambda}$. From Fig. 3 the scaling functions are simply written as

$$F_\lambda^B(\xi) = \sum_{j,k} \left[a_{jk}^{B\lambda} \right]^2 D_j(\xi) \tag{23}$$

We assume the partons to have only spin 0 and $\frac{1}{2}$. At high energy it is straightforward to check that

F_+ contains only right-hand partons and antipartons
F_- contains only left-hand partons and antipartons
F_0 contains only spin zero partons and antipartons.

As a first experimental result we shall discuss in the next lesson the transverse scaling functions dominate over the longitudinal ones both in electromagnetic and weak interactions. It is then legitimate to associate the interacting partons with spin $\frac{1}{2}$ quarks. Nevertheless we shall reserve the possibility of existence of integer spin gluons which have zero charges and which only carry energy momentum. The physical role attributed to gluons is then to bind the quarks into the hadron.

4) In weak interactions the partons may have different right-hand and left-hand couplings with the currents. We call a_{jk}^{BR} and a_{jk}^{BL} the corresponding charges. The indices j, k being from now positive we associate the distributions $D_j(\xi)$ to quarks and the distributions $D_{-j}(\xi)$ to antiquarks. Taking into account the symmetry relation

$$(a_{jk}^B)^2 = (a_{-k,-j}^B)^2$$

we easily compute the scaling functions associated to the current J^B

$$2\,F_+^B(\xi) = \sum_{j,k}\left[(a_{jk}^{BL})^2 D_{-k}(\xi) + (a_{jk}^{BR})^2 D_j(\xi)\right]$$

$$2\,F_-^B(\xi) = \sum_{j,k}\left[(a_{j,k}^{BL})^2 D_j(\xi) + (a_{jk}^{BR})^2 D_{-k}(\xi)\right] \tag{24}$$

For self Hermitian currents like J^Q and J^Z the charges are diagonal in the parton space: $a_{jk} = a_j \delta_{jk}$. Moreover, for the electromagnetic current parity is conserved and we simply have

$$a_j^{QR} = a_j^{QL} = Q_j$$

where Q_j is the electric charge of the parton j. The electroproduction scaling function is then written as

$$2\,F_T^Q(\xi) = \sum_j Q_j^2 \left[D_j(\xi) + D_{-j}(\xi)\right] \tag{25}$$

For a non Hermitian current like the weak charged current the scaling functions associated to its Hermitian conjugate $J^{\overline{B}}$ are easily deduced from equations (24) using the symmetry relation

$$(a_{jk}^B)^2 = (a_{kj}^{\overline{B}})^2$$

and the result is

$$2\,F_+^{\overline{B}}(\xi) = \sum_{j,k}\left[(a_{jk}^{BL})^2 D_{-j}(\xi) + (a_{jk}^{BR})^2 D_k(\xi)\right]$$

$$2\,F_-^{\overline{B}}(\xi) = \sum_{j,k}\left[(a_{jk}^{BL})^2 D_k(\xi) + (a_{jk}^{BR})^2 D_{-j}(\xi)\right] \tag{26}$$

5) Let us integrate the scaling functions with respect to ξ. Using the definition of the $\langle N_j \rangle$'s previously given we obtain two sum rules already derived from current algebra. Defining

$$2\,F_T = F_- + F_+ \qquad\qquad F_3 = F_- - F_+$$

we get

- a - the Adler sum rule for the weak charged current

$$\int_0^1 2\left[F_T^{\overline{W}}(\xi) - F_T^{W}(\xi)\right]d\xi = \tag{27}$$

$$= \frac{1}{2}\sum_{j,k}\left[(a_{jk}^{WL})^2 + (a_{jk}^{WR})^2\right]\left[\langle N_k - N_{-k}\rangle - \langle N_j - N_{-j}\rangle\right]$$

- b - The Gross-Llewellyn Smith sum rules for the weak currents

$$\int_0^1 \left[F_3^{W}(\xi) + F_3^{\overline{W}}(\xi)\right]d\xi = \tag{28}$$

$$= \frac{1}{2}\sum_{j,k}\left[(a_{jk}^{WL})^2 - (a_{jk}^{WR})^2\right]\left[\langle N_k - N_{-k}\rangle + \langle N_j - N_{-j}\rangle\right]$$

$$\int_0^1 F_3^{Z}(\xi)\,d\xi = \frac{1}{2}\sum_{j}\left[(a_j^{ZL})^2 - (a_j^{ZR})^2\right]\langle N_j - N_{-j}\rangle \tag{29}$$

The differences $\langle N_j - N_{-j}\rangle$ are linear combinations of conserved charges like the baryonic charge, the electric charge, the hypercharge, etc... These combinations and therefore the right-hand sides of the sum rules depend on the algebra of the quark model.

REFERENCES

T. D. Lee and C.N. Yang, Phys. Rev. Letters 4, 307 (1960)

M. Gourdin, Nuovo Cimento 21, 1094 (1961)

T. D. Lee and C. N. Yang, Phys. Rev. 126, 2239 (1962)

G. Charpak and M. Gourdin, Lectures at the Cargèse Summer School (1962)

M. Gourdin and A. Martin, CERN TH. 261 (1962)

S. L. Adler, Phys. Rev. 135B, 963 (1964); 143, 1144 (1966)

J. D. Bjorken, Phys. Rev. 148, 1467 (1966)

N. Christ and T. D. Lee, Phys. Rev. 143, 1310 (1966)

M. Gourdin, "Diffusion des Electrons de Haute Energie", Masson (1966)

N. Dombey, Rev. Mod. Phys. 41, 236 (1969)

D. J. Gross and C. H. Llewellyn Smith, Nucl. Phys. B14, 337 (1969)

J. D. Bjorken and E. A. Paschos, Phys. Rev. D1, 315 (1970)

J. D. Bjorken, Phys. Rev. D1, 1376 (1970)

M. G. Doncel and E. De Rafael, Nuovo Cimento 4A, 363 (1971)

M. Gourdin, Lectures given at CERN School, Grado (1972)

Physique des Neutrinos de Haute Energies, Colloque de Physique des Particules, Vittel, mai 1973.

PART B QUARK PARTON MODEL

I EXPERIMENTAL DATA ON ELECTROPRODUCTION

1) A systematic study of the electron deep inelastic scattering on hydrogen and deuterium is made at SLAC and DESY. The region of the q^2, W^2 plane where measurements have been performed is represented in Fig. 4. The two electroproduction structure functions on protons have been separated in the shaded region of Fig. 4 where data at three or more angles are available.

The results are generally presented in terms of the two quantities

$$F_2(q^2,\xi) = \frac{1}{\pi}\frac{\nu q^2}{\sqrt{\nu^2+q^2}}\left[\sigma_T(q^2,W^2) + \sigma_L(q^2,W^2)\right]$$

$$R(q^2,\xi) = \frac{\sigma_L(q^2,W^2)}{\sigma_T(q^2,W^2)}$$

If scaling holds

$$F_2(q^2,\xi) \rightarrow F_2(\xi) = 2\xi\left[F_T(\xi) + F_L(\xi)\right]$$
$$R(q^2,\xi) \rightarrow F_L(\xi)/F_T(\xi) \qquad (30)$$

2) The function $F_2^{ep}(q^2,\xi)$ for a proton target has been plotted in Fig. 5 versus ξ as usual. The results are compatible with a unique curve $F_2^{ep}(\xi)$ as suggested by the Bjorken scaling law (30) for values of q^2 larger than 1 GeV^2 and of W larger than 2.6 GeV. A typical example of scaling at $\xi = 0.25$ is shown in Fig. 6.

The ratio R^p is always smaller than 0.4 and its determination is considerably less accurate than that of $F_2{}^{ep}$. Various forms have been proposed for R^p and two possible fits having reasonable χ^2 values are

a - R constant with $R^p = 0.168 \pm 0.014$
b - $R^p = c\,\frac{M^2}{q^2}$ with $c = 0.35 \pm 0.05$

More sophisticated expressions between the forms a- and b- will obviously fit the data but a scaling of the quantity $\frac{\nu}{M} R^p$ as predicted by the U(3) quark model light cone algebra is certainly consistent with experiment.

Let us notice that a fit with a form $R^p = a\,\frac{q^2}{M^2}$ has a larger χ^2 value than fits of type a- and b-.

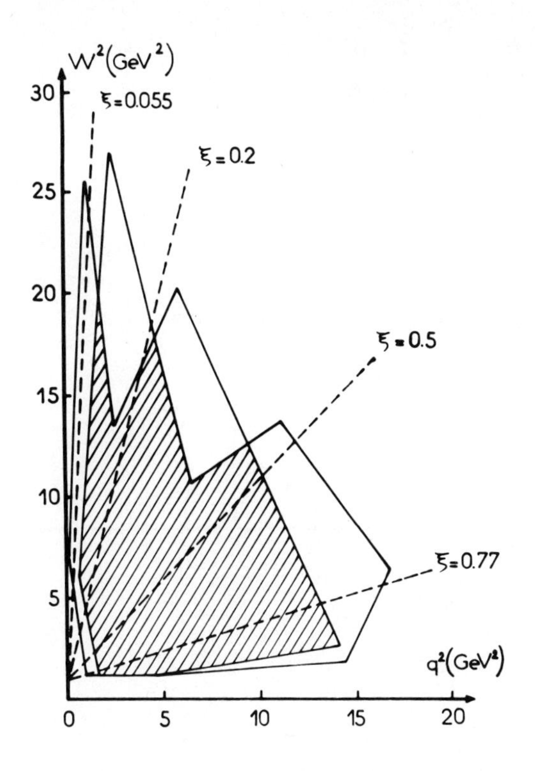

Fig. 4
Region of the q^2, W^2 plane studied at SLAC.

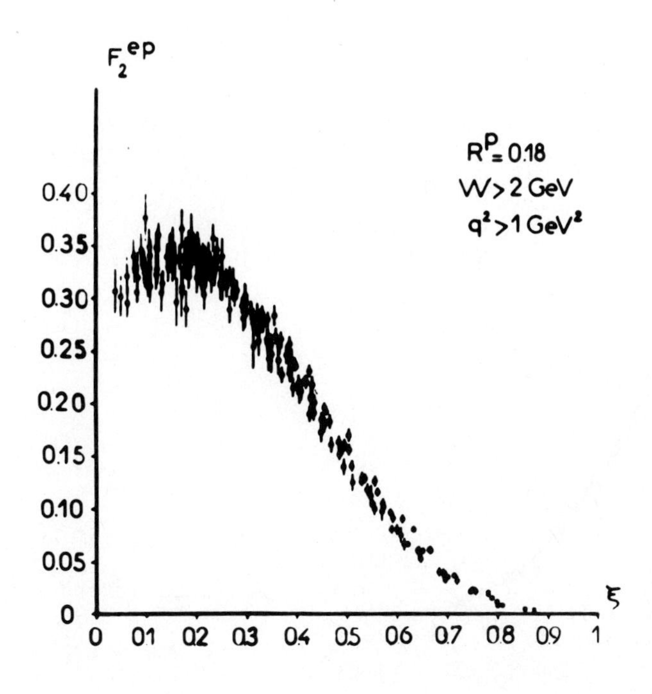

Fig. 5
$F_2^{ep}(\xi, q^2)$ versus ξ.

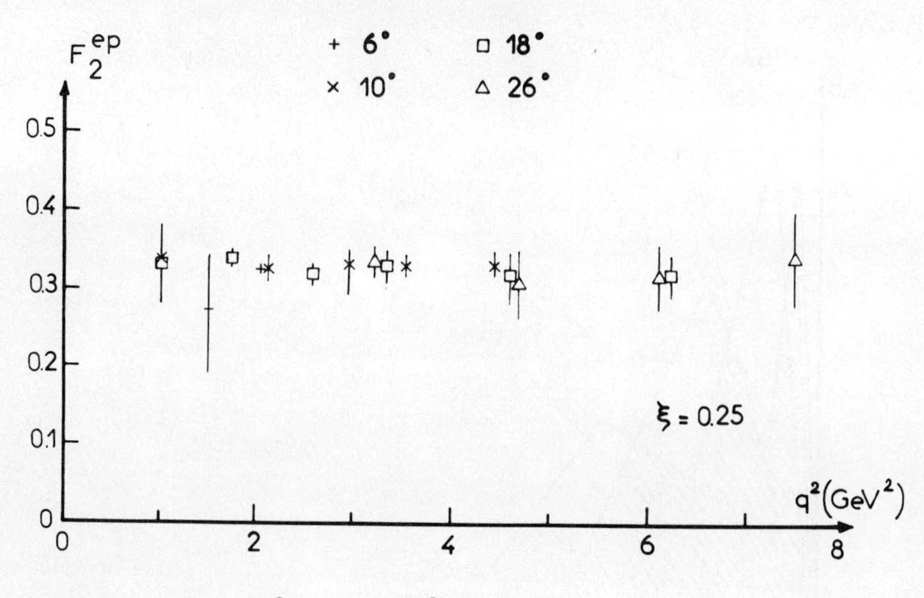

Fig. 6 $F_2^{ep}(\xi,q^2)$ versus q^2; ξ = 0.25

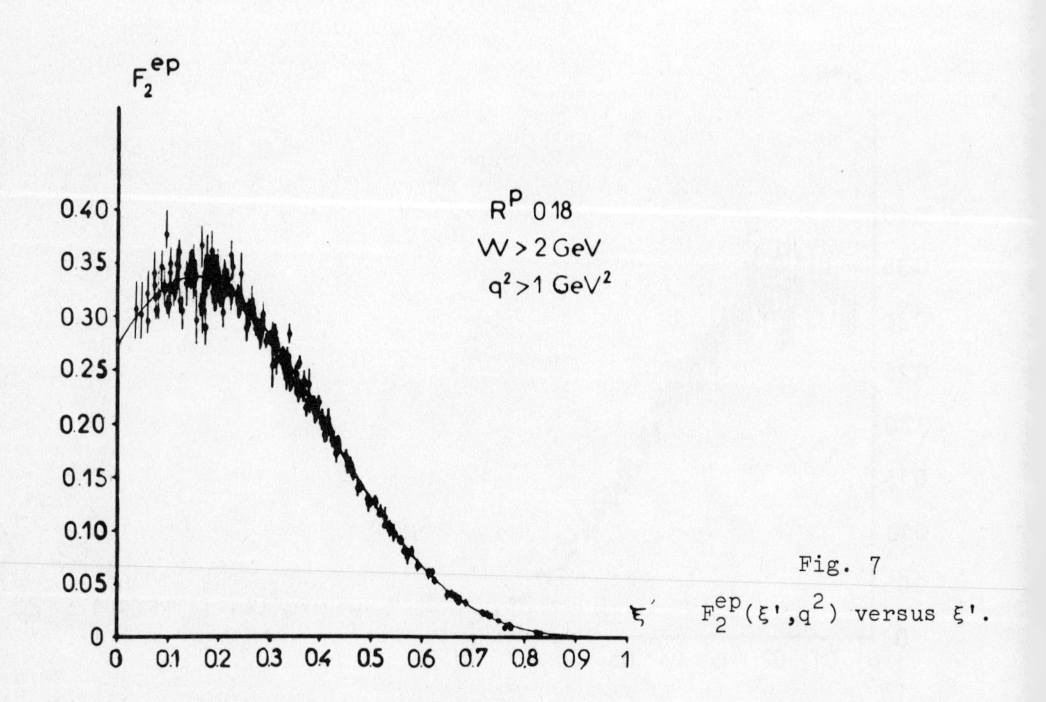

Fig. 7
$F_2^{ep}(\xi',q^2)$ versus ξ'.

3) Other scaling variables have been proposed in order to extend the region of the q^2, W^2 plane where the experimental data scale. Two well-known examples are the Bloom-Gilman variable

$$\xi' = \frac{q^2}{q^2 + W^2} \quad \text{or} \quad \omega' = 1 + \frac{W^2}{q^2} = \frac{1}{\xi} + \frac{M^2}{q^2}$$

and the Rittenberg-Rubinstein variable

$$\omega_W = \frac{2M\nu + M^2}{q^2 + a^2} = \frac{W^2 + q^2}{a^2 + q^2}$$

Fig. 7 represents a plot of F_2^{ep} versus ξ'. The dispersion in q^2 is somewhat less important than in Fig. 5.
Let us remark on the other hand that for resonances the quasi-elastic form factors exhibit similar shapes when plotted in the variable q^2/W^2. Therefore the variable ξ' has the advantage of nicely averaging the resonance contributions in a local way.

4) Experiments performed with a deuterium target have the same features: scaling for F_2^{ed} and smallness for R^d. A plot of F_2^{ed} versus ξ' is given in Fig. 8 and the shape of F_2^{ed} looks similar to that of F_2^{ep}. Moreover within errors $R^d = R^p$ as shown in Fig. 9.
After application of deuteron nuclear physics corrections due to the Fermi motion the neutron scaling functions are extracted by difference. But these corrections, very small for $\xi' < 0.65$, become more and more important when ξ' increase and also the uncertainty on these corrections and on the neutron scaling function F_2^{en}.

II EXPERIMENTAL DATA ON WEAK PROCESSES WITH CHARGED CURRENTS

1) The inclusive neutrino and antineutrino experiments with production of a charged final lepton or antilepton cannot be used to obtain an individual information about the structure functions. Most features observed in electroproduction like the scaling and the smallness of the longitudinal contribution in the deep inelastic region have not been directly checked but rather assumed in the analysis of data.

The experimental results are compatible with a linear rising with energy of the total cross sections and the ratio of antineutrino

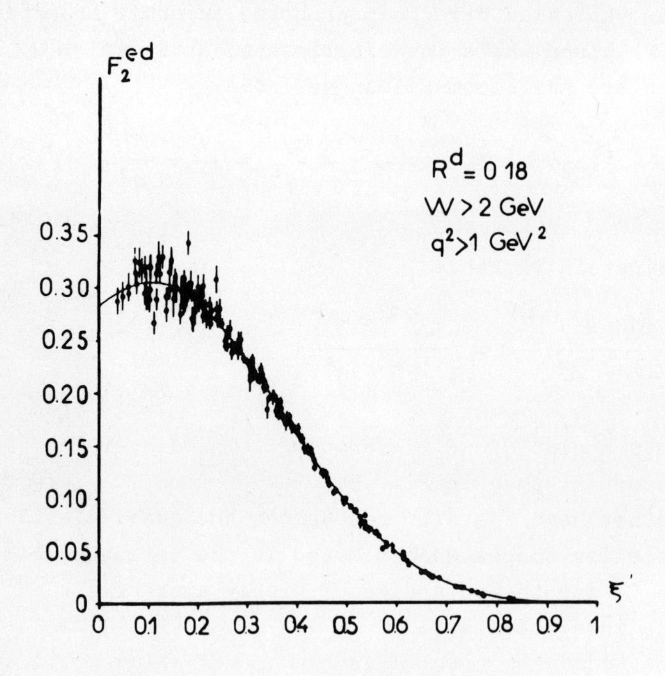

Fig. 8

$F_2^{ed}(\xi',q^2)$ versus ξ'.

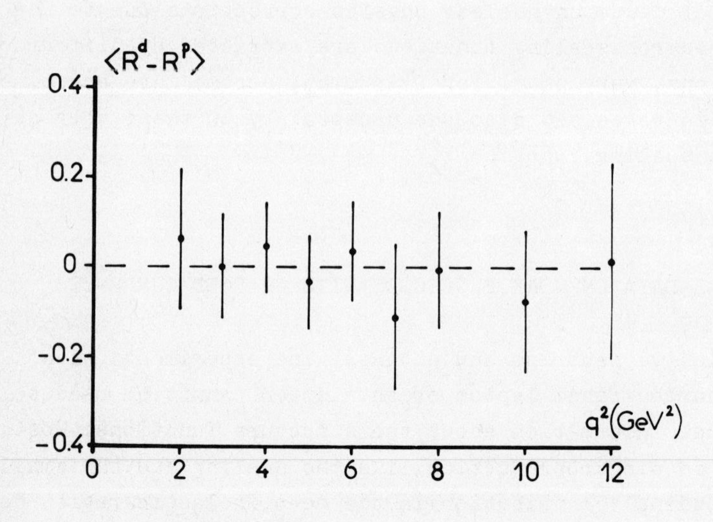

Fig. 9

$R^d - R^p$.

to neutrino total cross section is consistent with a constant value. The data is shown in Table 2 where the total cross sections are written

$$\sigma_{TOT}^{\nu,\bar{\nu}} = \alpha_{\nu,\bar{\nu}} E$$

with α given in units $10^{-38} cm^2\ GeV^{-1}$ per nucleon.

	Beam	α	$\sigma^{\bar{\nu}}\ \sigma^{\nu}$	Energy Range
CERN Propane	ν_μ	0.8 $\pm$ 0.2		1-12 GeV
CERN Gargamelle FREON	ν_μ	0.76 $\pm$ 0.08	0.38 $\pm$ 0.02	1-10 GeV
	$\bar{\nu}_\mu$	0.28 $\pm$ 0.03		
	ν_e	0.93 $\pm$ 0.17	0.40 $\pm$ 0.12	1 GeV
	$\bar{\nu}_e$	0.37 $\pm$ 0.09		
NAL Caltech IRON	ν_μ	0.83 $\pm$ 0.11	0.33 $\pm$ 0.08	40 110 GeV
	$\bar{\nu}_\mu$	0.28 $\pm$ 0.05		
NAL Harvard Pennsylvania Wisconsin	ν_μ	0.70 $\pm$ 0.18	0.41 $\pm$ 0.11	10-200 GeV
	$\bar{\nu}_\mu$	0.28 $\pm$ 0.09		
AVERAGE		0.78 $\pm$ 0.07		
		0.28 $\pm$ 0.025		

Table 2

The CERN Gargamelle data is shown in Figs. 10 and 11. The values of q^2 and W^2 brought into play are low and a large part of the experimental points in the q^2 - W^2 plane lies outside the scaling region obtained at SLAC $q^2 \gtrsim 1\ GeV^2$, $W \gtrsim 2.6$ GeV. Nevertheless linearity can be achieved for the total cross sections with reasonable χ^2 values.

Finite energy corrections are certainly important and they should be taken into account in a more refined analysis. Unfortunately these corrections depend on the choice of the scaling function and of the scaling variable and they cannot be uniquely predicted.

The N.A.L. high energy data for the neutrino total cross section

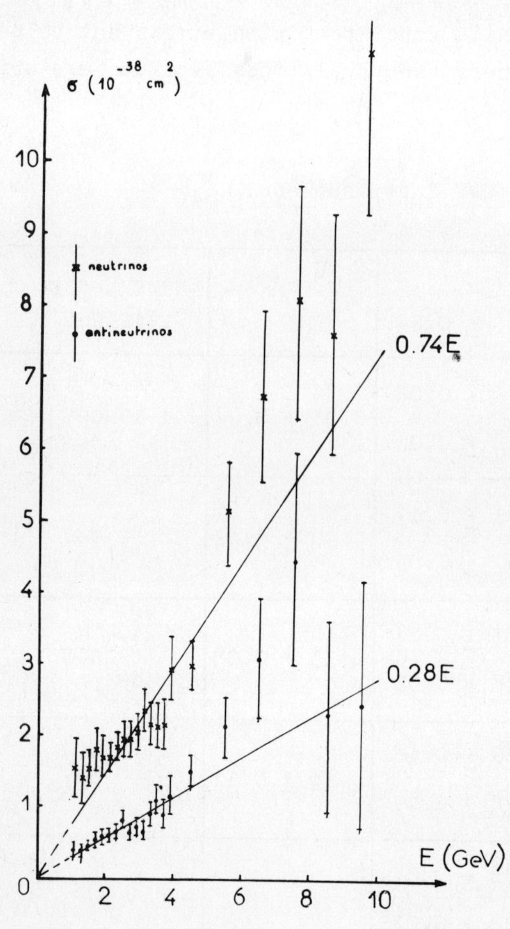

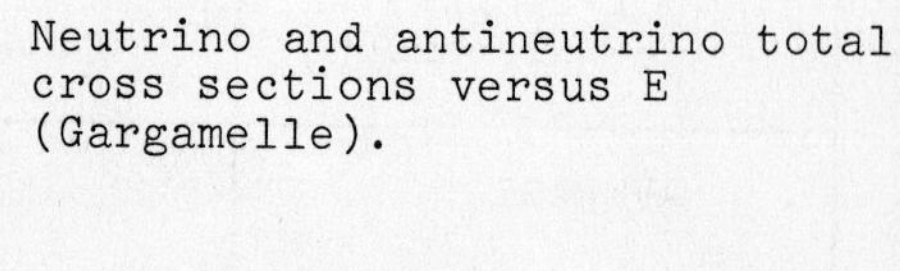

Fig. 10

Neutrino and antineutrino total cross sections versus E (Gargamelle).

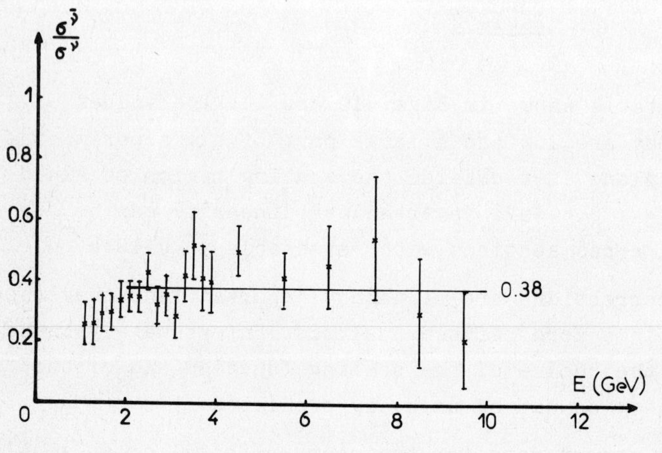

Fig. 11

Ratio of antineutrino to neutrino total cross sections versus E (Gargamelle).

and the antineutrino to neutrino ratio of total cross sections is consistent with the low energy one as shown in Fig. 12. In this case

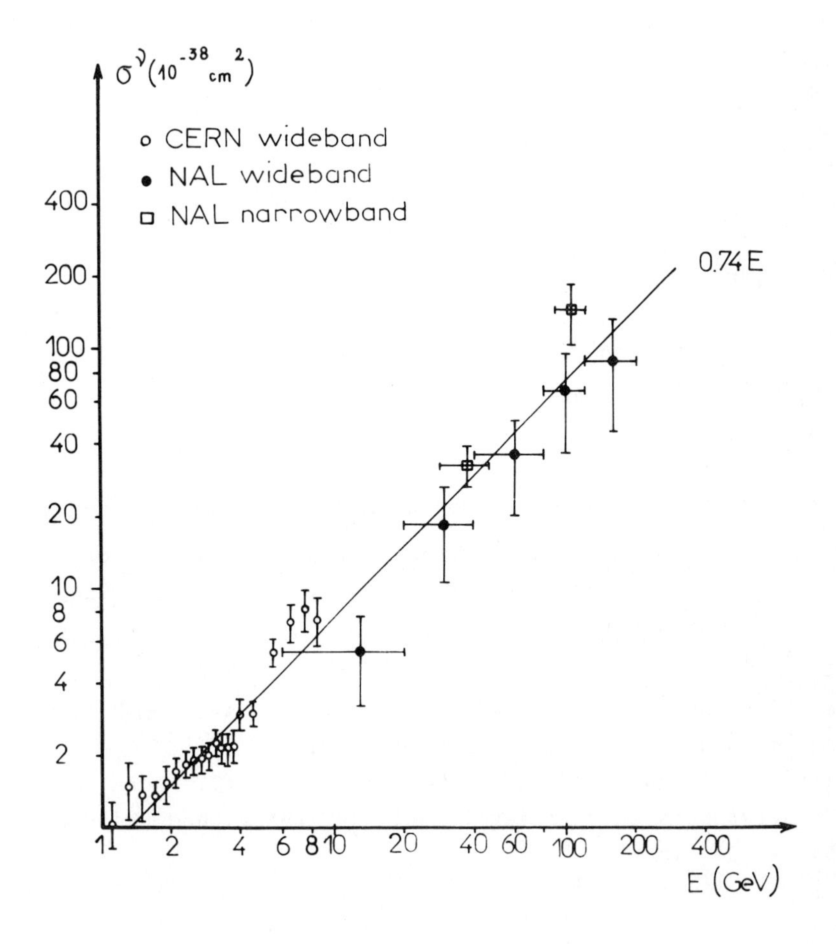

Fig. 12

Neutrino total cross section versus E (Gargamelle and NAL)

finite energy corrections are negligible but another type of phenomena may distort the linearity: the non locality due to an intermediate vector boson.

This effect cannot be measured with the present accuracy of the data and only a lower limit around 12 GeV can be put on the W meson mass.

2) We shall analyse the CERN-Gargamelle data in the units of G^2ME/π and the experimental value for the coefficients A^{ν} and $A^{\bar{\nu}}$ is

$$A^{\nu} = 0.483 \pm 0.051 \qquad A^{\bar{\nu}} = 0.178 \pm 0.019$$

In order to obtain the same quantities for an isoscalar target we must correct for the unequal number of protons and neutrons in the CF_3Br chamber where $N_n/N_p \simeq 1.19$. Anticipating the results of the quark parton model analyses we use

$$A^{\nu n}/A^{\nu p} \simeq 1.8 \qquad A^{\bar{\nu} p}/A^{\bar{\nu} n} \simeq 2$$

and we get

$$\begin{aligned} A^{\nu N} &\simeq 0.977 \ A^{\nu} \simeq 0.471 \pm 0.050 \\ A^{\bar{\nu} N} &\simeq 1.03 \ A^{\bar{\nu}} \simeq 0.183 \pm 0.020 \end{aligned} \qquad (31)$$

3) Other quantities have been measured in these experiments:

a- energy distributions of the final lepton or antilepton
b- averaged value of the final lepton or antilepton energy
c- averaged value of q^2
d- fixed ξ differential cross sections.

The points a-, b-, and d- will be studied in section V from both experimental and theoretical points of view.
For the averaged values of q^2 the CERN Gargamelle data has been fitted with a two parameter linear function of energy and the result for events with E>2 GeV is

$$\begin{aligned} \langle q^2 \rangle_{\nu} &= (0.21 \pm 0.02)E + (0.22 \pm 0.06) \\ \langle q^2 \rangle_{\bar{\nu}} &= (0.14 \pm 0.03)E + (0.11 \pm 0.08) \end{aligned}$$

The data is presented in Figs. 13 and 14. Let us remark that the dimensionless quantity $v = q^2/2ME$ is known from the final lepton or antilepton parameters

$$V = \frac{2E' \sin^2\frac{\theta}{2}}{M}$$

and therefore is independent of the incident spectrum. Theoretically the average values of q^2 involve the second moment of the scaling functions.

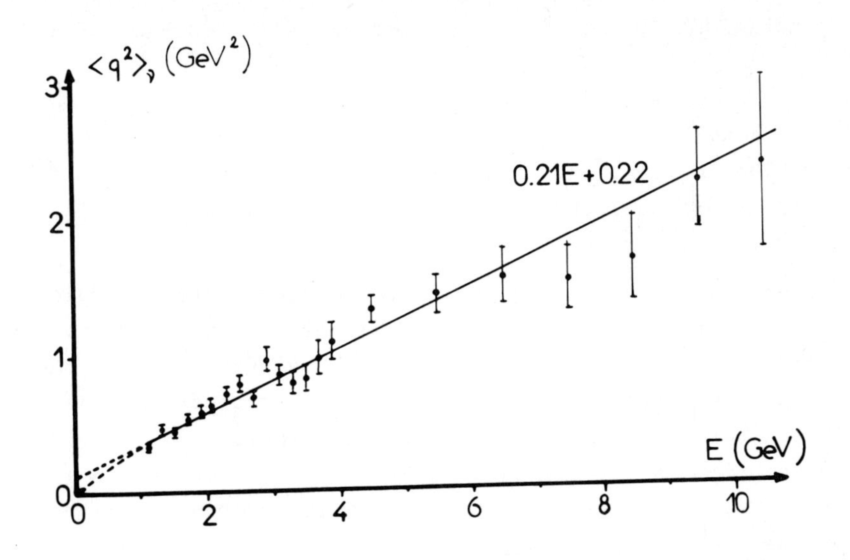

Fig. 13

Averaged q^2 for neutrinos versus E (Gargamelle).

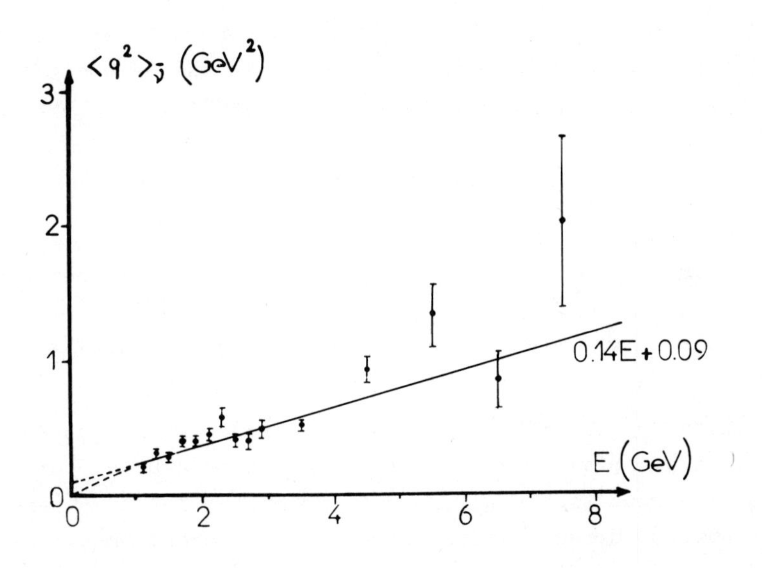

Fig. 14

Averaged q^2 for antineutrinos versus E (Gargamelle).

III EXPERIMENTAL DATA ON WEAK PROCESSES WITH NEUTRAL CURRENTS

1) The systematic research of neutral currents has been carried out at CERN with neutrino and antineutrino beam entering the Gargamelle chamber. The neutrino and antineutrino can scatter either off atomic electrons or off nucleons. Neutral currents have been studied in both cases and positive results found.

Here, we only discuss the hadronic case where the main characteristic of neutral current events is the absence in the final state of a charged lepton or antilepton trace. Such events have been observed and after a careful study of possible background sources they have been attributed for a large part to neutral currents.
The actual results for relative rates of neutral current events to charged currents events for interaction with hadron energy release larger than 1 GeV are as follows

$$\left(\frac{NC}{CC}\right)_{\nu} = 0.217 \pm 0.026 \tag{32}$$

$$\left(\frac{NC}{CC}\right)_{\bar{\nu}} = 0.43 \pm .12 \tag{33}$$

We notice that these quantities refer to numbers of events with identical cuts and not to total cross sections. In the same situation the ratio of antineutrino to neutrino for charged currents has been found to be

$$\frac{(CC)_{\bar{\nu}}}{(CC)_{\nu}} = 0.26 \pm 0.03 \tag{34}$$

2) Two different counter experiments have been performed at NAL by

- the Harvard-Pennsylvania-Wisconsin group (HPW)
- the Caltech group.

Data obtained by the HPW group has been presented in successive fluctuating steps but positive evidence for hadronic neutral currents is now claimed. The results are given in terms of two ratios R^{ν} and $R^{\bar{\nu}}$, comparing neutral current events with charged current ones. The mean incident energy is 50 GeV and only events carrying a total hadronic energy larger than 4 GeV have been retained. The average of successive experiments gives

$$R^{\nu} = 0.11 \pm 0.05$$

$$R^{\bar{\nu}} = 0.32 \pm 0.09$$

Another positive evidence is also claimed by the Caltech group with the following numerical estimates

$$R^{\nu} = 0.22$$
$$R^{\bar{\nu}} = 0.33$$

but the errors have not yet been computed.

3) It is clear that the results that come from the three experiments are consistent with each other. But we must keep in mind that there are no ratios of total cross sections but only of the number of events restricted with different cuts.

IV U(3) SYMMETRY GROUP

1) In the quark parton model based on SU(3) symmetry the interacting partons are 3 quarks and 3 antiquarks whose quantum numbers are given in Table 3

j	p 1	n 2	λ 3
Q	$\frac{2}{3}$	$-\frac{1}{3}$	$-\frac{1}{3}$
Y	$\frac{1}{3}$	$\frac{1}{3}$	$-\frac{2}{3}$
B	$\frac{1}{3}$	$\frac{1}{3}$	$\frac{1}{3}$

Table 3

The model is described by six distribution functions $D_j(\xi)$ and $D_{-j}(\xi)$ with $j = 1, 2, 3$ which are positive functions of ξ in the physical range $0 \leq \xi \leq 1$.
The conservation of the baryonic charge B, the electric charge Q and the hypercharge Y implies constraints on the mean values of quarks and antiquarks. From Table 3 we have

$$\langle N_1 - N_{-1} \rangle = B + Q \qquad \langle N_2 - N_{-2} \rangle = B + Y - Q \tag{35}$$
$$\langle N_3 - N_{-3} \rangle = B - Y$$

and for a given hadron only the mean numbers of antiquarks are free parameters.
The electromagnetic scaling function is immediately computed from Table 3:

$$2\,F_T^Q(\xi) = \frac{4}{9}\left[\mathcal{D}_1(\xi)+\mathcal{D}_{-1}(\xi)\right] + \frac{1}{9}\left[\mathcal{D}_2(\xi)+\mathcal{D}_{-2}(\xi)+\mathcal{D}_3(\xi)+\mathcal{D}_{-3}(\xi)\right] \quad (36)$$

2) The weak charged current is the Cabibbo current which in the quark language is simply written as

$$J_\mu^W = i\,\bar{p}\,\gamma_\mu(1+\gamma_5)\left[n\cos\theta_C + \lambda\sin\theta_C\right] \quad (37)$$

where θ_C is the Cabibbo angle.
The weak charges are purely left handed and they are given from equation (37) by

$$a_{21}^{WL} = 2\cos\theta_C \qquad\qquad a_{31}^{WL} = 2\sin\theta_C \quad (38)$$

It is convenient to separate in the scaling functions the contributions coming from the $\Delta Y=0$ and $\Delta Y=\pm 1$ transitions

$$F_\lambda = \cos^2\theta_C \; G_\lambda + \sin^2\theta_C \; H_\lambda$$

and the result is the following

$$G_+^W(\xi) = 2\mathcal{D}_{-1}(\xi) \qquad\qquad H_+^W(\xi) = 2\mathcal{D}_{-1}(\xi)$$

$$G_-^W(\xi) = 2\mathcal{D}_2(\xi) \qquad\qquad H_-^W(\xi) = 2\mathcal{D}_3(\xi)$$

$$G_+^{\bar{W}}(\xi) = 2\mathcal{D}_{-2}(\xi) \qquad\qquad H_+^{\bar{W}}(\xi) = 2\mathcal{D}_{-3}(\xi)$$

$$G_-^{\bar{W}}(\xi) = 2\mathcal{D}_1(\xi) \qquad\qquad H_-^{\bar{W}}(\xi) = 2\mathcal{D}_1(\xi)$$

We easily see that the splitting in states of definite helicity allows to isolate each quark and antiquark distribution function.
From electroproduction, neutrino and antineutrino processes on a given target one can measure nine structure functions. The number of differen types of quarks and antiquarks being six in this model we have at our

disposal only six distribution functions $D_j(\xi)$ so that the SU(3) quark parton model predicts three relations that one can write as

$$H_+^W(\xi) = G_+^W(\xi) \qquad\qquad H_-^{\overline{W}}(\xi) = G_-^{\overline{W}}(\xi)$$

$$F_T^Q(\xi) = \frac{1}{9}\left[G_+^W(\xi) + G_-^{\overline{W}}(\xi)\right] + \frac{1}{36}\left[G_-^W(\xi) + G_+^{\overline{W}}(\xi) + H_-^W(\xi) + H_+^{\overline{W}}(\xi)\right] \tag{40}$$

The relations between scaling functions are strict tests of the quark parton model.

The Adler and Gross-Llewellyn Smith sum rules (27) and (28) are obtained using equations (35) and (38)

$$\int_0^1 \left[F_T^{\overline{W}}(\xi) - F_T^W(\xi)\right] d\xi = (2B-Y)\cos^2\theta_c + (Q+Y)\sin^2\theta_c$$

$$\frac{1}{2}\int_0^1 \left[F_3^W(\xi) + F_3^{\overline{W}}(\xi)\right] d\xi = (2B+Y)\cos^2\theta_c + (2B-Y+Q)\sin^2\theta_c$$

In this quark parton model based on SU(3) $\otimes$ SU(3) algebra the sum rules have their original form.

3) When the target is a nucleon, charge symmetry relates the neutron and proton distributions

$$D_{\pm 1}^n(\xi) = D_{\pm 2}^p(\xi) \qquad D_{\pm 2}^n(\xi) = D_{\pm 1}^p(\xi) \qquad D_{\pm 3}^n(\xi) = D_{\pm 3}^p(\xi)$$

Only proton distributions will be used in what follows.
All the scaling functions on a neutron target are known from the scaling functions on a proton target. As an illustration let us give some examples of such relations: for neutrinos, antineutrinos

$$G_\pm^{\nu n}(\xi) = G_\pm^{\bar{\nu} p}(\xi) \qquad\qquad G_\pm^{\bar{\nu} n}(\xi) = G_\pm^{\nu p}(\xi)$$

$$H_-^{\nu n}(\xi) = H_-^{\nu p}(\xi) \qquad\qquad H_+^{\bar{\nu} n}(\xi) = H_+^{\nu p}(\xi) \tag{41}$$

and between electromagnetic and weak scaling functions

$$F_T^{ep}(\xi) - F_T^{en}(\xi) = \frac{1}{12}\left[G_+^{\nu p}(\xi) - G_-^{\nu p}(\xi) - G_+^{\nu n}(\xi) + G_-^{\nu n}(\xi)\right] \tag{42}$$

The situation is particularly simple for an isoscalar target N averaged over proton and neutron $N = \frac{p+n}{2}$. From equations (40) and (41) we get

$$G_{\pm}^{\nu N} = G_{\pm}^{\bar{\nu} N} = G_{\pm}^{N}$$
$$F_T^{eN} = \frac{5}{18} G_T^N + \frac{1}{36}\left[H_-^{\nu N} + H_+^{\bar{\nu} N}\right] \tag{43}$$

The equalities (42) and (43) are unambigous consequences of the quark parton model and positivity implies the simple inequality due to Llewellyn Smith

$$F_T^{eN} \geq \frac{5}{18} G_T^N \tag{44}$$

V QUARK PARTON MODEL FOR ELECTROPRODUCTION

1) The proton and neutron scaling functions are given by

$$2F_T^{ep}(\xi) = \frac{4}{9}\left[D_1 + D_{-1}\right] + \frac{1}{9}\left[D_2 + D_{-2}\right] + \frac{1}{9}\left[D_3 + D_{-3}\right]$$
$$2F_T^{en}(\xi) = \frac{1}{9}\left[D_1 + D_{-1}\right] + \frac{4}{9}\left[D_2 + D_{-2}\right] + \frac{1}{9}\left[D_3 + D_{-3}\right] \tag{45}$$

As a consequence of charge symmetry and of the positivity of the D_j's we obtain the inequalities

$$\frac{1}{4} \leq \frac{F_T^{en}(\xi)}{F_T^{ep}(\xi)} \leq 4 \tag{46}$$

or in terms of structure functions in the scaling region

$$\frac{1}{4} \leq \frac{\sigma^{en}(q^2, W^2)}{\sigma^{ep}(q^2, W^2)} \leq 4 \tag{47}$$

The experimental data shown in Fig. 15 are consistent with the theoretical bounds. Let us notice that the lower bound of 1/4 may eventually be reached at $\xi = 1$.

2) We now integrate the scaling function over ξ assuming these integrals to be convergent

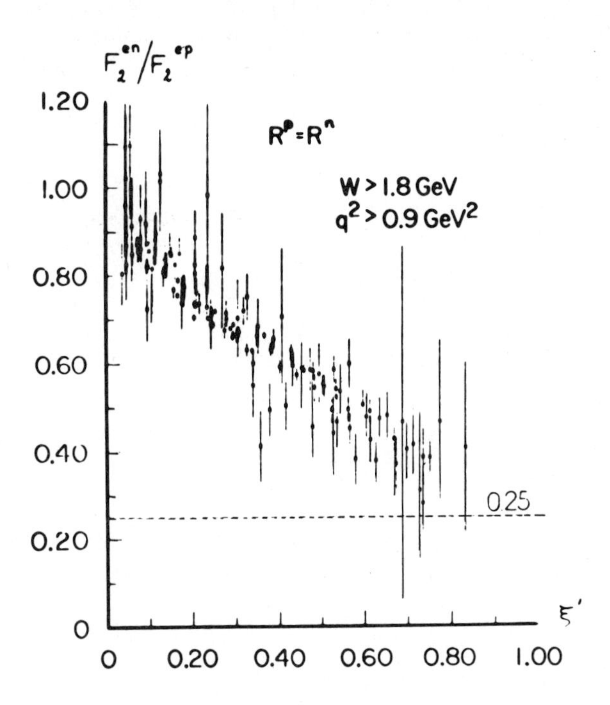

Fig. 15

σ_n/σ_p versus ξ'.

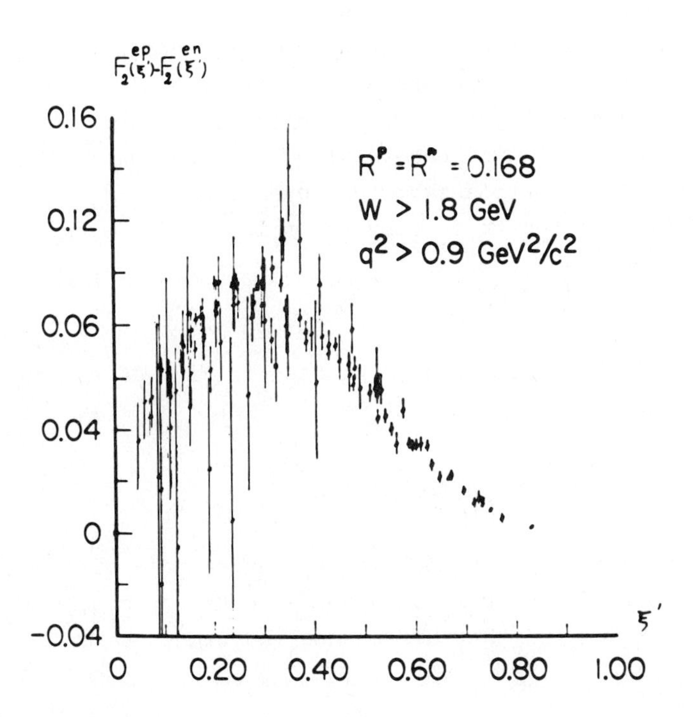

Fig. 16

$F_2^{ep} - F_2^{en}$ versus ξ'.

$$K^e = \int_0^1 2\, F_T^Q(\xi)\, d\xi$$

Using the equations (45) and the charge conservation relations (35) written for the proton we get

$$K^{ep} = 1 + \frac{8}{9}\langle N_{-1}\rangle + \frac{2}{9}\langle N_{-2}+N_{-3}\rangle$$

$$K^{en} = \frac{2}{3} + \frac{8}{9}\langle N_{-2}\rangle + \frac{2}{9}\langle N_{-1}+N_{-3}\rangle$$

The positivity of the mean number of antiquarks implies lower bounds for K^{ep} and K^{en}

$$K^{ep} \geq 1 \qquad\qquad K^{en} \geq \frac{2}{3} \tag{48}$$

The experimental situation is not very accurate. In fact the integrals K^{ep} and K^{en} look very dependent on the limits of integration. The most recent evaluation is

$$K^{ep} \simeq 0.81 \pm 0.04$$

$$K^{en} \simeq 0.65 \pm 0.03$$

the lower limit of integration being $\xi_m = 0.04$.
Obviously the lower bounds (48) are not violated. Unfortunately our information is unsufficient to decide whether the integrals K^e are convergent or not, or, in the parton language, whether the averaged number of partons is finite or not. The behaviour of the scaling function near $\xi = 0$ is obviously crucial to answer that question.
An interesting quantity expected to be convergent is the difference $K^{ep}-K^{en}$

$$K^{ep} - K^{en} = \frac{1}{3} + \frac{2}{3}\langle N_{-1} - N_{-2}\rangle$$

The structure function difference has been plotted in Fig. 16 and the experimental evaluation of $K^{ep}-K^{en}$ with $\xi_m = 0.05$ is

$$K^{ep}-K^{en} \simeq 0.18 \pm 0.04$$

Data below $\xi = 0.05$ are certainly crucial in evaluating this difference. In particular, the Gottfried sum rule which holds in parton models where $\langle N_{-1}\rangle = \langle N_{-2}\rangle$ predicts 1/3 for that difference and cannot be ruled out from experiment.

3) Let us now study the first moment of the quark and antiquark distributions

$$d_j = \int_0^1 \xi \, D_j(\xi) \, d\xi$$

These quantities are positive $d_j \geq 0$ and using energy momentum conservation we obtain

$$\sum_j (d_j + d_{-j}) = 1 - \varepsilon$$

where the parameter ε measures, in an averaged sense, the amount of gluons in the hadron

$$0 \leq \varepsilon \leq 1$$

By positivity, a non vanishing value for ε implies the existence of gluons in this model.
The first moment integrals for electroproduction are defined by

$$I^e = \int_0^1 \xi \, 2 F_T^Q(\xi) \, d\xi$$

For proton and neutron, using equations (45) we get

$$I^{ep} = \frac{1}{3}(d_1 + d_{-1}) + \frac{1}{9}(1-\varepsilon)$$

$$I^{en} = \frac{1}{3}(d_2 + d_{-2}) + \frac{1}{9}(1-\varepsilon)$$

Averaging over proton and neutron

$$I^{eN} = \frac{5}{18}(1-\varepsilon) - \frac{1}{6}(d_3 + d_{-3})$$

we deduce, by positivity, an absolute upper bound for the magnitude of electromagnetic scaling functions

$$I^{eN} \leq \frac{5}{18} \tag{49}$$

When the integral I^{en} is known from experiment the gluon parameter ε is restricted by

$$0 \leq \varepsilon \leq 1 - \frac{18}{5} I^{eN} \tag{50}$$

The most recent experimental evaluation of I^{eN} gives

$$I^{eN} \simeq 0.15 \pm 0.01 \tag{51}$$

The absolute bound (49) is satisfied and the limits for ε are

$$0 \leq \varepsilon \leq 0.46 \pm 0.04 \tag{52}$$

The proton-neutron difference is known with a poor accuracy and the result is

$$I^{ep} - I^{en} = 0.04 \pm 0.02 \tag{53}$$

VI QUARK PARTON MODEL FOR WEAK PROCESSES WITH CHARGED CURRENTS

A detailed analysis of weak processes can be done with the quark parton model starting from the set of expressions (39) for the scaling functions. Unfortunately our experimental information being extremely limited we had better concentrate over specific points where experimenta data are available.

1) The first of these points is the study of total cross sections The interesting quantities are the constants A^{ν} and $A^{\bar{\nu}}$ which govern the linear rising of the total cross sections in the local Fermi interaction Using the Cabibbo current the separation between strangeness conserving and strangeness changing transitions is achieved by putting

$$A^{\nu,\bar{\nu}} = \cos^2\theta_C \, B^{\nu,\bar{\nu}} + \sin^2\theta_C \, C^{\nu,\bar{\nu}}$$

From equation (20) these constants involve the first moments of the quark and antiquark distributions. The result is

$$\begin{aligned}
B^{\nu p} &= \tfrac{2}{3} d_{-1} + 2 d_2 & \qquad C^{\nu p} &= \tfrac{2}{3} d_{-1} + 2 d_3 \\
B^{\nu n} &= \tfrac{2}{3} d_{-2} + 2 d_1 & \qquad C^{\nu n} &= \tfrac{2}{3} d_{-2} + 2 d_3 \\
B^{\bar{\nu} p} &= \tfrac{2}{3} d_1 + 2 d_{-2} & \qquad C^{\bar{\nu} p} &= \tfrac{2}{3} d_1 + 2 d_{-3} \\
B^{\bar{\nu} n} &= \tfrac{2}{3} d_2 + 2 d_{-1} & \qquad C^{\bar{\nu} n} &= \tfrac{2}{3} d_2 + 2 d_{-3}
\end{aligned} \tag{54}$$

In order to compare these expressions with the Gargamelle results we first average over proton and neutron

$$B^{\nu N} = \frac{1}{3}(d_{-1}+d_{-2}) + (d_1+d_2) \qquad C^{\nu N} = \frac{1}{3}(d_{-1}+d_{-2}) + 2d_3$$

$$B^{\bar{\nu} N} = \frac{1}{3}(d_1+d_2) + (d_{-1}+d_{-2}) \qquad C^{\bar{\nu} N} = \frac{1}{3}(d_1+d_2) + 2d_{-3} \tag{55}$$

Let us recall that the quark parton model for the electroproduction integral I^{eN} is

$$I^{eN} = \frac{5}{18}(d_1+d_2+d_{-1}+d_{-2}) + \frac{1}{9}(d_3+d_{-3})$$

We solve these linear expressions and we get the theoretical expressions

$$d_1+d_2+d_{-1}+d_{-2} = \frac{3}{4[1-\frac{9}{2}\sin^2\theta_c]}\left[A^{\nu N}+A^{\bar{\nu} N} - 18\sin^2\theta_c I^{eN}\right] \tag{56}$$

$$d_3+d_{-3} = \frac{15}{8[1-\frac{9}{2}\sin^2\theta_c]}\left[\frac{24}{5}(1-\frac{3}{4}\sin^2\theta_c)I^{eN} - (A^{\nu N}+A^{\bar{\nu} N})\right] \tag{57}$$

and the numerical results with $\sin\theta_c = 0.23$ are

$$d_1+d_2+d_{-1}+d_{-2} = 0.505 \pm 0.054 \tag{58}$$

$$d_3+d_{-3} = 0.091 \pm 0.176 \tag{59}$$

For the gluon parameter ε the explicit expression is

$$\varepsilon = 1 - \frac{9}{8[1-\frac{9}{2}\sin^2\theta_c]}\left[8(1-\frac{9}{4}\sin^2\theta_c)I^{eN} - (A^{\nu N}+A^{\bar{\nu} N})\right] \tag{60}$$

and from experiment we obtain

$$\varepsilon = 0.40 \pm 0.13 \tag{61}$$

The quark parton model is consistent with electroproduction, neutrino and antineutrino data if and only if gluons are present.
It is now possible to have a first estimate of $\Delta S = 0$ and $\Delta S = \pm 1$ contributions using the expressions (54)

$$\frac{C^{\nu N}+C^{\bar{\nu} N}}{B^{\nu N}+B^{\bar{\nu} N}} = \frac{1}{4} + \frac{3}{2}\,\frac{d_3+d_{-3}}{d_1+d_2+d_{-1}+d_{-2}} \geq \frac{1}{4} \tag{62}$$

Taking into account the Cabibbo angle we get for total cross sections

$$\frac{(\sigma^{\nu N}+\sigma^{\bar{\nu}N})(|\Delta S|=1)}{(\sigma^{\nu N}+\sigma^{\bar{\nu}N})(\Delta S=0)} = \left(3\ {}^{+2}_{-1.4}\right)\% \qquad (63)$$

We now try to use the experimental information on the difference of neutrino and antineutrino total cross sections which in the quark parton model is written as

$$A^{\nu N}-A^{\bar{\nu}N} = \frac{2}{3}\left(1-\frac{3}{2}\sin^2\theta_c\right)\left[d_1+d_2-d_{-1}-d_{-2}\right]+2\sin^2\theta_c\left(d_3-d_{-3}\right) \qquad (64)$$

Taking into account the smallness of $\sin^2\theta_c$ and of the strange quark and antiquark moments from (59) we assume that the second term in the right-hand side of equation (64) is negligible as compared to the first one. We are now in position to separate the non strange quarks and non strange antiquarks contributions. The result is

$$d_1+d_2 = 0.487 \pm 0.052 \qquad (65)$$

$$d_{-1}+d_{-2} = 0.018 \pm 0.052 \qquad (66)$$

It clearly shows that the quark contributions dominate strongly over the antiquark ones and this feature will simplify the description of the nucleon in terms of quarks and antiquarks.
Let us emphasize that a ratio of antineutrino to neutrino cross-sections for the $\Delta S = 0$ part of 1/3 would imply the absence of antiquarks in the nucleon. The experimental value of that ratio is close to 1/3: 0.38 ± 0.02 and we immediately recover the previous result.
It is possible to compute the $\Delta S = 0$ part of the total cross section and to compare this contribution with the experimental value for all the events

$$B^{\nu N}\cos^2\theta_c = 0.467 \pm 0.051 \qquad A^{\nu N}_{exp} = 0.471 \pm 0.050$$
$$B^{\bar{\nu}N}\cos^2\theta_c = 0.170 \pm 0.051 \qquad A^{\bar{\nu}N}_{exp} = 0.183 \pm 0.020$$

By comparing these numbers we expect a very small $|\Delta S| = 1$ cross section induced by neutrinos and a measurable $|\Delta S| = 1$ cross section induced by antineutrinos

$$C^{\nu N} \ll B^{\nu N} \qquad C^{\bar{\nu}N} \simeq B^{\bar{\nu}N}$$

but a quantitative prediction is not possible because of the large experimental errors beside the antineutrino lower bound

$$C^{\bar{\nu}N} > \frac{1}{3}(d_1 + d_2)$$

which gives from equation (65)

$$C^{\bar{\nu}N} > 0.162 \pm 0.017$$

2) The proton and neutron total cross sections have not been separated in the CERN-Gargamelle experiment. Nevertheless in specific quark parton models and using the electroproduction result for $I^{ep}-I^{en}$ it is possible to make predictions. We shall give here two examples.

In the equipartition quark parton model the charge conservation relations (35) are assumed to be satisfied also by the first moment integrals and we have for a proton target

$$d_1 - d_{-1} = 2\langle \frac{1}{N} \rangle \qquad d_2 - d_{-2} = \langle \frac{1}{N} \rangle \qquad d_3 - d_{-3} = 0$$

where the parameter $<\frac{1}{N}>$ is interpreted as the averaged inverse number of partons in the nucleon. From equations (65) and (66) we compute a large value for that quantity

$$< \frac{1}{N} > = 0.156 \pm 0.029 \tag{67}$$

so that the first moments of the scaling functions can be described with a small number of partons. We obtain the predictions

$$\begin{aligned} A^{\nu p} &= 0.334 \pm 0.060 & A^{\bar{\nu}p} &= 0.246 \pm 0.032 \\ A^{\nu n} &= 0.607 \pm 0.060 & A^{\bar{\nu}n} &= 0.120 \pm 0.032 \end{aligned} \tag{68}$$

and for the neutron to proton ratios

$$\frac{A^{\nu n}}{A^{\nu p}} \simeq 1.8 \pm 0.3 \qquad \frac{A^{\bar{\nu}p}}{A^{\bar{\nu}n}} \quad 2 \pm 0.4$$

In the two component quark parton model we have also 4 independent parameters:

2 for the diffractive part represented by non strange valence quarks;

2 for the diffractive part associated to isoscalar quark-antiquarks seas.

The first moments are then written as

$$d_1 = v_1 + p \qquad d_2 = v_2 + p \qquad d_3 = q$$

$$d_{-1} = p \qquad d_{-2} = p \qquad d_{-3} = q$$

and from equations (65) and (66) we obtain

$$v_1 = 0.294 \pm 0.053 \qquad v_2 = 0.174 \pm 0.053 \tag{69}$$

Let us remark that the ratio v_1/v_2 is compatible with the value of 2 suggested by the naive quark model.
Again the proton and neutron total cross sections are separated

$$\begin{aligned} A^{\nu p} &= 0.357 \pm 0.080 & A^{\bar{\nu} p} &= 0.223 \pm 0.028 \\ A^{\nu n} &= 0.584 \pm 0.080 & A^{\bar{\nu} n} &= 0.143 \pm 0.028 \end{aligned} \tag{70}$$

and for the neutron-proton ratios we predict

$$\frac{A^{\nu n}}{A^{\nu p}} = 1.63 \pm 0.16 \qquad \frac{A^{\bar{\nu} p}}{A^{\bar{\nu} n}} = 1.56 \pm 0.14 \ .$$

The predictions of these two models are qualitatively the same.
In particular the neutron proton ratio of total cross sections induced by neutrinos is compatible with the value 1.8 ± 0.3 obtained in a previous CERN propane experiment.

3) The second point we wish to discuss here is the energy distribution of the final charged lepton or antilepton. We define a normalized distribution by

$$f(\rho) = \frac{1}{\sigma_{TOT}} \frac{d\sigma}{d\rho}$$

In the scaling region, from equations (19) and (21) they are independent of the incident energy E and given by

$$\begin{aligned} f^{\nu}(\rho) &= \frac{\rho^2 I_+^{\nu} + I_-^{\nu}}{\frac{1}{3} I_+^{\nu} + I_-^{\nu}} \\ f^{\bar{\nu}}(\rho) &= \frac{\rho^2 I_-^{\bar{\nu}} + I_+^{\bar{\nu}}}{\frac{1}{3} I_-^{\bar{\nu}} + I_+^{\bar{\nu}}} \end{aligned} \tag{71}$$

For an isoscalar target the first moment integrals $I_{\pm}$ have the following expressions

$$I_+^{\nu N} = d_{-1} + d_{-2} \qquad I_-^{\nu N} = \cos^2\theta_C (d_1 + d_2) + 2 \sin^2\theta_C \, d_3$$

$$I_+^{\bar\nu N} = \cos^2\theta_C (d_{-1} + d_{-2}) + 2 \sin^2\theta_C \, d_{-3} \qquad I_-^{\bar\nu N} = d_1 + d_2$$

Using the numerical results of the previous analysis of total cross sections we obtain

$$I_+^{\nu N} = 0.018 \pm 0.052 \qquad I_-^{\nu N} = 0.465 \pm 0.050$$
$$I_+^{\bar\nu N} = 0.021 \pm 0.050 \qquad I_-^{\bar\nu N} = 0.487 \pm 0.052 \tag{72}$$

Let us recall that in parton models with only left-hand (right-hand) couplings of partons (antipartons) with the weak current, the dominance of the helicity $\lambda = -1$ contribution over the helicity $\lambda = +1$ contribution is equivalent, at high energy, to the dominance of parton distributions over antiparton ones.
It is then convenient to rewrite the normalized energy distributions in the form

$$f^{\nu N}(\rho) = 1 + (3\rho^2 - 1) \frac{1}{3} \frac{I_+^{\nu N}}{A^{\nu N}}$$
$$f^{\bar\nu N}(\rho) = 3\rho^2 - (3\rho^2 - 1) \frac{I_+^{\bar\nu N}}{A^{\bar\nu N}} \tag{73}$$

The expected distribution for neutrino is essentially flat and the antineutrino one is very close to $3\rho^2$. The deviations from pure helicity $\lambda = -1$ shape are governed by the two coefficients

$$\frac{1}{3} \frac{I_+^{\nu N}}{A^{\nu N}} = 0.013 \begin{array}{l} + 0.037 \\ - 0.013 \end{array} \qquad \frac{I_+^{\bar\nu N}}{A^{\bar\nu N}} = 0.11 \begin{array}{l} + 0.28 \\ - 0.11 \end{array} \tag{74}$$

The predictions are shown in Figs. 17 and 18. The experimental data at high energy are in qualitative agreement with these results but they are not accurate enough to allow a quantitative comparison.

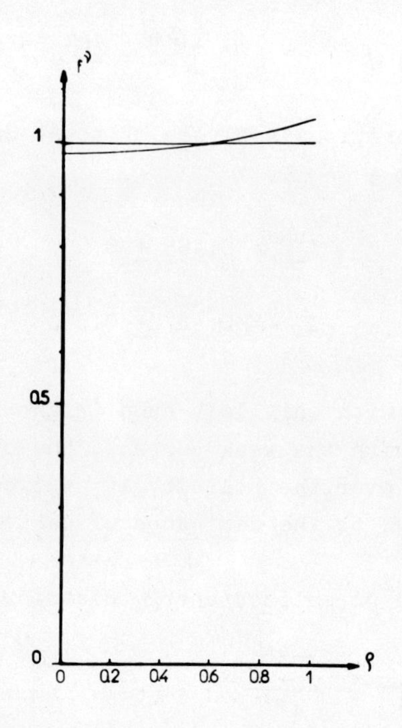

Fig. 17: Neutrino scattering: ℓ^- energy distribution

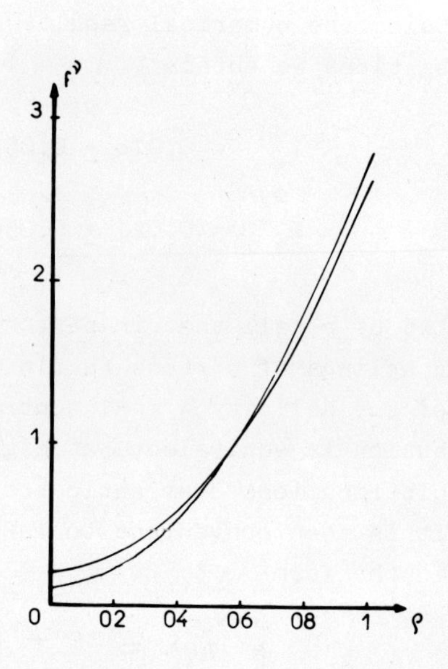

Fig. 18: Antineutrino scattering: ℓ^+ energy distribution

4) It is now straightforward to compute the averaged values of the final lepton and antilepton energies

$$\langle \rho \rangle = \int_0^1 \rho f(\rho) \, d\rho$$

By positivity of the I_λ's these quantities are bounded

$$\frac{1}{2} \leq \langle \rho \rangle \leq \frac{3}{4}$$

and from equations (73) and (74) we predict

$$\begin{aligned} \langle \rho \rangle_\nu &= 0.503 \begin{smallmatrix} + 0.009 \\ - 0.003 \end{smallmatrix} \\ \langle \rho \rangle_{\bar\nu} &= 0.722 \begin{smallmatrix} + 0.028 \\ - 0.071 \end{smallmatrix} \end{aligned} \qquad (75)$$

The results of the Gargamelle experiment

$$< \rho >_{\nu\ exp} = 0.54 \pm 0.04$$

$$< \rho >_{\bar{\nu}\ exp} = 0.72 \pm 0.05$$

are in good agreement with the quark parton model values based on total cross sections.

5) The weak scaling functions have not been experimentally separated and the only quantity we can discuss is the fixed ξ distribution which is written in the scaling limit as

$$\frac{d\sigma}{d\xi} \longrightarrow \frac{G^2 M E}{\pi} A(\xi)$$

The functions $A(\xi)$ for neutrino and antineutrino processes can be written as linear combinations of the quark and antiquark distributions $D(\xi)$. In the U(3) quark parton model and for an isoscalar target we get from equation (55)

$$A^{\nu N}(\xi) = \xi\left\{\frac{1}{3}\left[D_{-1}(\xi) + D_{-2}(\xi)\right] + \cos^2\theta_C\left[D_1(\xi) + D_2(\xi)\right] + 2\sin^2\theta_C\, D_3(\xi)\right\} \quad (76)$$

$$A^{\bar{\nu} N}(\xi) = \xi\left\{\frac{1}{3}\left[D_1(\xi) + D_2(\xi)\right] + \cos^2\theta_C\left[D_{-1}(\xi) + D_{-2}(\xi)\right] + 2\sin^2\theta_C\, D_{-3}(\xi)\right\} \quad (77)$$

Adding now the decomposition of the corresponding electroproduction function $F_2 = 2\xi F_T$

$$F_2^{eN}(\xi) = \xi\left\{\frac{5}{18}\left[D_1(\xi) + D_2(\xi) + D_{-1}(\xi) + D_{-2}(\xi)\right] + \frac{1}{9}\left[D_3(\xi) + D_{-3}(\xi)\right]\right\} \quad (78)$$

the system of equations (76), (77) and (78) can be solved as in the first paragraph of this section. With the present accuracy of experimental data the most interesting relation is that involving strange quark and antiquark distributions

$$\left[1 - \frac{9}{2}\sin^2\theta_C\right]\xi\left[D_3(\xi) + D_{-3}(\xi)\right] =$$

$$= 9\left(1 - \frac{3}{4}\sin^2\theta_C\right)F_2^{eN}(\xi) - \frac{15}{8}\left[A^{\nu N}(\xi) + A^{\bar{\nu} N}(\xi)\right] \quad (79)$$

Because of the positivity of the distribution functions the right-hand side of equation (79) must be positive for all values of ξ. This result which involves electromagnetic and weak functions is a non trivial and unambiguous test of the quark parton model. The comparison with experiment is shown in Fig. 19 where the variable ξ' is used for convenience. Positivity is satisfied within experimental errors and the quantity $\xi' \, |D_3(\xi') + D_{-3}(\xi')|$ is consistent with zero for $\xi' > 0.3$. This last result is expected in a two component model where the diffractive contributions are important only for small values of ξ'.

An analogous result is obtained by comparing the difference $A^{\nu N}(\xi) - A^{\bar{\nu} N}(\xi)$ with $F_2^{eN}(\xi)$ the coefficients being adjusted in order to eliminate the non strange quark distributions

$$\frac{12}{5}\left(1-\frac{3}{2}\sin^2\theta_c\right) F_2^{eN}(\xi) - \left[A^{\nu N}(\xi) - A^{\bar{\nu} N}(\xi)\right] =$$

$$= \frac{2}{3}\left(1-\frac{3}{2}\sin^2\theta_c\right) \xi \left[D_{-1}(\xi) + D_{-2}(\xi)\right] + \frac{4}{15}\left(1-9\sin^2\theta_c\right) \xi D_3(\xi) + \frac{4}{15}\left(1+\frac{3}{2}\sin^2\theta_c\right) \xi D_{-3}(\xi) \qquad (80)$$

The experimental situation is exhibited in Fig. 20. Positivity is consistent with experiment and the diffractive component of the right-hand side of eq. (80) is only sizeable for values of ξ' smaller than 0.4.

The Gross-Llewellyn Smith sum rule is easily translated into this language and from eqs. (76) and (77) we get

$$\int_0^1 \frac{d\xi}{\xi} \left[A^{\nu N}(\xi) - A^{\bar{\nu} N}(\xi)\right] = 2 - 3\sin^2\theta_c = 1.84$$

Moreover, taking into account the different numbers of neutron and protons in freon, the theoretical prediction becomes

$$\int_0^1 \frac{d\xi}{\xi} \left[A^{\nu}(\xi) - A^{\bar{\nu}}(\xi)\right] = 2.114 - 3.08\sin^2\theta_c = 1.96$$

Using the CERN-Gargamelle data an estimate of the integral has been done and the result 1.97 ± 0.20 is in excellent agreement with the theoretical prediction.

VII QUARK PARTON MODEL FOR WEAK PROCESSES WITH NEUTRAL CURRENTS

1) The simple quark parton model based on U(3) symmetry relates nicely electroproduction, neutrino and antineutrino data as shown in the previous section. Moreover the production of strange particles reduced by the Cabibbo angle remains small. It is then appealing to use for the neutral hadronic current a naive model proposed by Weinberg where strange

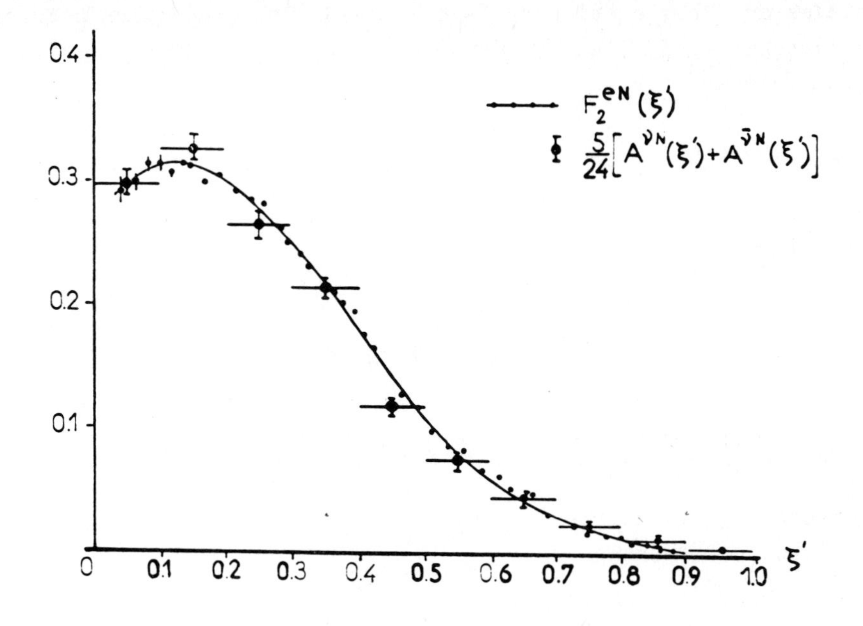

Fig. 19
Sum of neutrino and antineutrino cross sections compared to electroproduction.

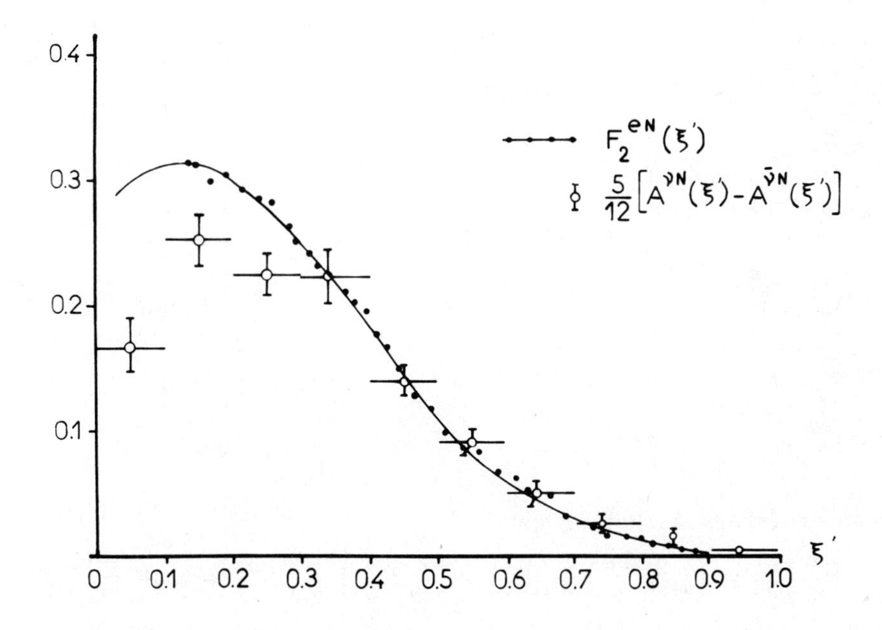

Fig. 20
Difference of neutrino and antineutrino cross sections compared to electroproduction.

particles and more exotic ones are ignored. The connection between the weak isotopic spin of gauge theories and the strong isotopic spin is made as follows

WEAK LEFT-HAND SU(2) → STRONG SU(2) ⊗ SU(2)

$$2 J_k^\mu = V_k^\mu + A_k^\mu$$

where as usual V means vector and A axial vector.
The neutral and charged weak current ΔS = 0 have very simple expressions

$$J_Z^\mu = (V_3^\mu + A_3^\mu) - 2 \sin^2\theta_W (V_3^\mu + V_S^\mu)$$

$$J_W^\mu = (V_1^\mu + A_1^\mu) + i (V_2^\mu + A_2^\mu)$$

the electromagnetic current being decomposed as usual into an isoscalar and isovector component

$$J_Q^\mu = V_3^\mu + V_S^\mu$$

The Weinberg mixing angle θ_W is a free parameter in the theory and we put $x = \sin^2 \theta_W$

2) The number of independent quark and antiquark distributions being six the scaling functions

$F_\pm^Z$ for $\nu_\ell(\bar{\nu}_\ell) + p \to \nu_\ell(\bar{\nu}_\ell)$ + HADRONS (ΔS = 0)

can be written as linear combinations of the scaling functions

$G_\pm^\nu$ for $\nu_\ell + p \to \ell^-$ + HADRONS (ΔS = 0)

$G_\pm^{\bar{\nu}}$ for $\bar{\nu}_\ell + p \to \ell^+$ + HADRONS (ΔS = 0)

F_T^Q for $\ell^\mp + p \to \ell^\mp$ + HADRONS

For an arbitrary target we get

$$F_+^Z(\xi) = \left(\frac{1}{4} - \frac{2x}{3}\right) G_+^\nu(\xi) + \left(\frac{1}{4} - \frac{x}{3}\right) G_+^{\bar{\nu}}(\xi) + 4x^2 F_T^Q(\xi)$$

$$F_-^Z(\xi) = \left(\frac{1}{4} - \frac{x}{3}\right) G_-^\nu(\xi) + \left(\frac{1}{4} - \frac{2x}{3}\right) G_-^{\bar{\nu}}(\xi) + 4x^2 F_T^Q(\xi)$$

In the particular case of an isoscalar target these relations become simpler

$$F_{\pm}^{ZN}(\xi) = \left(\frac{1}{2} - x\right) G_{\pm}^{N}(\xi) + 4x^2 F_T^{eN}(\xi) \tag{81}$$

Differential cross section relations can easily be obtained in the scaling region. Using equation (81) we get

$$d\sigma_{NC}^{\nu N} = \left(\frac{1}{2} - x\right) d\sigma_{CC}^{\nu N} + 4x^2 X \, d\sigma^{eN}$$

$$d\sigma_{NC}^{\bar{\nu} N} = \left(\frac{1}{2} - x\right) d\sigma_{CC}^{\bar{\nu} N} + 4x^2 \bar{X} \, d\sigma^{eN} \tag{82}$$

where $X = 2(G^2/e^4)q^4$. The indices NC and CC mean neutral current and charged current respectively.
In fact the relation between the differences $d\sigma^{\nu N} - d\sigma^{\bar{\nu} N}$ for neutral and charged current reactions is simply due to an isotopic spin rotation and it is a trivial consequence of the simple structure assumed for the weak currents.
Integrating the differential cross sections we obtain in the scaling limit

$$\int d^2\sigma^{\nu,\bar{\nu}} \rightarrow \frac{G^2 ME}{\pi} B^{\nu,\bar{\nu}}$$

$$\int X \, d^2\sigma^{e} \rightarrow \frac{G^2 ME}{\pi} \frac{2}{3} I^{e}$$

and from equations (82)

$$B_{NC}^{\nu N} = \left(\frac{1}{2} - x\right) B_{CC}^{\nu N} + \frac{8}{3} x^2 I^{eN}$$

$$B_{NC}^{\bar{\nu} N} = \left(\frac{1}{2} - x\right) B_{CC}^{\bar{\nu} N} + \frac{8}{3} x^2 I^{eN} \tag{83}$$

The ratio R^{ν} and $R^{\bar{\nu}}$ of neutrino and antineutrino total cross sections

$$R^{\nu} = \frac{\sigma_{NC}^{\nu N}}{\sigma_{CC}^{\nu N}} \qquad R^{\bar{\nu}} = \frac{\sigma_{NC}^{\bar{\nu} N}}{\sigma_{CC}^{\bar{\nu} N}}$$

become quadratic functions of x

$$R^{\nu} = \frac{1}{2} - x + \frac{8}{3} x^2 \frac{I^{eN}}{B_{CC}^{\nu N}}$$

$$R^{\bar{\nu}} = \frac{1}{2} - x + \frac{8}{3} x^2 \frac{I^{eN}}{B_{CC}^{\bar{\nu} N}} \tag{84}$$

The corresponding parabola have been represented in Figs. 21 and 22

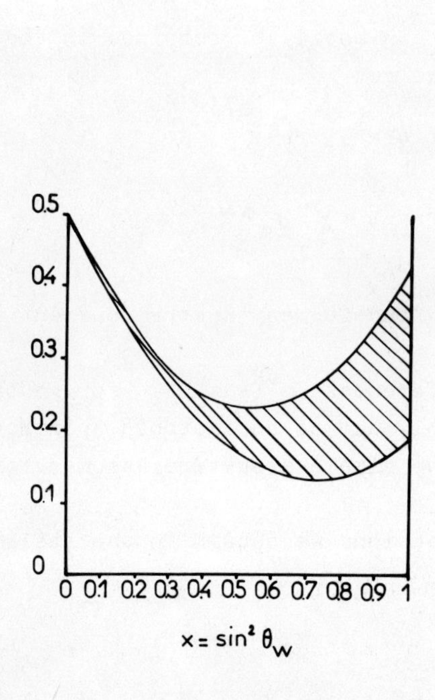

Fig. 21
Quark parton model prediction for the ratio of neutrino cross sections versus x.

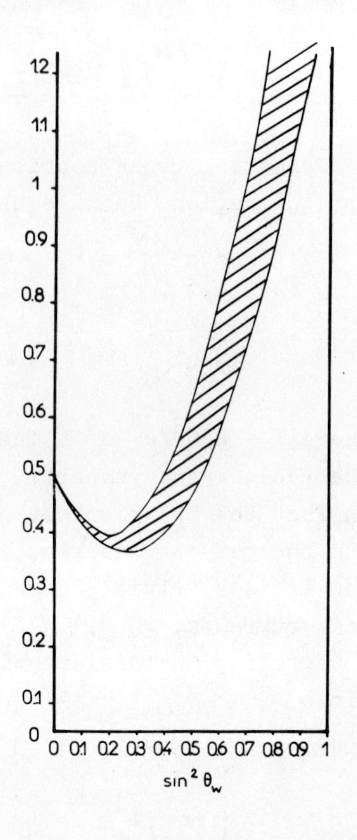

Fig. 22
Quark parton model prediction for the ratio of antineutrino cross sections versus x.

using the experimental data

$$B_{CC}^{\nu N} = 0.493 \pm 0.52 \qquad B_{CC}^{\bar{\nu} N} = 0.180 \pm 0.020$$

$$I^{eN} = 0.15 \pm 0.01$$

Lower bounds for the ratios R^{ν} and $R^{\bar{\nu}}$ are easily computed and in the one standard deviation limit we get

$$R^{\nu} > 0.14 \qquad R^{\bar{\nu}} > 0.37 \tag{85}$$

By eliminating x between the two equations (84) we obtain a relation between R^{ν} and $R^{\bar{\nu}}$

$$x = \frac{1}{2} + \frac{r_c R^{\bar{\nu}} - R^{\nu}}{1 - r_c} = \left\{ \frac{3}{8} \frac{1}{1-r_c} \frac{B_{cc}^{\bar{\nu}N}}{I^{eN}} \left[R^{\bar{\nu}} - R^{\nu} \right] \right\}^{1/2} \qquad (86)$$

where $r_c = B_{CC}^{\bar{\nu}N} / B_{CC}^{\nu N}$. The corresponding parabola in the R^{ν}, $R^{\bar{\nu}}$ plane has been drawn in Fig. 23 including the one standard deviation errors.

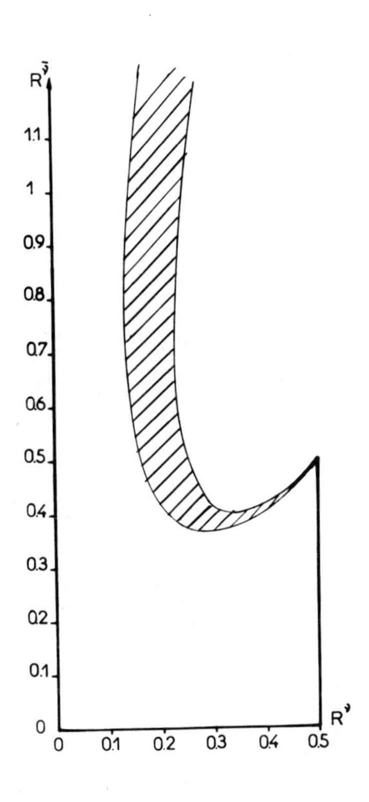

Fig. 23

Quark parton model prediction in the R^{ν}, $R^{\bar{\nu}}$ plane.

Let us write for completeness the Gross-Llewellyn Smith sum rule in this model

$$\int_0^1 F_3^{Z} d\xi = \left(\frac{1}{2} - x\right)\left(B + \frac{Y}{2}\right) - \frac{2x}{3} I_3$$

In particular for an isoscalar nucleon target we obtain

$$\int_0^1 F_3^{ZN} d\xi = 3\left(\frac{1}{2} - x\right) \qquad (87)$$

3) The Gargamelle data is now analysed in the framework of this simple quark parton model. We use as a first approximation*) numbers (32), (33) and (34) as ratios of total cross sections. The lower bounds (85) are satisfied and the Weinberg angle is computed from the two expressions of equation (86)**)

$$\sin^2 \theta_W = 0.36 \pm 0.06 \qquad (88)$$

$$\sin^2 \theta_W = 0.36 \pm 0.11 \qquad (89)$$

*) Without a good knowledge of the energy distributions it is not possible to compute the error made in replacing the ratio of total cross sections by that of the number of events. We expect the correction to be minimized by using, for the three ratios r_c, R^{ν} and $R^{\bar{\nu}}$ of eq. (86) the number of events with identical cuts.

**) Because of the existence of energy cut, the second expression of x in the right-hand side of eq. (86) can be computed in different ways and the central value may vary between 0.30 and 0.40. Such an uncertainty must be kept in mind when comparing the results (88) and (89).

The consistency of these results measures the consistency of the quark parton model with experiment. Moreover if the Weinberg current is replaced by a somewhat more general form

$$J_Z^\mu = \alpha(V_3^\mu + A_3^\mu) - 2x(V_3^\mu + V_S^\mu)$$

we easily check that a value of α close to unity can be found

$$\alpha = 1 \pm 0.10$$

the parameter x being given by equation (89).

REFERENCES

EXPERIMENTS

a) Electroproduction

E.D. Bloom et al., Phys. Rev. Letters 23, 930 (1969).

M. Breidenbach et al., Phys. Rev. Letters 23, 935 (1969).

M. Breidenbach, Ph. D. Thesis, M.I.T. (1970).

G. Miller, Ph. D. Thesis, SLAC-129 (1970).

E. D. Bloom et al., SLAC-PUB 796 (1970).

H.W. Kendall, Report to the Vth Intern. Symp. on Electron and Photon Interactions at High Energies, Cornell (1971).

E.D. Bloom and F. J. Gilman, Phys. Rev. D4, 2901 (1971).

A. Bodek, Ph. D. Thesis, M.I.T. COO3069-116 (1972).

G. Miller et al., Phys. Rev. D5, 528 (1972).

A. Bodek et al., Phys. Rev. Letters 30, 1087 (1973).

E. M. Riordan, Ph. D. Thesis, M.I.T. COO3069-176 (1973).

E. D. Bloom, Report to the VIth Intern. Symp. on Electron and Photon Interactions at High Energies, Bonn (1973).

J. S. Poucher et al., Phys. Rev. Letters 32, 118 (1974).

b) Weak processes with charged currents

I. Budagov et al., Phys. Letters 30B, 364 (1969).

T. Eichten et al., Phys. Letters 46B, 274 (1973).

T. Eichten et al., Phys. Letters 46B, 281 (1973).

B. C. Barish et al., Phys. Rev. Letters 31, 180 (1973); 31, 410 (1973); 31, 565 (1973).

A. Benvenuti et al., Phys. Rev. Letters 32, 125 (1974).

C. Franzinetti, Report to the Intern. Symp. on Electron and Photon Interactions at High Energies, Bonn (1973).

M. Haguenauer, Thèse, Paris (1973).
D. H. Perkins,Lectures at Hawaii Topical Conf. on Particle Physics(1973).
B. C. Barish et al., CALT preprints 68-452, 68-453 (1974).
D. Cundy, Report to the London Conf. (1974).
c) Weak processes with neutral currents
H.J. Hasert et al., Phys. Letters 46B, 138 (1973).
D. H. Perkins, Lectures at the Vth Hawaii Topical Conf. in Particle Physics (1973).
G. Myatt, Report to the Intern. Symp. on Electron and Photon Interactions at High Energies, Bonn (1973).
J.P. Vialle, Thesis, Orsay (1974).
A. Benvenuti et al., Phys. Rev. Letters 32, 800 (1974).
B. Aubert et al., Phys. Rev. Letters 32, 1454, 1457 (1974).
D. Cundy, Report to the London Conf. (1974).

N. B. The experimental data shown in the figures correspond to the situation in the spring 1974.

THEORY

J.D. Bjorken and E. Paschos, Phys. Rev. 185, 1975 (1969); Phys. Rev.D1, 3151 (1970).
C.H. Llewellyn Smith, Nucl. Phys. B17, 277 (1970).
M. Gourdin, Nucl. Phys. B29, 601 (1971).
J. Kuti and V. F. Weisskopf, Phys. Rev. D4, 3418 (1971).
M. Gourdin, Lectures at the Erice Summer School (1971).
C. H. Llewellyn Smith, Phys. Reports 3C, 263 (1972).
M. Gourdin, Nucl. Phys. B53, 509 (1973).
D. Cline and E. Paschos, Phys. Rev. D8, 984 (1973).
J.D. Bjorken, D. Cline and A. K. Mann, Phys. Rev. D8, 3207 (1973).
E. A. Paschos and V.I. Zakharov, Phys. Rev. D8, 215 (1973).
E. A. Paschos, Lectures at the Erice Summer School (1973).
V. Barger and R. J. N. Phillips, preprint (1973).
M. Gourdin, Invited talk at the IXth Rencontre de Moriond, Méribel (1974).
S. Weinber, Phys. Rev. D5, 1412 (1972).
S. L. Glashow, J. Iliopoulos, L. Maiani, Phys. Rev. D2 , 1285 (1970)
C. Bouchiat, J. Iliopoulos, Ph. Meyer, Phys. Letters 38B, 519 (1972).
A. Pais and S. B. Treiman, Phys. Rev. D6, 2700 (1972).
E. A. Paschos and L. Wolfenstein, Phys. Rev. D7, 91 (1973) .
C. H. Albright, Phys. Rev. D8, 3162 (1973);Nucl. Phys. B70,486 (1974).
L. M.Seghal, Nucl. Phys. B65, 141 (1973).

PART C FURTHER APPLICATIONS. BREAKING OF SCALING

I POLARIZATION EFFECTS IN ELECTROPRODUCTION

1) In the one photon exchange approximation polarization effects occur in electroproduction when the target is polarized. If time reversal invariance is assumed we have two new structure functions as seen in Table 1.
The polarized cross section is then generally written as

$$\frac{d^2\sigma_\eta}{dq^2\,dW^2} = \left(\frac{d^2\sigma}{dq^2\,dW^2}\right)_{UNP} \left[1 + \eta\, N\cdot\Delta\right] \tag{90}$$

where N is a unit space like polarization vector of the spin 1/2 target orthogonal to p

$$N^2 = 1 \qquad\qquad N\cdot p = 0$$

In thc laboratory frame N has only space components. The asymmetry vector Δ has no component orthogonal to the scattering plane if time reversal invariance holds. Then, with a polarization vector N in the scattering plane we can measure two independent asymmetries and therefore determine the two structure functions for polarization. Because of parity conservation in electromagnetic interactions there exist only three independent total cross sections for the photoabsorption reaction $\gamma + p \to$ HADRONS

$$\sigma^Q_{\lambda s}(q^2, W^2) = \sigma^Q_{-\lambda\, -s}(q^2, W^2)$$

Two of them have been associated to scattering on an unpolarized target and the third one describes the polarization effect in the $\vec{q}_{lab}$ direction

$$\Delta_q = -\frac{\sqrt{1-\varepsilon^2}}{2}\;\frac{\sigma_{1\,1/2} - \sigma_{1\,-1/2}}{\sigma_T + \varepsilon\,\sigma_L} \tag{91}$$

where the kinematical parameter ε is defined as usual by

$$\varepsilon^{-1} = 1 + 2\,\frac{\nu^2+q^2}{q^2}\,tg^2\frac{\theta}{2}$$

The second structure function for polarization is a transverse longitudinal correlation associated to total helicity $\pm\frac{1}{2}$.

2) An easily measurable quantity is the parallel-antiparallel asymmetry $\Delta_{\parallel}$ corresponding to a nucleon polarization vector $\vec{N}$ collinear to the incident beam momentum $\vec{k}$.
At high energy $\Delta_{\parallel} = \Delta_q$ and moreover if scaling holds for the structure functions we get scaling for the asymmetry $\Delta_{\parallel}$

$$\Delta_{\parallel} \rightarrow - \frac{1-\rho^2}{2} \frac{F^Q_{1\,1/2}(\xi) - F^Q_{1\,-1/2}(\xi)}{(1+\rho^2)F^Q_T(\xi) + 2\rho F^Q_L(\xi)} \tag{92}$$

where we have used the high energy limit

$$\varepsilon \rightarrow \frac{2\rho}{1+\rho^2}$$

From the positivity constraints on total cross sections and the relation $2\,F_T = F_{1\,1/2} + F_{1-1/2}$ we get an upper bound for the asymmetry

$$|\Delta_{\parallel}| \leq \frac{1-\rho^2}{(1+\rho^2) + 2\rho R(\xi)}$$

3) Let us define the polarization scaling function

$$2\,F^Q_{\parallel}(\xi) = F^Q_{1\,1/2}(\xi) - F^Q_{1\,-1/2}(\xi) \tag{93}$$

In the quark parton model we have an expression for $F^Q_{\parallel}(\xi)$ analogous to the equation (25) for $F^Q_T(\xi)$

$$2\,F^Q_{\parallel}(\xi) = \sum_{j,\sigma} Q_j^2\, \sigma \left[D_{j\sigma}(\xi) + D_{-j\,\sigma}(\xi) \right] \tag{94}$$

The $D_{j\sigma}$ are the distribution functions of the parton of type j in the hadron with momentum $\xi\vec{p}$ and spin parallel ($\sigma = +1$) or antiparallel ($\sigma = -1$) to the nucleon spin. We then have the obvious relation

$$D_j(\xi) = D_{j\,+1}(\xi) + D_{j\,-1}(\xi)$$

4) The distributions $D_{j\sigma}(\xi)$ are not known but we can have some information on their normalization integrals

$$\int_0^1 D_{j\sigma}(\xi)\, d\xi = \langle N_{j\sigma} \rangle \tag{95}$$

Let us define the quantity Z^Q by

$$Z^Q = \int_0^1 2\, F_1^Q(\xi)\, d\xi$$

From equations (94) and (95) we obtain

$$Z^Q = \sum_{j,\sigma} Q_j^2\, \sigma \left[\langle N_{j\sigma} \rangle + \langle N_{-j\sigma} \rangle \right] \tag{96}$$

or in an equivalent language

$$Z^Q = \sum_{j,\sigma} \left[\langle N_{j\sigma} \rangle \langle j\sigma | \sigma_3 Q^2 | j\sigma \rangle + \langle N_{-j\sigma} \rangle \langle -j\sigma | \sigma_3 Q^2 | -j\sigma \rangle \right] \tag{97}$$

It is clear, from equation (97) that Z^Q is the average value in the hadron of the operator $\sigma_3 Q^2$.
We now introduce the symmetric coefficients of the Lie algebra defined by the anticommutators

$$\{ X^a , X^b \}_+ = d^{ab}{}_c\, X^c$$

Equation (97) can be written in the form of the Bjorken sum rule

$$Z^Q = \frac{1}{2}\, d^{QQ}{}_c\, g_A^c \tag{98}$$

the axial vector coupling constant being defined in the quark parton model by

$$g_A^c = \sum_{j,\sigma} \left[\langle N_{j\sigma} \rangle \langle j\sigma | \sigma_3 X^c | j\sigma \rangle - \langle N_{-j\sigma} \rangle \langle -j\sigma | \sigma_3 X^c | -j\sigma \rangle \right] \tag{99}$$

In the U(3) algebra the symmetric coefficients computed in the 3 dimensional representation are given by

$$d^{QQ}{}_B = \frac{4}{3} \qquad\qquad d^{QQ}{}_Q = \frac{2}{3}$$

The nucleon belonging to an octet representation we have three reduced matrix elements

$$\begin{aligned} Z^{ep} &= \frac{2}{3} f_1 + \frac{1}{9} f_s + \frac{1}{3} f_a \\ Z^{en} &= \frac{2}{3} f_1 - \frac{2}{9} f_s \end{aligned} \tag{100}$$

The octet coefficients are known from an analysis of neutron and hyperon β decays with the Cabibbo current. From experiment we get

$$f_a + f_s = -\frac{G_A}{G_V} = 1.23 \qquad\qquad \frac{f_s}{f_a+f_s} \simeq 0.6$$

The difference $Z^{ep} - Z^{en}$ is therefore known from neutron decay only and we get the famous Bjorken relation

$$Z^{ep} - Z^{en} = \frac{1}{3}\left(-\frac{G_A}{G_V}\right) = 0.41 \tag{101}$$

The baryonic constant f_1 is not experimentally known. It can be computed in a model where the gluon spin is assumed to be uncorrelated in average with the nucleon one. Using equation (99) and spin conservation the result is simply $f_1 = 1/3$. In such a model we have the predictions

$$Z^{ep} \simeq 0.47 \qquad\qquad Z^{en} \simeq 0.06 \tag{102}$$

5) It is interesting to compare these predictions obtained in the $q^2 \to \infty$ limit with those of photoproduction at $q^2 = 0$. In this case the Drell-Hearn, Gerasimow sum rule derived on general grounds is written as

$$\frac{1}{\pi}\int_{M^2}^{\infty} \frac{\sigma_{1\,1/2}(0,W^2) - \sigma_{1\,-1/2}(0,W^2)}{W^2 - M^2}\, dW^2 = \frac{\mu^2}{2M^2} \tag{103}$$

In average the difference $\sigma_{1\,1/2} - \sigma_{1\,-1/2}$ is expected to be positive. On the other hand phenomenological analysis of photoproduction agrees with this qualitative statement and quantitatively the sum rule (103) seems to be in good shape.

On the basis of the quark parton model the situation looks radically different for large q^2. If one uses elementary arguments as those presented before we expect a positive asymmetry for the proton in the deep inelastic region and therefore a change of sign of the asymmetry $\Delta_{\parallel}^{ep}$ between the photoproduction region $q^2 = 0$ and the scaling region $q^2 > 1\ \text{GeV}^2$. There is no experimental evidence, at least for $q^2 \lesssim 0.6\ \text{GeV}^2$ for such a change of sign in the proton asymmetry and if one believes, following Bloom and Gilman, that the resonance region is well averaged using convenient variables by the deep elastic one we get a difficulty.

Of course if one accepts to leave to gluons a dominant role in polarization effects it is easy to obtain a negative asymmetry for the proton. In this case, from the Bjorken relation, the neutron asymmetry would become very large and negative.

An experiment is in progress at SLAC with a polarized electron beam. The target is a polarized proton target with proton polarization longitudinal to the direction of the incident electron beam. The parallel antiparallel asymmetry $\Delta_{\parallel}$ will be measured either reversing the electron polarization or the proton polarization.

II PARITY-VIOLATING EFFECTS IN ELECTROPRODUCTION

1) In inelastic scattering of charged leptons the one-photon exchange amplitude dominated the cross section at currently available energies. However an additional contribution due to neutral intermediate vector boson exchange may give peculiar effects we wish now to study. We compute the differential cross section in the inclusive case where only the final lepton is detected retaining the 1γ exchange term and the 1γ - $1Z$ interference contribution. The target is assumed to be unpolarized and we call η the longitudinal polarization of the incident beam. Details about the kinematics have been given in part A and the result has the following structure

$$\frac{d\sigma^{\mp}}{d\sigma_0} = 1 + \frac{\sqrt{2}\,G}{e^2} q^2 \left\{ (a \pm \eta b)\, \phi_1(q^2, W^2, E) - (a\eta \pm b)\, \phi_2(q^2, W^2, E) \right\} \tag{104}$$

The upper (lower) sign refers to lepton ℓ^- (antilepton ℓ^+) scattering and $d\sigma_0$ is the pure one-photon exchange cross section. The functions ϕ_1 and ϕ_2 are constructed from the interference structure functions $\tau_\lambda\,(q^2, W^2)$ and the electroproduction ones $\sigma_\lambda^Q(q^2, W^2)$

$$\phi_1(q^2, W^2, E) = \frac{\tau_T(q^2, W^2) + \varepsilon\, \tau_L(q^2, W^2)}{\sigma_T^Q(q^2, W^2) + \varepsilon\, \sigma_L^Q(q^2, W^2)}$$

$$\phi_2(q^2, W^2, E) = \frac{\sqrt{1-\varepsilon^2}}{2} \; \frac{\tau_{-1}(q^2, W^2) - \tau_{+1}(q^2, W^2)}{\sigma_T^Q(q^2, W^2) + \varepsilon\, \sigma_L^Q(q^2, W^2)}$$

where the kinematical quantity ε has been previously defined

$$\varepsilon^{-1} = 1 + 2\, tg^2 \frac{\theta}{2} \, \frac{\nu^2 + q^2}{q^2}$$

It is clear on these expressions that ϕ_1 is parity conserving and ϕ_2 parity violating.

2) Various asymmetries can be computed from equation (104). In order to eliminate higher order electromagnetic effects we only consider parity violating quantities and we define asymmetries for a given charge and opposite polarizations of the beam:

$$A_{\mp} = \frac{d\sigma_R^{\mp} - d\sigma_L^{\mp}}{d\sigma_R^{\mp} + d\sigma_L^{\mp}}$$

To lowest order in G we get from equation (104)

$$A_{\mp} = \frac{\sqrt{2}\, G}{e^2} q^2 \left[\pm \phi_1 - a\, \phi_2 \right] \tag{105}$$

In the Weinberg-Salam model the two parameters a and b are given by

$$a = 4 \sin^2 \theta_W - 1 \qquad\qquad b = 1$$

and the asymmetries are functions of the mixing angle θ_W. As previously we shall put $x = \sin^2 \theta_W$.

3) We now consider the simple model proposed by Weinberg for the hadronic current

$$J_Z^{\mu} = V_3^{\mu} + A_3^{\mu} - 2x\, J_Q^{\mu}$$

In the quark parton model based on U(3) symmetry the interference scaling functions $\tilde{F}_{\pm}(\xi)$ defined by

$$LIM \frac{M \sqrt{\nu^2 + q^2}}{\pi} \tau_{\pm 1}(q^2, W^2) = \tilde{F}_{\pm}(\xi)$$

can be expressed as linear combinations of the electroproduction neutrino and antineutrino strangeness conserving scaling functions.
The result is

$$\tilde{F}_{+}(\xi) = \frac{1}{6} G_{+}^{\nu}(\xi) + \frac{1}{12} G_{+}^{\bar{\nu}}(\xi) - 2x\, F_T^{Q}(\xi)$$

$$\tilde{F}_{-}(\xi) = \frac{1}{6} G_{-}^{\bar{\nu}}(\xi) + \frac{1}{12} G_{-}^{\nu}(\xi) - 2x\, F_T^{Q}(\xi)$$

Let us restrict now to the simple case where the proton and neutron scaling functions have been averaged

$$\tilde{F}^N_{\pm}(\xi) = \frac{1}{4}\ G^N_{\pm}(\xi) - 2x\ F^{eN}_T(\xi) \tag{106}$$

In the scaling limit the functions ϕ_1 and ϕ_2 are independent of the incident energy E and they can be computed in terms of electroproduction and weak scaling functions

$$\phi^N_1 \longrightarrow \frac{1}{4}\ \frac{G^N_T(\xi)}{F^{eN}_T(\xi)} - 2x$$

$$\phi^N_2 \longrightarrow \frac{1-\rho^2}{1+\rho^2}\ \frac{1}{8}\ \frac{G_{-1}(\xi) - G_{+1}(\xi)}{F^{eN}_T(\xi)}$$

By using the language of differential cross sections ($\Delta S = 0$) we get

$$\phi^N_1 = \frac{1}{8}\ \frac{d\sigma^{\nu N}_{cc} + d\sigma^{\bar{\nu}N}_{cc}}{X\ d\sigma^{eN}} - 2x \tag{107}$$

$$\phi^N_2 = \frac{1}{8}\ \frac{d\sigma^{\nu N}_{cc} - d\sigma^{\bar{\nu}N}_{cc}}{X\ d\sigma^{eN}} \tag{108}$$

where the notations are the same as in part B: $X = 2(G^2/e^4)q^4$.

Let us notice that the expression (108) for ϕ_2 is a trivial consequence of the simple isotopic spin structure assumed for the weak currents and it is independent of the quark parton model.

4) We first obtain a prediction for the sum of the asymmetries

$$A^N_- + A^N_+ = \frac{1-4x}{2}\ \frac{G}{\sqrt{2}\,e^2}\ q^2\ \frac{d\sigma^{\nu N}_{cc} - d\sigma^{\bar{\nu}N}_{cc}}{X\ d\sigma^{eN}} \tag{109}$$

Unfortunately the differential cross sections for neutrino and antineutrino are not well known and it is interesting to look at an averaged asymmetry defined as follows

$$\langle A^N_{\mp} \rangle = \int A^N_{\mp}\, X\, d\sigma^{eN} \Big/ \int X\, d\sigma^{eN} \tag{110}$$

It is clear, from equations (109) and (110) that averaged values of q^2 for neutrino and antineutrino reactions are involved and we obtain

$$\langle A_-^N + A_+^N \rangle = \frac{1-4x}{2} \frac{G}{\sqrt{2}e^2} \frac{\langle q^2 \rangle_\nu \sigma_{cc}^{\nu N} - \langle q^2 \rangle_\nu \sigma_{cc}^{\bar{\nu} N}}{\int X \, d\sigma^{eN}}$$

In the high energy limit we used scaling in the form

$$\frac{\sigma^{\nu,\bar{\nu}}}{\int X \, d\sigma^{eN}} \longrightarrow \frac{3}{2} \frac{B^{\nu,\bar{\nu}}}{I^{eN}}$$

all the parameters being defined in part B. The final result is

$$\langle A_-^N + A_+^N \rangle \longrightarrow \frac{1-4x}{2} \frac{3}{2I^{eN}} \frac{G}{\sqrt{2}e^2} \left[\langle q^2 \rangle_\nu B_{cc}^{\nu N} - \langle q^2 \rangle_{\bar{\nu}} B_{cc}^{\bar{\nu} N} \right]$$

Numerical estimates can be obtained using the CERN-Gargamelle data presented in part B

$$\langle q^2 \rangle_\nu \simeq (0.21 \pm 0.02)\, E \qquad \langle q^2 \rangle_{\bar{\nu}} \simeq (0.14 \pm 0.03) E$$

$$B_{CC}^{\nu N} \simeq 0.493 \pm 0.050 \qquad B_{CC}^{\bar{\nu} N} \simeq 0.180 \pm 0.020$$

and the electroproduction result

$$I^{eN} = 0.15 \pm 0.01$$

As expected the sum of averaged asymmetries increases linearly with the incident energy E

$$\langle A_-^N + A_+^N \rangle = (1-4x)(0.35 \pm 0.07)10^{-4} \; E \; GeV^{-1} \qquad (111)$$

5) An analogous treatment can be done for the difference of asymmetries $A_-^N - A_+^N$ and numerical estimatescan be computed for the averaged value of that difference. However it is interesting to remark that from the CERN-Gargamelle results presented in Fig. 19 the ratio of scaling functions $G_T(\xi)/F_T^{eN}(\xi)$ involved in the function ϕ_1 is practically constant at least for $\xi > 0.3$ and the constant turns out to be consistent with the value of 0.9 predicted by a pure three valence quark model. Therefore for $\xi > 0.3$

$$\phi_1 \simeq 0.9 - 2x$$

and we obtain a prediction for the difference of asymmetries

$$A_-^N - A_+^N = (0.9 - 2x) \; 3.6 \; 10^{-4} \; q^2 \; GeV^{-2} \qquad (112)$$

III BREAKING OF SCALING

1) The scaling à la Bjorken of the structure functions has been observed in electroproduction at SLAC and DESY in a limited range of values for q^2 and W^2

$$1\ \text{GeV}^2 < q^2 < 12\ \text{GeV}^2 \qquad\qquad 2\ \text{GeV} < W < 7\ \text{GeV}$$

The neutrino and antineutrino experiments performed at CERN with the Gargamelle bubble chamber cover an analogous range for q^2 and W^2. Indirect evidence for scaling has been obtained.
A possible physical interpretation of this fact is the parton model for hadrons: the elementary constituents have very small dimensions and appear as point like in their interactions with the electromagnetic and weak currents.

2) Even if the Bjorken scaling is an asymptotic theoretical statement we must ask the question: what will happen at higher values of q^2 and W^2? In principle the experiments performed at N.A.L. covering a more extended range of the q^2, W^2 plane will answer that question and we shall come back on this point later. On phenomenological grounds we have two possibilities

a- we are in an asymptotic region and nothing new will appear; we have reached the ultimate constituents of hadrons and life is simple

b- we are in a preasymptotic region and at larger values of q^2 and W^2 deviations of scaling will take place due for instance to the excitation of internal degrees of freedom of partons.

3) Let us first look at the possibility for partons to have a structure which can be represented by a form factor of the type propose by Chanowitz and Drell:

$$F(q^2) = \frac{1}{1 + q^2/m_G^2}$$

The new mass scale m_G which may be associated in a more or less effecti way to gluons is assumed to be very large as compared to the nulceon mass.

For instance the electroproduction structure function at large q^2 and fixed ξ will have the factorized form

$$F_2^{ep}(q^2,\xi) \rightarrow \frac{F_2^{ep}(\xi)}{[1+q^2/m_G^2]^2}$$

The parton structure has not yet been seen at SLAC. Although the analysis depends on the choice of the scaling variable the data put a lower bound on m_G of order 10 GeV. Crucial information will be provided by the NAL experiment with incident μ^- leptons and where values of q^2 as large as 40 GeV^2 can be reached.

For neutrino and antineutrino processes such a parton structure will compete with that due to the intermediate vector boson propagator. Unfortunately the breaking of scaling due to finite values of m_G and m_W produces analogous and indistinguishable effects.

An interesting consequence of the existence of a parton structure will also occur for time like photons in the annihilation process $e^+ + e^- \rightarrow$ HADRONS. The manifestation will now be an enhancement of the cross section of the resonance type. Something unexpected appears in the CEA and SPEAR experiments which may be associated to a parton structure or due to a totally different origin as for instance the production of new particles (heavy leptons or charmed and colored hadrons).

4) The results of the quark parton model for deep inelastic lepton scattering can equivalently be obtained in the framework of the light cone quark algebra supplemented by Wilson's operator product expansion. A better understanding of these simple results can be undertaken in a more systematic approach to asymptotic behaviour using the techniques of the renormalization group. Without giving any detail or proof we now briefly sketch some important steps of the method.

The hadronic tensor for inelastic lepton scattering is the Fourier transform of the one-particle matrix element of the product of two current operators. By taking advantage of the translational invariance equation (13) can be written as

$$M_{\mu\nu}^{\alpha\beta}(p,q) = \frac{M}{2\pi}\int e^{-iq\cdot x}\langle p|\, J_\nu^\beta\left(\tfrac{x}{2}\right) J_\mu^\alpha\left(-\tfrac{x}{2}\right)|p\rangle\, d_4x$$

We expand the hadronic tensor on a complete basis of covariants $I^j_{\mu\nu}$ the coefficients of that expansion being the structure functions

$$M_{\mu\nu}^{\alpha\beta}(p,q) = \sum_j I^j_{\mu\nu}\, F_j^{\alpha\beta}(q^2,\xi)$$

The $I^j_{\mu\nu}$'s are chosen so that to have simple properties for the structure functions the Bjorken conjecture about scaling holds. In this case we have

$$\lim_{\substack{q^2\to\infty \\ \xi \text{ fixed}}} F_j^{\alpha\beta}(q^2,\xi) = F_j^{\alpha\beta}(\xi)$$

We are interested in the behaviour of M(p,q) in the deep inelastic region for q spacelike and large with the target momentum p fixed. It is then convenient to expand the product of current operators near x = 0 by introducing an appropriate complete set of local operators O_n

$$J_\nu^\beta\left(\frac{x}{2}\right) J_\mu^\alpha\left(-\frac{x}{2}\right) = \sum_n C_{\mu\nu}^{\alpha\beta;n}(x)\, O_n(0) \tag{113}$$

In an analogous way the tensor $C_{\mu\nu}^{\alpha\beta;n}(x)$ is expanded on a Lorentz covariant basis and the Fourier transform $\tilde{C}(q)$ of the scalar coefficients C(x) can be studied by means of a generalization of the renormalization group equation of Gell-Mann and Low, the so-called Callan-Symanzik equation

$$\left[\mu\frac{\partial}{\partial\mu} + \beta(g)\frac{\partial}{\partial g} - \gamma_n(g)\right] \tilde{C}_j^{\alpha\beta;n}\left(q^2/\mu^2, g\right) = 0 \tag{114}$$

where μ is the subtraction point introduced in the renormalization and g a dimensionless coupling constant. For conserved or partially conserved currents the anomalous dimension of the current operators J vanishes so that $\gamma_n(g)$ is simply the anomalous dimension of the operator O_n. The solution of this Callan-Symanzik equation can be expressed in terms of an auxiliary function $\bar{g}(t,g)$ defined by

$$\frac{\partial}{\partial t}\bar{g}(t,g) - \beta(g)\frac{\partial}{\partial g}\bar{g}(t,g) = 0 \tag{115}$$

where

$$t = \frac{1}{2}\log\frac{q^2}{\mu^2}$$

with the initial condition $\bar{g}(o, g) = g$. The result is

$$\tilde{C}_j^{\alpha\beta;n}\left(q^2/\mu^2, g\right) = \tilde{C}_j^{\alpha\beta;n}\left[1, \bar{g}(t,g)\right] \exp\left[-\int_0^t \gamma_n\left[\bar{g}(\tau,g)\right] d\tau\right] \tag{116}$$

The connection between the structure functions $F_j^{\alpha\beta}$ and the Wilson coefficients $\tilde{C}_j^{\alpha\beta;n}$ is obtained at the level of the various moments and the result is simply

$$\int_0^1 \xi^{n-1} F_j^{\alpha\beta}(q^2,\xi) = \tilde{C}_j^{\alpha\beta;n}(q^2)\, M_n \tag{117}$$

where M_n is the one-particle matrix element of the operator O_n. The asymptotic behaviour of these moments is controlled by that of the Wilson coefficients $\tilde{C}(q)$ which is determined, from equation (116) by the large t behaviour of the function $\bar{g}(t,g)$. The result is

$$\lim_{q^2\to\infty} \tilde{C}_j^{\alpha\beta;n}(q^2/\mu^2, g) = \tilde{C}_j^{\alpha\beta;n}(1, g_0)\left[q^2/\mu^2\right]^{-\frac{1}{2}\gamma_n(g_0)}$$

where g_0 is the renormalization group fixed point

$$\beta(g_0) = 0 \qquad\qquad \lim_{t\to\infty} \bar{g}(t,g) = g_0$$

The condition for Bjorken canonical scaling is then simply

$$\gamma_n(g_0) = 0 \qquad \text{for all n's}$$

As shown in Part A in the case $\alpha=\beta$ the diagonal elements of the hadronic tensor in the helicity space are positive functions of their arguments. Therefore the corresponding momenta are positive functions of q^2 which, at fixed q^2, decrease when n increases. This positivity property enables to reduce the infinite number of constraints to two only.

It can be shown that the ultraviolet stable fixed point of the renormalization group must be at the origin $g_0 = 0$. This result is called asymptotic freedom because in this situation the strong interactions turn off for large space like momenta. Therefore if we insist to explain Bjorken scaling using the renormalization group approach the class of renormalizable theories for strong interactions one may consider is severely limited. Only gauge theories based on non Abelian gauge groups have the property of asymptotic freedom.

In an asymptotically free gauge theory the approach to asymptotic behaviour is not with a power law but with logarithmically vanishing correction terms. The functions $\beta(g)$ and $\gamma_n(g)$ are now expected to vanish around the origin according to

$$\beta(g) = -b_0 g^3 + O(g^5) \qquad\qquad \gamma_n(g) = \gamma_n g^2 + O(g^4)$$

Using equations (115) and (116) a straightforward computation gives the following result for the moments of the structure functions at large q^2

$$\int_0^1 \xi^{n-1} F_j^{\alpha\beta}(q^2,\xi)\, d\xi \longrightarrow \tag{118}$$

$$\longrightarrow \text{constant} \cdot M_n \left[\log \frac{q^2}{\mu^2}\right]^{-a_n} \left[\tilde{C}_j^{\alpha\beta;n}(1,0) + O\left(\frac{1}{\log q^2/\mu^2}\right)\right]$$

where $a_n = \gamma_n / 2b_0$ is a model dependent parameter and the rate of approach to this situation will obviously depend on the unknown scale μ.

Let us finally remark that the dependence on the indices α, β and j which is contained in the quantity $\tilde{C}(1,0)$ turns out to be the same as in free field theory. As a consequence the moments of the structure functions will satisfy all parton model relations and sum rules. The Adler sum rule is valid for all q^2 but the Gross-Llewellyn Smith sum rule is approached logarithmically

$$\int_0^1 \left[F_3^W(q^2,\xi) + F_3^{\bar{W}}(q^2,\xi)\right] d\xi = [GL]\ \mathcal{F}(q^2)$$

where the constant [GL] is its asymptotic value depending on the algebra and $\mathcal{F}(q^2)$ a function with the structure

$$\lim_{q^2\to\infty} \mathcal{F}(q^2) = 1 + O\left(\frac{1}{\log q^2/\mu^2}\right)$$

Analogously the dominance of spin $\frac{1}{2}$ partons is expressed by a Callan-Gross type relation

$$\frac{\int_0^1 \xi^n F_L(\xi,q^2)\, d\xi}{\int_0^1 \xi^n F_T(\xi,q^2)\, d\xi} \longrightarrow O\left(\frac{1}{\log q^2/\mu^2}\right)$$

REFERENCES

I. J. D. Bjorken, Phys. Rev. 148, 1467 (1966); D1, 1376 (1971).
J. Kuti and V. F. Weisskopf, Phys. Rev. D4, 3418 (1971).
M. Gourdin, Nucl. Phys. B38, 418 (1972).
J. Kuti, Invited talk at the VIIIth Rencontre de Moriond, Méribel (1973).
F. E. Close, Invited talk at the IXth Rencontre de Moriond, Méribel (1974).
A. Hey, Invited talk at the IXth Rencontre de Moriond, Méribel (1974).

II. A. Love, G.G. Ross and D.V. Nanopoulos, Nucl. Phys. B49,513 (1972).
E. Petronzio, Rome preprint 411 (1972).
E. Derman, Phys. Rev. D7, 2755 (1973).
S. M. Berman and J. R. Primack, SLAC-PUB 1360 (1973).
W. J. Wilson, Harvard preprint (1974).
C. P. Korthals-Altes, M. Perrottet and E. De Rafael, Marseille preprint 74/P595 (1974).
C. H. Llewellyn Smith and D. V. Nanopoulos, CERN preprint TH.1850 (1974).
M. Gourdin, Invited talk at the IXth Balaton Symp. (1974).

III. M. S. Chanowitz and S.D. Drell, Phys. Rev. Letters 30, 807 (1973).
S. D. Drell, Report to the Intern. Symp. on Electron and Photon Interactions at High Energies, Bonn (1973).
V. Barger, Wisconsin preprint (1973).
K. Wilson, Phys. Rev. 179, 1499 (1969).
M. Gell-Mann and F. Low, Phys. Rev. 95, 1300 (1954).
E.C.G. Stückelberg and A. Petermann, Helv. Phys. Acta 26,499 (1953).
C.G. Callan Jr., Phys. Rev. D2, 1541 (1970).
K. Symanzik, Commun. Math. Phys. 18, 227 (1970).
S. Coleman, Lectures at the Erice Summer School (1971).
C. G. Callan Jr., Phys. Rev. D5, 3203 (1972).
G. Parisi, Phys. Letters 50B, 367 (1974).
G.'t Hooft, unpublished.
D. J. Gross and F. Wilczek, Phys. Rev. Letters 30, 1343 (1973); Phys. Rev. D8, 3633 (1973); Phys. Rev. D9, 993 (1974).
H.D. Politzer, Phys. Rev. Letters 30, 1346 (1973).
T. Appelquist and H. Georgi, Phys. Rev. D8, 4000 (1973).
A. Zee, Phys. Rev. D, to be published.
H. Georgi and H. D. Politzer, Phys. Rev. D9, 416 (1974).
M. Veltman, Report to the Intern. Symp. on Electron and Photon Interactions at High Energies, Bonn (1973).
C. G. Callan Jr., Lectures given at the Cargèse Summer Institute (1973).

CONCLUSION

A complete study of the application of the quark parton model to electromagnetic and weak interactions would involve more topics than those considered here, namely the two sets of processes

(i) Semi-inclusive reactions where one or more hadrons in the final state is detected in coincidence with the final leptons

(ii) Electron-positron annihilation into hadrons.

The application of the parton model to semi-inclusive reactions implies new assumptions concerning the production mechanism and, to my point of view, the central point is to construct a model where the non observation of quarks and antiquarks as free particles in the final state - which is an experimental fact - appears as a natural consequence of the dynamics used to describe the production of hadrons. To my knowledge, the various proposals made are not totally satisfactory in this respect.

The description of annihilation processes with a quark parton model is generally made using a two-step mechanism; first a quark-antiquark pair is produced via one photon exchange or some less conventional way as that proposed by Pati and Salam, and then this quark-antiquark pair annihilates into hadrons in a way which again prevents the observation of a $q\bar{q}$ pair in the final state. The data produced by CEA and SPEAR lead a naive quark parton model into difficulties, the total cross section for $e^+e^- \rightarrow$ HADRONS being roughly constant between 9 and 25 GeV^2 for the squared total energy s. Therefore, the timelike region appears to behave differently from the spacelike one and satisfactory answers have not yet been given to this apparent contradiction, which is very important if experimentally confirmed.

THE RELATIVISTIC STRING

Norbert Dragon

Universität Karlsruhe, Germany

The relativistic string can be considered as a generalization of a free particle. It is well known, that the equations of motion of a free particle arise from a variational principle, taking the length of its path in space-time as the action. Enlarging the dimension of the zero dimensional particle to a one-dimensional curve which sweeps out a two-dimensional worldsheet in space-time, one gets the equations of the relativistic string, if one takes the area of the worldsheet as its action. There are different justifications for treating the relativistic string as a physical object: the dual models can be interpreted as strings; we will not pursue this point of view, however, but refer only to Ref.[2,3,4,6] and the references therein.

Instead, we will consider the relativistic string as a theoretical problem of its own and regain the results of Ref.[1]. Stress will be laid on the geometrical point of view as well as on a detailed treatment of the classical Hamiltonian formalism [5].

The worldsheet of the string is characterized by coordinate functions

$$x^\mu(\sigma,\tau)$$

where the parameters (σ,τ) vary in a domain D, which will be specified later. The parametrization of the string is to be regular, that is the natural tangent vectors

$$\dot{x}^\mu = \frac{\partial x^\mu}{\partial \tau} \quad , \quad x'^\mu = \frac{\partial x^\mu}{\partial \sigma}$$

have to be linearly independent. Furthermore, we will restrict ourselves to timelike surfaces, i.e. surfaces which cut the lightcone at each point.

Taking the area of the two-dimensional worldsheet as the action we have as Lagrangian

$$\mathcal{L} = -\frac{1}{\pi}\sqrt{(\dot{x}x')^2 - \dot{x}^2 x'^2} \quad (*) \tag{1.1}$$

where the factor $-\frac{1}{\pi}$ has been put in for convenience. In neglecting a dimensional constant in $\mathcal{L}$, we have adopted $\hbar = c = 1$ and a fundamental length. Varying $x^{\mu}(\sigma,\tau)$ in the action

$$\int_D \mathcal{L}\, d\sigma\, d\tau \tag{1.2}$$

we get the Euler-Lagrange equations

$$0 = \frac{\partial}{\partial\tau}\left(\frac{x'^{\mu}(\dot{x}x') - \dot{x}^{\mu}(x'^2)}{\sqrt{(\dot{x}x')^2 - \dot{x}^2 x'^2}}\right) + \frac{\partial}{\partial\sigma}\left(\frac{\dot{x}^{\mu}(\dot{x}x') - x'^{\mu}(\dot{x}^2)}{\sqrt{(\dot{x}x')^2 - \dot{x}^2 x'^2}}\right) \tag{1.3}$$

To obtain these equations, we had to integrate by parts. The boundary terms, which arise in this process cannot be neglected as usual, because we deal with a finitely extended string. They have to vanish separately and give rise to the boundary equations

$$0 = \left| m_0\left(\frac{x'^{\mu}(\dot{x}x') - \dot{x}^{\mu}(x'^2)}{\sqrt{(\dot{x}x')^2 - \dot{x}^2 x'^2}}\right) + m_1\left(\frac{\dot{x}^{\mu}(\dot{x}x') - x'^{\mu}(\dot{x}^2)}{\sqrt{(\dot{x}x')^2 - \dot{x}^2 x'^2}}\right)\right. \tag{1.4}$$

at boundary of D

Here (m_0, m_1) denotes the outward unit normal vector of the boundary of the parameterdomain D. We realize that adding total derivatives to the Lagrangian would have altered the boundary equations and left invariant the equations (1.3), so total derivatives represent interactions of the boundary of the string.

Multiplying (1.4) with $\dot{x}_{\mu}$, x'_{μ}, we get

$$\begin{pmatrix} m_0 \\ m_1 \end{pmatrix} \sqrt{(\dot{x}x')^2 - \dot{x}^2 x'^2}\,\Big|_{\text{at boundary}} = 0$$

that is, as $(m_0, m_1) \neq 0$

$$(\dot{x}x')^2 - \dot{x}^2 x'^2 = 0 \tag{1.5}$$

Writing this in the form

(*) We use the convention $g_{\mu\nu} = \text{diag}\,(-1,+1,+1,\ldots), \mu,\nu = 0,1,\ldots$

$$x'^2 \left(\dot{x}^\mu - \frac{(\dot{x}x')\, x'^\mu}{x'^2} \right)^2 = 0 \tag{1.6}$$

we see, that the boundary condition implies, that the transverse velocity of the boundary is the light velocity. In Euclidean space there is no solution to the boundary condition, a fact which is well known from everyday experience with soap bubbles. Returning now to equation (1.4), we realize, that the denominator vanishes, so we have to check, what additional conditions are implied by (1.4). Making use of (1.5) and the linear independence of $\dot{x}^\mu$, x'^μ we get

$$m_0 \, x'^2 - m_1 \, (\dot{x}x') \Big|_{\text{at boundary}} = 0 \tag{1.7}$$

as a necessary condition at the boundary. This condition has to be checked for consistency with the equations of motion (1.3), for (1.5) defines, where the boundary is, (1.3) where it will move to, so the direction of the boundary is fixed. One can prove, that (1.5) and (1.7) are sufficient for (1.4).

Let us now turn to the equations of motion (1.3). They do not determine uniquely the functions $x^\mu(\sigma,\tau)$ once the initial values $x^\mu(\sigma,\tau=0)$, $\dot{x}^\mu(\sigma,\tau=0)$ are specified. In fact, they allow an arbitrary reparametrization

$$\tilde{\sigma}(\sigma,\tau) \qquad \tilde{\tau}(\sigma,\tau) \qquad \tilde{x}^\mu(\tilde{\sigma},\tilde{\tau}) = x^\mu(\sigma,\tau)$$

So to get unique solutions of the initial value problem we have to impose additional gauge conditions. We will carefully choose such a parametrization, that the equations of motion can be solved.

Suppose, we are given a solution of (1.3), choose

$$\tilde{\tau} = c \cdot n_\mu x^\mu(\sigma,\tau) \tag{1.8}$$

as new parameter τ with an arbitrary constant vector n_μ, which is subject to

$$n^2 \leq 0$$

In the second step, we impose

$$\dot{x}\, x' = 0 \tag{1.9}$$

This can be done by solving the ordinary system of differential equations

$$\frac{d\tau(\lambda)}{d\lambda} = x'^2(\sigma,\tau) \qquad \frac{d\sigma(\lambda)}{d\lambda} = -\dot{x}x'(\sigma,\tau) \tag{1.10}$$

The solutions are paths $(\sigma(\lambda), \tau(\lambda))$. Now set $\tilde{\sigma}(\sigma,\tau)$ constant on these paths and let $\tilde{\tau} = \tau$. In this new parametrization, you can check, that (1.9) is fulfilled. Notice, that this condition fixes the form of the domain D. (1.7) yields

$$m_0 = 0 \quad , \quad m_1 = 1 \tag{1.11}$$

so the boundary is given by two curves

$$\sigma_{boundary} = const = \begin{cases} \sigma_{min} \\ \sigma_{max} \end{cases}$$

and we can choose $\sigma_{min} = 0$. The condition (1.9) could be met without spoiling (1.8), so we have

$$n\dot{x} = c \qquad n x' = 0$$

Multiplying the equations of motion (1.3) with n_μ, we get on account of (1.8) and (1.9)

$$\frac{\partial}{\partial\tau}\left(\sqrt{\frac{-x'^2}{\dot{x}^2}}\right) = 0$$

with the general solution

$$\dot{x}^2 + \lambda^2(\sigma)\, x'^2 = 0 \tag{1.12}$$

The first boundary condition (1.5) implies, that in a regular parametrization we must have

$$\lambda(0) = \lambda(\sigma_{max}) = 0 \tag{1.13}$$

We can however perform the singular reparametrization

$$\tilde{\sigma}(\sigma) = \int_0^{\sigma} \frac{1}{\lambda(\sigma')}\, d\sigma' \tag{1.14}$$

This is an integrable, finite reparametrization, if $\lambda^2(\sigma)$ has simple zeros at 0 and σ_{max}, and normalizes the function $\tilde{\lambda}(\tilde{\sigma})$ in (1.12) to unity, that is instead of (1.12) we have

$$\dot{x}^2 + x'^2 = 0 \tag{1.15}$$

and because of the singular parametrization

$$x'^\mu \Big|_{\text{at boundary}} = 0 \tag{1.16}$$

instead of (1.13). Odd numbers of partial derivatives with respect to σ have to vanish at the boundary. In a last step, we divide τ and σ by $\frac{\sigma_{max}}{\pi}$, thus normalizing the domain D to $D = \{ (\sigma,\tau): 0 \leq \sigma \leq \pi \}$. One can check, that each step was a invertible reparametrization of the timelike worldsheet of the string, so we didn't lose any solution of the equations of motion. (1.3) is now simply the wave equation in two dimensions

$$\ddot{x} - x'' = 0 \tag{1.17}$$

The gauge conditions (1.9) and (1.15) can be written

$$(\dot{x} \pm x')^2 = 0 \tag{1.18}$$

$$n x = c \cdot \tau \tag{1.18a}$$

The most general solution to (1.17) is given by

$$x^\mu(\sigma,\tau) = g^\mu(\tau+\sigma) + h^\mu(\tau-\sigma) \tag{1.19}$$

with arbitrary functions $g^\mu(t)$ and $h^\mu(t)$. Applying the boundary conditions (1.16) at $\sigma = 0$, we get

$$g^\mu(t) = h^\mu(t) \tag{1.19a}$$

and at $\sigma = \pi$ we have as result

$$g^\mu(t+2\pi) = g^\mu(t) + \pi \alpha_0^\mu \tag{1.20}$$

with an arbitrary constant α_0^μ . Making use of the periodicity of g^μ, which resulted from the finiteness of the string, we expand g^μ in discrete modes

$$g^\mu(t) = \frac{1}{2}\left(q^\mu + \alpha_0^\mu t - i\sum_{n=1}^{\infty}\frac{1}{\sqrt{n}}\left(a_n^{*\mu} e^{int} - a_n^\mu e^{-int}\right)\right) \tag{1.21}$$

with real constants q^μ, α_0^μ and complex a_n^μ. Notice, that $g^\mu(t)$ has a simple geometric interpretation. The boundary curve $x^\mu(\sigma = 0,\tau)$ is just

$$x^\mu(\sigma=0,\tau) = 2\, g^\mu(\tau)$$

and the world sheet is the mean value of positions of its boundary

$$x^\mu(\sigma,\tau) = \frac{1}{2}\left(x^\mu(0,\tau+\sigma) + x^\mu(0,\tau-\sigma)\right) \tag{1.22}$$

Inserting the expansion (1.21) of $g^\mu(t)$, we get

$$x^\mu(\sigma,\tau) = q^\mu + \alpha_0^\mu \tau - i\sum_{n=1}^{\infty} \frac{\cos n\sigma}{\sqrt{n}}\left(a_n^{*\mu} e^{in\tau} - a_n^\mu e^{-in\tau}\right) \tag{1.23}$$

The constants α_0^μ, a_n^μ are subject to the gauge conditions (1.18), or equivalently the condition, that the boundary curve is lightlike; this equivalence (in this gauge) is seen from

$$0 = (\dot{x} \pm x')^2(\sigma,\tau) = \Big(2g'(\tau+\sigma) + 2g'(\tau-\sigma) \pm 2\big(g'(\tau+\sigma) - g'(\tau-\sigma)\big)\Big)^2 =$$

$$= 4\left(g'(\tau\pm\sigma)\right)^2 = \dot{x}^2(0,\tau\pm\sigma) \tag{1.24}$$

So, using the expansion (1.21) and the more convenient coefficients

$$\alpha_n^\mu = \sqrt{n}\, a_n^\mu \quad , \quad \alpha_{-n}^\mu = (\alpha_n^\mu)^* \tag{1.25}$$

we get

$$\dot{x}^2(0,\tau) = \left(\sum_{n=-\infty}^{+\infty} \alpha_n^\mu e^{-in\tau}\right)^2 =$$

$$= \sum_{n=-\infty}^{+\infty} e^{-in\tau}\left(\sum_{m=-\infty}^{+\infty} \alpha_{n-m}^\mu \alpha_{\mu m}\right) \tag{1.26}$$

The gauge conditions thus imply

$$0 = L_n = \frac{1}{2}\sum_{m=-\infty}^{+\infty} \alpha_{n-m}^\mu \alpha_{\mu m} \tag{1.27}$$

$$\overset{n>0}{=} \sum_{m=1}^{\infty}\left(\sqrt{m(m+n)}\, a_m^{*\mu} a_{\mu\, m+n} + \sqrt{n}\,\alpha_0^\mu a_{\mu n} + \frac{1}{2}\sum_{m=1}^{n-1}\sqrt{(n-m)m}\, a_{n-m}^\mu a_{\mu m}\right)$$

For n=0 we have

$$0 = \frac{1}{2}\alpha_0^\mu \alpha_{\mu 0} + \sum_{n=1}^{\infty} n\, a_n^{*\mu} a_{\mu n} \tag{1.28}$$

and n < 0 follows from

$$L_{-n} = L_n^* \tag{1.29}$$

Exploiting the Poincare invariance of the Lagrangian (1.1) we get the conserved momentum and angular momentum of the string

$$\mathbb{P}^\mu = \int\left(\frac{\partial \mathcal{L}}{\partial \dot{x}^\mu} d\sigma + \frac{\partial \mathcal{L}}{\partial x'^\mu} d\tau\right)$$

$$M^{\mu\nu} = \int\left(x^\mu \frac{\partial \mathcal{L}}{\partial \dot{x}^\nu} d\sigma + x^\mu \frac{\partial \mathcal{L}}{\partial x'^\nu} d\tau\right) - \mu \leftrightarrow \nu \tag{1.30}$$

where the integral is taken along some arbitrary curve, which intersects the domain D. P^μ and $M^{\mu\nu}$ are conserved because of the equations of motion (1.3) and the boundary conditions (1.4). Choosing as path τ = const for convenience and inserting the expansion (1.23), we get

$$\mathbb{P}^\mu = \alpha_0^\mu$$

$$M^{\mu\nu} = \left(q^\mu \alpha_0^\nu - q^\nu \alpha_0^\mu\right) - i \sum_{n=1}^{\infty} \left(a_n^{*\mu} a_n^\nu - a_n^{*\nu} a_n^\mu\right) \tag{1.31}$$

Identifying α_0^μ with the momentum of the string and $a_n^{*\mu}$ with its excitations, we see that (1.28) is a spectrum condition, fixing the mass of the string in terms of its excitations.

To quantize the string, we have to go through a Hamiltonian formalism, and then substitute Poisson brackets by commutator, interpreting the dynamical variables as operators acting on the Hilbert space of states. Computing

$$\mathcal{H} = \frac{\partial \mathcal{L}}{\partial \dot{x}^\mu} \dot{x}^\mu - \mathcal{L} \tag{2.1}$$

we see, that $\mathcal{H}$ vanishes identically, furthermore we have the identities

$$p_\mu x'^\mu = 0 \qquad\qquad p^2 + \left(\frac{x'}{\pi}\right)^2 = 0 \tag{2.2}$$

where

$$p_\mu = \frac{\partial \mathcal{L}}{\partial \dot{x}^\mu} \tag{2.3}$$

So we see, that the phase space (p_μ, x^μ) is constrained and the usual Hamiltonian formalism cannot be applied. So we will gc through a generalized Hamiltonian formalism [5], and apply it to the string.

Let us be given a Lagrangian $\mathcal{L}(Q^A, Q^A_{,\mu})$ A = 1...N of fields Q^A (which for convenience are to be functions of space time coordinates x^μ and not of some parameters (σ,τ) as we will have to deal with) and its derivatives $Q^A_{,\mu} = \frac{\partial Q^A}{\partial x^\mu}$. The canonically conjugated momenta are

$$p_A = \frac{\partial \mathcal{L}}{\partial Q^A_{,0}} \tag{2.4}$$

The first step to Hamiltonian dynamics is to inverte this definition

$$Q^A_{,0} = Q^A_{,0}\left(p_A, Q^A, Q^A_{,\alpha}\right) \qquad \alpha \neq 0 \tag{2.5}$$

and substitute the time derivatives of the fields by these functions of the momenta, the fields and their space derivatives. What is the condition, for the inversion (2.5) to be possible?
By the implicit function theorem, the matrix

$$M_{AB} = \frac{\partial p_A}{\partial Q^B_{,0}} = \frac{\partial^2 \mathcal{L}}{\partial Q^A_{,0} \partial Q^B_{,0}} \tag{2.6}$$

has to be of maximal rank. If it is not, the Euler-Lagrange equations do not (in general) determine uniquely the initial value problem. This can be seen by writing them as

$$M_{AB}\, Q^B_{,00} + f_A = 0 \tag{2.7}$$

where f_A is a function of time derivatives of the fields up to first order at most. Now if M_{AB} is not of maximal rank, that is, if there exist vectors n^A_r with

$$n^A_r\, M_{AB} = 0 \tag{2.8}$$

then one cannot solve (2.7) for the second time derivatives. So given a set of initial values

$$\begin{aligned} Q^A(x^0 = 0, x^\alpha) &= u^A(x^\alpha) \\ Q^A_{,0}(x^0 = 0, x^\alpha) &= v^A(x^\alpha) \end{aligned} \tag{2.9}$$

the second time derivatives $Q^A_{,00}$ are determined only up to arbitrary combinations of n^A_r, if (2.7) admits a solution at all, more precisely the initial values are constrained by first order equations

$$n_r^A f_r^A = 0 \qquad (2.10)$$

It can be, that differentiating (2.10) gives new second order equations, independent from (2.7), thus removing some arbitrariness of the initial value problem.

On the other hand, if we have a gauge symmetry, then the solutions are determined only up to some arbitrary functions and the initial value problem is necessarily underdetermined, then we will not be able, to solve (2.4) for the time derivatives of the fields, because (2.6) is not of maximal rank. As a consequence, the phase space will be constrained by identities

$$\varphi_r\left(Q^A, Q^A_{,\alpha}, p_A\right)(x) = 0 \qquad \begin{matrix}\alpha \neq 0 \\ A = 1\ldots N \\ r = 1\ldots M\end{matrix} \qquad (2.11)$$

the number M of independent constraints being the deviation of the rank of (2.6) from its maximal value

$$\text{rank}\left(\frac{\partial p_A}{\partial Q^B_{,0}}\right) + M = N \qquad (2.12)$$

For convenience, we choose a functional parametrization of the constraints

$$\phi(\lambda) = \int d^3x\, \lambda^r(x)\, \varphi_r\left(Q^A, Q^A_{,\alpha}, p_A\right) = 0 \qquad (2.13)$$

with arbitrary functions $\lambda^r(x)$.

Variations of Q^A, $Q^A_{,0}$ give rise to variations of Q^A, p_A which preserve the constraints, they are thus subject to

$$0 = \int dx \left(\frac{\delta\phi(\lambda)}{\delta p_A(x)}\, \delta p_A(x) + \frac{\delta\phi(\lambda)}{\delta Q^A(x)}\, \delta Q^A(x)\right) \qquad (2.14)$$

The variational derivative of a functional

$$F[Q^A] = \int dx\, \mathcal{F}(Q^A, Q^A_{,\alpha})$$

is given by

$$\frac{\delta F}{\delta Q^A(x)} = \left(\frac{\partial\mathcal{F}}{\partial Q^A} - \partial_\alpha \frac{\partial\mathcal{F}}{\partial Q^A_{,\alpha}}\right)(x) \qquad (2.15)$$

(α is summed over space indices only)

Though we cannot substitute $Q^A_{,0}$ by a function of (Q^A, p_A), we can express H as a function of (Q^A, p_A), because variations of $Q^A{}_{,0}$ leave H invariant.

$$\delta H = \int \left(Q^A{}_{,0}\, \delta p_A + p_A\, \delta Q^A{}_{,0} - \frac{\delta L}{\delta Q^A}\, \delta Q^A - \frac{\partial \mathcal{L}}{\partial Q^A{}_{,0}}\, \delta Q^A{}_{,0} \right) d^4x \qquad (2.16)$$

The second and the fourth term cancel because of (2.4).

Varying Q^A and p_A we get from the variational principle

$$0 = \delta \int (p_A Q^A{}_{,0} - \mathcal{H})\, d^4x = \int d^4x \left(Q^A{}_{,0} - \frac{\delta H}{\delta p_A} \right) \delta p_A - \left(p_{A,0} - \frac{\delta H}{\delta Q^A} \right) \delta Q^A \qquad (2.17)$$

As the variations δp_A and δQ^A are only subject to (2.14), we get the following Hamiltonian equations

$$\begin{aligned} Q^A{}_{,0} &= \frac{\delta H}{\delta p_A} + \frac{\delta \phi(\lambda)}{\delta p_A} \\ - p_{A,0} &= \frac{\delta H}{\delta Q^A} + \frac{\delta \phi(\lambda)}{\delta Q^A} \end{aligned} \qquad \text{for some } \lambda \qquad (2.18)$$

Defining Poisson brackets

$$\{A, B\} = \int d^3x\, \frac{\delta A}{\delta Q^C(x)}\, \frac{\delta B}{\delta p_C(x)} \quad - \quad A \leftrightarrow B \qquad (2.19)$$

(2.18) can be cast into the form

$$\dot{G} = \{ G, H + \phi(\lambda) \} \qquad (2.20)$$

for an arbitrary functional $G\left[Q^A, p_A\right]$.

These equations of motion determine uniquely the development of the system, once the initial values have been specified. The constraints (2.13) therefore cannot be imposed as additional conditions but can be imposed at the initial time only. It has then to be checked, whether the evolution of the system, governed by $H+\Phi(\lambda)$ preserves the constraints. So for (2.13) to be valid, we must have

$$\dot{\phi}(\eta) = \{ \phi(\eta), H + \phi(\lambda) \} = 0 \qquad (2.21)$$

for any η and some fixed λ . It is sufficient that (2.21) vanishes on account of (2.13), that is

$$\{ \phi(\eta), H + \phi(\lambda) \} = \phi(\kappa(\eta, \lambda)) \qquad (2.21a)$$

It can be, that there is no λ, which fulfils (2.21) and that the Poisson brackets yield independent secondary constraints $\chi(\kappa)$ on the dynamical variables. Then these secondary constraints have to be imposed on the initial values of the system and its time derivatives have to vanish.

$$\dot{\chi}(\kappa) = \{\chi(\kappa), H + \phi(\lambda)\} = 0 \tag{2.22}$$

New constraints can arise, and we have to repeat the process. Let us suppose, that after a finite number of such steps, the equations (2.21), (2.22) are fulfilled, then we can look upon them as equations for the yet undetermined function λ. The equations form a linear inhomogeneous system of equations for λ, the most general solution of which is

$$\lambda = \bar{\lambda} + c^r \lambda_r \tag{2.23}$$

where $\bar{\lambda}$ is a particular solution, λ_r are solutions of the homogeneous equation (without H) and c^r are totally arbitrary functions, exhibiting the gauge symmetry of the system.

Let us apply this formalism to the string, where the coordinates x^μ take the role of fields and the parameters (σ,τ) correspond to the space-time coordinates. As $\mathcal{H}$ vanishes, the motion is generated by the constraint functions $\varphi_\pm$ alone

$$\varphi_\pm = (\pi p \pm x')^2 = 0 \tag{2.24}$$

As Hamiltonian, we can choose

$$\mathcal{H}(\lambda^+, \lambda^-) = \frac{\lambda^+}{2}\left(p + \frac{x'}{\pi}\right)^2 + \frac{\lambda^-}{2}\left(p - \frac{x'}{\pi}\right)^2 \tag{2.25}$$

The equations of motion are (corresponding to 2.18)

$$\begin{aligned} \dot{x}^\mu &= (\lambda^+ + \lambda^-)\, p^\mu + (\lambda^+ - \lambda^-)\frac{x'^\mu}{\pi} \\ \pi \dot{p}^\mu &= \frac{\partial}{\partial\sigma}\left[(\lambda^+ + \lambda^-)\frac{x'^\mu}{\pi} + (\lambda^+ - \lambda^-)\, p^\mu\right] \end{aligned} \tag{2.26}$$

and as we integrated by parts and have finite boundaries, we get the boundary conditions

$$\pi m_0 p^\mu = m_1\left[(\lambda^+ + \lambda^-)\frac{x'^\mu}{\pi} + (\lambda^+ - \lambda^-)p^\mu\right] \tag{2.27}$$

The phase space is constrained by

$$M^\pm(\lambda) = \pm\int_0^\pi d\sigma \frac{\pi}{4}\lambda(\sigma)\left(p \pm \frac{x'}{\pi}\right)^2 = 0 \tag{2.28}$$

The constraints obey the algebra

$$\{M^i(l), M^j(k)\} = \delta^{ij} M^i(lk' - l'k) \quad , \quad i,j = +,- \tag{2.29}$$

that is, they form a closed algebra, no secondary constraints arise. Defining

$$l(\sigma) = \begin{cases} \frac{\pi}{4}\left(p + \frac{x'}{\pi}\right)^2(\sigma) & \sigma \geq 0 \\ \frac{\pi}{4}\left(p - \frac{x'}{\pi}\right)^2(\sigma) & \sigma \leq 0 \end{cases} \tag{2.30}$$

and

$$L_n = \int_{-\pi}^{+\pi} d\sigma\, e^{in\sigma} l(\sigma) = M^+(e^{in\sigma}) - M^-(e^{-in\sigma}) \tag{2.31}$$

we get a discrete complete set of constraints. Their algebra is particularly simple

$$\{L_n, L_m\} = i(m-n)L_{n+m} \tag{2.32}$$

Making use of the constraints (2.2), one can cast the boundary condition into the form

$$\left.\begin{aligned} \lambda^+ + \lambda^- &= 0 \\ \pi m_0 &= (\lambda^+ - \lambda^-)m_1 \end{aligned}\right\} \text{at boundary} \tag{2.33}$$

Nevertheless, we will choose

$$\lambda^+ = \lambda^- = \frac{\pi}{2} \tag{2.34}$$

which corresponds to the singular parametrization, we dealt with above. We then have

$$H = \frac{\pi}{2} \int_0^\pi d\sigma \left(p^2 + \frac{x'^2}{\pi^2} \right) = L_0$$

$$\dot{x}^\mu = \pi p^\mu \qquad , \qquad \pi \dot{p}^\mu = x^{\mu\prime\prime} \tag{2.35}$$

or

$$\ddot{x} - x'' = 0 \tag{2.36}$$

The constraints give

$$0 = (\pi p \pm x')^2 = (\dot{x} \pm x')^2 \tag{2.37}$$

that is we get back the equations (1.16,1.17,1.18). They are invariant however under conformal transformations of the two-dimensional parameter space. This is, as one knows from the theory of functions e.g., a gauge group restricted by the "Cauchy Riemann differential equations" (with a different minus sign on account of the different metric). To be specific, we can allow for new parameters

$$\tilde{\sigma}(\sigma,\tau) \qquad , \qquad \tilde{\tau}(\sigma,\tau)$$

subject to

$$\frac{\partial \tilde{\sigma}}{\partial \sigma} = \frac{\partial \tilde{\tau}}{\partial \tau} \quad , \quad \frac{\partial \tilde{\sigma}}{\partial \tau} = \frac{\partial \tilde{\tau}}{\partial \sigma} \quad , \quad \left. \frac{\partial \tilde{\sigma}}{\partial \tau} \right|_{\text{at boundary}} = 0 \tag{2.38}$$

or

$$\ddot{\tau} - \tau'' = 0 \qquad , \qquad \tau' \big|_{\sigma = 0,\pi} = 0 \tag{2.39}$$

So we can choose

$$n_\mu x^\mu = c \cdot \tau \tag{2.40}$$

As $n_\mu x^\mu$ fulfils (2.39), n_μ is a constant vector with $n^2 \leq 0$ as above and the constant c turns out to be

$$c = n_\mu \alpha_0^\mu \tag{2.41}$$

the momentum in the direction of n. So we have reproduced all the equations of the Lagrangian treatment of the string, we have as the most general solution (1.23) and the constraints (1.27) and (2.31), defined in a different way, turn out to coincide, justifying the notation.

Using the expansion (1.23) and $p^{\mu} = \frac{\dot{x}^{\mu}}{\pi}$, we can solve for the modes.

$$\alpha_0^{\mu} = \int_0^{\pi} p^{\mu} d\sigma$$

$$q^{\mu} = \frac{1}{\pi}\int_0^{\pi} (x^{\mu} - \pi\tau p^{\mu}) d\sigma \tag{2.42}$$

$$\alpha_n^{*\mu} = \int_0^{\pi} e^{-in\tau} \cos n\sigma \left(\frac{p^{\mu}}{\sqrt{n}} + i\sqrt{n}\, x^{\mu}\right) d\sigma$$

From the definition of Poisson brackets

$$\{A, B\} = \int_0^{\pi} d\sigma \frac{\delta A}{\delta x^{\mu}(\sigma)} \frac{\delta B}{\delta p_{\mu}(\sigma)} \quad - \quad A \leftrightarrow B \tag{2.43}$$

we get

$$\{q^{\mu}, \alpha_0^{\nu}\} = g^{\mu\nu}$$

$$\{a_n^{*\mu}, a_m^{\nu}\} = i g^{\mu\nu} \delta_{n,m} \tag{2.44}$$

The other Poisson brackets of the modes vanish. The algebra of the constraints has been given already (2.32).

Quantization can proceed now along well known lines. Regard the dynamical variables x^{μ}, p_{μ} as operators acting on the Hilbert space of states and let the commutators be given by i times the Poisson bracket of the classical theory. The algebra of the modes becomes the algebra of annihilation-creation operators

$$[a_m^{\mu}, a_n^{\nu +}] = \delta_{m,n} g^{\mu\nu} \qquad n, m > 0$$

$$[q^{\mu}, \alpha_0^{\nu}] = i g^{\mu\nu} \tag{3.1}$$

It is also convenient to give the algebra of the α-operators (1.25)

$$[\alpha_m^{\mu}, \alpha_n^{\nu}] = n\, \delta_{m+n,0}\, g^{\mu\nu}$$

$$[q^{\mu}, \alpha_n^{\nu}] = i\, \delta_{n,0}\, g^{\mu\nu} \tag{3.1a}$$

Given the ground state, we can construct a basis for the Fock space

$$|\lambda, k\rangle = \prod_{i,\mu} \frac{1}{\sqrt{\lambda_{i,\mu}!}} \left(a_i^{+\mu}\right)^{\lambda_{i,\mu}} e^{ik_\mu q^\mu} |0\rangle \tag{3.2}$$

where the ground state has the properties

$$\alpha_m^\mu |0\rangle = 0 \tag{3.3}$$

The state (3.2) has momentum k_μ, its excitation is given by the multi-index $\lambda = (..\lambda_{\mu,i}...)$. The ordering of the creation-annihilation operators in those operators, containing $a_m^{+\mu}$ and a_n^ν in higher than first order, is either fixed by hermiticity (1.31) or is irrelevant (1.27) or gives rise to a yet undetermined c-number term. So we will define L_o as the normal ordered operator corresponding to (1.28). Because of the non-arbitrariness in the ordering of L_o, the algebra of the L_n operators is changed by an important dimension* dependent c-number. The number can be computed most easily be taking the groundstate expectation value of the L_n commutator. We have

$$\langle 0|[L_n, L_{-n}]|0\rangle \overset{n>0}{=} \langle 0| L_n L_{-n} |0\rangle =$$

$$= \frac{1}{4}\langle 0| \left\{ \sum_{\mu=0}^{D-1} \sum_{m=1}^{n-1} \sqrt{m(n-m)}\, a_m^\mu a_{\mu\, n-m} \right\} \left\{ \text{the same} \right\}^\dagger |0\rangle = \frac{D}{12}(n^3 - n)$$

So the constraints - or gauge operators obey the algebra (cf. 2.32)

$$[L_m, L_n] = (m-n) L_{m+n} + \frac{D}{12}(m^3 - m)\, \delta_{m,-n} \tag{3.4}$$

Because of these commutation relations, we cannot simply have $L_n = 0$ as constraint condition, but can only impose

$$\langle \psi | L_n + \delta_{n,0}\alpha | \varphi \rangle = 0 \tag{3.5}$$

thus singling out a subspace of the Hilbert space. (Here we allowed for the c-number term, arising from the arbitrariness of the ordering of the classical L_o.) Making use of $L_{-n} = L_n^+$ we get the sufficient condition for (3.5) to be valid as

$$(L_n + \delta_{n,0}\alpha)\, |\varphi\rangle = 0 \tag{3.6}$$

* Notice, that we didn't specify the dimension of space time up to now, so we can allow the index μ to vary from 0 to (D-1).

The set of all solutions to (3.6) will be denoted as the physical space the states $|\psi>$ as the physical states. One now has to show, that no negative norm states appear as solutions to (3.6), so that the gauge condition eliminates the negative norm states, generated by a_n^{+o}. Before we construct the physical states, some elementary remarks. We cannot expect to construct all solutions to (3.6) at once but want to make use of an inductive process rather. We notice, that we can diagonalize the level operator

$$R = \sum_{n=1}^{\infty} n \, a_n^{+\mu} a_{\mu n} \tag{3.7}$$

in the space of solutions to (3.6). This follows from the commutation relations

$$[R, L_m] = -m L_m \tag{3.8}$$

The eigenvalues of R, the level number M, are positive integers, the eigenspace at level number M will be denoted by R^M. The construction of solutions to (3.6) will make use of complete induction with respect to the level number.

To get rid of the continuum of momenta of the states, we would like to consider states only with one fixed momentum. The condition (3.6) for n = 0 however fixes the mass of the physical states

$$(L_0 + \alpha) |\varphi> = \left(\tfrac{1}{2} p^2 + R + \alpha\right) |\varphi> = 0 \tag{3.9}$$

so we cannot have one fixed momentum for physical states at different levels. (3.9 will be referred to as the mass-shell condition). So we will suppose the states, we are dealing with, as having the momentum

$$(p^\mu)_M = (p^\mu)_0 + M k^\mu \tag{3.10}$$

where $(p^\mu)_M$ is the momentum at level M, and k is restricted by

$$k^2 = 0 \;, \quad k^0 > 0 \;, \quad (p^\mu)_0 k_\mu = -1 \;, \quad (p)_0^2 = -2\alpha \tag{3.11}$$

We then have

$$(p_M)^2 = -2(\alpha + M) \;, \quad k_\mu (p^\mu)_M = -1 \tag{3.11a}$$

that is, (3.9) is fulfilled.
Before proceeding, we will introduce some notations:

The operators K_n will be of importance, they are defined by

$$K_n = k_\mu \alpha_n^\mu \quad , \quad K_{-n} = K_n^\dagger \tag{3.12}$$

and obey the commutation relations

$$[K_n, K_m] = 0, \; [L_n, K_m] = -m K_{n+m}, \; [R, K_m] = -m K_m \tag{3.13}$$

The transverse states $|t\rangle$ are the solutions of

$$K_n |t\rangle = L_n |t\rangle = 0 \qquad n > 0 \tag{3.14}$$

The space spanned by the transverse states at level M, $|t, M, \nu\rangle$ *) will be denoted by T^M. The following operator product, applied to transverse states will be important

$$L_{-1}^{\lambda_1} L_{-2}^{\lambda_2} \dots L_{-n}^{\lambda_n} K_{-1}^{\mu_1} \dots K_{-m}^{\mu_m} |t, M, \nu\rangle = \{\lambda, \mu\}_{M'} |t, M, \nu\rangle \tag{3.15}$$

where $\{\lambda,\mu\}$ denotes the multi-index $\lambda_1 \dots \lambda_n \; \mu_1 \dots \mu_m$ and

$$M' = \sum_{r=1}^{n} r \lambda_r + \sum_{s=1}^{m} s \mu_s \tag{3.16}$$

Using the commutation relations (3.13) and (3.8), one easily verifies, that the level number of $\{\lambda, \mu\}_{M'} |t, M, \nu\rangle$ is M'+M.

The importance of the states $|t, M, \nu\rangle$, which form a subspace of the physical space becomes transparent by the following lemma:
If $|t, M, \nu\rangle$ is a basis for the transverse space T^M at level M, then the states

$$\{\lambda, \mu\}_{N-M} |t, M, \nu\rangle \tag{3.17}$$

give a basis for the states at level number N, and as N varies, for the whole Hilbert space.

Using this lemma, we can write an arbitrary physical state as a linear combination of

---

*) (ν denotes a degeneration parameter, M the level number)

$$|\psi\rangle = K_{-1}^{\mu_1} \ldots K_{-n}^{\mu_n} |t\rangle + L_{-n} |s\rangle$$

and making use repeatedly of the commutation relations (3.4) and (3.13) we will show, that if the dimension D is 26 and $\alpha = -1$ then $|\psi\rangle$ is a linear combination of the form

$$|\psi\rangle = |t\rangle + L_{-n} |s\rangle \tag{3.18}$$

where $L_{-n}|s\rangle$ is physical and null,null meaning, that its scalar product with any physical state, including itself, vanishes. Then, the physical space is positive semidefinite. The choice of a specific vector k_μ turns out to be irrelevant, because, as $|t\rangle$ is a physical vector, it can be written as a sum of a transverse vector, constructed with a different k'_μ , and a null vector. The condition

$$K_n |\psi\rangle = 0 \qquad\qquad n > 0 \tag{3.19}$$

which corresponds to (2.30, 2.41) thus does not spoil Lorentz-covariance, but fixes only the null vector part of the physical states. It cannot be imposed however if $D < 26$, to fix a null vector part of the physical states.

Let us prove now the lemma. We show first, that the states

$$\{\lambda,\mu\}_{N-M} |t, M, \nu\rangle \qquad\qquad N-M > 0 \tag{3.20}$$

are linearly independent for fixed M and ν. Let us be given a linear combination $\sum_{\lambda,\mu} c[\lambda,\mu] \{\lambda,\mu\}_{N-M} |t, M, \nu\rangle = |f\rangle$
then for $|f\rangle$ to vanish, the terms with different excitations have to vanish separately. Consider the $a_1^{\dagger\mu}$ oscillators :
$\{\lambda\ ,\mu\}_{N-M}$ contributes terms of the form (cf. 1.27, 3.12)

$$(p a_1^\dagger)^{\lambda_1} (a_1^\dagger a_1^\dagger)^{\lambda_2} (a_1^\dagger a_2^\dagger)^{\lambda_3} \ldots (a_1^\dagger a_{n-1}^\dagger)^{\lambda_n} (k \cdot a_1^\dagger)^{\mu_1}$$

Consider the terms in $|f\rangle$, which maximize the number of $a_1^{\dagger\mu}$-oscillators; these terms have to cancel. But obviously, they can do so only if all these terms have the same $\lambda_1, \lambda_2, \ldots \lambda_n, \mu_1$. But then, for cancellation of the remaining excitations resulting from $K_{-s}^{\mu s}$ to be possible, we also have $\mu_2 \ldots \mu_n$ equal in all terms, so there is only one term, which has to vanish itself. So we conclude that for $|f\rangle$ to vanish, the coefficients $c[\lambda,\mu]$ have to vanish and the linear independence of (3.20) is shown.

We now prove that the states $|f\rangle$ do not contain transverse states. For this purpose, we first define an order of the multi-indices. We say $(\lambda_i) < (\lambda'_i)$ if $\lambda_1 = \lambda'_1, \dots \lambda_{m-1} = \lambda'_{m-1}, \lambda_m < \lambda'_m$, the same definition applying to (μ_i). Now assume $|f\rangle$ is transverse. Take those terms with smallest (μ_i) and choose from these the one with largest (λ_i). Suppose $(\lambda_i) \neq 0$ and especially $\lambda_1, \dots \lambda_{j-1} = 0, \lambda_j \neq 0$.
Apply K_j to $|f\rangle$. Using the commutation relation (3.13), one sees, that (μ_i) is increased and λ decreased. Because of $K_o|t,M,\nu\rangle = -|t,M,\nu\rangle$ (cf. 3.11a, 3.12), the only terms where (μ_i) is not increased have the structure $L_{-j}^{\lambda_j - 1} \dots |t\rangle$, they cannot cancel with other terms, arising in the commutation process, which have greater (μ_i), so for the smallest (μ_i) terms, the greatest (λ_i) is 0. Applying our consideration to the next larger (μ_i) and noting, that K_j annihilates the terms with smaller (μ_i) (no L_{-j} are present there), we get by induction, that all λ_j have to be zero. Now apply L_n to the vector, where $\lambda_j = 0$. Commuting L_n through K_{-i} operators either increases (μ_i) or gives rise to terms with $K_j, j > 0$, which are annihilated, or terms with K_o, which have a smaller (μ_i). Take the term with smallest non-zero (μ_i), let j be the minimal value, for which $\mu_j \neq 0$ and apply L_j. The smallest term which originated in commuting L_j to the right is $\mu_j - 1, \dots$, it cannot be cancelled by other terms. So there is no smallest non-zero term $\{0, \mu_i\}|t\rangle$ in the vector $|f\rangle$. It follows then, that the states (3.20) do not contain transverse states.
We now show, that the states (3.20) and the transverse states form a basis. Let G^N denote the space spanned by the states (3.20). T^N is orthogonal to it by definition (3.14) and because of $K_n = K_n^\dagger$, $L_{-n} = L_n^\dagger$. We prove by induction, that T^N is equal to the orthogonal complement of G^N (with respect to R^N), and as we know, that G^N and $T^N = (G^N)^\perp$ are disjoint, G^N and T^N span the whole R^N. It is easy to show the necessary conditions for N = 1. Let us be given, that up to N-1, $T^M \oplus G^M$ form a basis for R^M. Let $|\psi\rangle$ be from $(G^N)^\perp$, then it fulfils

$$\langle\psi| L_{-1}\{\lambda,\mu\}_{N-M-1}|t,M,\nu\rangle = 0$$

$$\langle\psi| L_{-2}\{\lambda,\mu\}_{N-M-2}|t,M,\nu\rangle = 0 \qquad (3.21)$$

$$\langle\psi| K_{-1}\{\lambda,\mu\}_{N-M-1}|t,M,\nu\rangle = 0$$

By induction hypothesis $\{\lambda,\mu\}_{N-M-1}|t,M,\nu\rangle$ and $\{\lambda,\mu\}_{N-M-2}|t,M\,\nu\rangle$ form a basis (N-M-1 and N-M-2 may be zero) for R^{N-1}, R^{N-2}.
As all scalar products of the vectors $L_1|\psi\rangle$, $L_2\,|\psi\rangle$, $K_1|\psi\rangle$ with a basis vanish, the states are zero themselves. From the commutation relations (3.4),(3.13) it then follows, that

$$L_n|\psi\rangle = K_n|\psi\rangle = 0$$

and $|\psi\rangle$ is a transverse state. Moreover, we can easily show now, that T^N is positive definite: it is positive semidefinite because $K_n|t\rangle = 0$ excludes timelike excitations, it is positive definite, because it is disjoint with its orthogonal complement G^N. Having established now the lemma it is easy to proceed to the main goal of determining the physical space. Any vector can be written as a linear combination of vectors $|t,M,\nu\rangle$, $K^{\mu_1}_{-1}\ldots K^{\mu_n}_{-n}\,|t,M,\nu\rangle$, $L_{-1}|a\rangle$, $\tilde{L}_{-2}|b\rangle$ because of the generating algebra of the L_n (3.4). $\tilde{L}_{-2}$ is defined by

$$\tilde{L}_2 = L_2 + \frac{3}{2}L_1^2 \qquad\qquad \tilde{L}_{-2} = \tilde{L}_2^{\dagger} \tag{3.22}$$

We only have to check

$$L_1|\psi\rangle = \tilde{L}_2\,|\psi\rangle = 0 \tag{3.23}$$

for $|\psi\rangle$ to be physical. Making use of the commutation relations (3.13) and especially

$$\begin{aligned}&[L_1, L_{-1}] = 2L_0\,, \quad [L_1, \tilde{L}_{-2}] = 6L_{-1}(L_0+1)\\ &[\tilde{L}_2, L_{-1}] = 6L_1L_0\,,\\ &[\tilde{L}_2, \tilde{L}_{-2}] = 13\left(L_0+\frac{D}{26}\right) + 18(L_{-1}L_1+L_0)(L_0+1)\end{aligned} \tag{3.24}$$

we see that if D = 26 and the groundstate mass squared $m^2 = 2\alpha$ is $m^2 = -2$ ($\alpha = -1$), then the states $L_{-1}|a\rangle + \tilde{L}_{-2}|b\rangle$ are mapped on states $L_{-1}|a'\rangle + \tilde{L}_{-2}\,|b'\rangle$ by L_1 and $\tilde{L}_2$. So for $|\psi\rangle$ to be physical, the states $K^{\mu_1}_{-1}\ldots K^{\mu_m}_{-m}|t\rangle$ and $L_{-1}|a\rangle + \tilde{L}_{-2}|\,b\rangle$ have to vanish separately on applying L_1 and $\tilde{L}_2$ to $|\psi\rangle$. Applying the same considerations as in the proof of the lemma, we get, that for $L_nK^{\mu_1}_{-1}\ldots K^{\mu_m}_{-m}\,|t\rangle$ to be zero all powers of K_{-j} have to vanish. So we get the result, which we indicated earlier: If the dimension of space time is 26 and $\alpha = -1$, all solutions of

$$L_n|\psi\rangle = (L_0+\alpha)|\psi\rangle = 0 \quad , n>0, \text{ with}$$

$$\langle\psi|\psi\rangle \neq 0$$

are of the form

$$|\psi\rangle = |t\rangle + |s\rangle \tag{3.25}$$

where $|t\rangle$ is a transverse state and $|s\rangle$ is a linear combination of states of the form $L_{-n}|\chi\rangle$, which is orthogonal to any physical state, including itself.

That 26 is an upper limit to the dimension of the string model is seen from the state

$$|\phi\rangle = \{\tilde{L}_{-2} + \frac{1}{4}(D-26)(K_{-2}+K_{-1}^2)\}|0\rangle \tag{3.26}$$

which is physical for any dimension D, but has norm

$$\langle\phi|\phi\rangle = \frac{1}{2}(26-D) \tag{3.27}$$

If D is smaller than 26, the transverse states do not span the physical space up to null vectors, as this example shows.

References

1) P. Goddard, J. Goldstone, C. Rebbi, C.B. Thorn, Nucl.Phys. B56, 109 (1973)
2) J. Scherk, An Introduction given to the Theory of Dual Models and Strings, Lectures given at New York University,NYU/TR3/74
3) C. Rebbi, The Physical Interpretation of Dual Models, CERN-Preprint TH-1691 (1973)
4) C. Rebbi, Dual Models and Relativistic Quantum Strings, Phys. Reports 12C , 1 (1974)
5) P.A.M Dirac, Generalized Hamiltonian Dynamics,Proc.Roy. Soc. A246, 326 (1958)
   P.A.M. Dirac, Lectures on Quantum Mechanics, Belfer Graduate School of Science, Monographs Series, New York 1964
6) J.H. Schwarz, Phys. Reports 8C,269 (1973)

FERMI - BOSE - SUPERSYMMETRY

J. Wess

Universität Karlsruhe, Germany

Introduction

Supersymmetries are based on algebras which are generalizations of Lie-algebras, i.e. there are commutators and anticommutators in the defining relations. They were first used by B. Zumino and the author[1)] to show that such symmetries can play an important role in the framework of renormalizable Lagrangian field theories.

The concept of supersymmetries has been abstracted from supergauge transformations in dual models[2)]. From a different point of view, Volkov and Akulov have arrived independently at the same algebra[3)].

The interesting and surprising features of supersymmetry are:

The powerful machinery of Lie algebras and their representation theory also works for supersymmetries. This has been very impressively demonstrated by the work of Salam and Strathdee[4)]. Under the name of "extended Lie algebras", supersymmetries have also been studied in the mathematical literature[5)].

Supersymmetry and Lagrangian field theory merge happily. Renormalization respects supersymmetry and supersymmetry leads to cancellations of divergencies[6)], rendering e.g. the least divergent quantum field theoretical Lagrangian model known up to now. Supersymmetry yields interesting relations among masses and coupling constants, involving Fermions and Bosons alike.

Supersymmetries contain the Poincaré group in a nontrivial way. They avoid difficulties with no go theorems[7)] and combine particles with different spin - Bosons and Fermions - in one supermultiplet. Relativistic models exist which contain SU(6) as a symmetry in the restframe.

It is this latter aspect of supersymmetry which we would like to emphasize in these lectures. To this end we develop the group theoretical methods, construct invariant Lagrangians and investigate the content of quantum numbers in supermultiplets.

A systematic review on supersymmetry was given by B. Zumino[8) in his Review talk at the XVII International Conference on High-Energy Physics. It is not our intention to duplicate this review here nor to give a systematic introduction to all what has been done with supersymmetry up to now.

Notation

$$\eta^{\mu\nu} = (-1, 1, 1, 1)$$

$$\sigma^{\mu} = (1, \vec{\sigma}) , \quad \bar{\sigma}^{\mu} = (1, -\vec{\sigma}) ,$$

$\vec{\sigma}$ are the 2 by 2 Pauli matrices.

$$\Theta Q = \Theta^{\alpha}{}_{i} Q^{i}_{\alpha} , \quad \bar{Q}\bar{\Theta} = \bar{Q}_{\dot{\alpha} i} \bar{\Theta}^{\dot{\alpha} i}$$

$$\varepsilon^{\alpha\beta} : \quad \varepsilon^{12} = 1 , \; \varepsilon^{21} = -1 , \; \varepsilon^{11} = \varepsilon^{22} = 0$$

$$\varepsilon^{\alpha\beta} \varepsilon_{\beta\gamma} = \delta^{\alpha}{}_{\gamma} .$$

$$g^{ij} \sim \varepsilon^{\alpha\beta} .$$

$$\sigma^{\mu\nu} = \frac{1}{4} (\sigma^{\mu}\bar{\sigma}^{\nu} - \sigma^{\nu}\bar{\sigma}^{\mu}) , \quad \bar{\sigma}^{\mu\nu} = \frac{1}{4} (\bar{\sigma}^{\mu}\sigma^{\nu} - \bar{\sigma}^{\nu}\sigma^{\mu}) .$$

$$\varepsilon_{0123} = 1 .$$

$$v_{\mu\nu} = \frac{\partial}{\partial x^{\mu}} v_{\nu} - \frac{\partial}{\partial x^{\nu}} v_{\mu} .$$

I Algebra

The algebra of supersymmetry transformations is defined through the following commutation and anticommutation relations:

$$\{ Q_\alpha^i , Q_\beta^j \}_+ = \{ \bar{Q}_{\dot{\alpha} i}, \bar{Q}_{\dot{\beta} j} \}_+ = 0$$

$$\{ Q_\alpha^i , \bar{Q}_{\dot{\beta} j} \}_+ = 2 \sigma_{\mu\, \alpha\dot{\beta}} P^\mu \delta^i{}_j \tag{1}$$

$$[P^\mu , Q]_- = [P^\mu , \bar{Q}]_- = [P^\mu , P^\nu] = 0$$

P^μ is the energy momentum operator, which generates four-dimensional translations. Q_α^i are constant Weyl spinors - α takes the values 1,2 and i the values 1...N .
$\bar{Q}_{\dot{\alpha} i}$ is the complex conjugate of Q_α^i. The defining relations (1) have the group SL(2C) ⊗ SU(N) as an automorphism. It is remarkable that this algebra can be represented in terms of field operators in the framework of a Lagrangian - field theory. The spinor operators are then integrals over conserved vector-spinor currents:

$$Q_\alpha^i = \int d^3x\, J_\alpha^{0i} \quad , \quad \partial_\mu J_\alpha^{\mu i} = 0$$

Theories, which possess such conserved currents have very interesting properties. In perturbation theory they are much less divergent then one might expect from power counting and there are relations among coupling constants and masses which remain true to all orders in perturbation theory - much the same as one is used to from ordinary symmetry groups, only that the symmetry relations involve particles with different spin. To construct such theories we have to find representations of the algebra (1).

II Representations

We shall follow the usual rules of the theory of representations of Lie algebras[9)], taking account of the anticommutators by using parameters which are elements of a Grassmann algebra, i.e. we use parameters with the properties:

$$\{ \Theta_i^\alpha , \Theta_j^\beta \} = \{ \bar{\Theta}^{\dot{\alpha} i}, \bar{\Theta}^{\dot{\beta} j} \} = \{ \Theta^\alpha_i , \bar{\Theta}^{\dot{\beta} j} \} = 0$$

as well as

$$\{\theta_i^\alpha , \theta_\beta^j\} = \{\theta_i^\alpha , \bar\theta_{\dot\beta j}\} = 0 \qquad \ldots$$

It then follows from (1) that:

$$[\theta Q , \theta Q] = [\bar Q \bar\theta , \bar Q \bar\theta] = 0$$

$$[\theta Q , \bar Q \bar\theta] = 2\, \theta \sigma_\mu \bar\theta P^\mu \quad .$$

A finite "group element", which depends on the parameters θ, $\bar\theta$ and x_μ is defined as follows:

$$G(\theta, \bar\theta, x) = e^{i\{\theta Q + \bar Q \bar\theta - x_\mu P^\mu\}}$$

We can multiply two such "group elements" and, using Hausdorf's formula we find:

$$G(\xi, \bar\xi, y) \cdot G(\theta, \bar\theta, x) =$$

$$= G(\theta + \xi , \bar\theta + \bar\xi , x_\mu + y_\mu - i\, \xi \sigma_\mu \bar\theta + i\, \theta \sigma_\mu \bar\xi)$$

The "group" induces a motion in the parameter space:

$$G(\xi, \bar\xi, y) : \{\theta, \bar\theta, x_\mu\} \longrightarrow \{\theta + \xi , \bar\theta + \bar\xi , x_\mu + y_\mu - i\, \xi \sigma_\mu \bar\theta + i\, \theta \sigma_\mu \bar\xi\}$$

A "<u>superfield</u>", introduced by Salam and Strathdee, is a function of the parameters θ, $\bar\theta$ and x_μ:

$$\phi(\theta, \bar\theta, x)$$

Under the "group" it transforms as follows:

$$G(\xi, \bar\xi, y) : \ \phi(\theta, \bar\theta, x) \longrightarrow$$

$$\longrightarrow \phi(\theta + \xi , \bar\theta + \bar\xi , x_\mu + y_\mu - i\, \xi \sigma_\mu \bar\theta + i\, \theta \sigma_\mu \bar\xi)$$

or, infinitesimally:

$$\delta \phi = \left(\xi \frac{\partial}{\partial \theta} + \bar\xi \frac{\partial}{\partial \bar\theta} - i(\xi \sigma_\mu \bar\theta - \theta \sigma_\mu \bar\xi) \frac{\partial}{\partial x_\mu} \right) \phi$$

It is easy to verify that we have found a representation of the algebra (1) in terms of the following differential operators:

$$Q^i_\alpha = \frac{\partial}{\partial \Theta^\alpha_i} - i\,\sigma^\mu_{\alpha\dot\alpha}\,\bar\Theta^{\dot\alpha i}\frac{\partial}{\partial x_\mu}$$

$$\bar Q_{\dot\alpha i} = -\frac{\partial}{\partial \bar\Theta^{\dot\alpha i}} + i\,\Theta^\alpha_i\,\sigma^\mu{}_{\alpha\dot\alpha}\frac{\partial}{\partial x_\mu}$$

$$P_\mu = i\frac{\partial}{\partial x_\mu}$$

The algebra (1) contains the algebra of the Q's <u>or</u> the $\bar{Q}$'s as <u>subalgebras</u>.

$$\{Q, Q\} = 0 \ , \quad \{\bar Q, \bar Q\} = 0$$

We can, therefore, study the motion of the "group" on the respective "co-sets".

We parametrise as follows:

$$\sigma_1(\Theta, \bar\Theta, x) = e^{i(\Theta Q - xP)}\, e^{i(\bar Q\bar\Theta)}$$

$$\sigma_2(\Theta, \bar\Theta, x) = e^{i(\bar Q\bar\Theta - xP)}\, e^{i(\Theta Q)}$$

and find, as before:

$$\sigma(\xi, \bar\xi, y)\,\sigma_1(\Theta, \bar\Theta, x) =$$

$$= \sigma_1(\Theta + \xi,\ \bar\Theta + \bar\xi,\ x_\mu + y_\mu + 2i\,\Theta\sigma_\mu\bar\xi + i\,\xi\sigma_\mu\bar\xi)$$

$$\sigma(\xi, \bar\xi, y)\,\sigma_2(\Theta, \bar\Theta, x) =$$

$$= \sigma_2(\Theta + \xi,\ \bar\Theta + \bar\xi,\ x_\mu + y_\mu - 2i\,\xi\sigma_\mu\bar\Theta - i\,\xi\sigma_\mu\bar\xi)$$

The transformation law of the corresponding superfields is, analogously:

$$\sigma(\xi,\bar{\xi},y): \quad \phi_1(\theta,\bar{\theta},x) \to \phi_1(\theta+\xi,\bar{\theta}+\bar{\xi}, x+y+2i\theta\sigma\bar{\xi}+i\xi\sigma\bar{\xi})$$

$$\phi_2(\theta,\bar{\theta},x) \to \phi_2(\theta+\xi,\bar{\theta}+\bar{\xi}, x+y-2i\xi\sigma\bar{\theta}-i\xi\sigma\bar{\xi})$$

or, infinitesimally:

$$\delta\phi_1 = \left(\xi\frac{\partial}{\partial\theta} + \bar{\xi}\frac{\partial}{\partial\bar{\theta}} + 2i\theta\sigma_\mu\bar{\xi}\frac{\partial}{\partial x_\mu}\right)\phi_1$$

$$\delta\phi_2 = \left(\xi\frac{\partial}{\partial\theta} + \bar{\xi}\frac{\partial}{\partial\bar{\theta}} - 2i\xi\sigma_\mu\bar{\theta}\frac{\partial}{\partial x_\mu}\right)\phi_2$$

We have found two other representations of Q and $\bar{Q}$:

$$Q = \frac{\partial}{\partial\theta}\ ; \quad \bar{Q} = -\frac{\partial}{\partial\bar{\theta}} + 2i\theta\sigma_\mu\frac{\partial}{\partial x_\mu}$$

and

$$Q = \frac{\partial}{\partial\theta} - 2i\sigma_\mu\bar{\theta}\frac{\partial}{\partial x_\mu}\ , \quad \bar{Q} = -\frac{\partial}{\partial\bar{\theta}}$$

It is easy to change the parametrisation of σ of σ_1 or σ_2. Using Hausdorf's formula we find:

$$\sigma(x,\theta,\bar{\theta}) = \sigma_1(x_\mu + i\theta\sigma_\mu\bar{\theta},\theta,\bar{\theta}) = \sigma_2(x_\mu - i\theta\sigma_\mu\bar{\theta},\theta,\bar{\theta})$$

We can, therefore, <u>shift</u> from one representation to the other by the corresponding change of variables:

$$\phi(x_\mu,\theta,\bar{\theta}) = \phi_1(x_\mu + i\theta\sigma_\mu\bar{\theta},\theta,\bar{\theta})$$

$$= \phi_2(x_\mu - i\theta\sigma_\mu\bar{\theta},\theta,\bar{\theta})\ .$$

To illustrate the meaning of this shift we calculate the following example:

$$\delta\phi_1(x'_\mu,\theta',\bar{\theta}') =$$

$$= \left\{\xi\frac{\partial}{\partial\theta'} + \bar{\xi}\frac{\partial}{\partial\bar{\theta}'} + 2i\theta'\sigma_\mu\bar{\xi}\frac{\partial}{\partial x'_\mu}\right\}\phi_1(x'_\mu,\theta',\bar{\theta}')$$

change of variables:

$$\begin{cases} x'_\mu = x_\mu + i\theta\sigma_\mu\bar\theta \\ \theta' = \theta \\ \bar\theta' = \bar\theta \end{cases} \qquad \begin{cases} \frac{\partial}{\partial x'_\mu} = \frac{\partial}{\partial x_\mu} \\ \frac{\partial}{\partial\theta'} = \frac{\partial}{\partial\theta} - i\sigma_\mu\bar\theta\frac{\partial}{\partial x_\mu} \\ \frac{\partial}{\partial\bar\theta'} = \frac{\partial}{\partial\bar\theta} + i\theta\sigma_\mu\frac{\partial}{\partial x_\mu} \end{cases}$$

$$\delta\phi_1(x_\mu + i\theta\sigma_\mu\bar\theta,\theta,\bar\theta) =$$

$$= \left\{ \xi\frac{\partial}{\partial\theta} + \bar\xi\frac{\partial}{\partial\bar\theta} - i\xi\sigma_\mu\bar\theta\frac{\partial}{\partial x_\mu} + i\theta\sigma_\mu\bar\xi\frac{\partial}{\partial x_\mu} \right\} \phi_1(x_\mu + i\theta\sigma_\mu\bar\theta,\theta,\bar\theta)$$

This shows that $\phi_1(x_\mu + i\theta q_\mu\bar\theta , \theta, \bar\theta)$ as function of x_μ, θ and $\bar\theta$ transforms exactly like $\phi(x_\mu,\theta,\bar\theta)$.

A <u>covariant derivative</u> can be defined such that it anticommutes with the variation. It is shown that

$$\frac{\partial}{\partial\bar\theta}\,\delta\phi_1 = \delta\frac{\partial}{\partial\bar\theta}\,\phi_1$$

and

$$\frac{\partial}{\partial\theta}\,\delta\phi_2 = \delta\frac{\partial}{\partial\theta}\,\phi_2$$

Through a shift we can write these covariant derivatives for any field, i.e.:

$$D_\alpha\,\phi = \left(\frac{\partial}{\partial\theta^\alpha} + i\sigma_{\mu\,\alpha\dot\beta}\,\bar\theta^{\dot\beta}\frac{\partial}{\partial x_\mu}\right)\phi$$

$$\bar D_{\dot\alpha}\,\phi = \left(-\frac{\partial}{\partial\bar\theta^{\dot\alpha}} - i\theta^\alpha\sigma_{\mu\,\alpha\dot\alpha}\frac{\partial}{\partial x_\mu}\right)\phi$$

$$D\,\phi_1 = \left(\frac{\partial}{\partial\theta} + 2i\sigma_\mu\bar\theta\frac{\partial}{\partial x_\mu}\right)\phi_1$$

$$\bar D\,\phi_1 = -\frac{\partial}{\partial\bar\theta}\,\phi_1$$

$$D\,\phi_2 = \frac{\partial}{\partial\theta}\,\phi_2$$

$$\bar D\,\phi_2 = \left(-\frac{\partial}{\partial\bar\theta} - 2i\theta\sigma_\mu\frac{\partial}{\partial x_\mu}\right)\phi_2$$

All these derivatives have the property that they anticommute with the respective variations.
The covariant derivatives allow us to impose covariant conditions on the superfields:

$$\bar{D}\phi = 0 \qquad \text{or} \qquad D\phi = 0$$

for example.
Superfields, satisfying one of these conditions are called <u>scalar superfields</u>. It is obvious that $\bar{D}\phi_1 = 0$ means that ϕ_1 does not depend on $\bar{\theta}$ and $D\phi_2 = 0$ means that ϕ_2 does not depend on θ.

III Scalar and vector superfields

We can make a power series expansion of a superfield in its variables θ and $\bar{\theta}$. This will always lead to a finite series. Let us take as an example the scalar field ϕ_1 ($\bar{D}\phi_1 = 0$). Throughout this and the following section we shall restrict ourselves to the case N = 1 (no internal symmetry).

$$\phi_1(x,\theta) = A(x) + \theta^\alpha \psi_\alpha(x) + \theta^\alpha\theta^\beta \varepsilon_{\alpha\beta} F(x)$$

The decomposition contains two complex scalar fields and one Weyl field. The transformation law of these fields can be obtained through:

$$\delta\phi_1(x,\theta) = \delta A + \theta\,\delta\psi + \theta\theta\,\delta F$$

$$= \left(\xi\frac{\partial}{\partial\theta} + 2i\,\theta\sigma_\mu\bar{\xi}\frac{\partial}{\partial x_\mu}\right)\phi_1$$

We find:

$$\delta A = \xi\psi$$

$$\delta\psi = 2F\xi + 2i\,\sigma_\mu\bar{\xi}\frac{\partial}{\partial x_\mu}A$$

$$\delta F = -i\frac{\partial}{\partial x_\mu}\psi\sigma^\mu\bar{\xi}$$

The complex fields F, ψ and A belong to the "scalar" representation of the algebra.

The <u>vector field</u> is a superfield which satisfies the reality condition $V^+ = V$. To find the vector representation we expand this field as follows:

$$V(x,\theta,\bar\theta) = \left(1 + \tfrac{1}{4}\,\theta\theta\bar\theta\bar\theta\,\Box\right)C$$

$$+ \left(i\theta + \tfrac{1}{2}\,\theta\theta\sigma^\mu\bar\theta\frac{\partial}{\partial x_\mu}\right)\chi + \tfrac{i}{2}\,\theta\theta(M+iN)$$

$$+ \left(-i\bar\theta + \tfrac{1}{2}\,\bar\theta\bar\theta\theta\sigma^\mu\frac{\partial}{\partial x_\mu}\right)\bar\chi - \tfrac{i}{2}\,\bar\theta\bar\theta(M-iN)$$

$$- \theta\sigma_\mu\bar\theta\, v^\mu + i\,\theta\theta\bar\theta\bar\lambda - i\,\bar\theta\bar\theta\theta\lambda$$

$$+ \tfrac{1}{2}\,\theta\theta\bar\theta\bar\theta D$$

We have not taken the monomials of θ and $\bar\theta$ as a basis for the expansion but have chosen more complicated combinations. This choice has the advantage that this basis transforms in a more "reduced" way than the monomials. To illustrate this we compute:

$$\delta\theta\theta = \left\{\xi\frac{\partial}{\partial\theta} + \bar\xi\frac{\partial}{\partial\bar\theta} - i\,\xi\sigma_\mu\bar\theta\frac{\partial}{\partial x_\mu} + i\,\theta\sigma_\mu\bar\xi\frac{\partial}{\partial x_\mu}\right\}\theta\theta$$

$$= 2\xi\left\{\theta - \tfrac{i}{2}\,\sigma_\mu\bar\theta\theta\theta\frac{\partial}{\partial x_\mu}\right\}$$

$$\delta\left\{\theta - \tfrac{i}{2}\,\sigma_\mu\bar\theta\theta\theta\frac{\partial}{\partial x_\mu}\right\} =$$

$$= \xi\left\{1 + \tfrac{1}{4}\,\theta\theta\bar\theta\bar\theta\,\Box\right\} + i\,\xi\theta\sigma_\mu\bar\theta\frac{\partial}{\partial x_\mu} + i\,\bar\xi\bar\sigma_\mu\frac{\partial}{\partial x_\mu}\theta\theta$$

$$\delta\left\{1 + \tfrac{1}{4}\,\theta\theta\bar\theta\bar\theta\,\Box\right\} = -i\,\xi\sigma^\mu\frac{\partial}{\partial x_\mu}\left\{\bar\theta + \tfrac{i}{2}\,\bar\theta\bar\theta\theta\sigma^\rho\frac{\partial}{\partial x^\rho}\right\}$$

$$+ i\,\bar\xi\sigma^\mu\frac{\partial}{\partial x^\mu}\left\{\theta - \tfrac{i}{2}\,\sigma^\rho\bar\theta\theta\theta\frac{\partial}{\partial x^\rho}\right\}$$

and we learn that the variation of $\theta\theta$ has, in this basis, no term $\theta\theta\bar{\theta}$. If we now compute the transformation law of the fields we find from

$$\delta V = \{ \xi \frac{\partial}{\partial\theta} + \bar{\xi} \frac{\partial}{\partial\bar{\theta}} - i \xi \sigma_\mu \bar{\theta} \frac{\partial}{\partial x_\mu} + i \theta \sigma_\mu \bar{\xi} \frac{\partial}{\partial x_\mu} \} V :$$

$$\delta C = i \xi \chi - i \bar{\xi} \bar{\chi}$$

$$\delta \chi = \xi (M + i N) + \sigma^\mu \bar{\xi} (\frac{\partial}{\partial x^\mu} C + i v_\mu)$$

$$\delta \bar{\chi} = \bar{\xi} (M - i N) + \xi \sigma^\mu (\frac{\partial}{\partial x^\mu} C - i v_\mu)$$

$$\delta N = \xi (i \lambda - \sigma_\mu \frac{\partial}{\partial x_\mu} \bar{\chi}) + \bar{\xi} (- i \bar{\lambda} + \bar{\sigma}_\mu \frac{\partial}{\partial x_\mu} \chi)$$

$$\delta M = \xi (\lambda + i \sigma_\mu \frac{\partial}{\partial x_\mu} \bar{\chi}) + \bar{\xi} (\bar{\lambda} + i \bar{\sigma}_\mu \frac{\partial}{\partial x_\mu} \chi)$$

$$\delta v_\mu = \xi \frac{\partial}{\partial x^\mu} \chi + \bar{\xi} \frac{\partial}{\partial x^\mu} \bar{\chi} + i \xi \sigma_\mu \bar{\lambda} + i \bar{\xi} \bar{\sigma}_\mu \lambda$$

$$\delta \lambda = \xi \sigma^{\mu\rho} v_{\mu\rho} + \xi D$$

$$\delta \bar{\lambda} = \bar{\xi} \bar{\sigma}^{\mu\rho} v_{\mu\rho} + \bar{\xi} D$$

$$\delta D = - \xi \sigma^\mu \frac{\partial}{\partial x^\mu} \bar{\lambda} + \bar{\xi} \bar{\sigma}^\mu \frac{\partial}{\partial x^\mu} \lambda$$

Through the choice of our basis we have achieved that D, λ, $\bar{\lambda}$ and $v_{\mu\rho} = \partial_\mu v_\rho - \partial_\rho v_\mu$ form a representation by themselves. This will have important consequences when we try to construct a Lagrangian.

A real field can also be constructed from a scalar field and its complex conjugate. If we take $\phi + \bar{\phi}$, $(D\bar{\phi} = \bar{D}\phi = 0)$ and shift it to the basis, in which we have considered our vector field we obtain a field that transforms like a vector field. Its components are:

$$(1+\tfrac{1}{4}\theta\theta\bar\theta\bar\theta\,\Box)(A+A^*) + i\,\theta\sigma_\mu\bar\theta\frac{\partial}{\partial x_\mu}(A-A^*)$$

$$+ (\theta - \tfrac{i}{2}\theta\theta\sigma^\mu\bar\theta\frac{\partial}{\partial x^\mu})\psi + (\bar\theta + \tfrac{i}{2}\bar\theta\bar\theta\theta\sigma^\mu\frac{\partial}{\partial x^\mu})\bar\psi$$

$$+ \theta\theta F + \bar\theta\bar\theta F^*.$$

We find that $\quad C = A + A^*\,,\ \chi = -i\psi\,,\ \bar\chi = i\bar\psi\,,$

$$v^\mu = -i\,\partial^\mu(A-A^*)\,,\ N = -(F+F^*)$$

$$M = -i(F-F^*)\,,\ \lambda = 0\,,\ D = 0$$

transform like a vector multiplet.

IV Lagrangians

The last member of a scalar or vector multiplet (F or D) transforms by a total derivative. A fourdimensional integral over such a quantity would yield an invariant action. The free Lagrangian has to be quadratic in the fields. We, therefore, have to combine two superfields to a vector or scalar multiplet.
For a scalar superfield we have two possibilities: $\phi\,\phi^+$ and $\phi^2 + \phi^{+2}$. These expressions will again transform like a superfield of the same type if ϕ and ϕ^+ are taken in the same basis.
Let us first study the expression $\phi\,\phi^+$:

$$\phi\phi^+ = \phi(x,\theta)\,\phi^+(x-2i\theta\sigma\bar\theta,\bar\theta)$$

$$= \phi(x,\theta)\,e^{-2i\theta\sigma^\mu\bar\theta\frac{\partial}{\partial x^\mu}}\,\phi^+(x,\bar\theta)$$

We expand in powers of θ and $\bar\theta$:

$$\phi\phi^+ = A(x)A^*(x) + \ldots$$

$$\ldots + \theta\theta\bar\theta\bar\theta\{F(x)F^*(x) - \tfrac{i}{2}\psi\sigma^\mu\partial_\mu\bar\psi + A\,\Box A^*\}$$

The action

$$\mathcal{L}_0 = \int d^4x \{ FF^* - \frac{i}{2} \psi\sigma^\mu \partial_\mu \bar{\psi} - \partial_\mu A \partial^\mu A^* \}$$

is invariant.
If we now expand $\phi\phi$ we obtain:

$$\phi(x,\theta)\phi(x,\theta) = A(x)A(x) + \dots$$

$$+ \theta\theta \{ 2AF - \frac{1}{2}\psi\psi \}$$

and analogously:

$$\phi^+(x,\bar{\theta})\phi^+(x,\bar{\theta}) = A^*(x)A^*(x) + \dots$$

$$+ \bar{\theta}\bar{\theta} \{ 2A^*F^* - \frac{1}{2}\bar{\psi}\bar{\psi} \}$$

The action:

$$\mathcal{L}_m = \frac{1}{2} m \int d^4x \{ 2(AF + A^*F^*) - \frac{1}{2}(\psi\psi + \bar{\psi}\bar{\psi}) \}$$

is invariant.

The action $\mathcal{L}_o + \mathcal{L}_m$ yields the equations of motion:

$$F + mA^* = 0, \quad \Box A + mF^* = 0, \quad i\sigma^\mu \partial_\mu \bar{\psi} + m\psi = 0$$

together with the complex conjugate equations. These are the free field equations for a complex scalar and a Weyl spinor field; they can also be written in the superfield notation:

$$-\frac{1}{4}\bar{D}\bar{D}\phi^+ + m\phi = 0$$

An interaction can easily be added. Interactions of the form $\phi^3 + \phi^{+3}$ and $\phi^4 + \phi^{+4}$ have been intensively investigated.
For a <u>vector superfield</u> we study the two invariants VV and $V\bar{D}\bar{D}DDV$.
If we expand VV in the variables θ and $\bar{\theta}$ we find:

$$VV = (1 + \frac{1}{4}\theta\theta\bar{\theta}\bar{\theta}\Box)C^2 + \dots$$

$$+ \frac{1}{2}\theta\theta\bar{\theta}\bar{\theta}\{ 2CD - \partial_\mu C \partial^\mu C - v_\mu^2 + M^2 + N^2$$

$$- 2\chi\lambda - 2\bar{\chi}\bar{\lambda} - 2i\chi\sigma^\mu\partial_\mu\bar{\chi} \}.$$

For $V\bar{D}\bar{D}DDV$ we obtain, up to terms which are total spacetime derivatives:

$$\frac{1}{4} V\bar{D}\bar{D}DDV = \ldots \theta\theta\bar{\theta}\bar{\theta}\{D^2 - 2i\lambda\sigma^\mu\partial_\mu\bar{\lambda}$$

$$-\frac{1}{2} v_{\mu\nu} v^{\mu\nu}\} + 2V\Box V$$

A suitable Lagrangian is obtained by taking the "D" member of

$$\frac{1}{2} V\{\frac{1}{4}\bar{D}\bar{D}DD - 2\Box\}V + m^2 VV.$$

This yields:

$$\mathcal{L} = \frac{1}{2} D^2 - i\lambda\sigma_\mu\partial^\mu\bar{\lambda} - \frac{1}{4} v_{\mu\nu} v^{\mu\nu}$$

$$+\frac{1}{2} m^2 \{2CD - \partial_\mu C\partial^\mu C - v_\mu^2 + M^2 + N^2$$

$$-2\chi\lambda - 2\bar{\chi}\bar{\lambda} - 2i\chi\sigma^\mu\partial_\mu\bar{\chi}\}.$$

The field equations are:

$$D + m^2 C = 0, \quad \Box C + D = 0,$$

$$\partial_\mu v^{\mu\nu} - m^2 v^\nu = 0,$$

$$M = N = 0$$

$$i\sigma^\mu\partial_\mu\bar{\lambda} + m^2\chi = 0, \quad i\partial_\mu\chi\sigma^\mu - \bar{\lambda} = 0.$$

They describe one vector field, one scalar field and <u>two</u> Weyl spinors (or <u>one</u> Dirac spinor), all with mass m. The same set of equations can be obtained from the covariant equation:

$$\frac{1}{16}(DD\bar{D}\bar{D} + \bar{D}\bar{D}DD)V - \Box V + m^2 V = 0.$$

V Internal symmetry

Here we consider the case N = 2 [10)] and we construct the scalar representation. As a basis for the decomposition of the superfield we choose the following:

$$\Theta_i^\alpha\,; \quad \tilde{\mathcal{C}} = \frac{1}{2}\,\Theta_i^\alpha \Theta_j^\beta \varepsilon_{\alpha\beta}\,(g\tilde{\tau})^{ij}$$

$$a^{\mu\nu} = \frac{1}{2}\,g^{ij}(\sigma^{\mu\nu}\varepsilon)_{\alpha\beta}\,\Theta_i^\alpha \Theta_j^\beta$$

$$\mathcal{U} = \Theta_1^1 \Theta_2^1 \Theta_1^2 \Theta_2^2$$

$$\chi_\alpha^i = \frac{\partial}{\partial\Theta_i^\alpha}\,\mathcal{U}\,.$$

We define:

$$\Phi = A + \Theta\Psi + \tilde{\mathcal{C}}\tilde{C} + a^{\mu\nu}F_{\mu\nu} + \phi\chi + F\mathcal{U}$$

and we derive the following transformation law for the fields:

$$\delta A = \xi_i^\alpha \Psi^i{}_\alpha$$

$$\delta\Psi_\beta^j = -\xi_i^\alpha\{\varepsilon_{\alpha\beta}(g\tilde{\tau})^{ij}\tilde{C} + (\sigma^{\mu\nu}\varepsilon)_{\alpha\beta}\,g^{ij}F_{\mu\nu}\} + 2i\,\sigma_{\varrho\beta\dot\beta}\,\bar{\xi}^{\dot\beta j}\,\partial_\varrho A$$

$$\delta C = \frac{1}{2}\,\xi_i^\alpha\,\varepsilon_{\alpha\beta}\,\phi_j^\beta(g\tilde{\tau})^{ij} + i\,\sigma_{\varrho\alpha\dot\beta}\,\bar{\xi}^{\dot\beta j}\,\varepsilon^{\alpha\beta}(\tilde{\tau}g)_{ij}\,\partial_\varrho\Psi_\beta^i$$

$$\delta F^{\mu\nu} = \frac{1}{2}\,\xi_i^\alpha(\sigma^{\mu\nu}\varepsilon)_{\alpha\beta}\,\phi_j^\beta g^{ij} + i\,\sigma_{\varrho\alpha\dot\beta}\,\bar{\xi}^{\dot\beta j}g_{ij}\,(\varepsilon\sigma^{\mu\nu})^{\alpha\beta}\,\partial_\varrho\Psi^i{}_\beta$$

$$\delta\phi_j^\beta = \xi_j^\beta F - 2i\,\sigma_{\varrho\alpha\dot\beta}\,\bar{\xi}^{\dot\beta i}(\varepsilon^{\alpha\beta}(\tilde{\tau}g)_{ji}\,\partial_\varrho\tilde{C} + (\varepsilon\sigma^{\mu\nu})^{\alpha\beta}g_{ji}\,\partial_\varrho F_{\mu\nu})$$

$$\delta F = 2i\,\sigma_{\varrho\alpha\dot\beta}\,\bar{\xi}^{\dot\beta i}\,\partial_\varrho\,\phi_i^\alpha\,.$$

This representation is reducible. We can impose the following supersymmetry covariant condition on the fields and their complex conjugates

$$\phi = -2i\bar{\sigma}_\mu \partial_\mu \bar{\Psi}, \quad F = 4\,\Box A^*, \quad \vec{\tilde{C}}^* = -\vec{\tilde{C}}$$

$$\partial_\mu F_{\varrho\sigma} + \partial_\varrho F_{\sigma\mu} + \partial_\sigma F_{\mu\varrho} = 0$$

This reduces the number of independent fields and we can find an invariant Lagrangian:

$$\mathcal{L} = \partial_\mu A^* \partial^\mu A - \frac{i}{2}\bar{\Psi}\bar{\sigma}^\mu \partial_\mu \Psi - \frac{1}{4}\vec{\tilde{C}}\vec{\tilde{C}} + \frac{1}{4}F_{\mu\nu}F^{\mu\nu}$$

This Lagrangian describes one complex scalar isoscalar; one real vector isoscalar; and one isodoublet Weyl spinor, all massless. From the following section - the representation in the rest frame - it will become clear that no mass term can be added for this set of fields. It is, however, possible to construct a Lagrangian with mass for the original set of fields. A proper choice of fields avoids all higher derivatives.

VI Rest frame representation

The operator P^2 commutes with the algebra (1). We are interested in a representation of (1) for which P^2 has the eigenvalue $M^2 > 0$ [11].

The subspace of states, for which P^μ has the eigenvalue (M, 0, 0, 0) (states at rest), is invariant under the algebra generated by Q and $\bar{Q}$. In this subspace the algebra (1) becomes:

$$\{Q^i_\alpha, Q^j_\beta\} = \{\bar{Q}_{\dot{\alpha}i}, \bar{Q}_{\dot{\beta}j}\} = 0$$

$$\{Q^i_\alpha, \bar{Q}_{\dot{\beta}j}\} = 2M\,\delta_{\alpha\dot{\beta}}\,\delta^i_j$$

which is the algebra of 2N Fermion creation and annihilation operators. It can be constructed by defining a "vacuum" state Ω through the condition:

$$Q^i_\alpha \Omega = 0$$

and by applying all the nonvanishing powers of $\frac{1}{\sqrt{2M}}\bar{Q}_{\dot{\alpha}}$

to the state Ω. These are exactly 2^{2N} states which can be labeled by the "quantum numbers": $(n_{1+}, n_{1-}, \ldots n_{N+}, n_{N-})$ where $n = 0$ or 1. The labels +, - correspond to the spin. It should be emphasized that all these states correspond to one particle states at rest. Let us consider the case $N = 1$ as an example. We obtain the four states $\Omega, \frac{1}{\sqrt{2M}} \bar{Q}_{\dot{\alpha}} \Omega, \frac{1}{4M} \varepsilon^{\dot{\alpha}\dot{\beta}} \bar{Q}_{\dot{\alpha}} \bar{Q}_{\dot{\beta}} \Omega$. The first and the last are spin zero states, the second one is a spin 1/2 state. These states correspond to the one particle states of our scalar field ϕ_1.
There we have one scalar, one pseudoscalar and one Weyl spinor. We can obtain other representations by assigning a spin index to the vacuum. A state $\Omega_{\dot{\alpha}_1 \ldots \dot{\alpha}_{2j}}$ corresponds to spin j if it is symmetric in all the α. The states created by the operators $\bar{Q}_{\dot{\alpha}}$ are now states with the spin $(j, j+1/2, j-1/2, j)$. With $j = 1/2$ we obtain two spin 1/2, one spin zero and one spin 1 state corresponding to our vector field V. If we want to describe particles and antiparticles we can double our algebra. The operators $Q_\alpha^{(1)}, Q_\alpha^{(2)}, \bar{Q}_{\dot{\alpha}}^{(1)}, \bar{Q}_{\dot{\alpha}}^{(2)}$ can be grouped to a Dirac spinor:

$$\Psi = \begin{pmatrix} Q^{(1)} \\ \bar{Q}^{(2)} \end{pmatrix}$$

in terms of which our algebra (1) can be written as:

$$\{\Psi, \Psi\} = \{\bar{\Psi}, \bar{\Psi}\} = 0$$

$$\{\Psi, \bar{\Psi}\} = 2\gamma^\mu P_\mu \qquad \bar{\Psi} = \Psi^+ \gamma^0$$

The one particle states at rest correspond exactly to the states which can be created by the creation operators of a Dirac particle at rest. If an internal symmetry is included we would call our Dirac particles quarks. The constant operators ψ^i play the role of constituent quarks. The one particle states of the supersymmetric theory correspond exactly to the states which can be created by powers of the static quark field ψ^i. At rest, these states span a representation of the group SU(4N). Especially it contains the group SU(2N) with the generators:

$$\frac{1}{2M} \bar{Q}^{(\ell)}_{i\dot{\alpha}} \vec{\sigma}^{\dot{\alpha}\beta} Q^{i(\ell)}_{\beta}; \quad \frac{1}{2M} \bar{Q}^{(\ell)}_{i\dot{\alpha}} \delta^{\dot{\alpha}\beta} T^{ri}{}_{j} Q^{(\ell)j}_{\beta}$$

$$\frac{1}{2M} \bar{Q}^{(\ell)}_{i\dot{\alpha}} \vec{\sigma}^{\dot{\alpha}\beta} T^{ri}{}_{j} Q^{(\ell)j}_{\beta}$$

T^r are the generators of SU(N).
For N = 3, these are the generators of SU(6). It is clear that the states of any quark model correspond to the one particle states of a super-

symmetric theory. To arrive at a reasonable model - whose states contain also the observed particles - it will be necessary to include colour in the internal symmetry.

The connexion between this representation theory approach and the superfield approach discussed in the previous sections can be established by expressing the generators Q in terms of differential operators and by solving the corresponding differential equations.

References

1) J. Wess and B. Zumino, Nucl. Phys. B70, 39 (1974)
Phys. Lett. 49B, 52 (1974)

2) A. Neveu and J. H. Schwarz, Nucl. Phys. B31, 86 (1971)
P. Ramond, Phys. Rev. D3, 2415 (1971)
Y. Aharnov, A. Casher and L. Susskind, Phys. Lett. 35B, 512 (1971)
J.L. Gervais and B. Sakita, Nucl.Phys. B34, 633 (1971)

3) D.V. Volkov and V.P. Akulov, Phys. Lett. 46B, 109 (1973)

4) A. Salam and J. Strathdee, Nucl. Phys. B76, 477 (1974)
Trieste preprints IC/74/42 (1974)
IC/74/85 (1974)
IC/74/36 (1974)

5) F.A.Berezin and G.I. Katz, Mathemat. Sbornik (USSR) 82, 343 (1970), English translation Vol. 11

6) J. Wess and B. Zumino, Phys. Lett. 49B, 52 (1974),
Nucl. Phys. B78, 1 (1974)
J. Iliopoulos and B. Zumino, Nucl. Phys. B76, 310 (1974)
S. Ferrara, J. Iliopoulos and B. Zumino, Nucl. Phys. B77,413(1974)
W. Lang and J. Wess, University of Karlsruhe Preprint (1974) to be published in Nucl. Phys.

7) L. O'Raifeartaigh, Phys. Rev. Lett. 14, 575 (1965),
Phys. Rev. 139B, 1052 (1965)
S. Coleman and J. Mandula, Phys. Rev. 159, 1251 (1967)

8) B. Zumino, Review talk given at the XVII International Conference on High-Energy Physics, 1-10 July 1974, Imperial College,London
CERN preprint Ref. TH 1901-CERN

9) S. Ferrara, J. Wess and B. Zumino, Phys. Lett.61B, 239 (1974)

10) P. H. Dondi and M. Schnius, University of Karlsruhe, Preprint (1974) to be published in Nucl. Physics

11) A. Salam and J. Strathdee, Trieste preprint IC/74/80 (1974)

STRINGS, CONDUCTIVE AND OTHERWISE

Jorge Willemsen

Stanford Linear Accelerator Center
Stanford University, Stanford, California 94305

1. The Physical Nature of the String

In his lectures, Dr. Dragon has reviewed for us the theory of the relativistic string, discussing the formal problems associated with incorporation of the constraints of the theory, and the unphysical features encountered in attempting a canonical quantization.

In the course of this seminar I want to describe a model worked out in collaboration with C. Carlson, L. N. Chang, and F. Mansouri, which was motivated by an attempt to circumvent some of these difficulties in quantization by identifying canonical variables different from the coordinates of the string, already at the classical level.

Before I get to our model, however, I would like to take time to remind you of some of the physical pictures which have been proposed as candidates for the underlying structure of the string. The "string model" itself is, after all, an attempt to understand what physical structures can give rise to Veneziano-type scattering amplitudes, and it is natural to proceed a step further and ask, "What makes up the string, and how does it hold together?" Hopefully, this discussion will provide physical motivation for our "conductive string" model.

The first conceptual problem one is faced with is in visualizing how a hadron could look like a one-dimensionally extended object at all. It is reasonable to expect that what something looks like depends upon how we look at it, and I want to remind you that field-theoretic models have already been around for a long time, in which the hadrons <u>do</u> appear as one-dimensionally extended matter distributions. The hadrons are supposed to "look" this way to the extent that in a high energy, low momentum transfer collision, the inclusive particle distribution in the central region provides us with a "snapshot" of the hadrons' constituent matter distribution. A convenient way to display this information is by drawing a $p_\perp - p_\parallel$ phase space plot

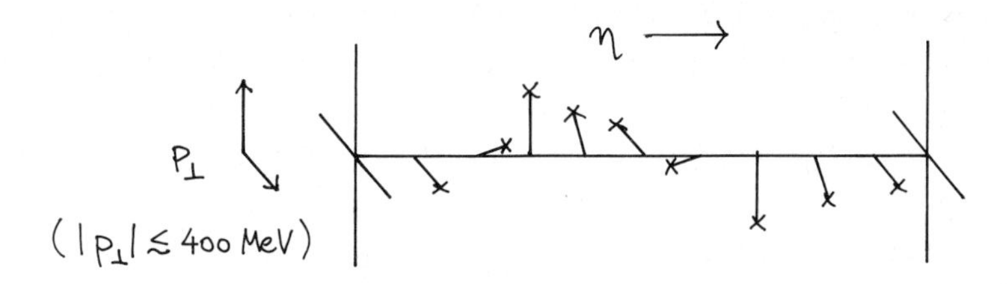

Figure 1

If, as the number of partons increases, we can smoothly interpolate between the points on the plot with a single curve, we will see the hadron as a string of partons.

Does this parton string have anything to do with the dualists' string? If we transform $p_\perp$ to $x_\perp$, but keep the "length" axis as the longitudinal momentum fraction, we do indeed get just the picture that emerges from the Goddard, Goldstone, Rebbi, and Thorn (GGRT) parametrization of the Nambu string action. The original papers on the subject should be consulted to see that this is so (here and elsewhere), because I will not have time to show you the details. But this very nice mesh between the intuitive parton picture and the actual mathematical formalism of the string model should be kept in mind when thinking about the physics of the string.

This simple physical picture is not without its faults, of course. As indicated in the drawing, we cannot really make contact with what is happening to the "leading particles" in the hadrons. Phenomenologically, current induced reactions suggest that partons with the quantum numbers of the quarks carry the most significant fractions of the hadron's momentum. Although Harari-Rosner diagrams suggest a picture where the ends of the string are like quarks and antiquarks, with an essentially "neutral" middle, the mathematical string formalism does not genuinely reflect this fact in any way.

In addition, this picture reflects only one component of the transverse momentum distribution, the exponentially bounded part dying off at $\simeq$ 400 MeV. We now know that there are other parts of the hadronic wavefunction, extending out much further than this with a power-law behavior. The string picture is no more able to account for this region than the simple "wee", "soft" parton picture that motivates the Feynman-Wilson phase space plot.

I want to milk this soft parton picture for two more features.

The first is that, as the energy increases, the number of produced particles increases logarithmically. On a rapidity plot, this means, of course, that the central region is filled uniformly. We expect, in the snapshot, that there will be a small likelihood for multiple occupancy at a given rapidity value. This is necessary for our hadron to extend out in one dimension, rather than filling up like a three-dimensional jelly. However, this still leaves a question about whether the hadron is one big string, or a series of short strings, i.e., whether there is any correlation in $x_\perp$ space between neighboring points in longitudinal momentum. I will return to this question momentarily.

My second point is not really derived from the parton picture, but rather an input to it originating from the notion of duality. Duality tells us there is an intimate relationship between the behavior in the Regge region, which the soft parton concepts describe, and the resonance region. That is, there should be a connection between the extended, "wee sea" component of the hadronic wave-function, and the spectrum of hadrons that is observed. It is natural to anticipate that, if there is such a relationship, it arises from the collective behavior of the wee partons in the sea. The collective excitations of the sea give rise to the observed spectrum of hadrons. To do this, of course, one cannot adhere too closely to the strictly "free" notion of partons. Allowance must be made for parton-parton interactions, which can give rise to sequences of excitations, but do not destroy the basic linear extension of the hadrons.

At this point, it is natural to wonder if there is any model, no matter how simple, that can actually give rise to behaviors of the type we have been discussing. Many of the concepts of the parton model itself are realized in the field theoretic multiperipheral model, and this provides a natural point of departure for a first attempt to describe strings. To go from an initial Regge picture to a dual picture, one should generalize the class of graphs that are considered:

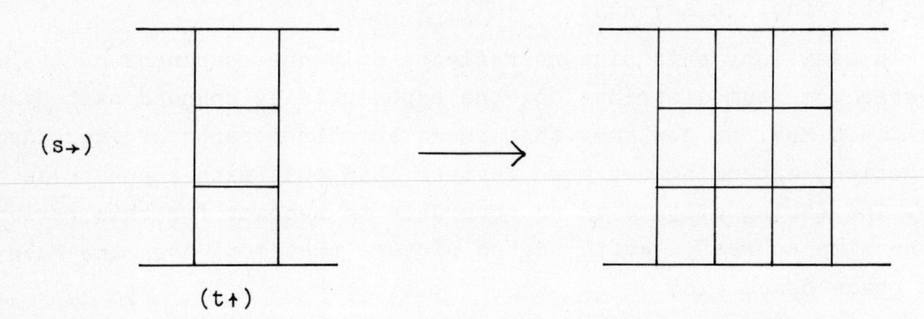

Figure 2

Kraemmer, Nielsen, and Susskind, and separately Gervais, Sakita, and Virasoro studied the properties of such "fishnet" diagrams, and elaborated the approximations under which such diagrams give rise to Veneziano-type scattering amplitudes. Not unexpectedly, this kind of behavior is obtained under the assumption that momentum flows smoothly and uniformly throughout the graph, so that propagators can be sensibly approximated by Gaussians. By so doing, one loses information about the short-distance structure of the constituent interactions, and this is already a strong hint as to the inherent limitations of the string picture.

The physical assumptions involved in jumping from a quantum field theory to a string picture are even more vividly portrayed in an old model due to Bjorken (Tel-Aviv lecture). This model is also discussed in detail by Kogut and Susskind in their Physics Report about partons. The basic idea is that one can examine the hadronic wavefunction in terms of the constituents, at a given time, in ϕ^3 theory,

$$\psi_{(n)}(t) \sim < \eta_1, \underset{\sim}{K}_{\perp,1}; \ldots \eta_n, \underset{\sim}{K}_{\perp,n} | u(t,-\infty) \quad \psi_0 | 0 > \tag{1a}$$

$$\sim \prod_{j=1}^{n-1} \left[\frac{M^2}{2} - \sum_{i=1}^{j} \frac{K_i^2 + m^2}{2\eta_i} + \frac{L_j^2 + m^2}{2\beta_j} \right]^{-1} \left[\frac{\beta_{n-1}}{\beta_j} \right] \quad . \tag{1b}$$

Here η_j is the longitudinal momentum fraction of the jth parton, $(p_{\parallel})_j / \Sigma_j (p_{\parallel})_j$; and $(K_{\perp,j})$ is the transverse momentum of the jth parton. The quantities β_i are the sequential longitudinal momentum transfers down the chain, and $(L_{\perp,i})$ are the transverse momentum transfers. Equation (1b) is obtained using the rules of old-fashioned perturbation theory, for a single time ordering in which the cascade occurs as one long sequence. (Further, the calculation is performed in the infinite momentum frame. Alternately, one may work directly using "light-cone quantization". I will use the languages of these treatments interchangeably where there is no real ambiguity.)

Now, if we further <u>order</u> the momentum flow as in the multiperipheral model, $\beta_j << \eta_j$, and Fourier transform to transverse configuration space, we obtain

$$\psi_{(n)}(t) \to \left[\prod_{j=1}^{n-1} K_0(m|\underset{\sim}{x}_{\perp,j} - \underset{\sim}{x}_{\perp,j+1}|) \right] F(\eta) \tag{2}$$

Here K_0 is a Bessel function, which dies exponentially for transverse separations on the order of the length associated with a parton mass.

Let us examine the important qualitative features of this result:

1) Near neighbors in η are nearby in $\perp$ configuration space. The second parton "orbits" about the first, the third "orbits" about the second, etc., with the net result that the whole configuration random walks out in transverse configuration space.

This simple calculation provides, therefore, a justification, albeit loose, for the notion that the hadron is a single long string, since indeed the transverse coordinates of neighboring elements on the string are tightly correlated.

2) The transverse and longitudinal momentum dependence of the wavefunction factorize. This is a dynamical result, strongly dependent on the trivial nature of the ϕ^3 coupling. It is nonetheless a useful notion to hang onto, since the longitudinal momentum can then serve purely as a label for the points along the string, and one can as well write the transverse coordinates $x_{\perp,j}$ as $x_{\perp}(\eta_j)$.

3) As already noted, these results are obtained from looking at a special graph under a special approximation. To "derive" stringlike behavior from any field theory, one must of course make some special approximations. The value of having a simple model is, equally obviously, that one can see clearly just what the nature of the special approximations is. In addition to the approximations above, one needs still further assumptions plus a "leap of faith" to finally arrive at the string. However, these further steps suggest themselves naturally from the qualitative picture we've been pursuing. They are that:

a) Since the near-neighbors in rapidity are close together in configuration space, the residual soft interactions needed for the system to produce a spectrum are short range, and in fact between nearest neighbors;

b) Since the graph that motivates the thing neglects virtual pair formation in any link along the chain (these processes are actuall down for the usual reasons in the $p_z \to \infty$ frame), and since the characteristic property of the partons that appears is the transverse coordinate, one can think of the $x_{\perp}(\eta_j)$ as the relevant dynamical variables in terms of which the residual dynamics can be described. We will work in first-quantization.

c) In the Regge region, it should be really the infinite "wee sea" that mediates the dynamics. We shall neglect, therefore, any

$[\Sigma_n P(n)]$, i.e., assume the parton configuration with an infinite number of partons is in some sense "dominant", so that a continuum approximation is possible. This is really a terrible assumption on two counts.

First of all, we seem to close the door on being able to go back and account sensibly for the "valence" configurations, where an infinite sea plays no role. Field theoretically, the wavefunction must be a sum over resolutions into configurations with all possible numbers of partons compatible with the overall quantum numbers of the hadron. Actually, the "conductive string" model suggests a way to deal with this situation, and I will discuss it further in Section III.

The second problem does not relate to whether we are doing sensible physics vis-a-vis field theory, but is a problem of internal consistency. It is that if we random-walk out in $x_\perp$-space with an infinite number of steps, we fill out all of the space, i.e., the hadron is infinitely big. Scaling the momentum from 0 to π, $[\theta/\pi = \int_0^\theta d\theta' P_\parallel(\theta')/\int_0^\pi d\theta' P_\parallel(\theta')]$, one finds $<(x_\perp(\pi) - x_\perp(0))^2>$ diverges logarithmically. Going backwards now, one naturally says any physical string is made out of constituents, and is not a true continuum. The paradox is that if we ask how fine-grained we should make our string so that the size of a "hadron" turns out to be the size of a hadron, we come up with a spatial cut-off so small as to be physically meaningless. That's why I say this is a problem of consistency.

Nevertheless, this problem is not troublesome for hadron-hadron-scattering. As I will remind you again later, scattering is involved with the overlap of the $P_\parallel$ ends of separate strings, and it is only when we really try to look inside the string that we find the inside is infinitely big. To discuss interactions with currents, for example, one has to subtract off $[<x_\perp^2(\pi)> + <x_\perp^2(0)>]$, each of which are also logarithmically divergent, to obtain finite results. Once this is done, the effective size of the hadron reduces to something like $[\ell n\ 2]$. This is just an aside, but again points to the limitations we must remain aware of.

d) The final thing we must do is make the dynamics stringlike. Using the null-plane Hamiltonian to describe the residual interactions, we have

$$H_{eff} \sim \sum_i \frac{p_{\perp,i}^2 + m^2}{2\eta_i} + (x_{\perp,i} - x_{\perp,i+1})^2$$

$$\to \int_0^\pi d\theta \left[\left(\frac{\partial x_\perp}{\partial \theta}\right)^2 + \left(\frac{\partial x_\perp}{\partial \tau}\right)^2 \right] + \text{(const.)} \qquad (3)$$

provided that the density $d\eta/d\theta$ = const. Only then do the (η_i) come out in the proper fashion to give a uniform string Hamiltonian in the continuum limit. It is gratifying that this requirement is also a mathematical property of the GGRT treatment of the string.

I stress once again that it is because of the rather precise manner in which the GGRT results fit the physical picture suggested by the parton model that I have dwelled on this model for so much time. Progress in attempting to understand hadrons as one dimensionally extended objects has not ended here, of course, and a lot of effort is currently going into incorporating more physics (such as hard, short range forces) into the structure of the theory from the very beginning. I prepared a set of (hopefully) pedagogic notes on the string model for the SLAC Summer Institute, entitled "The Beginner's String", in which references to the stimulating works of numerous authors may be found. I will not be able to go into details of these works here.

II. Lorentz Invariance

The parton model discussions may be helpful in providing some basis for insight into how a hadron can be a string, but in its mathematical formulation it hardly looks like it could be a Lorentz covariant theory. One logical possibility, the one initially explored by Nambu, is to complete the process of abstraction by postulating an action principle for the string dynamics that incorporates simultaneously the 1d extension and the Lorentz invariance of the system. The string is, after all, imbedded in the four-dimensional Minkowski space. Dr. Dragon has been lecturing on the consequences of this elegant postulate.

With the benefit of hindsight, however, we are now in a position to ask whether a set of ten Poincare generators for the dynamical string system could have been guessed if one had been very clever. The motivation for attempting to invent the generators rather than derive them is what is lost in elegance may be made up for in flexibility.

I will now discuss one route to guessing the desired generators, because this is the way we constructed them in the conductive string model.

Actually, the method is not really too much guesswork, since the approach was discussed in detail by Bacry and Chang, and by Bardakci and Halpern in their works on light-cone quantization. More recently, other relevant articles have appeared in Phys. Rev. by Biedenharn and van Dam, and by Staunton. Rather than trudging through the formal arguments, however, I would like to give you a simple mnemonic device which conveys the idea.

Recall one nice thing about the l.c. quantization is that the dynamics has a non-relativistic appearance to it. The Hamiltonian is $(p_\perp^2 + m^2/2\eta)$, two of the boosts are transverse Galilei boosts, a longitudinal boost is a scaling operation, etc. But simply using the metric $A_\mu B^\mu = A_+ B_- + A_- B_+ - A_\perp B_\perp$ does <u>not</u> give this simple structure to the Dirac system. The non-relativistic structure only emerges if we first decompose the Dirac field as $\psi = \psi_+ + \psi_-$, using projectors $(\gamma^\pm \gamma^\mp)$, and then observe that the Dirac equation for ψ_- involves only a "spatial" derivative. The components ψ_- are not canonical dynamical variables, but can be eliminated in favor of the true independent degrees of freedom ψ_+.

If, further, we prudently choose the Bjorken, Kogut, Soper representation for the γ^μ, we obtain their expression for ψ_+ (free),

$$\psi_+(x) \sim \sum_{\lambda=\pm 1/2} \int d^2 p_\perp \int \frac{d\eta}{\sqrt{\eta}}$$

$$\left[b(p_\perp, \eta; \lambda)\, e^{-ipx}\, \omega(\lambda) + d^\dagger(p_\perp, \eta; \lambda)\, e^{ipx}\, \omega(-\lambda)\right] , \qquad (4)$$

where $\omega(1/2) = \binom{1}{0}$, $\omega(-1/2) = \binom{0}{1}$. We can use two component spinors with no loss of generality.

I've gone into these elementary results to remind you that once ψ_- is eliminated, and Eq. (4) used for ψ_+, the ten Poincare generators of the free Dirac theory may be written as follows:

$$G \sim \int dx\, \psi^\dagger(x)\, g\psi(x) \qquad , \qquad (5)$$

where the first-quantized forms for the generators, g, are:

$$p_\perp = -i\partial_\perp, \qquad P^+ \equiv \eta, \qquad P^- \equiv H = \frac{p_\perp^2 + m^2}{2\eta} \quad ; \tag{6a}$$

$$K_3 = \frac{i}{2}\{\eta, \partial\eta\}; \tag{6b}$$

$$J_3 = \varepsilon_{ab}\, x_a\, p_b + \frac{\sigma_3}{2} \quad ; \tag{6c}$$

$$B_\perp = \eta x_\perp \quad ; \tag{6d}$$

$$S_k = \frac{1}{2}\{\, x_k, H \,\} - \frac{1}{2}\{\, \frac{1}{\eta}, K_3 \,\} P_k + \frac{\varepsilon_{k\ell}}{2\eta}\left[\frac{\sigma_3}{2} P_\ell - m\frac{\sigma_\ell}{2}\right] \tag{6e}$$

These generators obey the Poincare algebra under the first-quantization canonical commutation rules $[x_a, P_b] = i\delta_{ab}$.

For N free particles, we have $g(N) = \sum_i^N g_i$. It is convenient to use CM and relative coordinates, e.g.,

$$H_{1+2} = \frac{p_\perp^2}{2M} + \frac{\underline{\pi}^2 + m^2}{2\mu} .$$

If, to the two free particle terms, we add an interaction term between them, it is convenient to introduce a $(\mathrm{mass})^2$ operator in which the interaction is buried. What follows is simply a definition:

$$H = H_{1+2} + V_{12} \equiv \frac{p_\perp^2 + \mathcal{M}^2}{2M} , \tag{7}$$

with

$$\mathcal{M}^2 = 2M\, V_{12} + \frac{M}{\mu}(\underline{\pi}^2 + m^2)$$

Now, the mnemonic is quite simple, and consists of making the following replacements in the generators Eq. (6):

$$m^2 \text{ (parameter)} \to \mathcal{M}^2 \text{ (operator)} \tag{8a}$$

$$\frac{\sigma_i}{2} \text{ (Pauli)} \quad \to \quad j_i \text{ (operators)} \tag{8b}$$

Also $\eta \to M$; and $x_\perp$, $p_\perp$ are C.M. operators which commute with $\mathcal{M}^2$ and j_i. All the algebraic properties of Eq. (6) are to be preserved. Thus the j_i satisfy the spin algebra, and $\mathcal{M}^2$ must be a rotational scalar. The idea is that the mass, which is a parameter in an elementary particle theory, becomes an operator in a composite particle theory. Similarly the spin is not an intrinsic property, but arises from the dynamical configuration of the system. All of the information of the state of internal excitation of the particle is carried by $\mathcal{M}^2$ and the j_i, which are to be expressed in terms of some set of appropriate internal degrees of freedom of the system. In the string, these degrees of freedom are $x_\perp(\theta,\tau)$ and $p_\perp(\theta,\tau)$, or equivalently, their Fourier coefficients, the boson operators a_n and $a_n^\dagger$:

$$x_\perp(\theta,\tau) = x_\perp^{(0)} + p_\perp^{(0)}\tau + \sum_n \sqrt{\frac{2}{n}}\left[\cos n\theta \, (a_{n\perp}(\tau) + a_{n\perp}^\dagger(\tau))\right] ;$$

$$p_\perp(\theta,\tau) = \partial x_\perp/\partial\tau; \tag{9}$$

$$\left[a_{n_i}, a_{m_j}^\dagger\right] = \delta_{nm}\,\delta_{ij} .$$

As it turns out, this is not quite right for the string model. Following Gursey and Orfanidis, and Ramond, introduce operators which <u>transform</u> as $m\,\sigma^i$

$$T^i \sim \sqrt{\mathcal{M}^2}\,\frac{\sigma^i}{2} \text{ , i.e.,} \tag{10a}$$

$$[T^i, T^j] = i\,\varepsilon^{ij}\,\mathcal{M}^2\,J^3 ; \tag{10b}$$

$$[J^3, T^i] = i\,\varepsilon^{ij}\,T^j . \tag{10c}$$

The combination $(m\sigma^i)$ appears in the generators S_k, Eq. (6e), and from GGRT we learn that the structures that emerge in the string model in those generators has the algebra of the T^i. Dr. Dragon has discussed for us the difficulties with 26 dimensions and tachyons that arise from the structure of those generators.

III. The Conductive String

I have gone to some length to provide you with a non-formal background on strings, most of which is well known to specialists, because the conductive string model does not really follow from any pretty formalism, but rather arose as a tentative step away from the rather close-knit formal structure of string theory.

One specific mathematical motivation for this particular way to move away from the string model arises from asking why the canonical quantization procedure fails for this theory. Why is it that we run into these troubles with dimension and with tachyons? Perhaps one has not chosen the dynamical variables properly, for which quantization rules are to be prescribed. At the classical level, another choice for the variables suggests itself quite naturally. As Dr. Dragon has noted, the equations of motion become the string equations, $\ddot{x} = x''$, if the coordinate conditions

$$\left(\frac{\partial x^\mu}{\partial u^\pm}\right)^2 = 0, \tag{11}$$

with $u^\pm = \tau \pm \theta$, are imposed. That is, $(\partial x^\mu/\partial u^+)$ and $(\partial x^\mu/\partial u^-)$ are null vectors.

Now, any such null vector may be represented as

$$\partial x^\mu/\partial u^\pm = \xi_\pm^\dagger \, \sigma^\mu \, \xi_\pm \tag{12}$$

already at the classical level. We shall try, therefore, to take the pair of two component spinors $\xi_\pm$ as our basic dynamical variables.

The string equation of motion will then be satisfied if

$$\partial_{u_\pm} \xi_\mp = i \, B_\mp \, \xi_\mp \tag{13}$$

where $B_\mp$ are arbitrary Hermitian functions. However, by Eq. (12) the $\xi_\pm$ enjoyed a phase invariance under $\xi \to e^{i\lambda(\theta,\tau)}\xi$. This invariance can be preserved in Eq. (13) if the $B_\mp$ simultaneously transform as gauge fields,

$$\begin{aligned} &\xi \to e^{i\lambda}\,\xi \\ &B \to B - \partial\lambda \end{aligned} \tag{14}$$

We now depart from our strict adherence to the string model by treating $B_\pm$ as gauge fields. Specifically, this means that we will derive Eq. (13) from a new <u>effective</u> Lagrangian, and include kinetic energy terms for $B_\pm$ as well.

Before displaying this effective Lagrangian and plunging into the details of the spectrum, etc., I want to jump the gun a little and confirm your suspicion that we will be doing two-dimensional electrodynamics.

The point I want to make right now is that there are physical motivations for doing this. The argument regarding the choice of proper classical variables could not guide one into making $B_{\pm}$ gauge fields, but it is reasonable to try this nonetheless, for different reasons.

These physical motivations stem in part from work done by Nielsen and Olesen, who observed that the electrodynamics of scalar fields, cum Higgs mechanism, could give rise to filamentary solutions at the classical level. These filaments are analogous to trapped magnetic flux lines in a type II superconductor. Nambu has argued that if these flux lines terminate on (abelian) magnetic poles, the static, classical expression for the energy contains two pieces,

$$E \sim a\,L + b\ (\text{Yukawa}). \tag{15}$$

The first piece is proportional to the length of filament between the poles, and should represent something like the ground state energy of of the unexcited string. The second piece contributes for short wavelengths, and is desirable for producing power-law fall-offs in form factors.

In addition, we have heard Professor Susskind's lecture on the hadron-wurst picture he has been working on with J. Kogut. Following K. Wilson, one examines the current-current correlation function

$$\langle J_\mu(x)\ J_\nu(o)\rangle \sim \int DA\ D\psi\ D\psi^+\ J_\mu(x)\ J_\nu(o)\ \exp i \int d^4x\mathcal{L}(x;A,\psi,\psi^+).$$

The factor $[\exp i \int dx_\mu\ A^\mu]$ leads one to believe that if the q and $\bar{q}$ in a loop exchange photons in a fairly uniform manner,

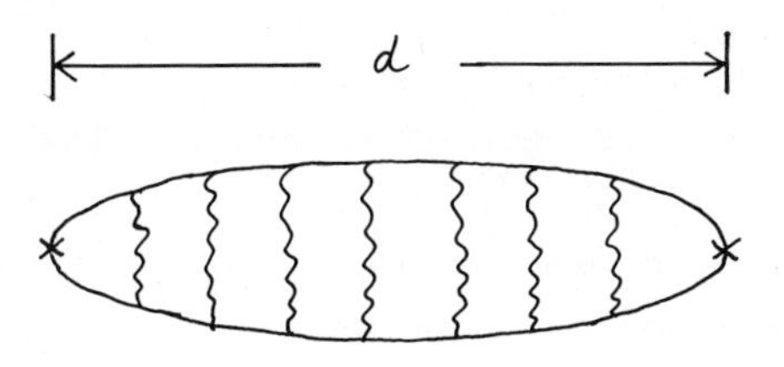

Figure 3

the contribution to the action will go as $[2d\langle A\rangle]$, where (2d) is the perimeter of the loop.

But if for some reason the exchanges do not work out this way, but rather conspire to make the effective action proportional to the <u>area</u>, something interesting occurs. Let me try to give a crude argument for how this works.

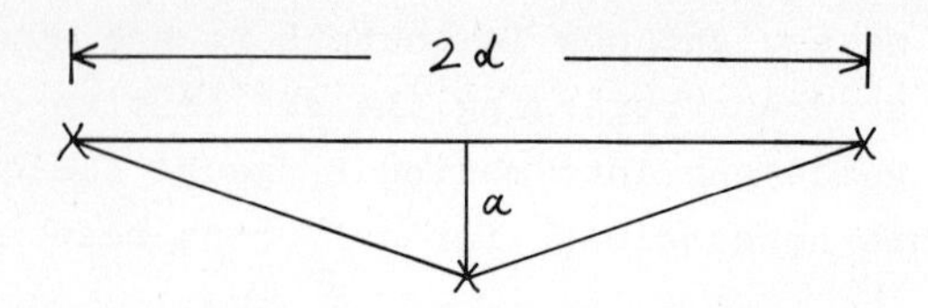

Figure 4

First, if the action $I \sim d + \sqrt{d^2 + a^2}$, clearly $dI/da \sim a/\sqrt{d^2 + a^2} \sim a/d$. It does not cost much to separate the quarks further and further apart ($d \to \infty$). On the other hand, if $I \sim da$, $dI/da \sim d$, and we lose a great deal. This was why Wilson wanted to get the action to go like the area.

In addition, however, in a static situation $L = -H$, and so $I \propto E$, whence $E \propto ad$. Here "a" is, in some frame, the spatial separation between the members of a pair created at the origin and moving toward the point x. So again we have a situation where the potential energy grows linearly with the separation of the pair.

There are, to summarize, various ways in which conventional field theories may support solutions in which the effective interaction is just like the interaction in two-dimensional electrodynamics, or perhaps like its non-Abelian brother. One may view the conductive string either as an abstraction from such models, on the same footing as deciding on harmonic forces between partons in Bjorken's illustrative model or as an approximation to the full field theory which may be appropriate for studying a special class of properties of the hadron. Let me now continue with the main line of development of the model.

We have

$$\mathcal{L} = i\left[\bar{\psi}\Gamma^a \; (\partial_a - ig\, B_a)\psi\right] - \frac{1}{4}\, F_{ab}\, F^{ab}, \tag{16}$$

with "a" and "b" running over 0,1. The spinor ψ has four components, consisting of the two components each of ξ_+ and ξ_-. The 4x4 matrices Γ^a satisfy the algebra $\{\Gamma^a,\Gamma^b\} = 2\eta^{ab}$, with $\eta^{oo} = -\eta^{11} = 1, \eta^{o1} = 0$. In terms of the usual Dirac matrices, we may write $\Gamma^0 = i\gamma^o\gamma^5, \Gamma^1 = i\gamma^5$. As usual, $F_{ab} = \partial_b B_a - \partial_a B_b$, where the two components B_a are linear combinations of B_+ and B_-. It is easily seen that with these choices, variation of $\mathcal{L}$ gives the Eqs. (13), plus the Maxwell equations for B.

Working in analogy with the string model, we confine our spatial domain to $[0,\pi]$, and impose boundary conditions on ψ such that $(\partial x^\mu/\partial\theta)$ vanishes at the boundaries, using Eq. (12). These boundary conditions also lead to the conditions $j'(0) = j'(\pi) = (\partial j^o/\partial\theta)|_o = (\partial j^o/\partial\theta)|_\pi = 0$, where the currents

$$j^a \equiv \bar{\psi}\Gamma^a\psi. \tag{17}$$

For our purposes we are not interested in the Green's functions of the theory, but rather in the physical spectrum of excitations supported by the system. As is well known, TDQED has no genuine radiation field, and in the gauge $B_1 = 0$, the timelike field can be solved for in terms of the charge density,

$$B_0 = - \frac{g}{2} \int_0^{\pi} d\theta' |\theta-\theta'| \, j^0(\theta',\tau). \tag{18}$$

Forming the Hamiltonian, then, we have

$$H = -i \int d\theta \bar{\psi} \Gamma^1 \partial_1 \psi - \frac{g^2}{4} \int\int d\theta \, d\theta' \, j^0(\theta,\tau) |\theta-\theta'| \, j^0(\theta',\tau). \tag{19}$$

The idea is to diagonalize this Hamiltonian, and display the energy eigenstates. Our task is somewhat simplified by the consistency condition on TDQED first discussed by Zumino, which says that all the physical states of the system must be neutral.

To perform the diagonalization, it is useful to introduce a set of coupled fermion operators which satisfy Bose commutation relations. These "plasmons" are the Fourier components of the vector current,

$$\rho(p) = \frac{1}{\sqrt{2p}} \int_0^{\pi} d\theta : [j_0 \cos p\theta + i j_1 \sin p\theta] : \tag{20}$$

It is easily seen that the δ' Schwinger term in the equal time j_0, j_1 commutator provides

$$[\rho(p), \rho^+(q)] = \delta_{pq}. \tag{21}$$

Also, $[\rho(p),Q] = 0$, so acting with plasmons does not destroy the neutrality of a state. Inverting Eq. (20) for j_0 and inserting into (19), one obtains

$$H = H_0 + \frac{\mu^2}{4} \sum_{n=1}^{\infty} \frac{1}{n} [2\rho_n^+ \rho_n + \rho_n \rho_n + \rho_n^+ \rho_n^+], \tag{22}$$

where $\mu^2 \equiv 2g^2/\pi$. In this form, it is straightforward to diagonalize H by means of a Bogoliubov transformation. The details are presented in SLAC-PUB-1418, to be published in Phys. Rev., and I will give only the results:

1) The ground state is

$$|\Omega\rangle = \exp [- \sum_{n=1}^{\infty} (\tfrac{1}{2} \tanh^{-1} \frac{\mu^2}{\mu^2+2k^2})(\rho_k^+ \rho_k^+ - \rho_k \rho_k)] |0\rangle \tag{23}$$

$$\equiv e^{-iS} |0\rangle.$$

This state clearly contains indefinite numbers of quarks and anti-quarks, and it is because of this that it acts like a "conductive" medium, any excess test charge being screened in the interior and only reappearing on the boundary.

Notice the $q\bar{q}$'s do not pair off to make bosons localized in space. Rather, there are correlated pairs of fixed total momentum. The ground state $|\Omega\rangle$ has a finite negative definite energy with respect to the Fock vacuum $|0\rangle$,

$$\varepsilon_0 = \frac{1}{2} \sum_n [\varepsilon_n - n - \frac{\mu^2}{2n}], \tag{24}$$

with

$$\varepsilon_n = \sqrt{\mu^2 + n^2}. \tag{25}$$

The fact this energy is negative reflects, of course, that the correlated state is favored over the no-particle state.

We can recognize in this result a possible answer to the problem posed earlier as to whether a string is really only the $N = \infty$ parton configuration in the wave function. In this type of model, where correlations are extremely important, the stationary states will project onto all possible (neutral) bare parton states.

2) The presence of doubled spinors allows for the presence of neutral "filled Fermi sea" states in the spectrum. These are of the form

$$|F\rangle = \prod_{n=1}^{F} b_i{}^+(n)\, c_j{}^+(n)|0\rangle, \quad (i \neq j), \tag{26}$$

where b^+ and c^+ are particle-antiparticle creation operators. The essential property of these states is that

$$\rho(n) \quad |F\rangle = 0, \text{ (all } n).$$

Then

$$H(e^{-is}|F\rangle) = \varepsilon_F(e^{-is}|F\rangle),$$

where

$$\varepsilon_F = 2 \sum_{n=1}^{F} (n - \frac{1}{2}) = F^2$$

comes entirely from the non-interacting part of the Hamiltonian. (We are now shifting the energy by ε_0 so $|\Omega\rangle$ has zero energy.)

3) Plasmons may be added onto the filled sea states,

$$|N_p, P; F\rangle = \prod_{m=1}^{P} \frac{[\rho^+(m)]^{N_m}}{\sqrt{N_m!}} |F\rangle. \qquad (27)$$

These states are also energy eigenstates, the plasmons contributing as massive bosons, by Eq. (25),

$$H[e^{-is}|N_p, P; F\rangle] = (\varepsilon_F + \varepsilon_p)[e^{-is}|N_p, P; F\rangle],$$

$$\varepsilon_p = \sum_{m=1}^{p} N_m \sqrt{m^2 + \mu^2}.$$

In this model, then, we see clearly that the states of excitation consist of collective excitations of the constituent fermions.

To complete the story, we have to construct the Poincare generators, and discuss how physical states transform in the full four-dimensional Minkowski space. We can do this with the machinery erected in Section II, in two steps.

First, we can identify the dynamical $(\text{mass})^2$ operator to be some function of the Hamiltonian of our theory above. It turns out by a simple counting argument that the Hagedorn degeneracy is reproduced if we choose

$$\mathcal{M}^2 = H_{TDQED} \qquad (28)$$

We will need center of mass four-momenta for our particles, subject to the mass-shell conditions

$$P_\mu P^\mu = \mathcal{M}^2. \qquad (29)$$

This of course relates the "true" Hamiltonian P^- to the spectrum of internal excitations, as in Eq. (7).

The state of a single free particle in motion may then be labelled

$$|k^+, \underset{\sim}{k}^\perp, \mathcal{M}^2, J, \lambda\rangle = e^{ikx}|N_p, P; F\rangle. \qquad (30)$$

The second stage in our construction is to interpret the spin and helicity labels J and λ by producing operators j_i. This can be done very naturally in our model, because the TDQED with four component spinors enjoys an extra SU(2) symmetry generated by

$$j^k = \frac{1}{2} \int d\theta : \psi^+ \begin{pmatrix} \sigma^k & 0 \\ 0 & \sigma^k \end{pmatrix} \psi :.$$

One then finds that the plasmons are Lorentz scalar excitations,

$$[j^i, \rho(m)] = 0, \qquad (31a)$$

while

$$j^3 | F> = (\pm)F \mid F >.$$

Only the "filled sea" states carry spin, and by means of the ladder operators $(j^1 \pm ij^2)$ one can complete the multiplets of spin F. Said differently, one easily checks that

$$W_\mu W^\mu |F\rangle = m^2 F(F+1)|F\rangle ,$$

where W_μ is the Pauli-Lubanski vector.

The net result of all this is that the model describes a system of parabolic trajectories, $J = \sqrt{m^2}$, with sea states providing the leading trajectory. The plasmons then shift this trajectory to the right to form an infinite family of particles. Unfortunately, this spectrum does not appear to be particularly realistic, although we are always in four dimensions and we have no tachyon and no ghosts. It has to be admitted that our Lorentz generators are constructed in an ad hoc fashion, and other ways of proceeding may exist that would lead to a different trajectory structure. For the present, however, I have no light to shed on this question.

IV. Speculations

The great virtue of the string model, as compared with other theories of composite hadrons, is that scattering amplitudes already exist, and a great deal is known about the structure of these amplitudes. Especially in the last year, a lot of progress has been made in formulating the theory of interacting strings, both as a particle theory in the sense of Feynman path integrals, and as an interacting multilocal field theory.

While I think that the conductive string model illustrates many good features that a more realistic model of this genre should possess, it remains to be seen whether enough can be learned from the string model to be able to discuss interactions of conductive strings. These closing remarks are speculations on approaches to this problem.

To describe string-string interactions as a second quantized field theory, Kaku and Kikkawa have introduced master fields describing entire strings, which are functionals of the first quantized coordinates $x_\perp(\theta,\tau)$,

$$\psi = \psi[x_\perp(\theta,\tau)] \ .$$

Now, in addition to the original string model, there are other string models in which the constituents are endowed with intrinsic spins, the Neveu-Schwarz and Ramond models. The master field would then have to be a functional of two sets of "fields" on the two-dimensional submanifold,

$$\psi = \psi[x_\perp(\theta,\tau),\ \phi(\theta,\tau)]$$

where $\phi(\theta,\tau)$ is a spinor field describing the spin excitations.

Structures like this are reminiscent of the supergauge fields that Professor Wess told us about, with $x_\perp$ and ϕ playing the role of the gauge parameters. Here these fields themselves obey free equations of motion, Klein-Gordon and Dirac respectively. What we would want for our conductive string model is to extend this even further and allow the functional arguments of the master field to be interacting quantities. The speculation consists of the conjecture that if we hold on to the guiding principle of gauge invariance, the consistent formulation of such a theory will already contain the allowed forms of the master field interactions. This hope is bolstered by the observation that there are, in fact, residual vortex-vortex interactions in type II superconductors, whose form is determined from the original Landau-Ginzburg equations.

In any case, the study of multilocal field systems is just beginning, and it may be valuable to investigate master fields with quite general field arguments, $\psi = \psi(x,\phi_j(x))$, where "j" is any space-time and/or internal symmetry index, and x are the Minkowski space coordinates, in addition to the forms suggested by the string models. Such master fields would represent "particles" whose constituents' dynamics is itself field-theoretic in nature. One intuitively expects the particles' dynamics to follow from the form of the constituents' dynamics, and an interesting problem is how much of this can be deduced from general considerations such as gauge invariance.

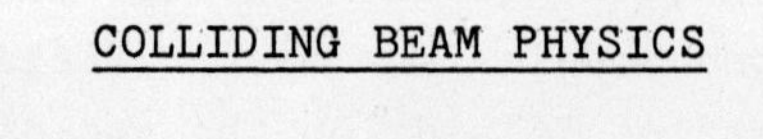

HADRON PHYSICS AT ISR ENERGIES

M. Jacob

CERN, Geneva, Switzerland

INTRODUCTION

One year ago I gave a series of lectures on the same topic at the Louvain Summer Institute of Theoretical Physics. There is no point in repeating what was discussed then and the more so because new developments provide quite enough material for a new series of lectures. Minimizing overlap with what was discussed one year ago, I would therefore advise the reader not already familiar with physics at the ISR to first read the Louvain lecture notes[1)] or the relevant rapporteur's talks at the Aix-en-Provence Conference. This should provide him with a detailed introduction to the new points discussed here and also with references to papers where the important discoveries made so far with this unique instrument were first reported. I shall merely recall that the range of energies covered by the ISR corresponds to the 250-2000 GeV domain, when translated into terms of standard machine energy with a stationary target. There is, therefore, a sizeable overlap with NAL, which currently works at energies ranging up to 400 GeV. Indeed, many of the topical questions to be discussed here borrow from results at NAL as well as from results at the ISR. There is an obvious complementary role for the two machines (and later on the SPS) to play. The ISR reaches much higher energies. Nevertheless, its two main weaknesses as compared to NAL (SPS) are, firstly, the lack of variety in the type of reactions which can be studied[2)] -- one is limited to proton-proton reactions -- and, secondly, the impossibility to actually look at the reaction vertex. The vacuum pipe generates spurious secondaries, which are difficult to eliminate. For these reasons, the detailed exploration of phenomena first discovered at the ISR will probably soon shift to NAL (SPS), the ISR quickly losing competitivity. In particular, this is the case whenever the observed effect shows only a logarithmic energy behaviour, which is a very common property for hadronic reactions. It remains true, however, that some processes -- this is for instance the case for large transverse momentum phenomena and lepton (pair?) production -- have a marked energy dependence over the ISR energy range.

The corresponding yields rise by an order of magnitude as opposed to a gain by a factor of 1.5 for the available rapidity range (a typical logarithmic effect). Their study should call for an extensive research programme at the ISR, irrespective of progress at NAL (SPS) and this for at least a few years

At present, research at the ISR has two main aspects:

i) Search for as yet unobserved effects which experimentation at very high energy could detect. Classified in such a category are the new particle search experiment in I4 and the monopole and the multigamma event research experiment in I1. Also classified in this category are the two approved major projects for lepton pair search, namely the CERN-Columbia-Rockefeller solenoid and the MIT-Pisa muon pair detector. In all cases, thresholds could be such that, at present, the ISR could be the only instrument able to detect as yet unknown effects. One may soon have to add search for charmed particles to the above list.

ii) The further exploration of hadronic phenomena which are already known. In most cases, the specific effects under study were first discovered at the ISR. One should then quote the following:
   a) the rising total cross-section;
   b) evidence for scaling properties in hadronic production (at the single-particle and two-particle levels);
   c) evidence for an important single-diffractive excitation mechanism, extending up to high masses;
   d) evidence for prominent and specific effects associated with the observation of a large transverse momentum particle.

This last point is probably the most exciting topic in hadron physics at present, to the extent that one could thus see effects similar to those observed in deep inelastic electron scattering at SLAC, with, at the origin, an effective point-like structure within the proton.

The properties listed under these four points are all very important discoveries which make present understanding of hadronic phenomena very different from what it was before the ISR started. It should be stressed, however, that these discoveries could have all been made at NAL had the ISR not been as good a machine as it is, or had experimentation at the ISR not been as active and successful as it has been. This point should now have important implications, carrying options for the future research programme at the ISR. Competition with NAL has to be kept in mind. As already mentioned, there are, however, specific points for which the detailed study of what happens between 400 and 2000 GeV appears as of paramount importance. At present, it is in particular the case for the analysis of large transverse momentum phenomena, which show very important variations with increasing energy. As mentioned earlier, these variations are very much stronger than the logarithmic behaviour which seems to prevail in most other cases, such as those listed under (a), (b) and (c) above. It remains that in all cases co-operation with research at NAL (SPS) is of great importance for the further exploration of all the discovered phenomena. As discussed later, this is already particularly the case for the study of rising cross-sections. Confirmation from NAL of predictions based on ISR results (the vanishing of the real part of the forward elastic scattering amplitude, for instance) has been among the important news of the past year.

Looking back over the past year, one can say that what refers to (i) has been at the setting up, or conception stage. In particular, preparation for a lepton pair search programme, which includes the analysis of the associated hadrons, is now going ahead. It should make 1976 (let us hope) a very important year for research at the ISR. With respect to (ii), the past year has also been primarily a setting-up period for sophisticated devices which now allow correlation studies on all intersections[3]. The setting up of such extensive and sophisticated devices has taken time. As a result, the analysis of the many pieces of data which could already be collected has hardly started. This explains why, physics-wise, the past year has not brought up as many new results as the preceding one. Yet there is much to be mentioned. This is dicussed in the following, where we take in turn the list of topics listed under (ii). In all cases, we bring together results from the ISR and results from NAL, whenever they pertain to the same physics.

2. THE TOTAL CROSS-SECTIONS

The situation, as it was at the time of the Aix-en-Provence Conference, is displayed in Fig. 1. The ISR results, showing the rise of the proton-proton total cross-section over the 250-2000 GeV range, are almost as they were early in 1973[1].

The situation as it is now, after the recent NAL results, is shown in Fig. 2. The proton-proton cross-section already starts to rise

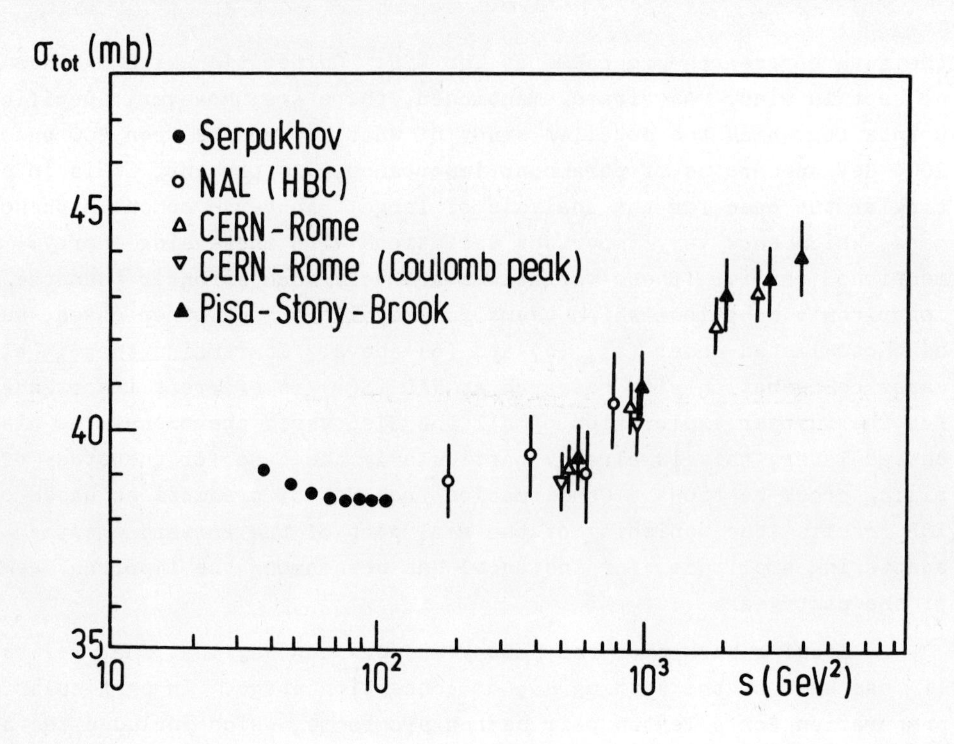

Fig. 1 Proton-proton total cross-section. The situation as it was one year ago.

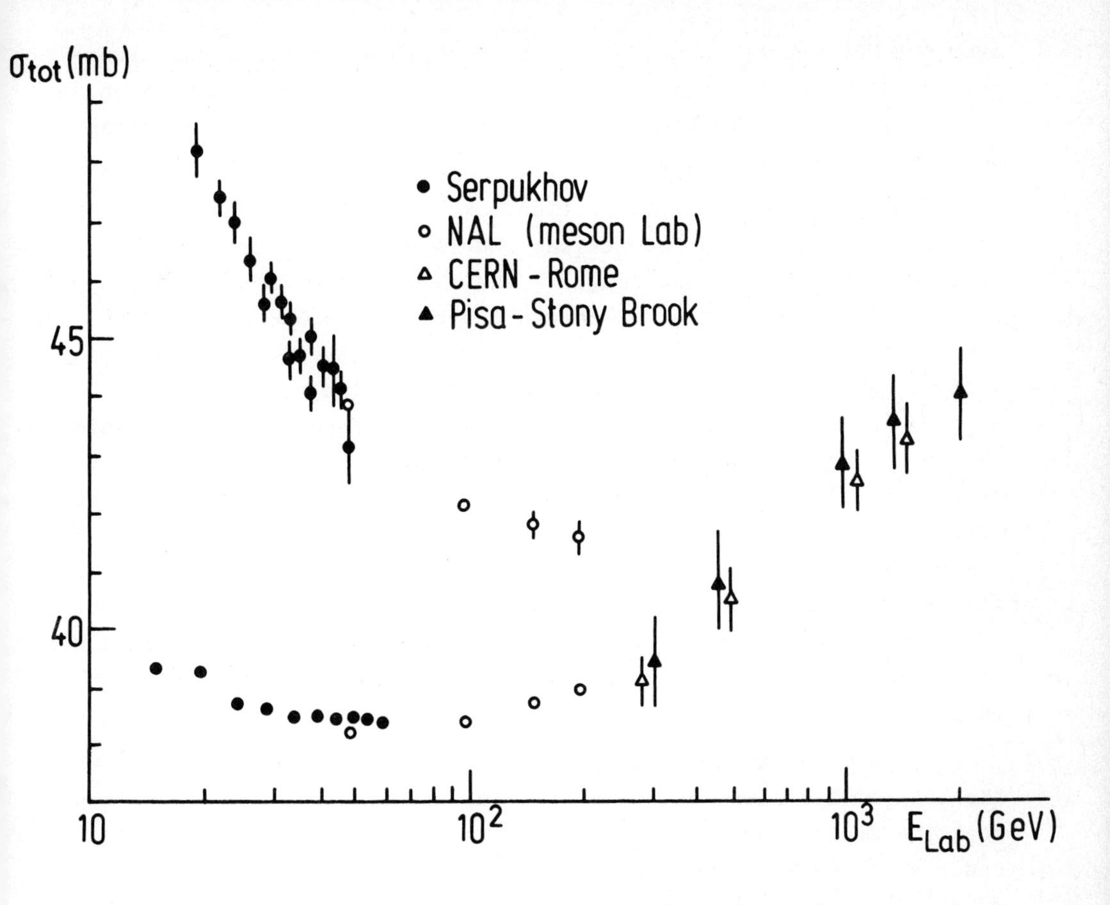

Fig. 2 The situation at present. The NAL results confirm the rise of σ_{tot} for pp and are compatible with the presence of a minimum of σ_{tot} for $\bar{p}p$.

over the NAL energy range, while the antiproton-proton total cross-section appears to level off. This is not an isolated phenomenon, as shown for instance by the behaviour of the K^+p and K^-p total cross-sections, shown in Fig. 3.

A rising total cross-section came somewhat as a surprise. This is now a solid fact. Theoretical models did allow for a rising cross-section. This was, in particular, the case in the Cheng and Wu approach where an input amplitude, which would violate the Froissart bound, is limited so as to stay within the proper unitarity limits. As a result, a $\log^2 s$ behaviour is obtained. This was also the case in the Gribov Reggeon calculus, where the driving term (Pomeranchon pole) is shielded at non-asymptotic energies (Pomeranchon cuts). As a result, one obtains a slow (logarithmic) approach of the limiting value of the total cross-section. However, as dicussed in detail elsewhere[4)], the observed rise is either too modest, or too big, to fit naturally within one of the two aforementioned schemes. With respect to the behaviour of the total cross-section, the ISR energy range appears as a huge transition domain where no sign of any simple asymptotic behaviour yet appears.

One is, however, not limited to such evasive statements. When the ISR results first became available, it was still natural to expect that the scattering amplitude, which had then to be asymptotically even under crossing, should be becoming even (to a good approximation) already over the ISR energy range. In such a limit, the forward elastic amplitude at centre-of-mass energy squared $s + i\varepsilon$ (the physical amplitude) is set equal to the complex conjugate of its value at $-s + i\varepsilon$ and then also equal to the $\bar{p}p$ elastic amplitude at the same energy. A logarithmic rise of the cross-section, and therefore of the imaginary part of the forward elastic amplitude (which is the fastest one tolerable asymptotically), has then important and simple consequences since, as one goes from $s + i\varepsilon$ to $-s + i\varepsilon$, $\log s$ whenever present in the amplitude, goes into $\log s + i\pi$, which is different from $(\log s)^*$. An easy trick to maintain an even amplitude is therefore to replace $\log s$ by $\log s - (i\pi/2)$, whenever it would appear in an expression for the elastic amplitude. A candidate amplitude for a total cross-section rising as $\log^\alpha s$ should then be is $[\log s - (i\pi/2)]^\alpha$ instead of is $\log^\alpha s$, which, through the optical theorem, would _a priori_ most simply give the required behaviour.

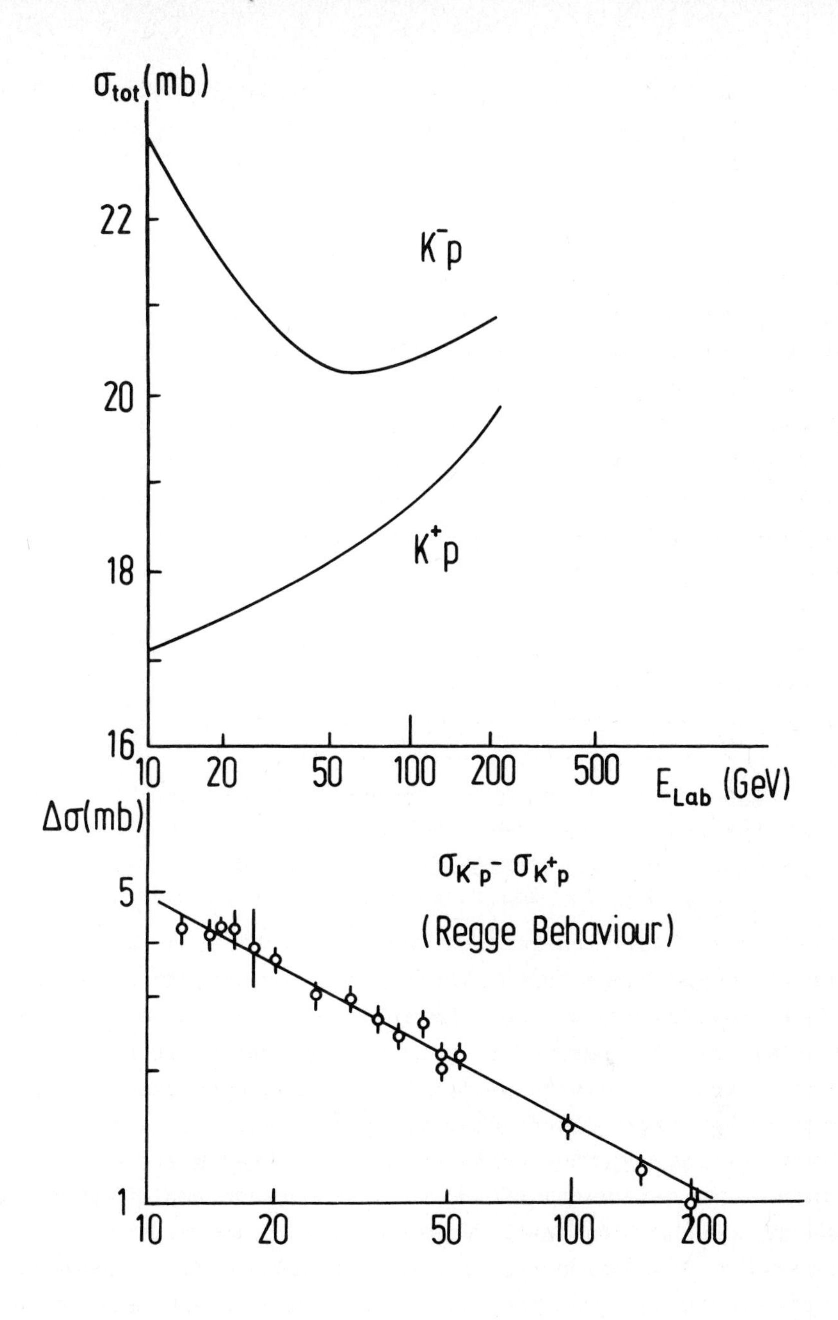

Fig. 3 The rising cross-section in the Kp channel. The rise of σ_{tot} for K^+p and the minimum for K^-p are found at a lower energy. The difference has a simple Regge behaviour.

In the complex energy plane we have the following relation between different limits:

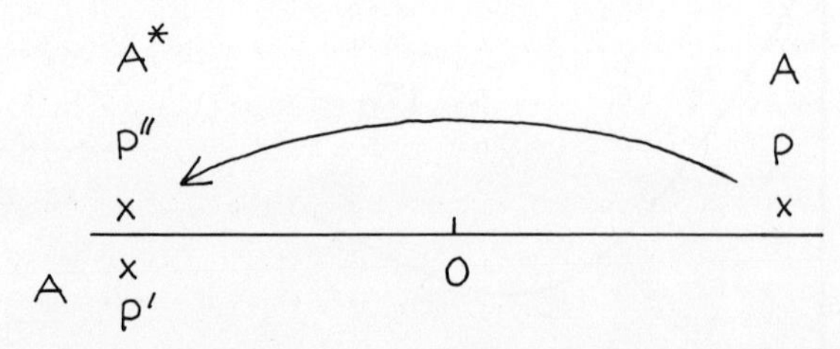

An even amplitude A is the same at points P and P'. The value at P'' (symmetric to P' across the unitarity cut) is equal to A^*. One reaches P'' from P through a path drawn in the upper half plane. The argument of A at <u>P</u>, i.e. s, is therefore multiplied by $e^{i\pi}$ in order to correspond to <u>P</u>''. With

$$A = i\beta s\left(\log s - \frac{i\pi}{2}\right)^{\alpha} \qquad (1)$$

$$\sigma_{tot} = \frac{\text{Im}\, A}{s}$$

One has $\sigma_{tot} \sim (\log s)^{\alpha}$ asymptotically, but it follows that the ratio ρ of the real part to the imaginary part of the elastic forward amplitude should eventually be <u>positive</u> and decrease only as $(\log s)^{-1}$, and this, as soon as an even amplitude becomes a good approximation.

This was very interesting in view of the fact that, over the Serpukhov energy range, ρ was known to be negative and decreasing as $s^{-1/2}$, a behaviour most simply described in terms of secondary-Regge contributions. The parameter ρ had then to vanish in between its known Serpukhov values and its expected ISR values. The exact location of the zero of ρ has, however, also much to do with the assumed respective behaviour of the pp and $\bar{p}p$ total cross-sections, which measures how the imaginary part (at least) of the elastic forward amplitude becomes eventually even. "Reasonable" assumptions led to expect (through a then standard dispersion relation calculation) that ρ should vanish through the NAL energy range but that, at the same time, the $\bar{p}p$ total cross-section should sharply level off at NAL energies in order not to fall below the value of the pp total cross-section.

The same simplifying assumptions were thus implying, both the vanishing of ρ (while becoming clearly positive) and the minimum of σ_{tot} for $\bar{p}p$ scattering. The odd part of the forward amplitude was accordingly expected to continue its power law (Regge) fall with increasing energy. This corresponds to the linear behaviour for $(\sigma_{\bar{A}A} - \sigma_{AA})$ as a function of log s shown in fig. 3. All points have now been verified. Evidence for the latter two is displayed in Figs. 2 and 3. In order to understand the fall of σ_{tot} for pp over the Serpukhov energy range, it was necessary to call for non-vanishing Regge contributions, irrespective of the exotic quantum numbers of the corresponding channel. It was, however, expected that for the Kp channel, which is the showcase for exoticity[5)], the corresponding behaviour, namely the rise of the K^+p cross-section and the minimum of the K^-p cross-section, should be seen at lower energies. This is now verified (Fig. 3).

The vanishing of ρ has also been ascertained[6)] as shown in Fig. 4. Measurements of ρ at the ISR (22 and 30 GeV) were indeed compatible with such behaviour. Figure 4 shows the ISR points together with the new NAL results (APS Chicago meeting). The expected behaviour is beautifully confirmed. The NAL results on $\bar{p}p$ (Fig. 2) also show that, as expected, the vanishing of ρ has actually to do with the rising pp cross-section and not with a peculiar relative behaviour of the pp and $\bar{p}p$ cross-sections. The observed rise of the cross-section is now associated with the even diffractive part of the amplitude. It is a feature common to all particles, while differences between particle-particle and particle-antiparticle total cross-sections indeed appear to decrease very fast with increasing energy (inverse power).

The situation about the correlated behaviour of the slope parameter B(s) has not changed since one year ago[1)]. The success met with the behaviour of ρ(s) should not undermine the fact that the rise of the cross-section is, however, far from being understood. As was already stressed, as soon as the ISR results became available, the effect is far too small as compared to what to expect if the proton were to become opaque at such high energies. (It is a hundred times smaller than what is possible within a range defined by the pion Compton wavelength). On the other hand, it is far too large to fit easily into a scheme where, despite the observed rise, we would be near an asymptotic regime with eventually a constant cross-section. With log s as the key variable, the asymptotic cross-section would have to be at least 1.5 times larger than the now observed value. We are still where we were a year ago[1)]. Asymptopia is but an elusive concept. This in itself is, however, a very important finding at the ISR.

It is probable that information on a variety of channels will be more efficient at probing models than the knowledge of what occurs for pp only, even if it is over a wider energy range and, the more so,

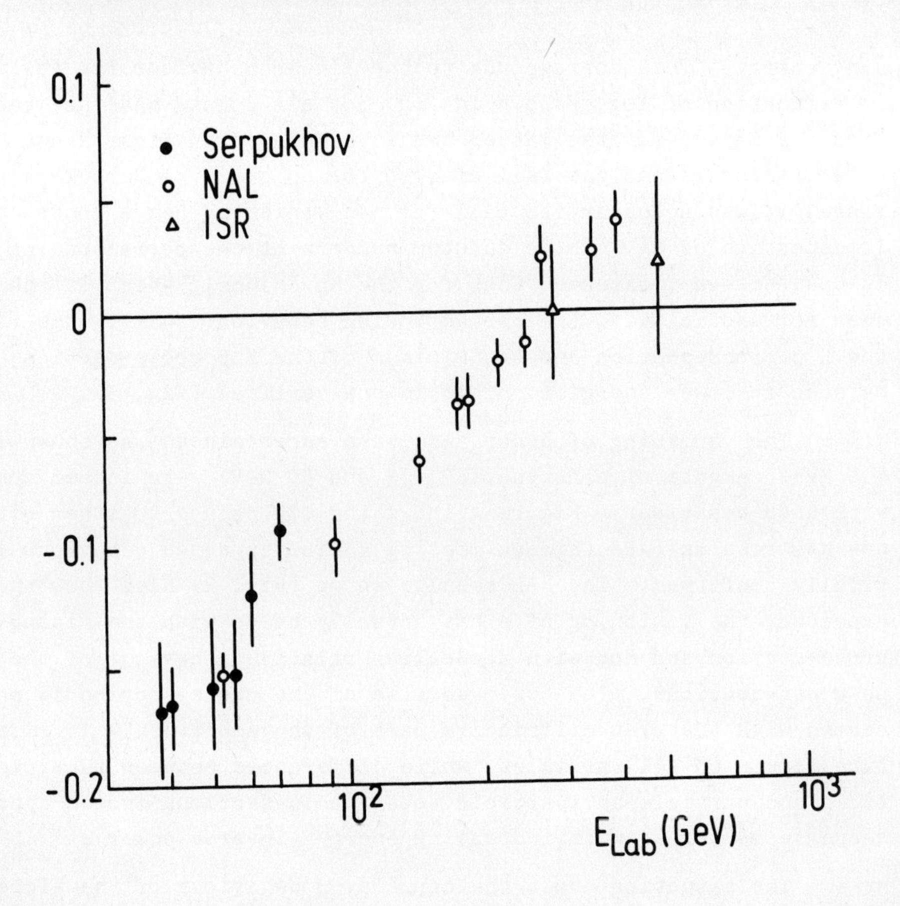

Fig. 4 The behaviour of ρ (s). It is expected to have a very wide maximum, with a value of 0.05 to 0.15, depending on hypothesis abaout the detailed behaviour of σ_{tot}, and then to decrease very slowly towards zero, i.e. as $(\log s)^{-1}$. Measuring ρ up to 2000 GeV will be attempted next year by the CERN-Rome Collaboration.

since the behaviour of σ_{pp} is already known rather well. Nevertheless, only the ISR can provide two further important contributions. The first one is a better measurement of σ_{tot}, which should be achieved through a combination of the global counting (Pisa-Stony Brook) and optical theorem (CERN-Rome) approaches, thus making an independent measurement of the luminosity superfluous. This should bring down present error bars by a significant amount. The second one is a measurement of ρ over the full ISR energy range, its measurement presently stopping at 30 GeV with large error bars. As previously discussed, it is predicted that ρ should rise to a very wide maximum. Checking this qualitative prediction is important. Quantitatively speaking, whether it rises to 5% or to 10% (typical predicted values at 2000 GeV) will help in contriving models for the asymptotic behaviour of σ_{pp} and $\sigma_{p\bar{p}}$. The first programme is under way (CERN-Rome-Pisa-Stony Brook). The second one is at present the object of a proposal (CERN-Rome).

3. THE SHAPE OF THE PROTON

The modest rise of σ_{pp}, as compared to what could be allowed by unitarity, shows something which is still far from a black disc. The combined study of σ_{tot} and of the differential elastic cross-section indeed suggests an object which, if rather dark at the centre, becomes quickly grey as one moves away from it, with a Gaussian-like absorption factor extending over a zone 0.9 fm in radius. The observed rise of the cross-section is then mainly due to a rather modest increase in opacity on the outer side. This is discussed in some detail below. The experimental information has not changed appreciably over the past year. The Aachen-CERN-Genova-Harvard-Torino Collaboration had then data showing a beautiful diffractive pattern with the elastic cross-section dropping almost exponentially by over 6 orders of magnitude to a dip followed by a secondary maximum. The dip is located at $|t| \simeq 1.4\ (\text{GeV/c})^2$. The analysis of the corresponding data, however, developed from these data and those of the CERN-Rome Collaboration. At present, the CHOV (CERN-Hamburg-Orsay-Vienna) Collaboration is obtaining data on elastic scattering at the Split Field Magnet. These data already show that the dip moves towards lower $|t|$ values with increasing energy, as if the proton would "swell" with increasing energy without changing its shape appreciably. When discussing the shape of the proton, it can be convincingly argued that the elastic scattering amplitude should be mainly imaginary. One may then <u>define</u> the proton shape as the Fourier transform of the observed diffraction pattern[4).

An exponential differential cross-section with slope B(s) thus gives a Gaussian impact parameter profile, namely:

$$A(s,t) = i s \sigma_{tot} e^{-\frac{B|t|}{2}} \sim a(r,s) = \frac{i}{8\pi} \sigma_{tot} \frac{e^{-\frac{r^2}{2B}}}{B} \qquad (2)$$

A typical slope of 10 $(GeV/c)^{-2}$ thus corresponds to a radius $\sqrt{2B}$ of 0.9 fm. The presence of the dip and of the second maximum (they are at the 10^{-6} level!) gives a very small flattening of the primarily Gaussian shape around r = 0. The presence of the steeper slope at low $|t|$ [$|t| < 0.1\ (GeV/c^2)$] gives some widening in the edge from the same primarily Gaussian profile. With measured values of σ and B, one gets $a(0) \approx 0.36i$ as opposed to 0.5i, which full absorption (unitarity) would impose. The proton then appears as a grey object. One may, however, take a different definition for the proton shape, taking instead of a(r) the inelastic cross-section at impact parameter r. With a purely imaginary amplitude, one has

$$H(r) = \frac{1}{2\pi r} \frac{d\sigma_{in}(r)}{dr} = 4|a(r)|\left(1-|a(r)|\right) \qquad (3)$$

Pure absorption now corresponds to H(r) = 1. With a(0) = 0.36i, one gets H(0) = 0.92. The proton may thus look darker at the centre; yet substituting eq. (1) into eq. (2), one quickly gets a Gaussian fall-off and basically a soft grey object again. The difference in the appreciation of darkness at the centre, which is reached at first sight from either relation, stems from the fact that H(r) and a(r) depend, respectively, quadratically and linearly, upon the inelasticity parameter, which is small, but not zero (as full absorption would impose). Getting full absorption would mean a sizeable relative rise of the contribution of the central region to elastic scattering, while its contribution to inelastic scattering (its absorbing power) would rise by less than 10%. Using eqs. (2) and (3) for different energies (different values of σ and B) one may indeed locate the rise of σ_{tot} in the outer edge (peripheral region). It is very small as compared to reaching H(r) = 1 there would imply. The relative transparency of the proton, even at such energies, appears as a prominent (and puzzling) fact. This is illustrated in Fig. 5. The change with energy is shown in Fig. 6 [6,7].

Here again model testing should be most efficient when information from a variety of channels is at hand (differential cross-sections from NAL). Yet a better knowledge of the differential cross-sec-

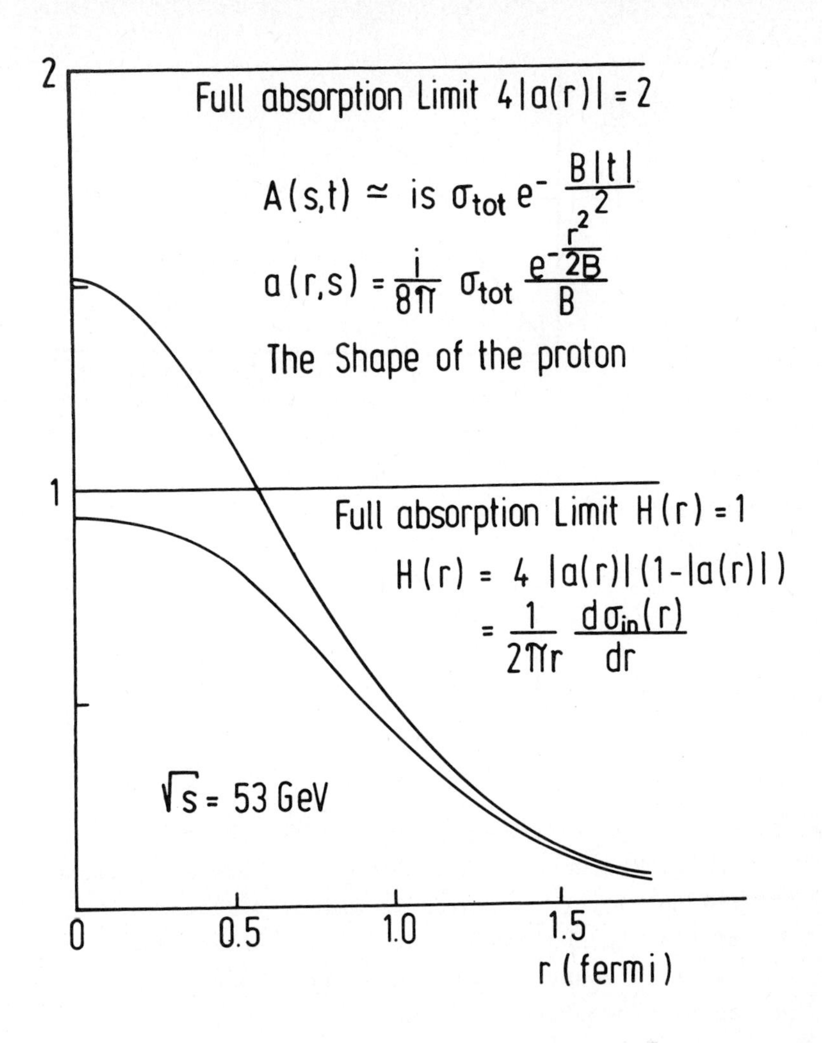

Fig. 5 The proton shape, as defined by the Fourier transform of the elastic diffraction peak [a(r)] and through the inelastic cross-section at fixed impact parameter [overlap function, $\sigma_{in}(r)$]. The proton is rather dark at the centre but quickly grey as one moves out. It appears as an extended "soft" object as opposed to a black disc with a well defined edge.

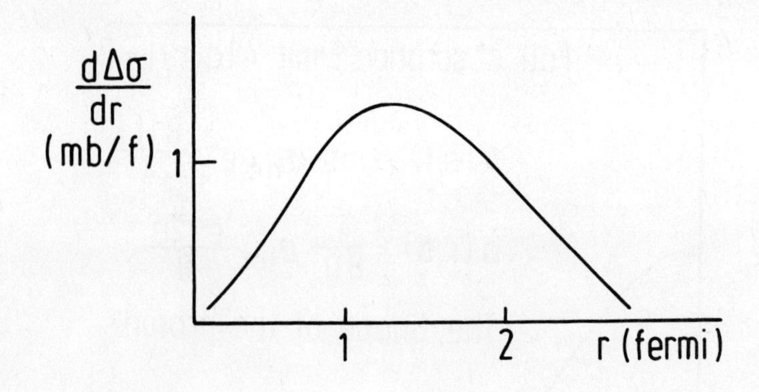

Fig. 6 The change in the inelastic cross-section as a function of energy (between 31 and 53 GeV). There is a modest rise of the opacity (σ_{in}) which is mainly peripheral[6,7]. The motion of the dip indicates that the radius increases, while the over-all shape remains the same. This is generally referred to as geometrical scaling.

tion over the ISR energy range should be urged. The impact parameter parametrization of the rise of σ_{tot} presented here should be improved upon. In particular, the localization of the dip as a function of energy is now being ascertained. The ratio σ_{el}/σ_{tot}, which differs in different models, should be known as a function of s (at present it is constant within errors). The structure of $d\sigma/dt$ at low $|t|$, which corresponds optically to the outer edge of the proton, should be better studied. This is the object of part of the present programme under way at I6 and at the SFM.

4. THE QUESTION OF SCALING

Evidence for scaling of single-particle distributions in the fragmentation region and for the approach to a scaling limit in the central region also (the rapidity plateau) was a very important discovery at the ISR. It is indeed what gave the corresponding theoretical concepts their needed tests. Experimentally, the situation is almost identical to what it was a year ago. Data have gained in precision though. There are indications that scaling may not hold when one probes for deviations of a few per cent. Nevertheless, it should be stressed that there are at present no theoretical motivations for scaling to be better an approximation than constant cross-sections are. There is still no point in repeating what was already known and discussed a year ago[1,8]. Further research on inclusive distributions at the ISR could orient itself towards the analysis of secondary particle yields at very large energy ($x > 0.9$, say), where the triple-Regge formalism motivates a large demand for new data. The corresponding yields are small. Nevertheless, this is very interesting as a means of studying reaction amplitudes off the mass shell. This is particularly the case for baryon exchange, namely the following amplitude:

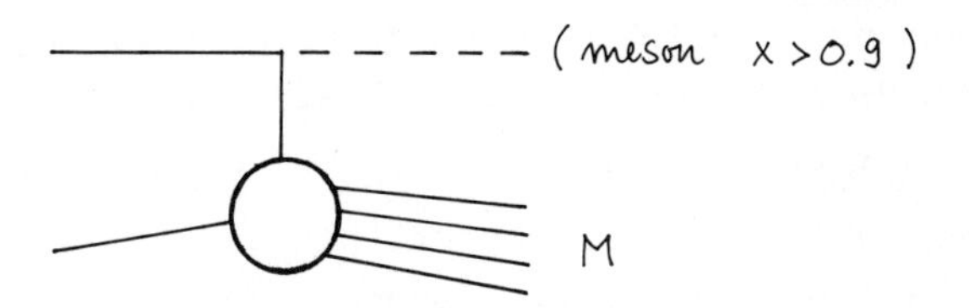

where the missing mass to the forward meson corresponds to centre-of-mass energies at the PS or Serpukhov. Production of exotic mesons, if they exist at all, should also be allowed[5]. As already mentioned, the cross-sections are very small, but the corresponding trigger is simple.

Further research on inclusive distributions could also improve upon the study of yields at low transverse momentum ($p_T < 0.1$ GeV/c), which should be better known. The fact that one has not access to the reaction vertex has raised so far paramount background problems for their measurement. An experiment in I8 is now trying to tackle the latter question (Scandinavian-MIT Collaboration).

Most of the progress over the past year went into the study of correlations. Correlations are labelled as short-range and long-range, rapidity-wise, whether they refer to effects pertaining to the same region of phase space (nearby rapidities) or to effects which interest the reaction process as a whole. This was discussed in detail one year ago[1]. One can describe part of what happens in a very high energy collision in terms of the fragmentation of the impinging particles into several secondaries. Their transverse momentum distribution, with $\langle p_T \rangle = 0.35$ GeV/c leads one to expect that such an obvious fragmentation region should interest at least two units of rapidity on either side of the over-all rapidity interval. With four units of rapidity typically available at PS energies, it was practically impossible to see anything else. At the ISR, the 8 units which are available open a wide region of phase space where secondaries, if present at all, are clearly dissociated kinematically from the obvious fragmentation of either proton. This is the central region. An important discovery at the ISR was to find that this region was highly populated with a density rising only extremely slowly with energy (the central rapidity plateau). Another important discovery was to find that these secondaries have important short-range correlations among themselves. The observation of a secondary, even when it is slow in the centre-of-mass system, and hence not associated with the several fragments of either proton, does make more probable the observation of another one in the same region of phase space. Everything looks as if particles were emitted in <u>clusters</u>. The dynamical meaning of such clusters is still open (giant or correlated resonances, droplets of hadronic glue, shock waves in hadronic matter, etc.) Their phenomenological study, however, has made some progress.

The existence of a central plateau is most generally understood in terms of a multiperipheral production amplitude[9]. One may, however not consider any longer the production of individual pions along a multi-exchange chain. The observed correlations call for important grouping of particles. Resonance formation (ρ, ω, ...) should be present and contribute to at least part of the observed correlations.

This, however, does not appear to be enough.

Our information about correlations is at present almost limited to two-body correlations. Such correlations are, furthermore, often averaged over the transverse momentum distribution and, hence, refer to rapidity correlations. Usually, one defines a correlation function as

$$C(y_1, y_2) = \frac{1}{\sigma_{in}} \frac{d^2\sigma}{dy_1 dy_2} - \frac{1}{\sigma_{in}} \frac{d\sigma}{dy_1} \frac{1}{\sigma_{in}} \frac{d\sigma}{dy_2} \qquad (4)$$

when $\gamma(y) = (1/\sigma_{in})(d\sigma/dy)$ is the rapidity density averaged over all inelastic collisions. The rapidity y cannot often be measured directly. One has to make do with the variable $\eta = -\log \mathrm{tg}(\theta/2)$. In the following, we will use either η or y for the actual rapidity. For charged particles one finds that $\gamma(0)$ is of the order of two. The density does not vary much over a sizeable range of rapidity which increases in dimension with energy. For obvious background reasons (spurious generation in the vacuum chamber walls), a safer quantity to focus upon at the ISR is often the quantity:

$$R(y_1, y_2) = \frac{C(y_1, y_2)}{\gamma(y_1)\gamma(y_2)} = \frac{\sigma_{in} \frac{d^2\sigma}{dy_1 dy_2}}{\frac{d\sigma}{dy_1} \frac{d\sigma}{dy_2}} - 1 \qquad (5)$$

All this was discussed in detail in Ref. 1. With these definitions at hand, one may examine Fig. 7, which shows important features of two-body correlations as they are known at present. The data are from the Pisa-Stony Brook Collaboration. They have been confirmed by results at NAL (bubble chamber) and at ISR (streamer chamber). Figure 7a shows a two-dimensional plot of R as a function of y_1 and y_2. The key feature is the high "ridge" at $y_1 \simeq y_2$, which extends over the whole central region. It falls sharply as one moves away from the equal rapidity line as a witness to the short-range nature of these positive correlations. Indeed, in the central region, R depends mainly on Δy and not separately on y_1 and y_2. This is typical of a short-range effect. With increasing energy, the ridge extends in length but the value of the correlation does not change appreciably. These two features are also seen clearly but in a slightly different way in Fig. 7b. The value measured for R(0,0), which is of the order of 0.7, should be considered as <u>large</u>. This should be compared with the correlation which the production of uncorrelated ρ mesons instead of uncorrelated

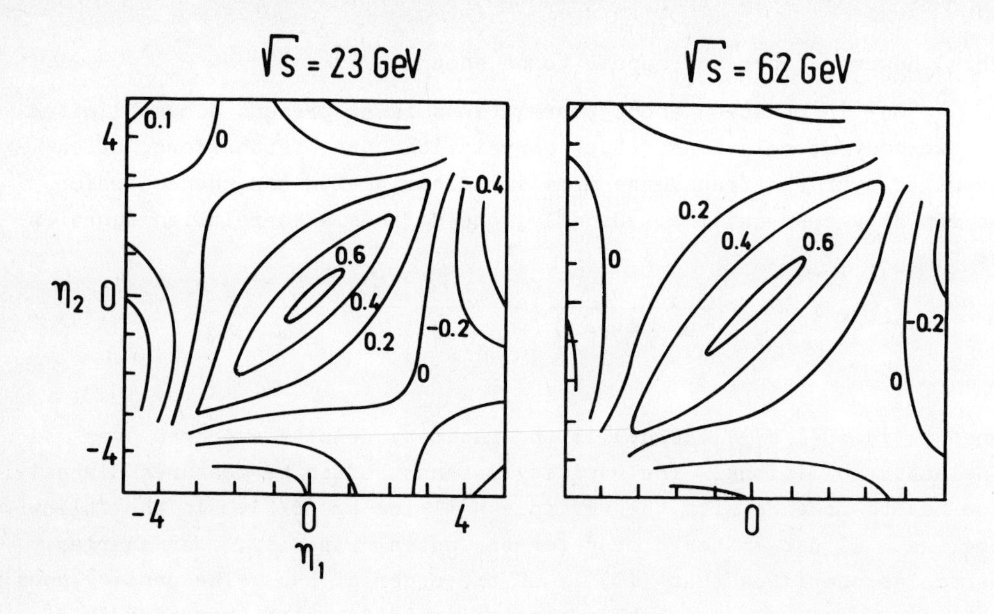

Fig. 7a

Evidence for short-range correlations in the central region. R depends mainly on $|\eta_1-\eta_2|$ and more weakly on η_1 and η_2 separately. R (0,0) does not depend appreciably on s. The central ridge extends with energy but only slowly. One, therefore, does not gain very much when going from NAL to top ISR energies, when spurious secondaries are difficult to eliminate in ISR experiments.

pions would give, which is at most of the order of 0.2.

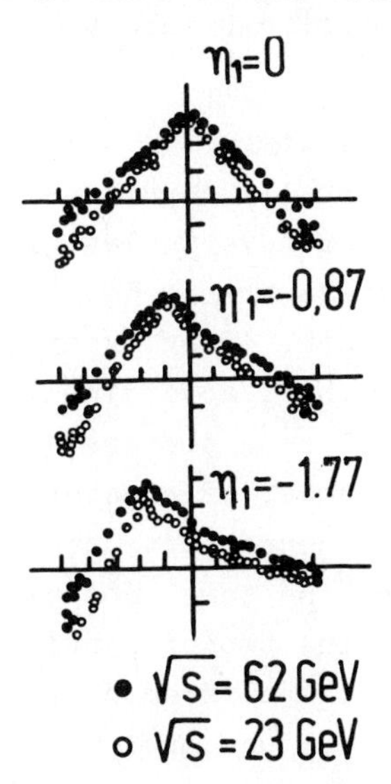

Fig. 7b Correlation profile for fixed values of η_1. One also sees clearly the short-range property and the energy independence of the effect.

The relatively low correlation effect associated with a typical resonance is due to the fact that secondaries originating from the same resonance are spread over two units or so of rapidity range (typical momentum of 0.35 GeV/c measured in the rest frame of the resonance) and that, over these two units, one expects an average of 4 charged particles anyway. Generalizing, however, from what is expected from resonances, the existence of important positive correlations, together with their energy independence, naturally led to the cluster picture of particle production. What is important is its simplicity and of course the fact that it easily meets all the effects observed so far. The cluster picture borrows a lot from the concept of resonances which, once formed, decay into particles which automatically cluster in phase space, or rapidity-wise. For instance, a group of particles isotropically distributed in their centre-of-mass system, and showing the typical transverse momentum cut-off, spreads over two units of rapidity. This is conversely what defines a typical short-range effect or a typical cluster size.

If one speaks about clusters rather than directly about resonances, it is because the meson multiplicity which one is led to attribute to clusters calls for bigger objects than the well-known resonances.

At present, one may <u>tentatively</u> quote a mean pionic multiplicity of 4 to 5 (mean charged multiplicity about 3) and a mean mass of the order of 2 GeV.

The existence of clusters may translate the fact that resonances, if produced, are generally correlated among themselves for mere quantum number reasons. It nevertheless looks an interesting approach to assume that such clusters are formed in the first place and fairly independently of one another. They eventually resolve themselves into particles (pions), which show, accordingly, important short-range correlations. According to such a picture, increasing the energy allows an increase in the mean number of clusters, but does not modify appreciably their properties. Hence correlations among pions remain of short-range nature and of the same intensity (scaling at the two-particle level). These are the two properties, clearly seen in Fig. 7, which actually motivated the cluster picture in the first place[10). As already stressed one year ago, determining cluster properties from the observed values of R is, however, difficult. Even in the central region, correlations are far from being only short-range in nature. This is partly due to the importance of diffractive excitation processes which provide a sizeable cross-section for rapidity contributions with practically nothing in the central region. This already gives a positive contribution to R (0,0) as defined in eq. (5).

A popular, but oversimplified, approach is to assume that an inelastic event is either diffractive, with no secondary in the central region (probability α) or of a different nature, with a mean rapidity density $\gamma_{ND}(y)$, and probability $\beta = 1 - \alpha$. One may then consider separately correlations proper to non-diffractive production and define new correlation functions $C_{ND}(y_1,y_2)$ and $R_{ND}(y_1,y_2)$, specific to those events. Diffractive excitation is providing the positive correlations also seen at $\eta_1 = -\eta_2$ and both close to their maximum absolute value in Fig. 7a. It will be discussed in detail in the following section. At present, one considers that α is of the order of 0.2.

Such a dichotomic separation allows the separation of a fixed long-range contribution in the correlations measured in the central region, namely

$$R(y_1,y_2) \approx \frac{1}{\beta} R_{ND}(y_1,y_2) + \frac{\alpha}{\beta} \tag{6}$$

while associating correlations in the central region to cluster formation only leads to expressing R_{ND} (mainly) as a function of Δy (γ_{ND}

constant) as:

$$R_{ND}(\Delta y) \approx \frac{\langle K(K-1)\rangle}{\langle K\rangle} \frac{D(\Delta y)}{\rho_{ND}} \qquad (7)$$

In this expression K is the number of relevant particles (those involved by the correlation which is being studied) which are to be associated with a single cluster and therefore localized in rapidity according to a function of Δy, D. Accordingly, $D(\Delta)$ is a function normalized to 1, describing how particles form a cluster spread in rapidity. A Gaussian distribution with $D(0) \approx 0.3$ will do. This is satisfactory for a mere generalization of resonance formation and decay with $\langle K\rangle >$ 2. $D(0) \approx 0.3$ is probably too large if one has to consider looser correlations among resonances. The other quantities in expression (7) are the global non-diffractive particle yield per unit of rapidity γ_{ND}, and $\langle K(K-1)\rangle/\langle K\rangle$ which is the only parameter directly accessible. Defining $\langle K\rangle$ calls for a special assumption about the multiplicity distribution.

Testing relations (6) and (7) against data on inclusive two-body correlations allows the determination of cluster parameters which, at present, summarize, in a simple way, all available facts about correlations.

Present conclusions about cluster properties call for three remarks.

i) The relatively large multiplicity attributed to each cluster leads one also to expect sizeable short-range correlations among two π^-'s or two π^+'s. They should, of course, be smaller than those observed among charged particles, since only 4 to 5 pions are involved. They should nevertheless stand clearly above the value of 0.25 tentatively attributed to the long-range effect (common to all) imposed by the presence of diffractive configurations. At the same time, charged-neutral, charged-charged, and neutral-neutral correlations should be of similar values. Results from NAL (Aix-en-Provence Conference) confirm this. The value of R(0,0) for charged-charged, charged-neutral, $\pi^+\pi^-$ and $\pi^-\pi^-$ are, respectively, 0.65 ± 0.06 (ISR), 0.65 ± 0.08 (ISR), 0.75 ± 0.1 (NAL) and 0.4 ± 0.1 (NAL).

It is certainly worth getting more extensive track chamber data. This should allow semi-inclusive tests, such as those discussed later which are still too crude.

ii) With a rather large number of particles in each cluster, varying the transverse momentum of one particle should not affect appreciably the transverse momentum distribution of its "fellow particles" at the same rapidity. Recent results from the Saclay double-arm spectrometer experiment at 90° are compatible with that. The relevant distributions for $\pi^{+}\pi^{-}$ pairs and $\pi^{-}\pi^{-}$ pairs are shown in Fig. 8 [11]. This is, however, but a passive test. Nevertheless, the data are good evidence for the lack of strong azimuthal correlations, which is a general ingredient in such analyses.

One should also consider that two a priori compensating effects are at work. This is probably why the observed effect is so weak. If particles are produced in clusters, the cluster transverse motion provides an over-all transverse momentum for each particle (0.20 of the cluster momentum for a cluster mass of 2 GeV). Hence, observing one particle in one direction with $p_T > \langle p_T \rangle = 0.35$ GeV/c is a bias in favour of a cluster moving in that direction in the first place. The other cluster particles have therefore a smaller mean momentum in the opposite direction than what would correspond to a local balance of p_T in rapidity.

iii) Finally, attributing three or so charged particles to each cluster, and giving each of them two units of rapidity to spread over, provides but a somewhat uniform rapidity distribution. It then seems hopeless to clearly separate out clusters in an event-to-event analysis of all configurations with multiplicity close to or larger than the mean value. This is a serious difficulty in looking for definitive tests of the cluster picture.

Figure 8 indicates a local charge balance. The cluster model considers a neutral cluster as an input compatible with present data. Correlations among two π^{-}'s remain sizeable as a result of the mean cluster multiplicity. These general conclusions about the validity of the cluster picture as an easy and efficient way to summarize present data about correlations in the central region have not changed appreciable over one year[1]. The parameters which have been mentioned were already available one year ago.

What has become available since is a better knowledge of diffractive configurations relevant, as stressed above, to the determination of short-range correlations (Section 5) and a new test of the cluster picture which worked in a satisfactory way. If clusters exist at all as dynamical entities, their properties should be fairly independent of the over-all conditions under which one looks for them. We already stressed the importance of energy independence which motivated the idea.

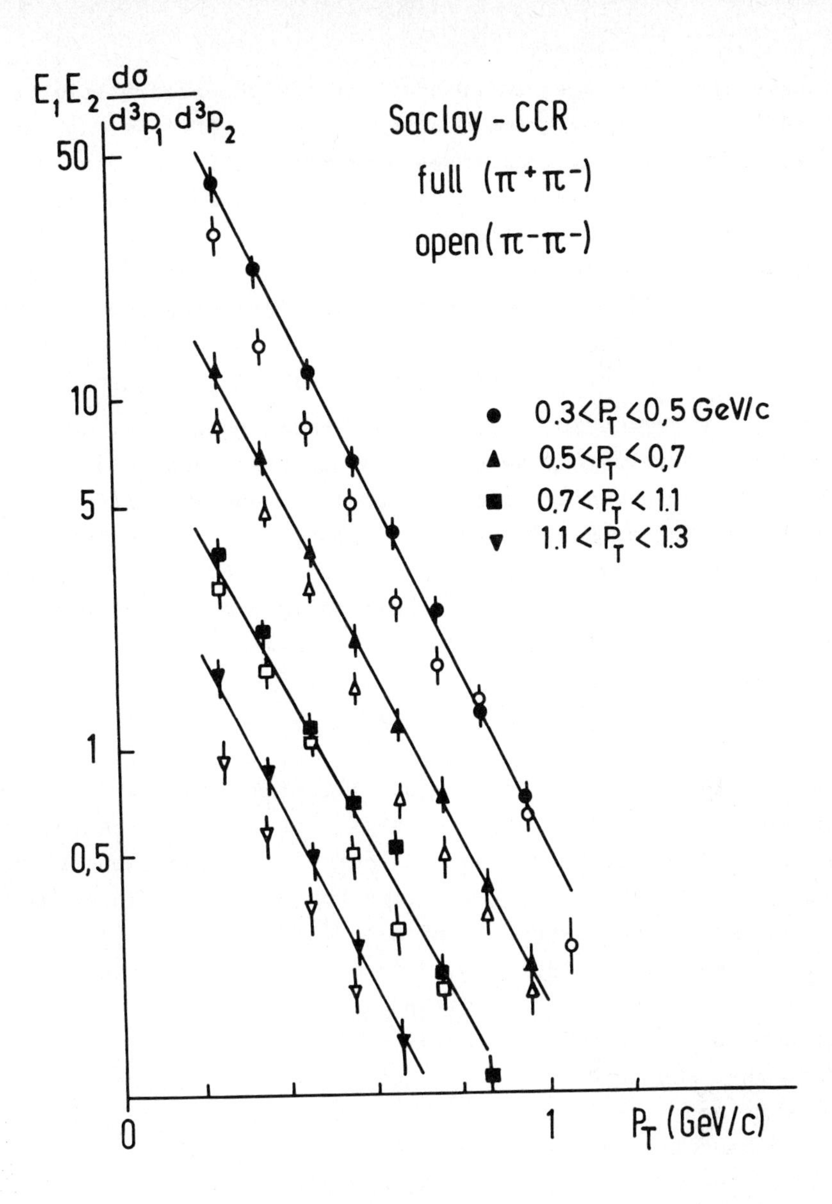

Fig. 8 Back-to-back correlations at $\theta = 90^{\circ}$. Saclay-CERN-Columbia-Rockefeller double-arm spectrometer. Note that $p_T \gtrsim \langle p_T \rangle$.

One should also find them independent of the observed multiplicity, provided it does not vary too much around $\langle n\rangle$. If $n \ll \langle n\rangle$ at ISR energies ($n < \langle n\rangle/2$ say), diffraction configurations become dominant. If $n \gg \langle n\rangle$, energy-momentum conservation will force the observed particles into a giant cluster anyway. This is, indeed, what occurs in the case of large multiplicity configurations at typical PS energies. Yet, at a fixed energy events with widely different multiplicities ($\langle n\rangle/2 < n < 2\langle n\rangle$ say) should show the same type of clusters, more numerous of course in the one case than in the other.

One may define still other correlation functions, one for each observed multiplicity n (the global one or that seen in a fixed solid angle), with the advantage that, for n large enough, diffractive excitation can be neglected. One uses σ_n instead of σ_{in} and can then forget about the $1/\beta$ and α/β terms in expression (6). One may again isolate a short-range term in the corresponding correlation function and express it in terms of cluster decay. It now reads

$$\frac{\langle K(K-1)\rangle_n}{\langle K\rangle_n} \gamma^{(n)} D(\Delta y) \tag{8}$$

This is similar to expression (7). The rapidity density $\gamma^{(n)}$ is assumed to be constant over the region of interest (central plateau). The function D should be the <u>same</u> as before. However, observing a fixed number of particles gives a bias for or against large multiplicity clusters and adds a subscript n to $\langle K(K-1)\rangle$ and $\langle K\rangle$. Such quantities are generally calculable in terms of the assumed generating function for the multiplicity distribution, which easily incorporates the assumption of independent emission for the clusters. As a matter of fact, they should not vary much with n as it varies around $\langle n\rangle$. On the other hand, $\gamma^{(n)}$ should almost increase linearly with n if one fixes the rapidity interval which is to be used (in the central region) to ΔY or increase linearly with $1/\Delta Y$ at fixed n. The multiplicity refers to any type of particles selected. The two-particle density and correlation function at fixed n have, of course, another term besides expression (8), which corresponds to secondaries originating from different clusters. It is proportional to $[\gamma^{(n)}]^2$ (or $n^2/\Delta Y$ in the central plateau) and can be calculated so that $C_2^{(n)}(y_1,y_2)$ is normalized to the proper value (-n) after integration over ΔY. The function D is normalized to one except for edge effects.

One may fit two-particle distributions at fixed n using such expressions and extract $[\langle K(K-1)\rangle/\langle K\rangle]\, D(0)$ as the relevant parameter[12]. One may also try to isolate a term independent of n and ΔY in $C_2^{(n)}(\Delta Y/n)$ or in $R_2^{(n)}\ n/\Delta Y$ and identify it with expression (8). This may be a better procedure since the remaining terms, to be added to expression (8) in order to get $\gamma_2^{(n)}(y_1,y_2)$, are sensitive to long-range correlations of many kinds[13]. The two-particle density is denoted by γ_2.

If one considers the full rapidity interval, but accordingly does not consider $\gamma^{(n)}$ as constant, one gets a short-range term which should be proportional to n and inversely proportioanl to log s. These features are beautifully verified by the Pisa-Stony Brook Collaboration[12]. The result of their recent analysis is shown in Fig. 9. The shape of the short-range term of the correlation function, which is thus isolated for different multiplicities, and different energies, is remarkably stable. Its Gaussian shape stands for isotropic cluster decay. The value of $\langle K(K - 1)\rangle/\langle K\rangle$, which one extracts from such data,is compatible with that obtained from a global analysis, namely about 3 charged particles per cluster. The cluster picture thus stands successfully an important test. One can refer to mean cluster properties irrespective of the over-all multiplicity. Such an analysis should be applied to different sets of data. It is only by varying the required conditions and still getting the same parameters that one may gain confidence in the identification of the important short-range correlations which are observed with the primary production of hadronic clusters. Further progress in probing the cluster picture should involve analysis of clustering properties when particles other than the overwhelmingly dominant pions are produced and in particular K or $\bar{p}$. Whether one can define a correlation length in such cases and, if it is the case, how it varies with the kind of particle considered, are at present open questions. One would thus like to differentiate a cluster formation and decay approach, with two successive steps, from what could still be final-state interactions among pions. The present programme with the SFM should eventually answer many of these questions. As of now, information is becoming available from the Saclay-CERN-Columbia-Rockefeller double-arm spectrometer at 90°. This cannot yet be discussed fully. Still, a prominent feature is that the observation of a proton at 90° boosts the probability of finding another one at 90° by a factor of 2 as compared to what is expected in a (typical non diffractive) collision with a pion observed at wide angle. Furthermore, this new probability is a factor of 2 larger than that of observing a $\bar{p}$. Hence, it is often not a $p\bar{p}$ which is produced at low centre- of- mass

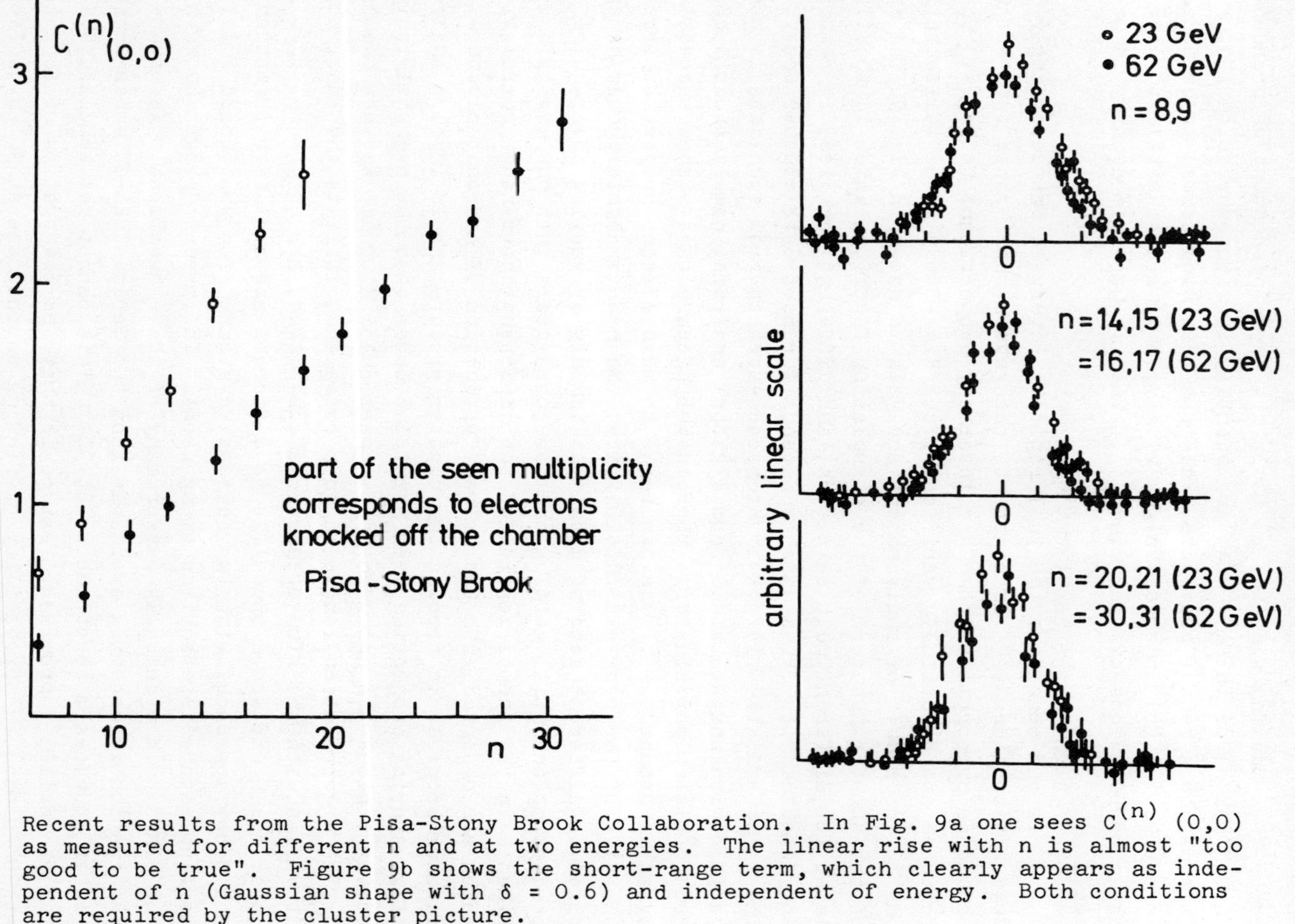

Fig. 9 Recent results from the Pisa-Stony Brook Collaboration. In Fig. 9a one sees $C^{(n)}$ (0,0) as measured for different n and at two energies. The linear rise with n is almost "too good to be true". Figure 9b shows the short-range term, which clearly appears as independent of n (Gaussian shape with $\delta = 0.6$) and independent of energy. Both conditions are required by the cluster picture.

momentum in association with a slow proton, but rather the two incident protons which happen to be slowed down to rest through the reaction. The further analysis of such a reaction is very interesting. This may correspond to a new class of event with well-defined properties. There is evidence for such configurations in PS energy track-chamber results[14]. The observation of a slow proton highly favours the observation of another slow one.

The further study of reactions where two $\bar{p}$'s are observed at wide angle (there is some evidence for that) is also very interesting. This may correspond to reaction mechanisms which are quite different from the typical one, with its independent cluster production picture, so far conceived. This may also still fit within this general pattern. It would then be probing the cluster picture in its higher mass behaviour.

It will be interesting to have results over a sizeable rapidity range, whereas those so far available refer to rapidity zero only. From the information available from the Saclay experiment and the British-Svandinavian Collaboration, one may tentatively conclude that $p\pi$ correlations at wide angle are similar, in intensity and range, to the $\pi\pi$ ones. This is, however, not very much.

We have so far considered particle production in the central region. It should be stressed that irrespective of the study of diffraction excitation described later, the description of reaction mechanism is also progressing from the analysis of collisions involving a fast secondary, yet not fast enough to signal an obvious diffractive excitation, as discussed in Section 5. This allows the exploration of long-range effects so far neglected in the global description advocated for non-diffractive events with basically short-range correlations among secondaries.

Figure 10, which corresponds to recent results of the CERN-Holland-Lancaster-Manchester Collaboration, illustrates such an effect. Observing a rather fast proton $1.8 < y_p < 3$ is a bias against observing pions at wide angles and the more so the more energetic the proton is. This is an obvious long-range effect. The measurement of the proton momentum provides information about the reaction process as a whole. A very similar effect is seen when a fast pion is observed in the first place. The energy behaviour of such double fragmentation configurations, with depletion in the central region, should be an important piece of information. Understanding such long-range effects has much

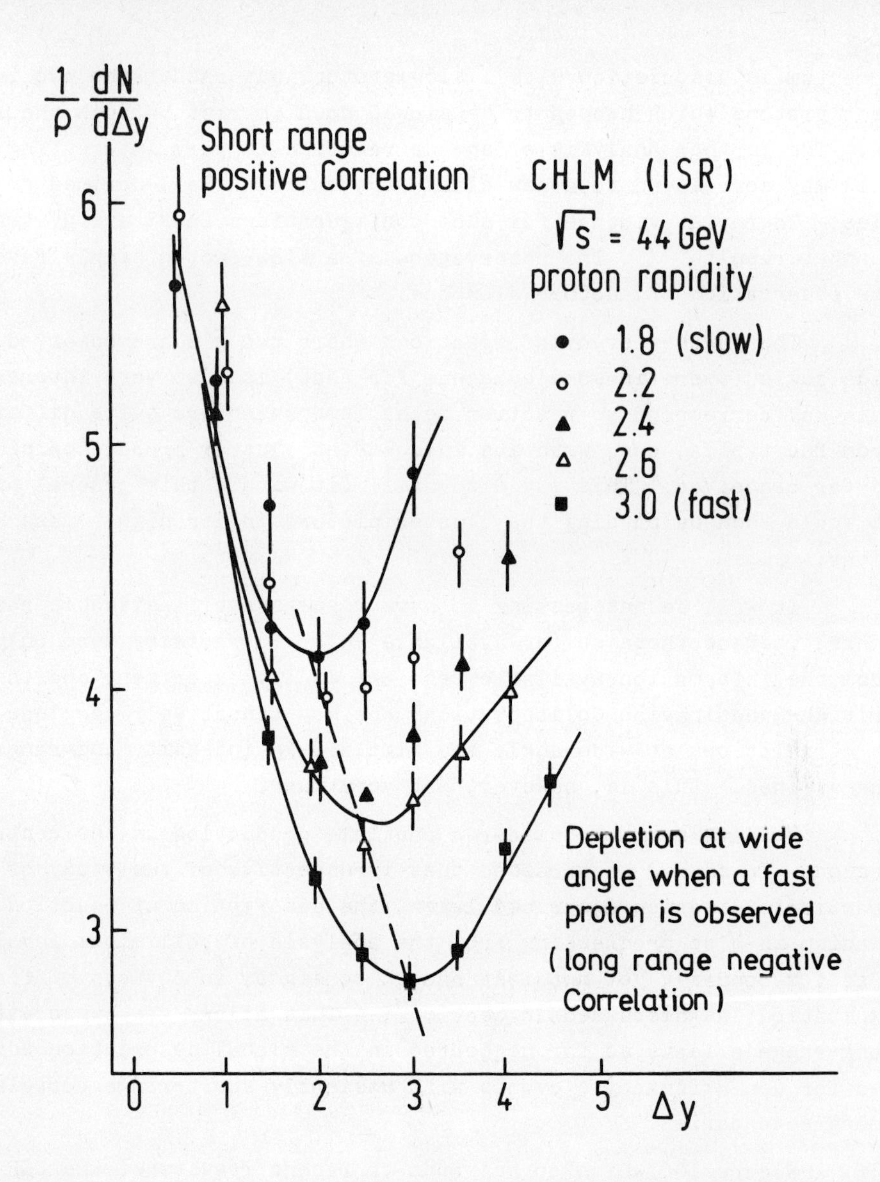

Fig. 10 Correlations involving a fast proton. Very similar correlations are observed with a fast pion (of similar rapidity). The dashed line indicates rapidity zero.

to do with our reliable parametrization of clustering effects proper. However, the depletion in the central region is still weak as compared to what corresponds to diffractive configurations. It thus corresponds to a correction only to the previous description of typical non-diffractive configurations.

At present the British barrel detector, used in coincidence with the CERN-Holland-Lancaster-Manchester small-angle spectrometer, should provide a rather thorough analysis of such effects. Information will also come from the forward negative spectrometer of the CERN-Rome Collaboration, used in coincidence with the Pisa-Stony Brook detector[3].

The study of correlations in particle production has unravelled very interesting features. The concept of cluster, which got its momentum from the important short-range effects first found at the ISR, is worth a thorough probe. An important set of correlation experiments are now under way in I1, I2 and I8[3]. Several more are to be performed at the SFM. This is a very interesting field of research at the ISR. It should provide a much more refined picture for many-particle configurations (we still know practically nothing about three-body correlations!) than what is presently available. Nevertheless, it should come a time where what is gained in energy (the extension of the central ridge in Fig. 7a illustrates it very well) should be weighed against the non-accessibility of the reaction point proper in its observation through a background generating pipe.

The ISR has given, or will still provide, the gross (but key) features, in particular correlations. The finer work will certainly have to be carried out at NAL and SPS.

Reviewing correlations as studied at the ISR, one should also mention the observation of well-known resonances through the measurement of several particles in a wide acceptance magnet. Evidence for Δ production has been reported by the Aachen-CERN-UCLA Collaboration. The Δ appears often as part of a $\Delta\pi^-$ (diffractive?) cluster. Inclusive analysis of resonance production and its association with a possible Deck effect are interesting programmes to follow. In particular, the important muon yield at large p_T makes it very interesting to measure the vector meson yield as a function of p_T.

5. DIFFRACTIVE EXCITATION

Diffraction excitation has long been known as a production mechanism for some of the prominent low mass resonances. Its key features are:

i) a sharp t dependence (as expected for a coherent effect);

ii) no quantum number exchange;

iii) an energy behaviour similar to that of elastic scattering and thus very different from the power law decrease of all other quasi-two-body cross-sections.

These three features can conversely be used for defining diffractive excitation. Easier to detect is single-diffractive excitation, whereby one proton (at the ISR) is quasi-elastically scattered, while the other one flares into several hadrons.

In this domain, the very important contribution of the ISR has been twofold. The first point is that single diffractive excitation is an important mechanism. It is not limited to a few leading resonances and involves about 1/5 of all inelastic processes. The second point is that diffractive excitation extends up to very high masses (10 GeV at least) with a diffractive cross-section behaving as M^{-2} at large mass. This should provide a logarithmically rising contribution to the total cross-section. These two points are illustrated by Fig. 11, which gives results of the CERN-Holland-Lancaster-Manchester Collaboration. The diffractive cross-section $d\sigma/dM^2dt$ (at fixed t) shows a sharp peak (most of the cross-section refers to $M < 5$ GeV), but also a high mass tail (Fig. 11a). Evidence for such a tail is also provided by the scaling property of the quasi-elastic peak in the inclusive distribution (Fig. 11b)[1]. The kinematical separation of diffractive excitation depends on x, and is thus possible up to masses increasing linearly with the centre-of-mass energy. This explains the success of the ISR at separating out diffractive excitation as a specific mechanism. One has (x close to 1):

$$s \frac{d\sigma}{dM^2 dt} \simeq \frac{d\sigma}{dx\, dp_T^2} \quad \text{with} \quad x = 1 - \frac{M^2}{s} \qquad (9)$$

The sizeable single-diffractive cross-section, as well as the large mass tail, have been recently confirmed at NAL in several experiments, but particularly in the gas-jet experiment on deuterium. The results (APS Chicago meeting) are shown in Fig. 12[6,15].

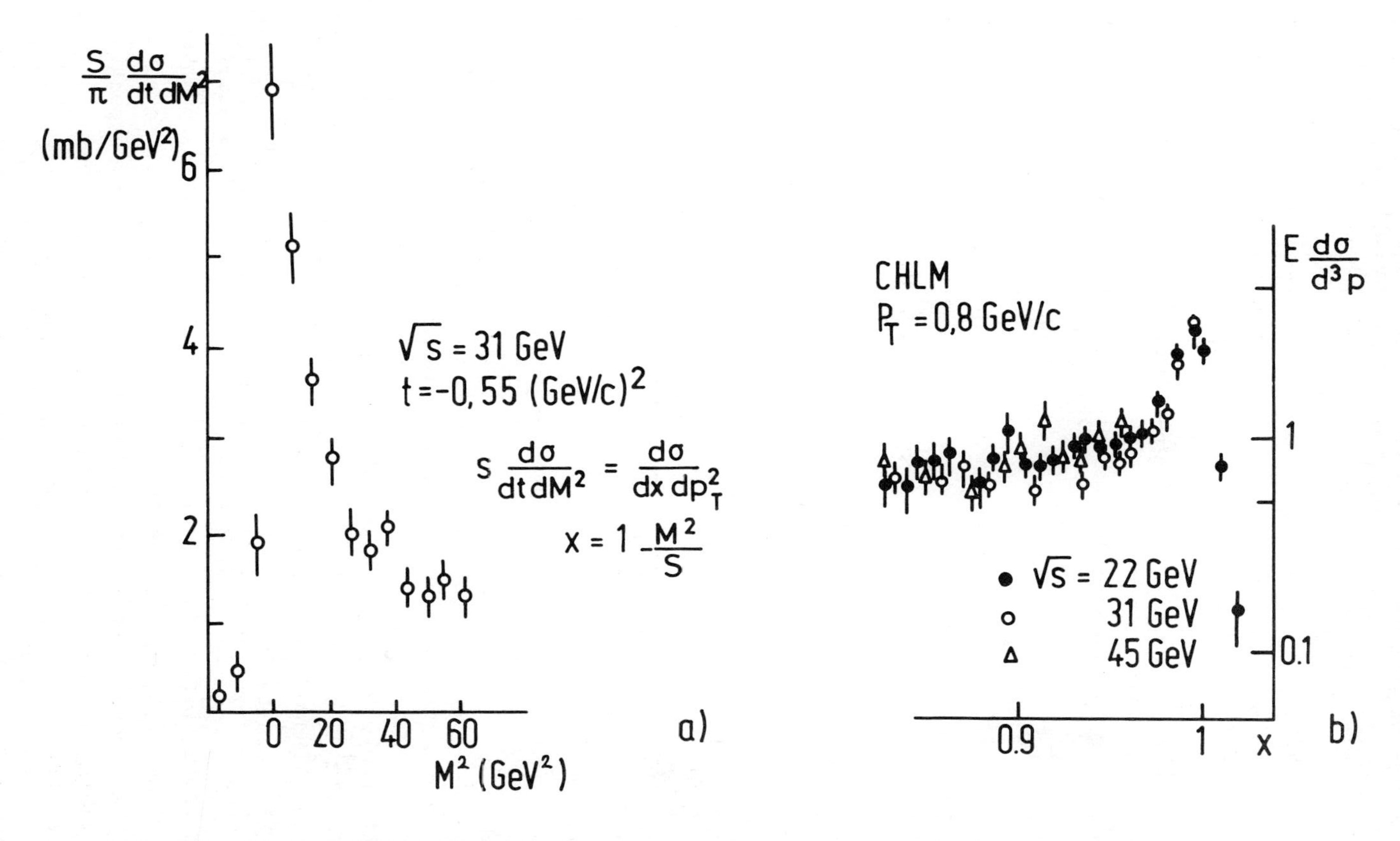

Fig. 11 Single-diffractive excitation as observed by the CHLM Collaboration. a) Missing-mass distribution. b) Inclusive proton distribution at large x and evidence for scaling.

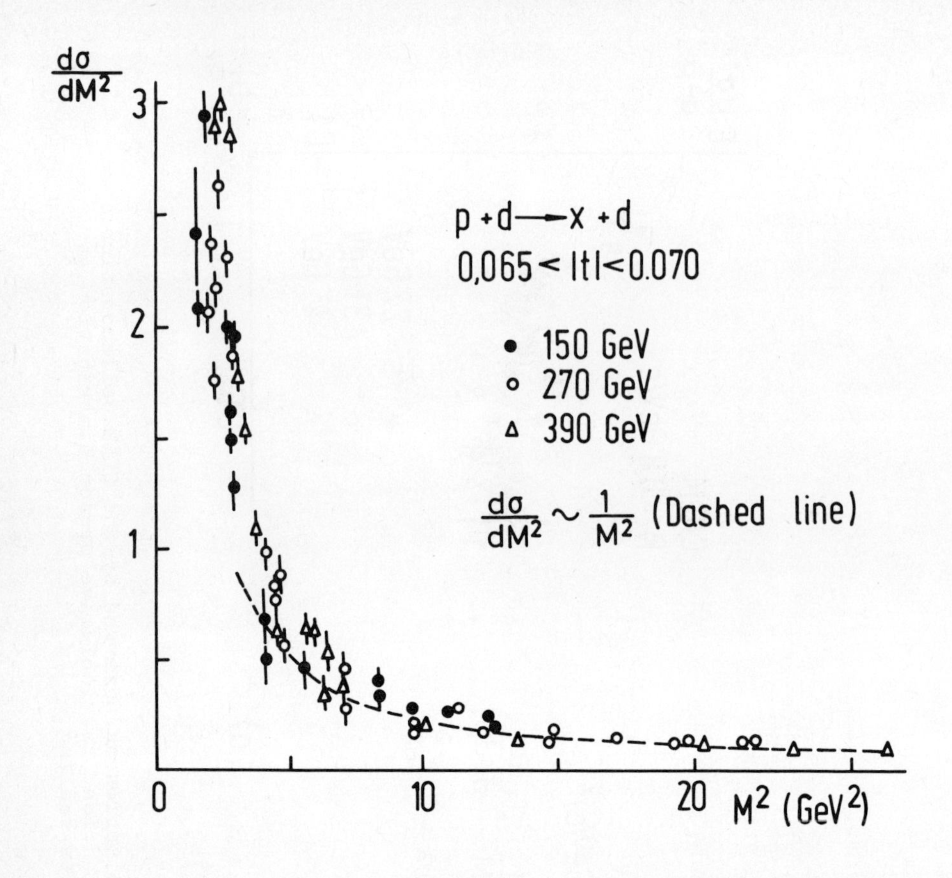

Fig. 12 Diffractive excitation as observed at NAL. Most of the cross-section still corresponds to low excitation masses. But there is evidence for a large M tail.

A weak point of the ISR results (so far) came from the difficulty of observing protons at very low $|t|$ when this is where most of the cross-section is. Indeed, estimates for the cross-section have to combine ISR and NAL results in order to reach a global picture of the t dependence. This is illustrated by Fig. 13 [16]. It should obviously be improved upon. Diffractive excitation should be measured down to very low $|t|$ values ($|t| \lesssim 0.05$) at the ISR, and also done with a better missing -mass resolution. From present results, one may expect a logarithmically rising contribution to the cross section. This is a very important point to check and it should be possible with the ISR.

A typical single-diffractive event has <u>a priori</u> a very specific rapidity configuration with a wide gap between the quasi-elastically scattered proton at $y \sim y_{max}$ and the remaining particles which are asso-

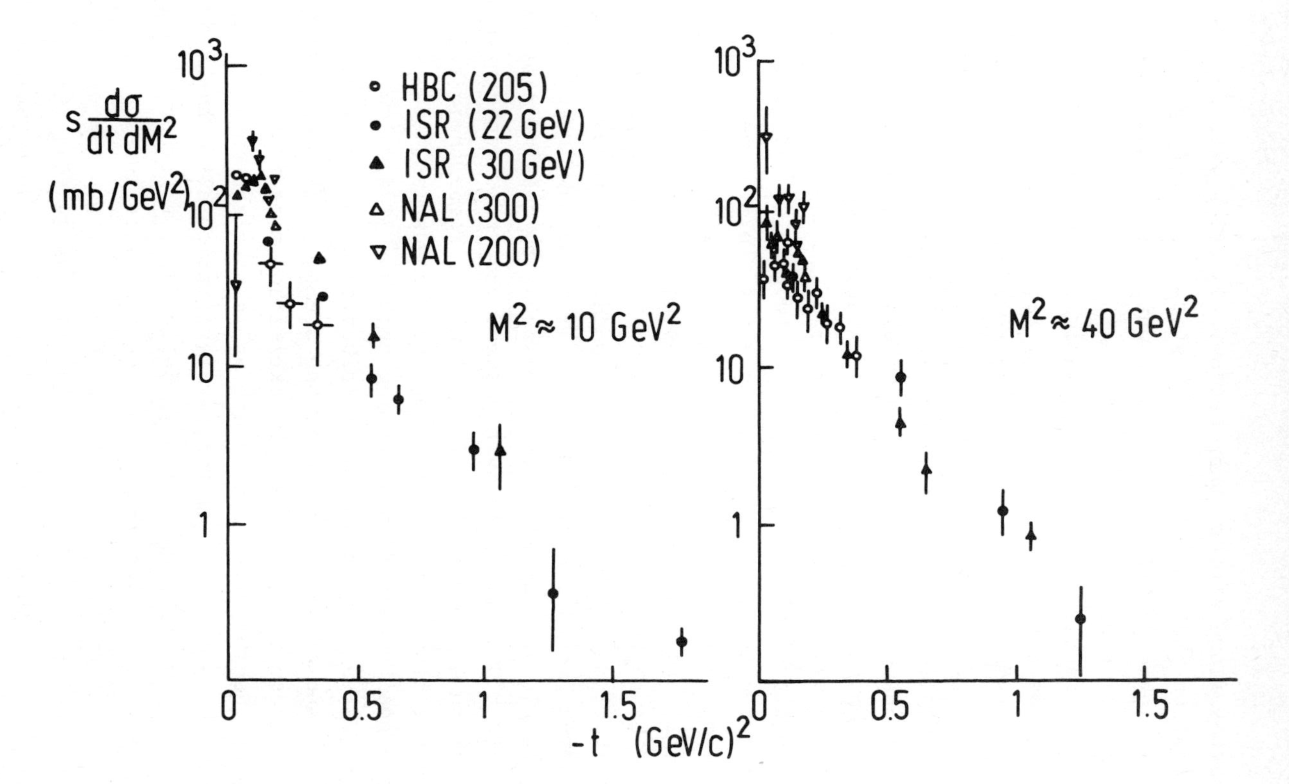

Fig. 13 The shape of the quasi-elastic peak. A compilation of data from NAL and the ISR.

ciated with the flare of the other proton. Indeed a year ago, there was an indication that diffractive excitation was imposing a very low yield at wide angles. This is a feature which was actually used in the discussion of correlations at wide angles in Section 4. Such a rapidity gap is merely implied by kinematics for x close enough to 1. How it sets in as a function of x (or M^2) is, however, an important question. The relative behaviour of the many hadrons coming from the flaring proton (often referred to as a hadronic nova) also opens up interesting questions. Diffractive excitation (with, by definition, Pomeranchon exchange) can be, in principle, summarized in terms of a Pomeranchon-particle collision amplitude at centre-of-mass energy equal to the observed excitation mass (missing mass in the case of single diffractive excitation). Probing such a new type of interaction, with important bearing on model building, is a new field of research[17). Improvements on what was known a year ago (very little) are illustrated by Figs. 14 and 15, which both correspond to new results of the CERN-Holland-Lancaster-Manchester Collaboration. The rapidity and angular distribution of the associated secondaries with the quasi-elastically scattered proton are given in Figs. 14a and 14b, respectively. The quasi-elastic peak of Fig. 11b corresponds to missing-mass values up to 7 GeV ($x = 0.95$). Up to such masses, one sees indeed a very strong depletion at wide angles (Fig. 14 uses a logarithmic scale when Fig. 11a uses a linear scale), most of the secondaries being in the hemisphere opposite that of the leading proton. If for low missing mass ($M < 3$ GeV,say) the rapidity distribution of the secondaries may correspond to one of the clusters of the preceding section (almost isotropic decay), one sees that, as the excitation mass increases, the over-all distribution extends in width while it rises only slowly. This, of course, shows that diffractive excitation also involves rather large muliplicities (though typically below $\langle n \rangle$) together with large missing masses. This also shows that the global features of the Pomeranchon-proton reaction are not very different from those of a typical hadron-hadron reaction. The secondaries eventually fill the available rapidity interval which extends on both sides of a mean rapidity $\bar{Y} = \log \sqrt{s}/M$. The inclusive yield then eventually becomes a slowly rising function of M. These features are typical of hadron collisions past the low-mass resonance region.

A further point in the analogy between Pomeranchon-proton and hadron-hadron amplitudes should be the observation of a leading proton effect within the hadronic nova. Looking back at the rapidity distri-

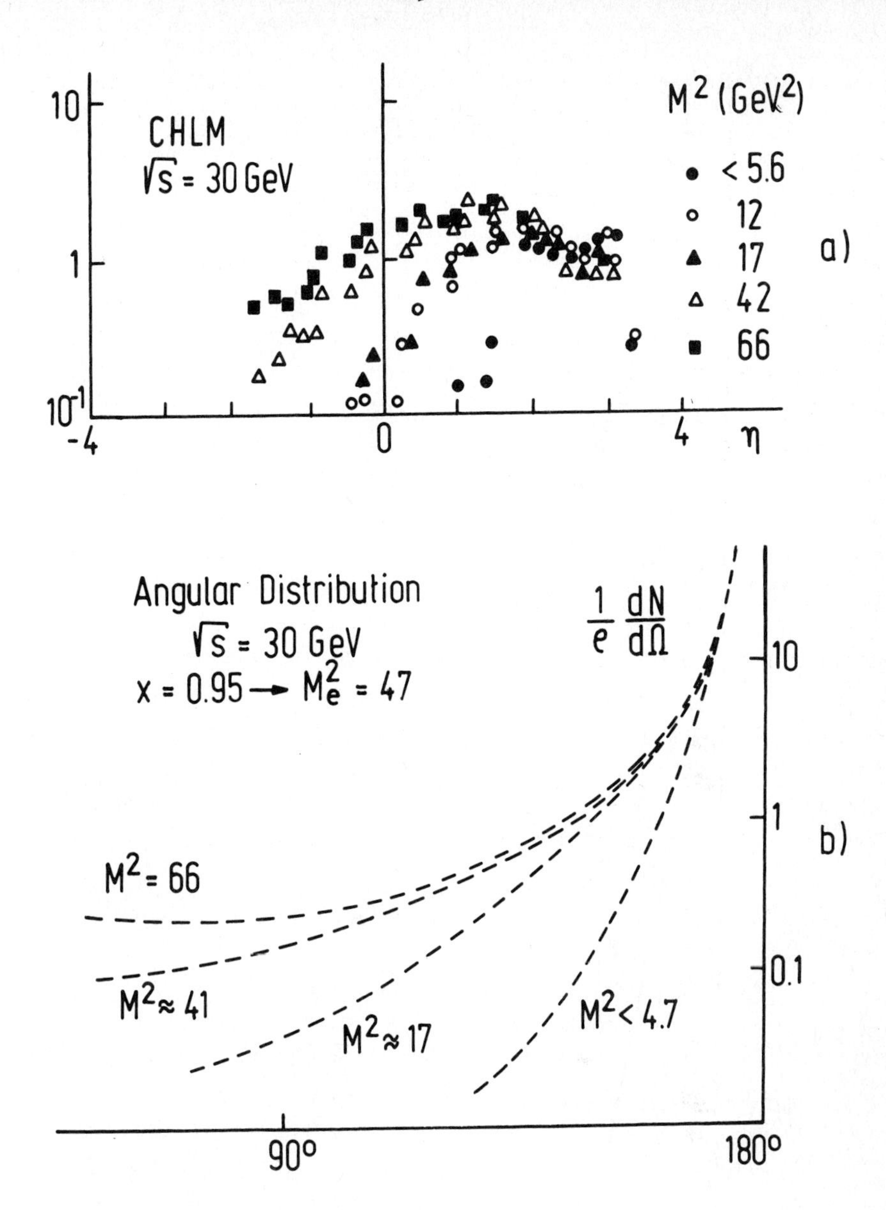

Fig. 14 Associated multiplicity with a quasi-elastically scattered proton as studied by the CHLM Collaboration. a) Rapidity distribution. b) Angular distribution. The curves are normalized to the inclusive proton yield.

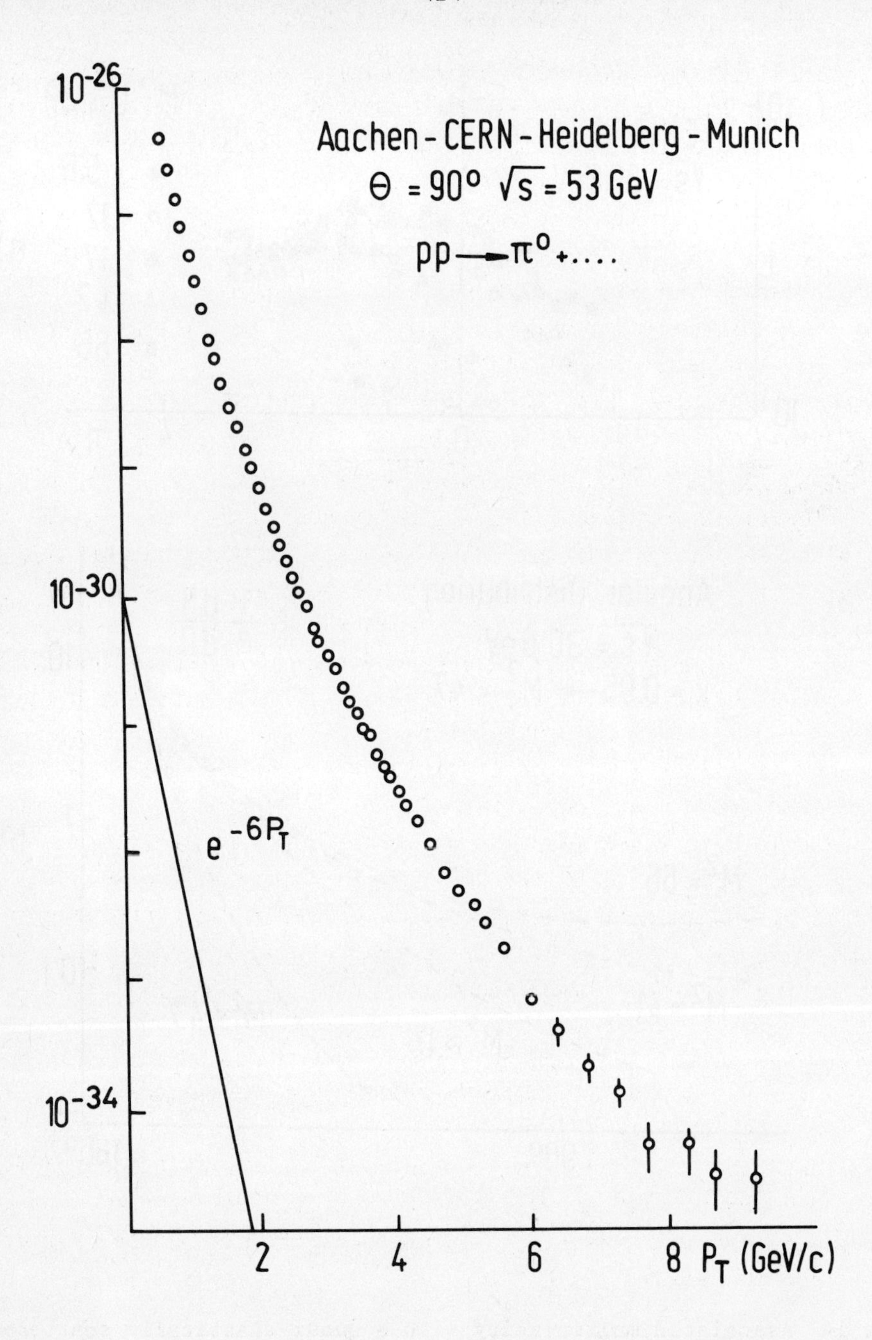

Fig. 15 Inclusive π^0 distribution at $\theta = 90^0$ ($\sqrt{s} = 53$ GeV).

bution in Fig. 14a, the proton should take a somewhat asymmetrical stand towards higher rapidity values and the more so the larger is M. This effect has again important dynamical consequences, one of which is the existence of double-Pomeranchon exchange in production processes, which is still an open question. This is something which should be studied at the ISR, measuring the proton/pion ratio over the rapidity distribution for different M values. This also calls for measurements at low $|t|$ values, where actually most of the cross-section is. One could eventually isolate a term where the vacuum flares into pions. Its energy dependence over the ISR energy range may give it a big enough cross-section within conservative kinematical limits.

Another point is the existence of a definite skewness in the rapidity distribution of Fig. 14a, which present data do not allow one to detect. In photoproduction one observes an important photon fragmentation effect with a pionic inclusive distribution asymmetrical at $x = 0$ and peaking on the proton side. Whether there is also an important Pomeranchon fragmentation is interesting. Low statistics results from NAL indicate that it could well be there[18]. This last point may be hard to check at the ISR. Nevertheless, one should stress that research at the ISR has brought diffractive excitation into a completely new perspective. Some, among the many, questions which are still open can probably find satisfactory answers in a few ISR experiments still to be done. Later on, the two weak points of the ISR, namely the limited missing-mass resolution and the spurious effects produced in the vacuum tube, will have to be weighed against the limited energy range (and hence accessible missing-mass range) of NAL (SPS). The detailed study of correlations among the nova particle should rather belong to the SPS than to the ISR. The energy behaviour of the diffractive cross-section and the possible leading-proton effect should be settled at the ISR. There should also be double-diffractive excitation (though with a smaller cross-section). The analysis of the Aachen-CERN-UCLA experiment may provide satisfying evidence for it. The SFM early programme may also bring interesting results about it. It should be there at the 1 to 2 mb level.

6. LARGE TRANSVERSE MOMENTUM PHENOMENA

A year ago, reaction processes with a secondary of large transverse momentum, could already be distinguished from typical events in several different ways. The contrast between what is associated with a particle of low transverse momentum (0.3 GeV/c, say) and what is associated with a particle of large transverse momentum (3 GeV/c say) is obvious from Table 1. All these points have now been further confirmed by new results at the ISR and by results at NAL. There are two important points to stress though:

i) Practically all the information available so far pertains to wide-angle secondaries. There is evidence that, at the inclusive level at least, nothing changes dramatically with rapidity (within one unit on either side of $y = 0$ say). Nevertheless, information on large p_T phenomena at $x \neq 0$ would be very useful. Indeed one of the most important recent pieces of data involves correlations with a large p_T triggering particle at $y \neq 0$. It remains that, at present, when one discusses data for large p_T phenomena, one actually means large p_T phenomena at wide angles.

ii) Large p_T phenomena clearly distinguish themselves from the rest only as one gets beyond $p_T = 1$ GeV/c. By then, the cross-section is typically 3 orders of magnitude down from that of a typical process (5 to 6 orders of magnitude for $p_T = 4$ GeV/c). Even if the observed yields are very much larger than what the naive extrapolation of the low p_T behaviour would give, one is still faced with a relatively rare phenomenon. As a result, many an _a priori_ unlikely effect can be called upon in order to interpret what is observed. If several points of view are still acceptable, the association of such effects with the hard incoherent interaction of proton constituents is at present the only one for which all the observed effects are straightforward consequences. It seems fair to say that, in all other approaches considered so far, meeting experimental results requires _ad hoc_ choices of parameters. The hard scattering picture considers large p_T phenomena as another facet of those effects responsible for the key features of deep inelastic electron scattering. They are then a further hint at a parton structure[19]. Anomalously large yields at large p_T at the ISR could be postulated in analogy with what is observed in electron scattering at large momentum transfer[20]. Nothing yet contradicts such a relation. Even though several other approaches should be kept in mind, we will use the hard-scattering picture as a reference picture when discussing data[21,22].

The semi-quantitative statements of a year ago which are summarized in Table 1 [1] have been much improved upon quantitatively speaking over the past year. The conclusions which could be drawn then have been confirmed. Far from all questions have been answered.

Table 1

Key features which distinguish configurations with a large transverse momentum secondary from typical ones. The change occurs as p_T becomes larger than 1 GeV/c.

Feature ($y \sim 0$)	Low p_T $p_T \sim 0.3$ GeV/c	Large p_T $p_T \sim 3$ GeV/c
$\frac{d\sigma}{dp_T^2}$	e^{-6p_T}	$\sim p_T^{-8}$
Scaling over the ISR energy range	Good (within 10%)	Wrong (rise by a factor of 10)
$\frac{\text{positive}}{\text{negative}}$	Compatible with 1	Positive excess ~ 1.4
$\frac{\text{pion}}{\text{heavy particle}}$	Large (6)	About 1
Associated multiplicity towards an observed π^0	Important short-range positive correlations	Same as those observed with a low p_T π^0
Associated multiplicity away from an observed π^0	No strong azimuthal effect	Very important positive correlations increasing with p_T

As an example of the inclusive distributions now available, Fig. 15 gives the π^0 distribution just reported by the Aachen-CERN-Heidelberg-Munich Collaboration. It corresponds to the yield at 90^0 at $\sqrt{s} = 53$ GeV. The obvious departure from the low p_T exponential approximation does not result in a simple inverse power behaviour. The type of expression suggested by the hard-scattering approach (and in particular parton models), namely

$$E\frac{d\sigma}{d^3p} = \frac{A}{p_T} f(x_T) \qquad x_T = \frac{2p_T}{\sqrt{s}} \qquad \text{at } y \approx 0 \tag{10}$$

with, tentatively and <u>empirically</u>

$$f(x_T) \approx e^{-b\frac{p_T}{\sqrt{s}}} \tag{11}$$

is, however, not contradicted by the data. The new Aachen-CERN-Heidelberg-Munich results, as previously the CERN-Columbia-Rockefeller results[1], are amenable to such a parametrization. Relations (10) also apply to the NAL results of the Princeton-Chicago Collaboration[23]. However, the values of n claimed by the different groups are not the same. This is a question which has been clarified recently and one may stress that:

i) There is a perfect consistency between the NAL and ISR results, which do not always overlap in p_T (and x_T).

ii) The best value of n (at fixed p_T) changes with energy (or as one covers the p_T x_T plot). It shows that even the ISR energy range is a transition domain. One may even say that even if $n = 8$ may be favoured at the ISR[24], $n = 4$ which would correspond to full scaling[25] is not excluded asymptotically.

In order to support these two points, one may display equal rate curves on a log p_T log x_T plot as shown in Fig. 16a[26]. The open (full) dots indicate curves with yields in the ratio of 10^2. The separation between two curves at fixed x_T is inversely proportional to n. Phase-space limits force all curves to jam and fall as x_T reaches one. One sees that the NAL results, which partly correspond to large x_T, still indicate this limitation with values of n larger than what is found for the ISR results ($x_T < 0.3$). The equal rate curves open up as one moves toward lower x_T (higher energy at fixed p_T) and, despite the local validity of a parametrization of the type given by eqs. (10) and (11), the ISR energy range appears yet as far from asymptotic. According to eq. (10), one would expect the equal rate curves to eventually become parallel. They may do so, but n could well reach a lower value ($n = 4$) before. This is further illustrated by Fig. 16b, which shows the same data on a log p_T x_T plot, with $x_T = 0$ being the asymptotic limit at fixed p_T. One again sees that there is a good agreement between the NAL and the ISR results. Yet it would be meaningless to try to fit a form such as eq. (11) to the NAL results alone or to try to come to a conclusion on the value of n relevant to a fit of the type of eq. (10) from any set of data taken separately. There is, of course, evidence for an opening of the equal rate curves with increasing energy. The ISR can be considered as providing the highest possible energy for several years. Yet one sees that there is obviously a need for better measurements and a consistency check at the ISR[27]. Reaching final agreement between the CERN-Columbia-Rockefeller and Aachen-CERN-Heidelberg-Munich Collaborations could clearly provide (or elimi-

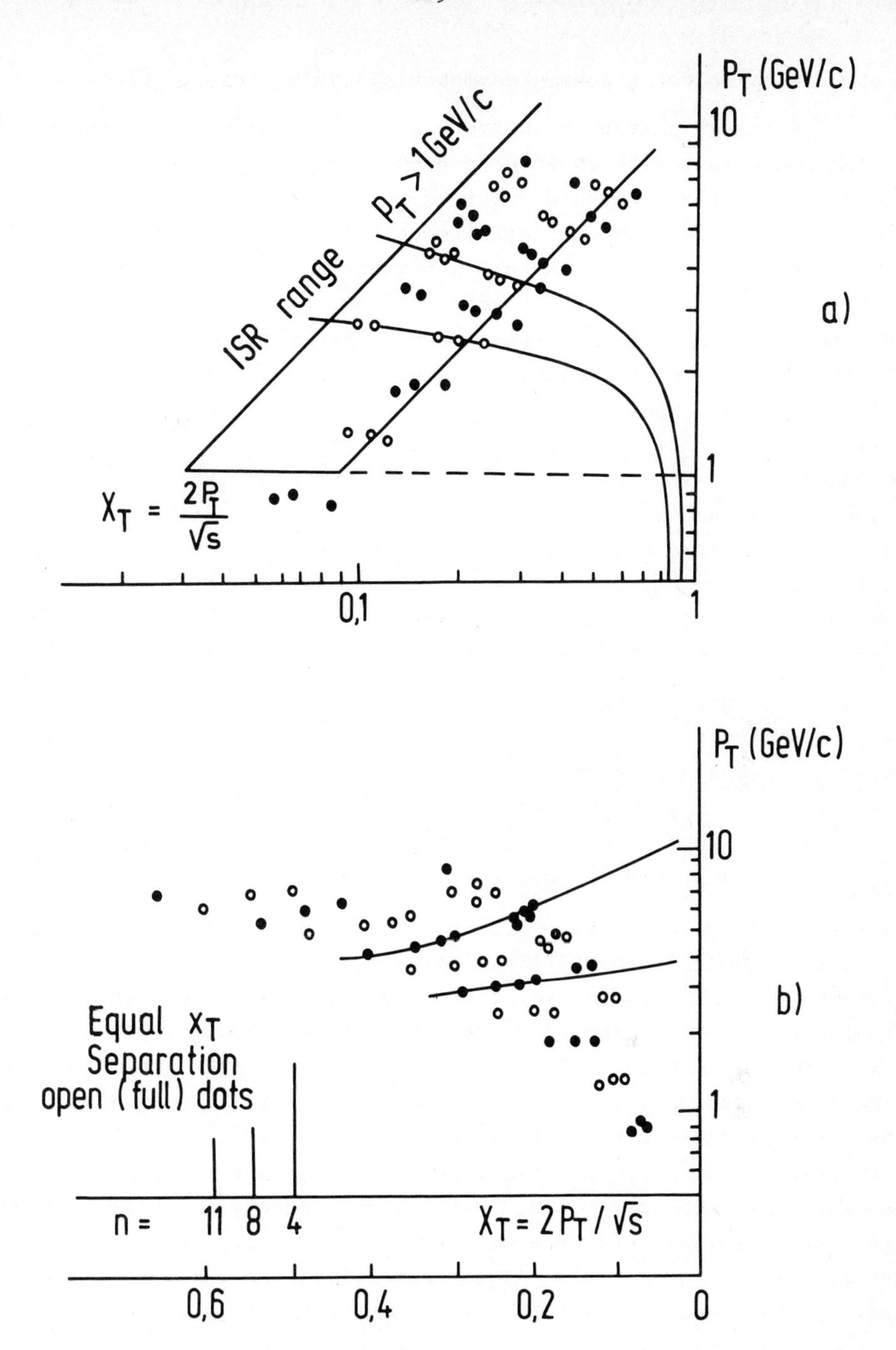

Fig. 16 Separation of equal rates by factors of 10 on p_T, x_T plots.

nate) evidence for a lower (asymptotic?) value for n.

At large p_T the inclusive distribution clearly indicates a behaviour compatible with an inverse power. This is supporting evidence for a point-like structure. The relevance of a fit using eq. (10) is rather impressive, but no definite conclusion can be drawn at present as to what should be the actual behaviour. Full scaling (n = 4) is not excluded, but it is also certainly not relevant yet. As often advocated, hard scattering between proton constituents should rather give p_T^{-4} at least asymptotically. A larger power, and in particular p_T^{-8}, is not excluded, but it would then indicate that one pion only collects all the transverse momentum due to the primary parton collision, as opposed to its spreading over a "yet" of secondaries. It may be considered a good point that the analysis of the associated charged multiplicity indicates that this seems to be the case. There are, practically speaking , no other large p_T particles in the direction of the neutral pion[28]. It remains a puzzle, though, that such configurations dominate. Using different triggering techniques may settle the issue.

The extra dimensions correspond to imposing the production of a single pion from a parton (quark) scattered at wide angle with large momentum. The amplitude for its sticking as a pion with another parton (anti-quark) is weighed down by the pion form factor at large t, namely p_T^{-2}. One would then expect the p_T^{-8} behaviour [typically $p_T^{-4}(|p|^{-4})$] to break down at small angle ($|\theta| < 20^o$ say) and a steeper fall-off to be found.

The question of particle ratios has much improved over one year. For instance, Fig. 17 shows the fraction of all charged particles as a function of p_T as observed at $\sqrt{s}$ = 53 GeV[29]. These results have been obtained by the British-Scandinavian Collaboration. The pions quickly lose their overwhelming majority. Even more striking is the stabilization of the different ratios beyond p_T = 1 GeV/c. The p/π^+ ratio is of the order of 0.3. The NAL results show a much larger ratio (as large as one at 200 GeV) which, however, decreases with increasing energy (Aix-en-Provence Conference at 200 and 300 GeV and the London Conference at 400 GeV) (Fig. 18). At the same time, the $\bar{p}/\pi^-$ ratio increases between NAL energy and 53 GeV. The latter point is expected on quite general grounds. The former one is rather peculiar[30]. It indicates that large p_T protons at wide angles may originate from quite different mechanism, the relative importance of which change with energy. As a matter of fact, the proton yield increases with energy. It, however, does not increase as fast as the pion one [relations (10) and (11)].

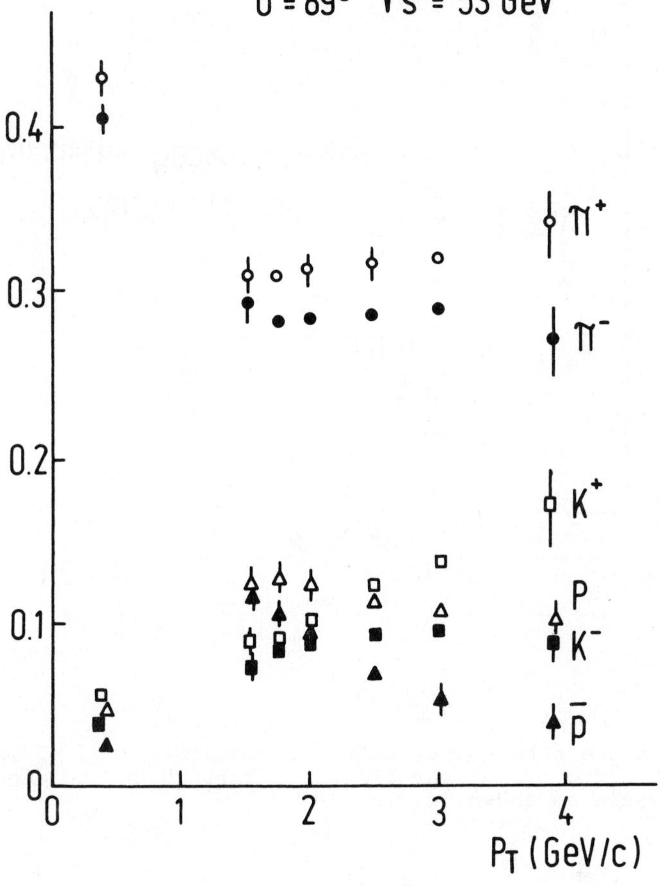

Fig. 17 Particle ratios. Their levelling as p_T exceeds 1 GeV/c is the prominent fact next to the relative importance of heavy particles.

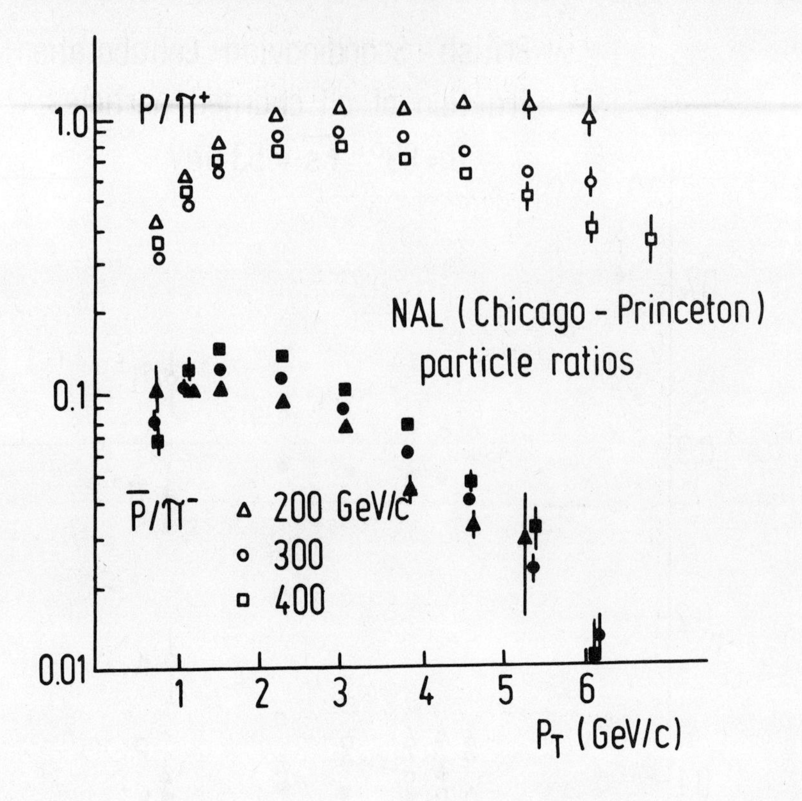

Fig. 18 Part of the discrepancy with the results of Fig. 17 is probably due to nuclear effects. Part is due to energy dependence as shown in Fig. 19

This energy behaviour of the p/π^+ and $\bar{p}/\pi^-$ ratios has now been measured by the British-Scandinavian Collaboration (Fig. 19). These ratios give an altogether positive excess, which is already present at the pion level. This is of course what is expected from an incoherent scattering among proton constituents. This is, however, not specific of it[31].

All ratios are compatible with asymptotic limits. With $3 > p_T > 1$ GeV/c one, however, explores a region where the relevant parton would have only a small fraction of the total momentum. Valence quark effects are present if one wishes to see them. They are, however, not overwhelming.

It would be interesting to have data about charge correlations at large p_T and eventually correlations among particles with opposite

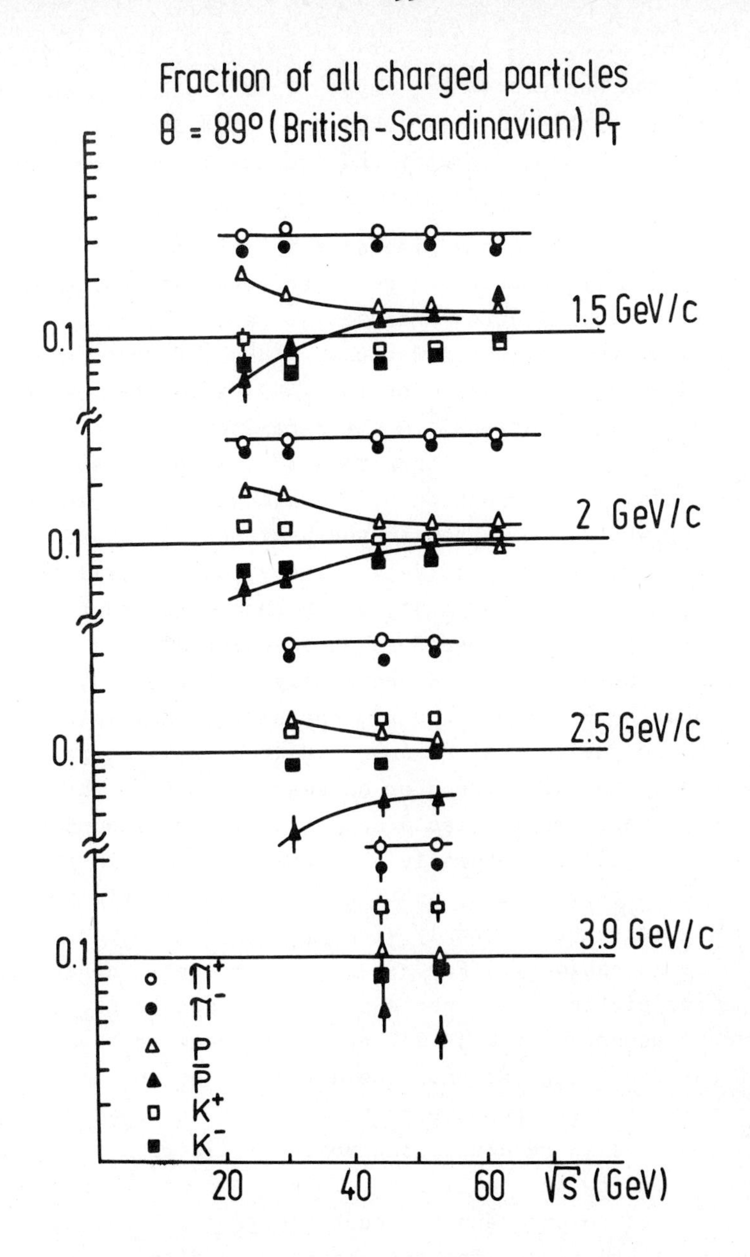

Fig. 19 Particle ratios as functions of energy.

(different) quantum numbers (K^+K^-, say). This would help enormously to devise models. Nothing is available now. The Saclay-CERN-Columbia-Rockefeller double-arm spectrometer and eventually the SFM should provide important pieces of information in that domain.

The most interesting new pieces of information about large p_T phenomena are probably those which refer to correlations involving a large p_T particle (in all cases analysed so far, a large p_T π^0). This refers to the last two points of Table 1. This information was then separately available from results of the CERN-Columbia-Rockefeller and Pisa-Stony Brook Collaborations[1]. The Pisa-Stony Brook people have now a comprehensive set of results with a π^0 trigger at 90°. Figure 20 gives the rapidity distribution of the associated charges multiplicity, separating what is measured in the hemispheres towards the observed π^0 and away from it, respectively. The yields are all normalized to those observed with a low p_T π^0, which then includes a positive short-range correlation and therefore an associated yield 60 to 70% above its mean value as discussed previously in Section 4. One sees that in the hemisphere towards the π^0, the charged multiplicity is independent of p_T in the central region, but drops with increasing p_T for large rapidity in the forward or backward regions. The observation of a large p_T particle implies a depletion in the amount of energetic secondaries. This is naturally also what is found in the hemisphere away from the observed pion, but the associated multiplicity within short-range now increases strongly with p_T. The corresponding peak is relatively wide in rapidity. For reference purposes, one should say that in a parton picture one expects a two-unit spread due to a "typical" transverse momentum of 0.35 GeV/c with respect to the over-all momentum of the recoiling parton. The direction of this momentum is furthermore only roughly opposite to that of the detected π^0 (rapidity zero here or 90°) as a result of the momentum distribution of the primary partons suffering the collision. This is to say that even in a hard collision picture one cannot expect too good a localization of the associated secondaries. The one observed is fine. One may however say that this is yet another passive test. The increase in multiplicity at low rapidity, and the decrease in multiplicity at large rapidity, altogether result in an over-all increase, and at fixed p_T the more so the larger is the energy. Increasing the energy reduces the relative depletion in energetic particles, which for p_T = 3 GeV/c, say, results in an important variation of the associated multiplicity over the ISR energy range. In the opposite hemisphere, the multipli-

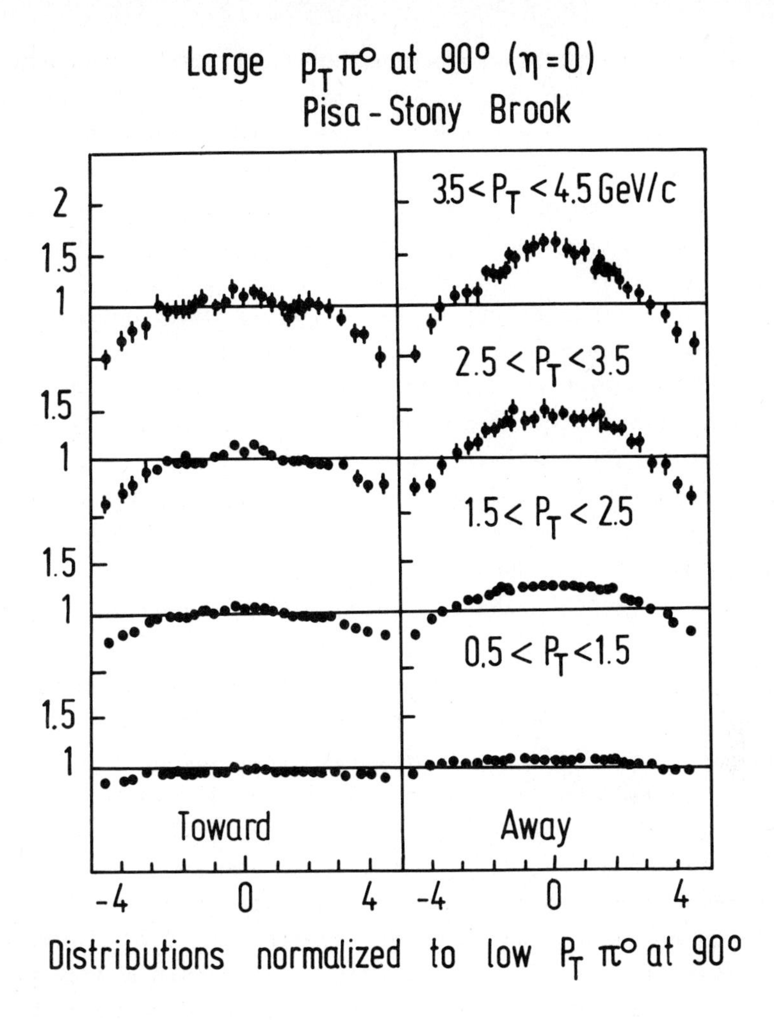

Fig. 20 Associated charged multiplicity with a trigger π^0 as observed in the hemispheres towards the π^0 and away from it. $\sqrt{s}$ = 53 GeV.

city increase is essentially a function of p_T and not of energy. The large increase at low rapidity strongly dominates over the depletion at large rapidity. These effects have been observed by the Pisa-Stony Brook Collaboration and their results for the over-all multiplicity in both hemispheres are shown in Fig. 21. The yields are normalized to those observed triggering on a low p_T (typical) π^0. The rise of the multiplicity at fixed p_T observed in the hemisphere toward the π^0 is due to the decreasing effect of depletion at small angles (energetic secondaries). The relative yields at wide angle do not change with energy. The very marked difference between what happens in the two hemispheres thus separated corresponds to a strong azimuthal variation of the associated multiplicity. This is shown in Fig. 22. The direction of the triggering π^0 defines 0°. The yields are again normalized to what is measured triggering on a low p_T π^0. The multiplicity is almost independent of p_T when measured in the direction of the π^0, while it increases almost linearly when it is measured in the opposite direction.

This defines what one may wish to call a jet of secondaries. Looking back at the parton picture for reference purposes, this would correspond to the free flare of particles resulting from a parton scattered at 90°, whereas the actual π^0 trigger would select a jet with essentially one hard particle only in it. The angular spread of Fig. 22 may also suggest a jet, but a rather wide one indeed. A large fraction of the associated secondaries are produced at rather large angles ($\theta > 45^o$) with respect to the mean jet direction. This suggests that they are basically soft ($|p| \sim 0.4$ GeV/c), while only a very small fraction of the associated multiplicity may correspond to rather hard particles ($|p| \gtrsim 1$ GeV, say). From Figs. 20 and 22, one may draw an identity kit portrait for the typical large p_T event selected triggering on a 4 GeV/c π^0 at 90°. The key feature is an increase of the associated multiplicity at wide angles. Within two units of rapidity centred on 0, one typically finds (on the average) two charged secondaries in each hemisphere (3 pions, say). In a large p_T event, the corresponding multiplicity in the same hemisphere rises to 5.5 (one for the triggering π^0 and one to two more rather soft secondaries). The multiplicity in the other hemisphere shows a number of soft particles which strongly increases with p_T, while the over-all momentum distribution extends gradually toward higher p_T. The multiplicity within two units of rapidity may now reach 8 at p_T = 4 GeV/c, most of the particles though (6, say) being rather soft secondaries.

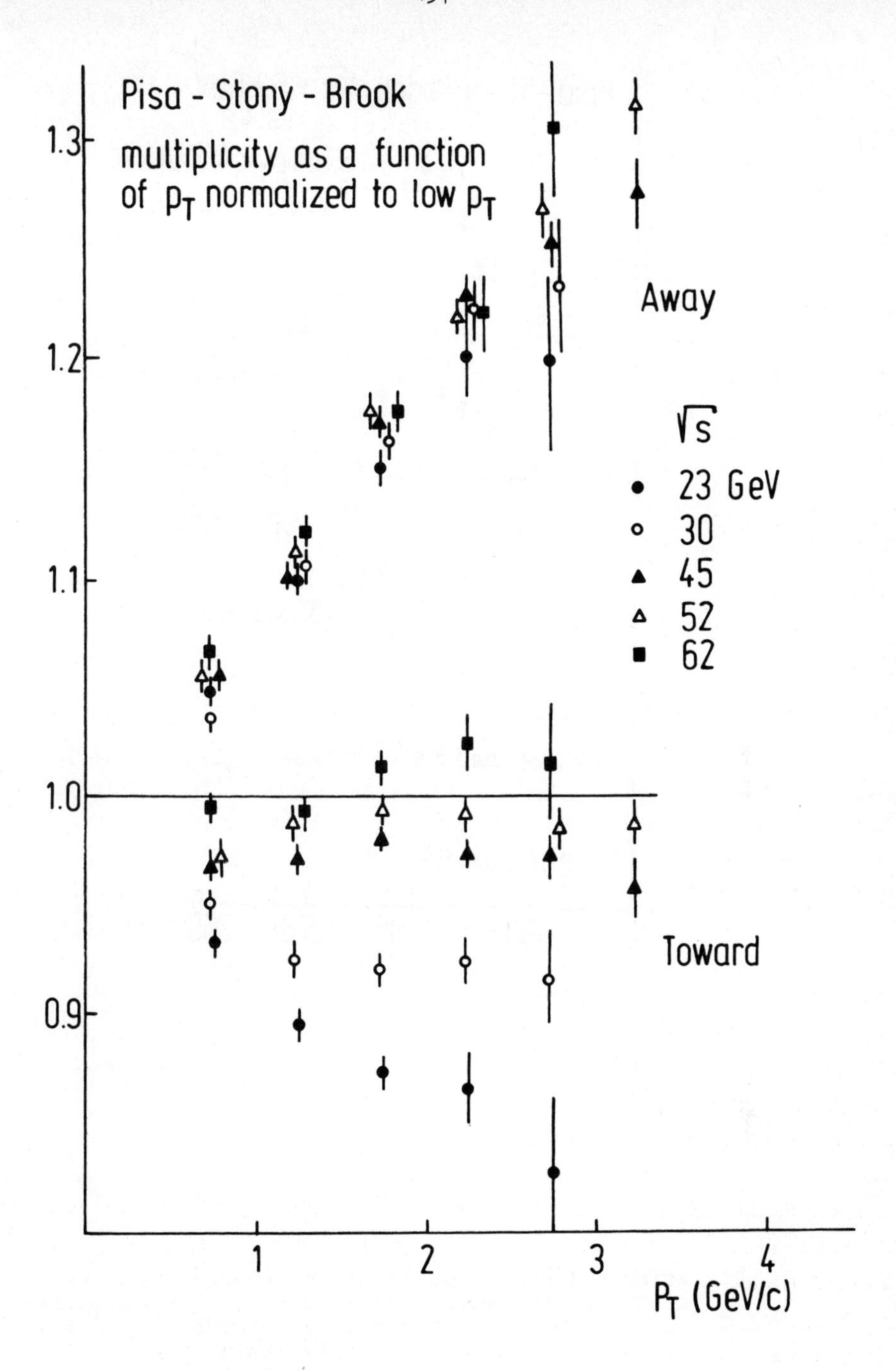

Fig. 21 Multiplicity as a function of p_T (normalized to low p_T value). Shown are the charged multiplicities in both hemispheres, away and towards.

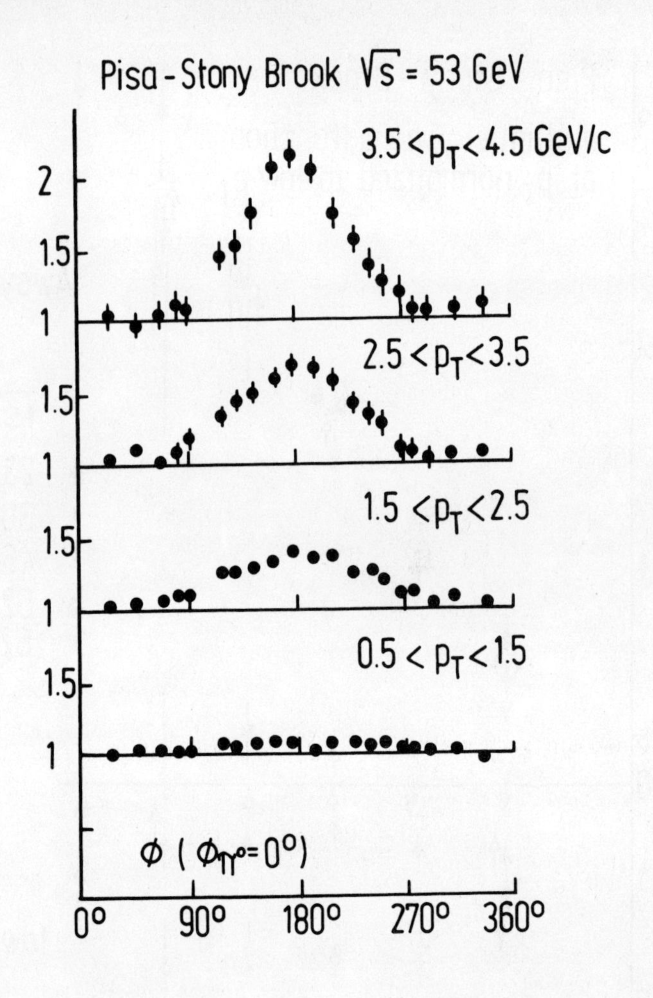

Fig. 22 Azimuthal distribution of the charged particles seen together with a π^0 at $\phi = 0^\circ$. This corresponds to a 1.5 unit rapidity interval centred at $\eta = 0$. One observes a coplanatary effect. The distribution remains rather wide in ϕ .

The separation between hard and soft particles is likely to be smooth. Harder secondaries are associated here with a specific dynamical effect. Part of the effect may, however, be due to mere momentum conservation. Detailed model calculations are still lacking for a conclusion[31], Nevertheless, one may say that the mean p_T value of the associated secondaries in the opposite hemisphere ($\langle p_T \rangle \sim 0.8$ GeV/c) is definitely larger than the typical p_T value of 0.35 GeV/c. One does not have merely more typical soft pions in order to balance an unlikely one with a large p_T value. One also has harder ones. Yet the ϕ distributions show that most of them are soft anyway and, as a result, the probability of getting real hard ones ($p_T > 1$ GeV/c) should rise in a very important way from what it is for all inclusive distributions.

If I stress this picture it is again because of its possible connection with the p_T^{-n} behaviour which favours $n = 8$ at ISR energies. With a large value of p_T on one single pion, one may favour configurations where one hard particle only occurs on one side (peculiar jet, with corresponding amplitude behaving as p_T^{-2}), while an uncontrived jet, with several particles sharing the over-all value of p_T, would develop on the other side. One would then expect, however, that, with increasing energy, at fixed p_T, the corresponding bias should weaken. As a result, hard particles could also be found on the same side, while the best value for n would decrease, accordingly. Triggering on energy (calorimeter) rather than on a single pion should give a different power.

This description is extremely tentative. It should be soon ruled out or confirmed by the CERN-Columbia-Rockefeller-Saclay experiment[32]. It, however, just passed an important test. Rotating the direction of the π^0 trigger, one was led to expect that, to a first approximation, the identity kit portrait would just rotate. Such a rotation should not much affect all soft secondaries, but should bring the harder ones toward the direction opposite to that of the large p_T π^0 triggered upon. The rapidity distribution of the associated secondaries should accordingly become skewed, still important at rapidity zero, whatever the rapidity of the π^0 trigger is, it should shift to the other side. Recent results from the Pisa-Stony Brook Collaboration with a trigger at 20° confirm this. The rapidity of the π^0 may be too large not to give also an important role to mere energy-momentum conservation. The new results of the Aachen-CERN-Heidelberg-Munich Collaboration, shown in Fig. 23, correspond to a relatively small shift in rapidity[33]. They are obtained with a π^0 trigger at 90° in the

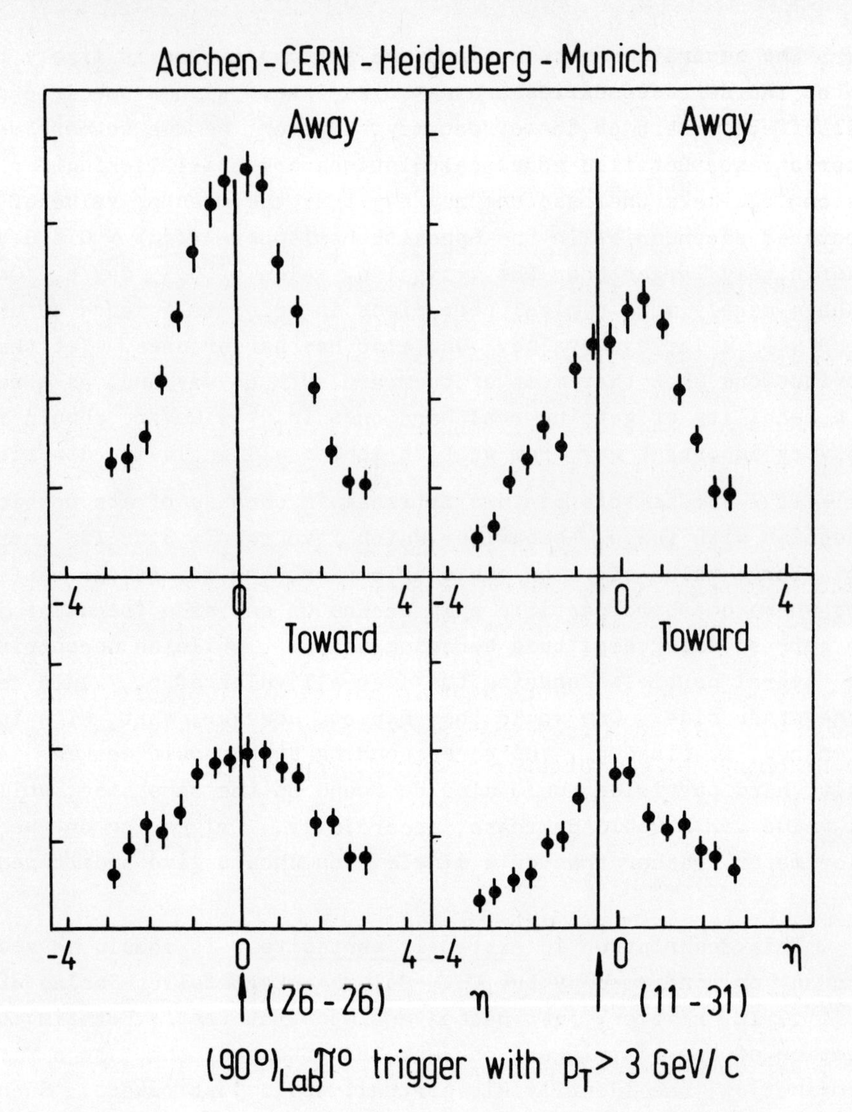

Fig. 23 Associated charged multiplicity with a π^0 with $p_T \gtrsim$ 3 GeV/c at rapidity 0 and rapidity 0.6.

laboratory but with asymmetric beam momenta, with accordingly a rotation of the selected direction in the centre-of-mass system. The distribution of the associated charged particles is now skewed in the expected direction. It is, as expected, mostly the case in the opposite hemisphere. In the same hemisphere, the distribution still peaks at rapidity zero.

Advocates of a close relationship between large p_T phenomena and deep inelastic electron scattering are far from being able to point at angular (or better momentum) distributions, which would be clearly "jet-like". Nevertheless, none of the straightforward effects expected from hard collisions between proton constituents has yet been ruled out by the data, while the picture certainly offers a simple self-consistent description of what is happening.

The important rise of the inclusive rate with increasing energy should make large p_T phenomena a very important topic at the ISR, irrespective of the much larger yields at NAL. Very important results from NAL should, however, be those corresponding to incident mesons. They should differ much from what is observed with incident protons[22]. In any case this is a domain where covering the 400-2000 GeV range will be enough of an advantage over NAL to consider a long-range programme at the ISR[34].

REFERENCES AND FOOTNOTES

1. Proc. 1973 Lovain Summer Institute in Theoretical Physics (Plenum Press, New York, 1974). The abbreviations used throughout are:
   ISR (Intersecting Storage Rings, CERN);
   NAL (National Accelerator Laboratory, Batavia, Illinois, now FermiLab);
   SPS (Super Proton Synchrotron, CERN, presently being constructed).

2. Studies for antiproton storage give a luminosity of the order of 10^{25}, which is very low for experimentation.

3. A guided tour around the ISR, with 6 intersections used for physics now includes the following set-ups:
   I1 Saclay double-arm spectrometer used in coincidence with the CERN-Columbia-Rockefeller lead-glass detector.
   I2 British barrel detector used in coincidence with the CERN-Holland-Lancaster-Manchester small-angle spectrometer and the former British-Scandinavian wide-angle spectrometer.
   I4 Extensive research programme with the split field magnet detector (four different experiments with machine time so far).

I6 Analysis of several particles at the same time through a wide acceptance magnet (Aachen-CERN-UCLA) and then more precise measurements of $d\sigma/dt$.
I7 Aachen-CERN-Heidelberg-Munich streamer chamber used in coincidence with a wide-angle lead-glass detector.
I8 Pisa-Stony Brook 4π detector used in coincidence with a lead-glass detector, the CERN-Rome forward spectrometer, the Scandinavian-MIT wide angle spectrometer, etc.

4. See also J.D. Jackson, in Proc. Scottish Universities Summer School (1973).

5. For a discussion of phenomenological duality see K. Igi, Phenomenological duality, to be published in Phys. Reports; M. Jacob, Regge models and duality, in Proc. Brandeis University Summer School (1970).

6. A. Melissinos, Invited paper, APS Chicago meeting (1974).
A. Melissinos, to be published in Phys. Reports.
The ISR results are reviewed by U. Amaldi, Rapporteur's talk, Aix-en-Provence Conference, 1973.

7. H. Miettinen, Rencontre de Moriond, CERN preprint (1974).

8. L. Foà, Rapporteur's talk, Aix-en-Provence Conference, 1973.

9. For a review, see S. Fubini, Multiperipheral Dynamics, Scottish Universities Sommer School (1963).

10. The definition of cluster properties from correlations in the central region is discussed in Ref. 1. For a detailed review of clustering effects and their observation through statistical analysis see R. Slansky, to be published in Phys. Reports 11 C.

11. Saclay-CERN-Columbia-Rockefeller Collaboration, Contribution to 17th Int. Conf. on High-Energy Physics, London, 1974.

12. Pisa-Stony Brook Collaboration, Contribution to the 17th Int. Conf. on High-Energy Physics, London 1974. For a detailed discussion see ISR summary 9, CERN internal report (1974). Similar results have since been obtained by the Aachen-CERN-Heidelberg-Munich streamer chamber experiment.

13. One may generalize this procedure and try to isolate a term independent of n (varying ΔY) and ΔY (varying n) in $C_3^{(n)}(\Delta Y/n)$ or $R_3^{(n)}(n/\Delta Y)$ and identify it with

$$\frac{\langle K(K-1)(K-2)\rangle n}{\langle K\rangle_n} D^{(3)}(v_1^2, v_2^2, v_3^2)$$

with

$$v_i = Y - y_i \quad \text{and} \quad Y = \frac{1}{3}(y_1 + y_2 + y_3)$$

This is very interesting in view of the value tentatively attributed to $\langle K\rangle$. This, however, requires very large statistics. The procedure is to isolate the short-range correlation term and identify it with what is expected from particles produced from the same cluster.

14. Scandinavian Collaboration, K. Hansen, private communication.

15. US-USSR Collaboration at NAL, Contribution to 17th Int. Conf. on High-Energy Physics, London, 1974.

16. M. Jacob and R. Stroynowski, The shape of the quasi-elastic peak, to be published in Nuclear Phys. Of particular interest is the presence of a structure at $|t| \sim 0.3$ (GeV/c)^2, which would be associated with an important peripheral no-helicity flip contribution. This could correspond to most of the non-resonant "background" when coherently produced.

17. The ISR can be considered as a "Pomeranchon accelerator" in the same energy range as SLAC is a "photon accelerator".

18. J. Whitmore and T. Ferbel, private communications.

19. R.P. Feynman, Photon-hadron interactions, Benjamin Frontiers series (Benjamin, NY, 1971).
J. Kogut and L. Susskind, Physics Reports 8 C, 75 (1973)
S.D. Ellis and M.B. Kisslinger, to be published in Phys. Rev.

20. S. Berman, J. Bjorken and J. Kogut, Phys. Rev. D4, 3388 (1971).
M. Jacob and S. Berman, Phys. Rev. Letters 25, 1683 (1970).
R. Blankenbecler, S. Brodsky and F. Gunion, Rev. D6, 2652 (1972); D8, 287 (1973).
P. Landshoff and J. Polkinghorne, as reviewed by J. Polkinghorne, Aix-en-Provence Conference, 1973.

21. There is no theory for large p_T phenomena. There are many models which are not reviewed here. One should consult J. Bjorken, Rapporteur's talk, Aix-en-Provence Conference, 1973; P. Landshoff, Rapporteur's talk, 17th Int. Conf. on High-Energy Physics, London, 1974;
S. Ellis and R. Thun, Rencontre de Moriond, 1974.

22. The theoretical aspect of large transverse momentum phenomena in the parton picture is described by J. Polkinghorne in his series of lectures (Bonn Institute).

23. Chicago-Princeton Collaboration, Contribution to 17th Int. Conf. on High-Energy Physics, London, 1974.

24. This was the situation at the time of the 1973 Aix-en-Provence Conference (results from the CCR Collaboration as discussed by J. Bjorken).

25. The inclusive distribution is then written as $d\sigma/dp_T^2 = (1/s^2)\, G(x_T)$, since s is the only quantity with a dimension. This can of course be rewritten as $d\sigma/dp_T^2 = (1/p_T^4)\, F(x_T)$. A different function F should appear for different production angles. Since all available data are at 90° this is often omitted.

26. This was emphasized by P. Darriulat in his invited paper at the Trieste Topical Conference on Intersecting Storage Rings Physics, 1974.

27. Confusion between η and π^0 (γ-rays are detected) with different bias in the two experiments may explain, at least partly, the actual discrepancy. To the extent that there is an important K/π ratio at large p_T (0.3 say), one may expect also a sizeable η/π ratio.

28. What is meant by "no" is that in most cases, all the secondaries seen on the same side are soft ($p_T \sim 0.4$ GeV/c). This does not exclude low probability configurations with two (several) large p_T particles. They, however, do not contribute much to the inclusive yield (Fig. 15).

29. British-Scandinavian Collaboration, Contribution to the 17th Int. Conf. on High-Energy Physics, London, 1974.

30. The p/π^+ is found to decrease with the atomic number. Chicago-Princeton Collaboration, Contribution to the 17th Int. Conf. on High-Energy Physics, London, 1974.

31. Another possible picture is that of a heavy central fireball with approximate isotropic decay. It would trap the two positive charges and give accordingly a positive excess. One would not expect charged yields at large p_T to factorize in opposite directions as one would expect in the parton picture.

32. Results reported at the London Conference indicate a large increase of the number of soft secondaries with increasing p_T, while the probability of finding harder ones increases strongly. This occurs on both sides. Most of the associated secondaries are soft.

33. Aachen-CERN-Heidelberg-Munich Collaboration, ISR summary 10, CERN internal report (1974). I am indebted to P. Darriulat for many discussions about this experiment.

34. Two topics, large transverse momentum phenomena and diffractive excitation are covered in more detail in M. Jacob, Erice Lecture Notes (1974). The discussion on particle correlation given here follows what was presented at the Trieste Topical Conference on Intersecting Storage Rings Physics, unpublished, 1974.

EXPERIMENTAL REVIEW

REVIEW OF RESULTS OBTAINED AT THE CERN INTERSECTING STORAGE RINGS.

Ugo Amaldi
CERN, Geneva, Switzerland

1. INTRODUCTION

In this experimental review, after a short presentation of the Intersecting Storage Rings (Section 2), a summary is given in Section 3 of the information obtained at this machine during the years 1971-1973, i.e. in the first-generation experiments. Section 4 is devoted to some of the latest experimental results, while Section 5 contains a list of experiments which are at present under way or will be running in 1975.

2. THE INTERSECTING STORAGE RINGS (ISR)

The construction of the ISR started in 1966 and the first proton-proton collisions were observed on 27 January 1971 [1]. The diameter of the ISR is 300 m and the beams cross at an angle of 14.8° in eight intersection regions[2]. The layout of one octant is shown in Fig. 1. At present the average vacuum is a few 10^{-11} Torr, the stacked currents for physics runs are as high as 15 A, and the fractional losses are of the order of 10^{-6} min^{-1}.

The motivation of colliding beam machines lies in the very large centre-of-mass energies which can be obtained using beams of "conventional" momenta. However, this simple idea could not be applied until in 1956 a group at the Midwestern Universities Research Association put forward the idea of particle stacking in circular accelerators[3]. The ISR are fed with protons accelerated by the CERN Proton Synchrotron (PS), and the stacking procedure is such that the ISR available phase space is filled with almost the same phase-space density as that of the PS beam. After stacking, a d.c. current of several amperes circulates in each of the two rings, and the protons of each beam have a momentum spread $\Delta p/p$ which is of the order of 2%. (This corresponds to $\Delta p \simeq 500$ MeV/c, to be compared with $\Delta p \simeq 7$ MeV/c in the PS.) The horizontal width of the two beams is ~ 40 mm and the vertical height $\simeq 3$ mm, so that they look like two ribbons crossing at an angle of about 15°.

The project luminosity $L \simeq 4 \times 10^{30}$ cm^{-2} sec^{-1} has been reached at the end of 1972, and now normal physics runs are performed with luminosities of the order of 5×10^{30} cm^{-2} sec^{-1}. By definition the luminosity L is the factor which relates the total cross-section σ_{tot} to the total interaction rate R_{tot} measured in one intersection region:

$$R_{tot} = L \cdot \sigma_{tot} \quad (1)$$

At the ISR the luminosity is measured by applying the method suggested in 1968 by Van der Meer[4], which is based on displacing the two beams vertically in small and known steps and on measuring the rate in a system of monitor counters which detects beam-beam events[5]. In 1972 the accuracy of the method has been pushed by the CERN-Roma and the Pisa-Stony Brook Collaborations to the level of ±2%, and at present it is approaching ±1%.

Future developments of the ISR include the focalization of the two beams in one intersection region (in a so-called low-β section), which should improve the luminosity by a factor of about three. The possibility of stacking antiprotons is also under study: calculations show that by using a 200 GeV proton from the SPS to produce the antiprotons, it should be possible to obtain an antiproton-proton luminosity of the order of 5×10^{25} cm^{-2} sec^{-1}.

3. RESULTS OF FIRST-GENERATION EXPERIMENTS

Since large transverse momenta phenomena are discussed at this School by M. Jacob, I shall consider only large cross-section experiments, i.e. experiments which study the bulk of the proton-proton interaction.

In inclusive experiments the variables are, as usual, the longitudinal and transverse momenta p_L and p_T; the Feynman variable

$$-1 \leq x = \frac{p_L}{p_{L_{max}}} \leq 1 \; ;$$

the centre-of-mass rapidity

$$y = \frac{1}{2} \ln \frac{E + p_L}{E - p_L} \; ;$$

and the pseudorapidity

$$\eta = -\ln \mathrm{tg} \frac{\theta}{2} \; ,$$

which coincides with y for $p_T^2 >> m^2$.

The ISR run at five beam momenta. The corresponding values of the c.m. energy $\sqrt{s}$ and of the proton rapidity y_p appear in Table 1.

Table 1

p_{ISR} (GeV/c)	$\sqrt{s}$ (GeV)	y_p
11.8	23.2	3.2
15.4	30.5	3.5
22.5	44.5	3.9
26.6	52.5	4.0
31.5	62.0	4.1

It is seen that at the ISR the so-called "central region", defined as $|y| \lesssim 2$, is separated by the so-called "central region", for which $|y - y_p| \leq 2$, and that the full rapidity range increases by $2 \times 0.8 \simeq 1.6$ units when the ISR momentum passes from 11.8 to 31.5 GeV/c.

The quantities measured in first-generation experiments were the total, elastic, and inelastic cross-sections σ_{tot}, σ_{el}, and σ_{in}, the elastic differential cross-section $d\sigma/dt$, the invariant cross-section for the inclusive production of various types of particles

$$f(y,s,p_T) = E \frac{d^3\sigma}{d^3p} ,$$

and the normalized two-body correlation function

$$R(y_1,y_2) = \frac{\sigma_{in}(d^2\sigma/dy_1\ dy_2)}{(d\sigma/dy_1)(d\sigma/dy_2)} - 1 . \tag{2}$$

In the last equation $d\sigma/dy$ is the single-particle y-distribution and $d^2\sigma/dy_1dy_2$ is the joint distribution for finding two particles, one at rapidity y_1 and the other at rapidity y_2.

The most important results of the ISR first-generation experiments can be summarized under the following points.

3.1 Scaling of the inclusive spectra

Many of the relevant data on the invariant cross-section $f = E(d^3\sigma/d^3p)$ at $p_T = 0.4$ GeV/c are summarized in Fig. 2, which shows that scaling, i.e. independence of f from s, is valid to $\sim$ 10% and is reached first in the fragmentation region ($y_{LAB} \lesssim 2$) then in the central region. Note also that lighter particles reach the limiting distributions at lower energies than do the heavier ones.

3.2 Existence of a large diffractive component

The proton inclusive distributions measured at small angles with respect to the beams, by the CERN-Holland-Lancaster-Manchester Collaboration[6], showed a large peak close to the kinematic limit x = 1 (Fig. 3). (The relation between the x of the forward-going proton and the mass M of the object in which the other proton is supposed to be excited reads

$$x \cong 1 - \frac{M^2}{s} .) \tag{3}$$

The data of Fig. 3 show that in this process masses as high as $\sim$ 7 GeV are excited. The measured x-distributions agree with a $1/(1 - x)$ dependence, i.e. the mass spectra have a $1/M^2$ behaviour. The inelastic cross-sections of this "single dissociation" process is $\sim$ 5 mb, a relevant fraction of the total inelastic cross-section $\sigma_{in} \simeq 35$ mb. It is usually assumed that the physical origin of these single dissociation events is inelastic diffraction, so that the process is named "diffraction dissociation in large masses".

3.3 Scaling of the two-particle correlation function

Measurements of correlations between photons and charged secondaries by the CERN-Hamburg-Vienna Collaboration[7)] and between charged particles by the Pisa-Stony Brook Collaboration[8)] have revealed the existence of strong positive correlation between particles of equal rapidity. Some of the data obtained in the central region by the Pisa-Stony Brook Collaboration appear in Fig. 4. They show that not only the value of the normalized correlation function for equal pseudo-rapidities $\eta_1 = \eta_2$ is energy independent in the central region, but also that the correlation length $\Delta\eta \simeq 2$ is almost energy independent, so that the two-particle correlation function scales. This experimental fact is most easily interpreted by assuming that groups of hadrons (clusters) are produced independently with a uniform distribution in rapidity and then isotropically decay[9)]. Fitting the data, it is found that at ISR energies there are 3-4 hadrons per cluster and the cluster mass is of the order of 2-3 GeV. Of course this is only one particularly simple model, and it may very well be that clusters are nothing more than a convenient way of describing the experimental results.

3.4 Diffractive nature of elastic scattering

The Aachen-CERN-Genova-Harvard-Torino Collaboration[10)] has measured proton-proton elastic scattering for momentum transfers $|t| \lesssim 5$ GeV^2 and found a clear minimum at $|t| \simeq 1.4$ GeV^2 (Fig. 5). The presence of such a minimum is typical of diffraction scattering, so that the experiment confirms the expectations: at these large energies elastic scattering is mainly due to diffraction, i.e. to the absorption of the incoming waves induced by the inelastic processes. If this is the case, the behaviour of elastic scattering in the forward region is simply related, in an optical picture, to the radius of interaction of the two protons. In Fig. 6 are plotted the data at present available on the slope b of the differential cross-section (b is obtained by writing, in a small interval of $|t|$, $d\sigma/dt \propto e^{-b|t|}$). The ISR and FNAL data show that b is roughly proportional to ln s and that it increases by $\sim$ 8% in the ISR energy range.

3.5 Increasing total cross-sections

The CERN-Roma and Pisa-Stony Brook Collaborations have shown, by applying two different methods, that the total cross-section σ_{tot} increases by $\sim$ 10% in the ISR energy range[11,12)]. As obtained by the CERN-Roma Collaboration[11)], the same happens also for the elastic and inelastic cross-sections σ_{el} and σ_{in} (Fig. 7).

Leaving out the very important, but also very rare, phenomena in which particles with large transverse momenta are produced, the following picture emerged from first-generation experiments at the ISR. Two types of inelastic processes dominate the inelastic cross-section. A cross-section of about 5 mb is due to single diffraction dissociation events in which, together with the known low mass resonances, a continuum of masses up to $\sim$ 10 GeV are produced; these events have a relatively small multiplicity. The rest of the inelastic cross-section, about 30 mb, corresponds to events in which many particles (90% of which are pions) are produced. These "pionization" events fill up the central region of the rapidity plot and cause a multiplicity which rises as ln s and very strong short-range correlations. The amount of correlation can be explained by assuming the independent emission of clusters and their successive isotropic decay. The protons show the so-called "leading particle" effect, i.e. in pionization events they tend to keep their momentum so that their inclusive distribution is virtually y-independent (see Fig. 2), apart from the peak which is due to the diffractive events. At these energies elastic scattering is mainly due to diffraction, i.e. it is the shadow of the two types of inelastic mechanisms described above: diffraction dissociation and pionization. This fact implies that the proton-proton elastic amplitude f(t) is mainly imaginary, so that from the knowledge of the elastic cross-section $d\sigma/dt$ one can deduce, by a simple bidimensional Fourier transform, the amplitude f(a) as a function of the proton-proton impact parameter a. This has been done by various authors[13)], and in Fig. 8 the inelastic overlap function $G_{in}(a)$ is plotted versus the impact parameter. This function is defined as the contribution of all inelastic processes to the unitarity relation in impact parameter space[14)]:

$$\mathrm{Im}\, f(a) = |f(a)|^2 + G_{in}(a)$$

and can be obtained from the knowledge of Im f(a) if the elastic amplitude is essentially imaginary. Figure 8 shows that the inelastic overlap function is not a simple Gaussian in impact parameter space and, more important, that the increase ΔG_{in} of this quantity in the ISR energy range is mainly peripheral, i.e. that the increase of σ_{in} and thus of σ_{tot} is concentrated around $a \simeq 1$ fm. A point of current interest is how to explain this fact in the framework of a two-component (diffraction and pionization) picture of inelastic processes[15)].

4. SOME SECOND-GENERATION EXPERIMENTS

4.1 Diffraction dissociation

Last year, data on diffractive events were gathered by the Aachen-CERN-UCLA (ACU), the CERN-Hamburg-Orsay-Vienna (CHOV), and the CERN-Holland-Lancaster-Manchester (CHLM) Collaborations. The ACU Collaboration has collected data on the inclusive production of states which decay in $p\pi^+\pi^-$, detecting the production of well-defined resonances already seen at lower energies[16]. The CHOV Collaboration has been working with the Split Field Magnet (SFM), a general magnetic facility which contains a large number of multiwire proportional chambers (Fig. 9). For this experiment four neutron impact detectors were added to the SFM, so that the events $pp \rightarrow pn\pi^+$ could be completely reconstructed. Figure 10 shows the measured angular distributions at $\sqrt{s} = 53$ for various masses M of the $n\pi^+$ system[17].

The small t-region can be fitted with an exponential, and its slope decreases rapidly with increasing mass. Comparing, for a fixed value of M, the measured slope with lower energy data, it is concluded that all slopes increase with s approximately at the same rate as in elastic scattering. Thus, also in this respect, diffraction dissociation seems to behave as elastic diffraction.

With the apparatus shown in Fig. 11, the CHLM Collaboration has studied the decay distribution of the masses produced in single dissociation events[18]. In the small-angle spectrometer (which comprises two setpum magnets and three momentum measuring magnets) the undissociated proton is detected, while a set of counters around the intersection region detects the other charged particles. Since only angles are measured, the results are plotted as a function of the pseudorapidity η of the charged tracks accompanying the quasi-elastically scattered proton, which has x close to 1. Some of the results are shown in Fig. 12. The horizontal scale is the rapidity of the charged tracks, while on the vertical scale their inclusive distribution is plotted. This distribution is normalized to the proton yield. The various figures refer to two different centre-of-mass energies, to various masses M, and to slightly different values of the transverse momentum p_T of the proton. The rapidity of the proton is indicated by an arrow, and it determines kinematically the mean rapidity y_m of the distribution of the remaining tracks $[y_m = \log(\sqrt{s}/M)]$. Figure 12 shows that the width Δy of this distribution increases with the mass M roughly as $\ln M^2$. This is a relevant fact, because it tells us that the mass M is not a single cluster decaying isotropically, thus implying $\Delta y \simeq 2$. On the contrary, the data indicate that, at least for large masses, the "decay" of the diffractionally excited mass fills a rapidity interval $\Delta y \simeq \ln M^2$, as it happens in proton-proton collision at an energy $\sqrt{s} = M$.

4.2 Correlations in the fragmentation region

In intersection region 8 the CERN-Roma group has mounted a spectrometer which detects negative particles emitted at zero degrees and are deflected outwards by the magnetic field of the ISR (Fig. 13). The spectrometer has been used to measure inclusive spectra of π^-, K^-, and $\bar{p}$ in the fragmentation region ($0.3 \leq x \leq 0.9$). The CERN-Roma group together with the Pisa-Stony Brook Collaboration have collected correlation data, using the PSB omni-directional counter hodoscope system.

The forward-going negative particle is recognized and momentum-analysed, while only the angles of the tracks of all the other particles are measured in the Pisa-Stony Brook counter system[19]. Figure 14 shows the measured normalized correlation function R as a function of the pseudorapidity η in the counter system for $\bar{p}$, K^-, and π^- of $x = 0.4$ detected in the spectrometer (which sits at positive values of η). The pion data show an enhancement for large negative values of η and a broad structure at positive η on top of a steadily decreasing continuum. This continuum is probably a consequence of the fact that the momentum taken away by the spectrometer reduces the probability of particle emission in the same hemisphere. The structure on top of the continuum indicates that the detected π^- belongs to a forward-going cluster, which gives rise to the typical short-range correlations. Monte Carlo calculations show that the mass of this cluster must be of the order of 2 GeV. In Fig. 15 the pion data have been subdivided into four classes according to the multiplicity detected in the counter system. It is seen that the strongest correlations are found in small multiplicity events, indicating that the leading cluster, which gives rise to the measured pion, is mainly decaying into a few bodies. By comparing the graphs relative to antiprotons and K^-'s in Fig. 14 with that of the π^- one, it is seen that the short-range correlation bump is very similar in the three cases. This seems to indicate that also K^- and $\bar{p}$ of this relatively large x are produced in clusters. Monte Carlo calculations show that the mass of the leading cluster decaying into $\bar{p}$ is of the order of 3 GeV.

4.3 Correlations at fixed values of the total multiplicity

The correlative data plotted in Fig. 4 have been obtained by the Pisa-Stony Brook Collaboration by taking all the events, irrespective of their multiplicity. These inclusive data have clearly demonstrated the existence of strong short-range correlations in the pionization component, but cannot be unambiguously interpreted because this would require a knowledge of the long-range correlations which are due to single and double dissociation events. Since these dissociation events have an average multiplicity which is smaller than the multiplicity in the pionization events, the correlations in events of multiplicities n larger than the average value $\langle n \rangle$ are dominated by the pionization component and their study allows a better understanding of the clustering effect observed in the central region.

Preliminary data on semi-inclusive correlations have been presented by L. Foà at the Aix-en-Provence Conference[20]. The final analysis is almost completed and the observed correlations as a function of n give new information on the decay distribution of the central clusters[21].

5. AN OUTLOOK ON THE FUTURE

In spite of the many experiments, up to now scaling has been checked with typical errors of 10-15%, and this is not enough to see a possible $\sim$ 10% effect in the central region related to the 10% increase of the total cross-section. At present in intersection region I8 the British-Scandinavian-MIT Collaboration is performing an accurate test of scaling; the experiment will be completed in 1974.

In the Split Field Magnet two groups are taking, or about to take, data on two-body correlations. The fragmentation region is being studied by the CERN-College de France-Heidelberg-Karlsruhe Collaboration, while the MIT-Orsay-Scandinavian Collaboration is interested in particle correlations at large angles. In 1975 the Adelphi-Brookhaven-Roma Collaboration will mount in intersection I1 a large coverage gamma detector to study multigamma correlations. Further in the future will be the installation, in the same intersection, by the CERN-Columbia-Rockefeller Collaboration, of a superconducting solenoid to search for electron pairs, and to study multipion correlations. Muon pairs will be detected in 1976 by the set-up of the Genova-Harvard-MIT-Pisa Collaboration.

Diffraction dissociation at small angles will be measured in 1975 by the CERN-Holland-Manchester Collaboration, while an experiment on double diffraction dissociation in the channel $(p\pi^+\pi^-)(p\pi^+\pi^-)$ is at present under way by the Pavia-Princeton Collaboration in the Split Field Magnet.

Some elastic scattering data have been presented by the CHOV Collaboration at the London Conference, and many more should be produced by the end of 1974, while in 1975 the CERN-Genova-Harvard-Munich-Northwestern-Riverside Collaboration will detect in intersection region I6 elastic scattering events at large momentum transfer by using both magnets and hadron calorimeters.

Measurements of total cross-sections which do not rely on the Van der Meer method for the machine luminosity are under way both by the CERN-Pisa-Roma-Stony Brook and the Aachen-CERN-Genova-Harvard-Torino Collaborations. In 1975 the CERN-Roma Collaboration will extend the measurement of the forward real part of the proton-proton scattering amplitude at the highest ISR energies.

Finally, in the field of large transverse momenta, not discussed in this review, data are expected to come from the Split Field Magnet (Aachen-CERN Collaboration), from the solenoid of the CERN-Columbia-Rockefeller Collaboration and from the set-up of the Brookhaven-CERN-Saclay-Syracuse-Yale Collaboration, which uses calorimeters and transition radiation detectors.

REFERENCES

1) ISR staff, Phys. Letters 34 B, 425 (1971).

2) K. Johnsen, Nuclear Instrum. Methods 108, 205 (1973).

3) D.W. Kerst et al., Phys. Rev. 102, 590 (1956).
K.R. Symon and A.M. Sessler, Proc. Int. Conf. on High-Energy Accelerators and Instrumentation (CERN, Geneva, 1956), p. 44.

4) S. Van der Meer, CERN Int. Report ISR-PO/68-31 (1968).

5) For a discussion of the method see, for instance, U. Amaldi, Proc. Roy. Soc. London A335, 431 (1973).

6) CERN-Holland-Lancaster-Manchester Collaboration, M. Albrow et al., Nuclear Phys. B51, 388 (1973) and M. Albrow et al., Nuclear Phys. B54, 6 (1973).

7) CERN-Hamburg-Vienna Collaboration, H. Dibon et al., Phys. Letters 44 B, 313 (1973).

8) Pisa-Stony Brook Collaboration, R. Amendolia et al., Phys. Letters 48 B, 359 (1974).

9) An incomplete list of references in 1973 is:
P. Pirilä and S. Pokorski, Phys. Letters 43 B, 502 (1973) and Nuovo Cimento Letters 8, 141 (1973).
J. Ranft and G. Ranft, Phys. Letters 45 B, 43 (1973).
A. Białas, K. Fiałkowski and K. Zalewski, Phys. Letters 45 B, 337 (1973).
W. Schmidt-Parzefall, Phys. Letters 46 B, 399 (1973).
C. Quigg and G.H. Thomas, Phys. Rev. D7, 2752 (1973).
E.L. Berger and G.C. Fox, Phys. Letters 47 B, 162 (1973).
S. Pokorski and L. Van Hove, CERN preprint TH.1772 (1973).
C.B. Chiu and K.H. Wang, Phys. Rev. D8, 2929 (1973).

10) Aachen-CERN-Genova-Harvard-Torino Collaboration, A. Böhm et al., Phys. Letters 49 B, 491 (1974).

11) CERN-Roma Collaboration, U. Amaldi et al., Phys. Letters 44 B, 112 (1973).

12) Pisa-Stony Brook Collaboration, R. Amendolia et al., Phys. Letters 44 B, 119 (1973).

13) U. Amaldi, Proc. 2nd Aix-en-Provence Conf. on Elementary Particles, J. Phys. (France) 34, 241 (1973).
R. Henzi and P. Valin, Phys. Letters 48 B, 119 (1974).
H.I. Miettinen, Invited talk at the 9th Rencontre de Moriond, CERN preprint TH.1864 (1974).
F.S. Henyey, R. Hong Tuan and G.L. Kane, University of Michigan report UMHE 73-18 (1973).

14) L. Van Hove, Rev. Mod. Phys. 36, 655 (1964).

15) N. Sakai and J.N.J. White, Nuclear Phys. B59, 511 (1973).
F.S. Henyey, Elastic scattering from the two-component picture, Max-Planck preprint (1974).
T. Inami, R.J.N. Phillips and R.G. Roberts, Phys. Letters 52 B, 355 (1974).

16) Aachen-CERN-UCLA Collaboration, L. Baksay et al., Measurement of $pp \to (p\pi^+\pi^-) + X$ at the CERN ISR; scaling test and evidence for double excitation, Contribution to the 17th Int. Conf. on High-Energy Physics, London (July 1974).

17) CERN-Hamburg-Orsay-Vienna Collaboration, E. Nagy et al., Experimental results on inelastic diffraction scattering in pp collisions at the ISR, Contribution to the 17th Int. Conf. on High-Energy Physics, London (July 1974).

18) CERN-Daresbury-Holland-Lancaster-Manchester Collaboration, M.G. Albrow et al., Phys. Letters 51 B, 424 (1974).

19) CERN-Pisa-Roma-Stony Brook Collaboration, J.V. Allaby et al., Correlations between charged particles and one momentum-analysed forward negative particle at the ISR, Contribution to the 17th Int. Conf. on High-Energy Physics, London (July 1974).

20) Pisa-Stony Brook Collaboration, L. Foà, Proc. 2nd Aix-en-Provence Conf. on Elementary Particles, J. Phys. (France), 34, 317 (1973).

21) For a recent paper, see E.L. Berger, Rapidity correlations at fixed multiplicity in cluster emission models, CERN/D.Ph.II/PHYS 74-17, and reference therein, to be published in Nuclear Phys. B.

Figure captions

Fig. 1 : Layout of the magnets of one octant in the ISR.

Fig. 2 : Invariant cross-section for the production of $\pi^{\pm}$, $K^{\pm}$, $p^{\pm}$ plotted versus the rapidity in the laboratory y_{LAB} at a fixed transverse momentum p_T = 0.4 GeV/c. This quantity is defined as the rapidity of the particle measured in the reference system of the proton which was initially going in the same direction as the detected particle. It is particularly useful to study scaling in the fragmentation region.

Fig. 3 : Spectra of the quasi-elastically scattered protons at small angles as measured by the CHLM Collaboration as a function of $M^2 \simeq s(1 - x)$.

Fig. 4 : Two-particle correlation in the central region measured by the Pisa-Stony Brook Collaboration.

Fig. 5 : Proton-proton elastic cross-section as a function of the momentum transfer at various c.m. energies (Aachen-CERN-Genova-Harvard-Torino Collaboration).

Fig. 6 : Slope b of the forward elastic scattering. The data have been obtained by the CERN-Roma and the Aachen-CERN-Genova-Harvard-Torino Collaborations at the ISR, and by the USSR-USA Collaboration at FNAL.

Fig. 7 : Total, inelastic and elastic cross-sections measured at the ISR.

Fig. 8 : Proton-proton inelastic overlap function computed by various authors as a function of the square of the impact parameter a^2 [13]. Note that in this graph a Gaussian in a is represented by a straight line.

Fig. 9 : A sketch of the Split Field Magnet and its multiwire proportional chambers.

Fig. 10 : Differential cross-section of the reaction $p + p \rightarrow p + (n\pi^+)$ as measured by the CHOV Collaboration for various masses M of the $(n\pi^+)$ system.

Fig. 11 : Apparatus used by the CHLM Collaboration to study the decay distribution of the masses produced in single dissociation events.

Fig. 12 : Normalized inclusive distributions of the charged secondaries which are produced in coincidence with a forward-going proton (CHLM Collaboration).

Fig. 13 : The CERN-Roma forward magnetic spectrometer and the PSB counter hodoscope system.

Fig. 14 : Normalized correlation functions R versus the pseudorapidity η of the particles detected in the PSB hodoscope system in coincidence with a π^-, a K^- and an antiproton of $x = 0.4$ detected in the CERN-Roma spectrometer.

Fig. 15 : Correlation function of a π^- of $x = 0.4$ emitted at 0° with the particles emitted in the whole solid angle, for various ranges of the multiplicity n detected by the PSB hodoscope system.

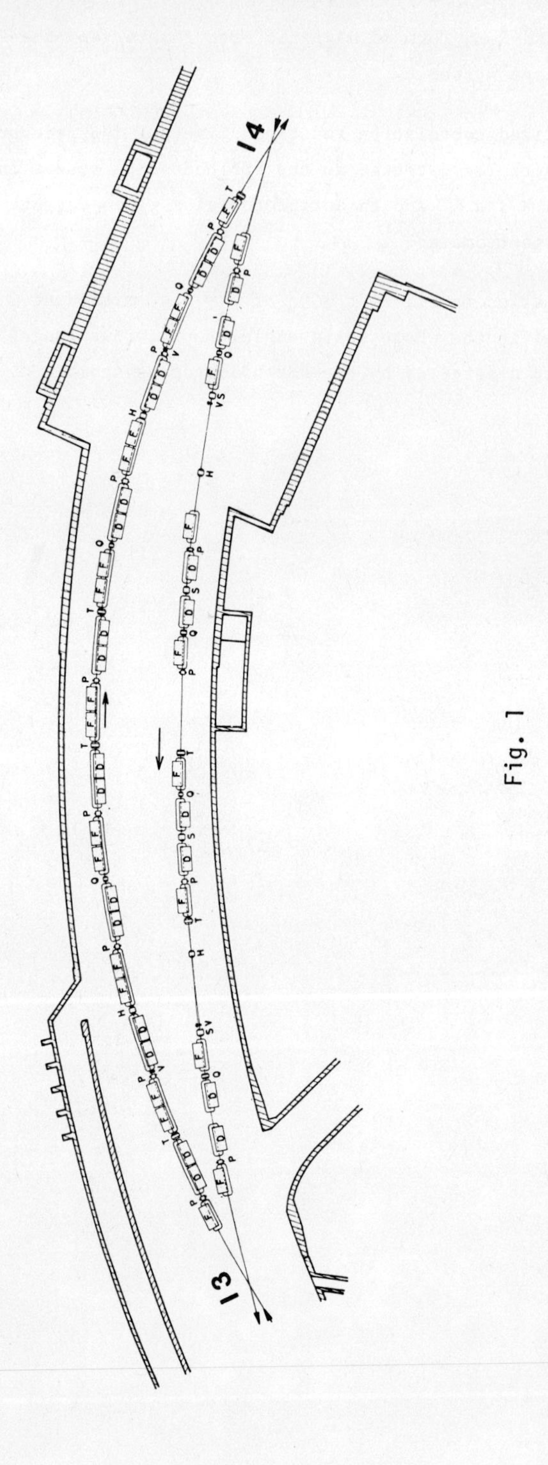

Fig. 1

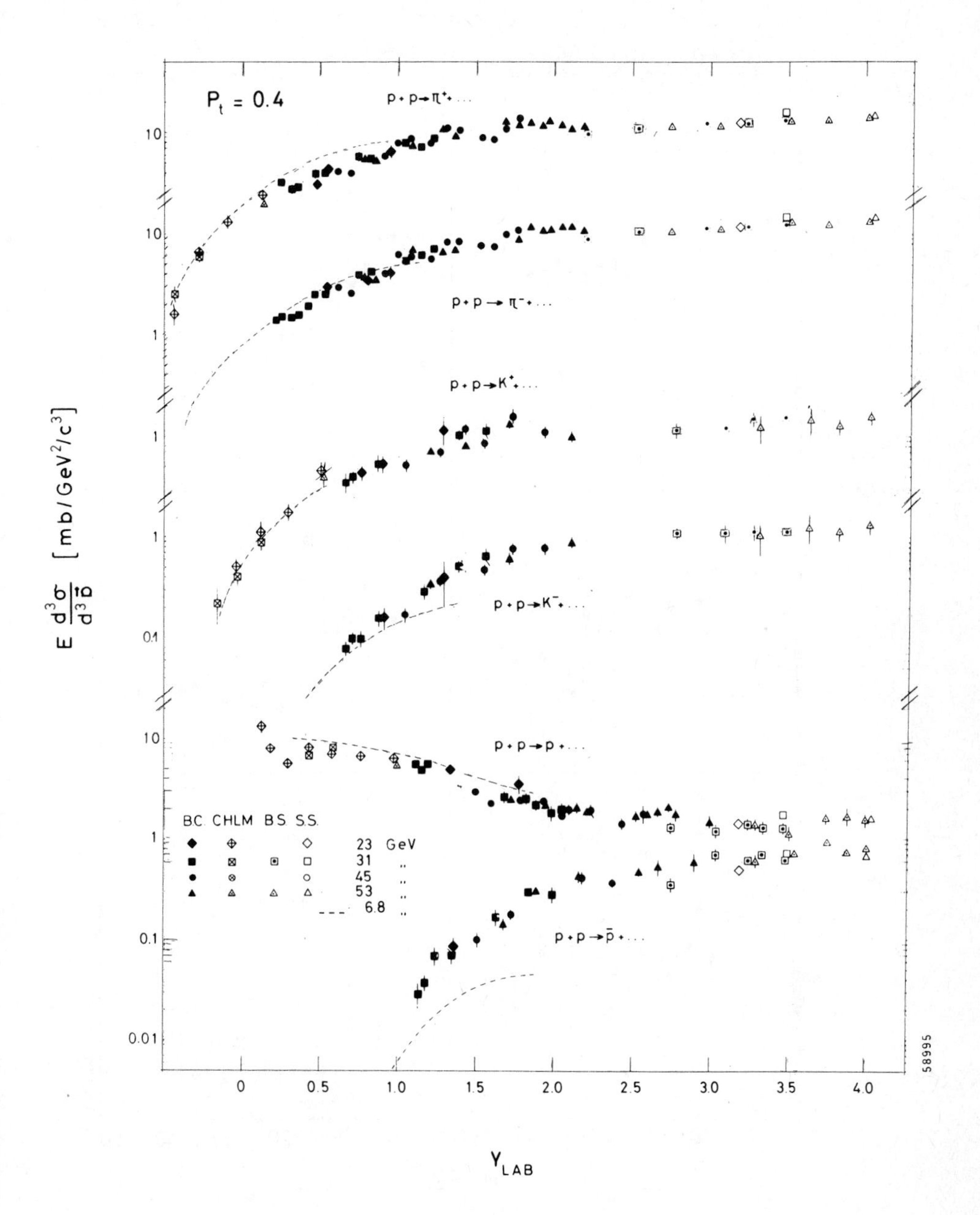

P_t = 0.4
p + p → π⁺ + ...
p + p → π⁻ + ...
p + p → K⁺ + ...
p + p → K⁻ + ...
p + p → p + ...
p + p → p̄ + ...
BC. CHLM B.S. S.S.
23 GeV
31
45
53
6.8
$E\frac{d^3\sigma}{d^3\vec{p}}$ [mb/GeV²/c³]
Y_{LAB}
58995

Fig. 2

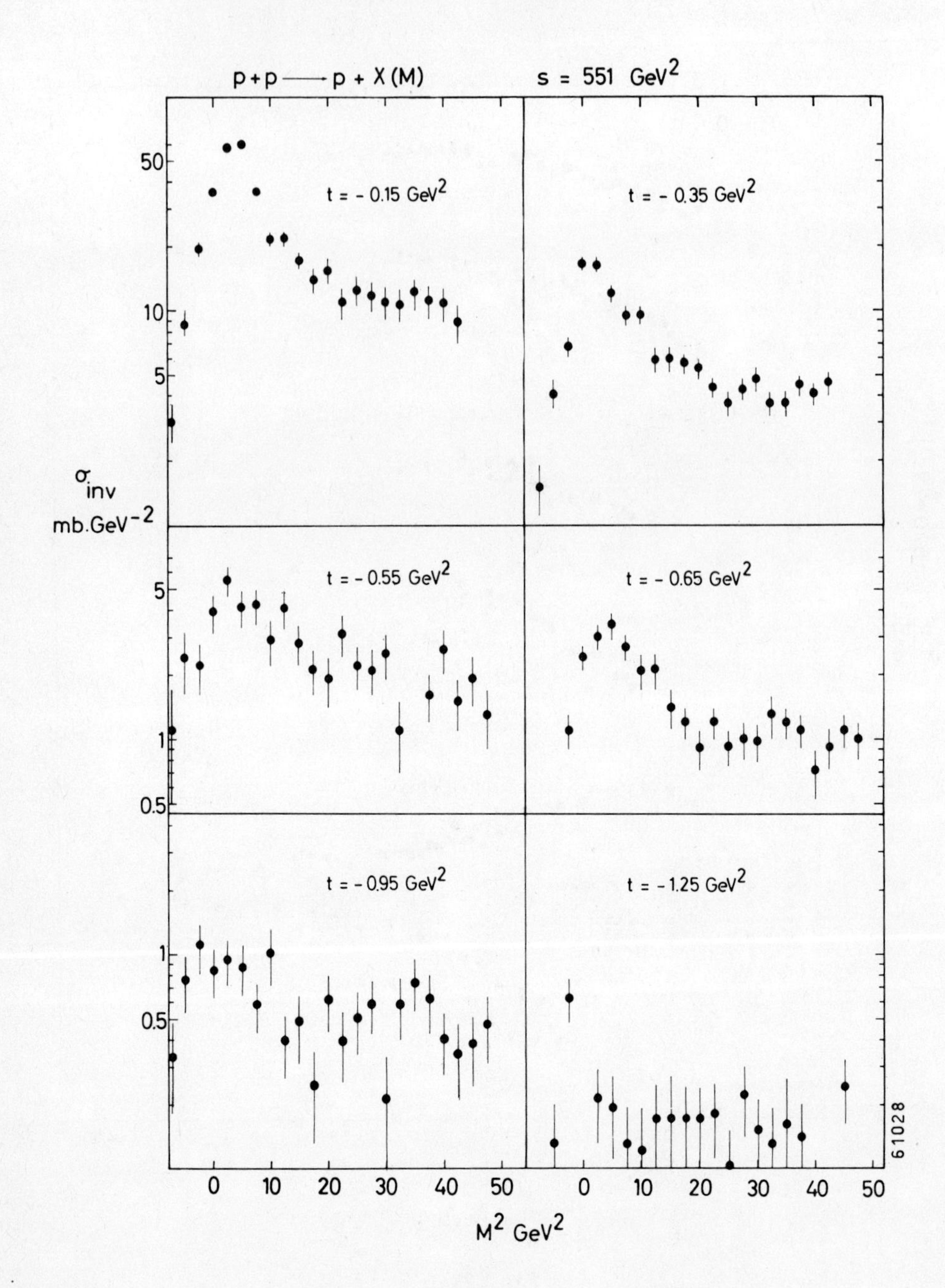
p + p ⟶ p + X (M)
s = 551 GeV²
t = - 0.15 GeV²
t = - 0.35 GeV²
t = - 0.55 GeV²
t = - 0.65 GeV²
t = - 0.95 GeV²
t = - 1.25 GeV²
σ inv
mb.GeV⁻²
M² GeV²
61028

Fig. 3

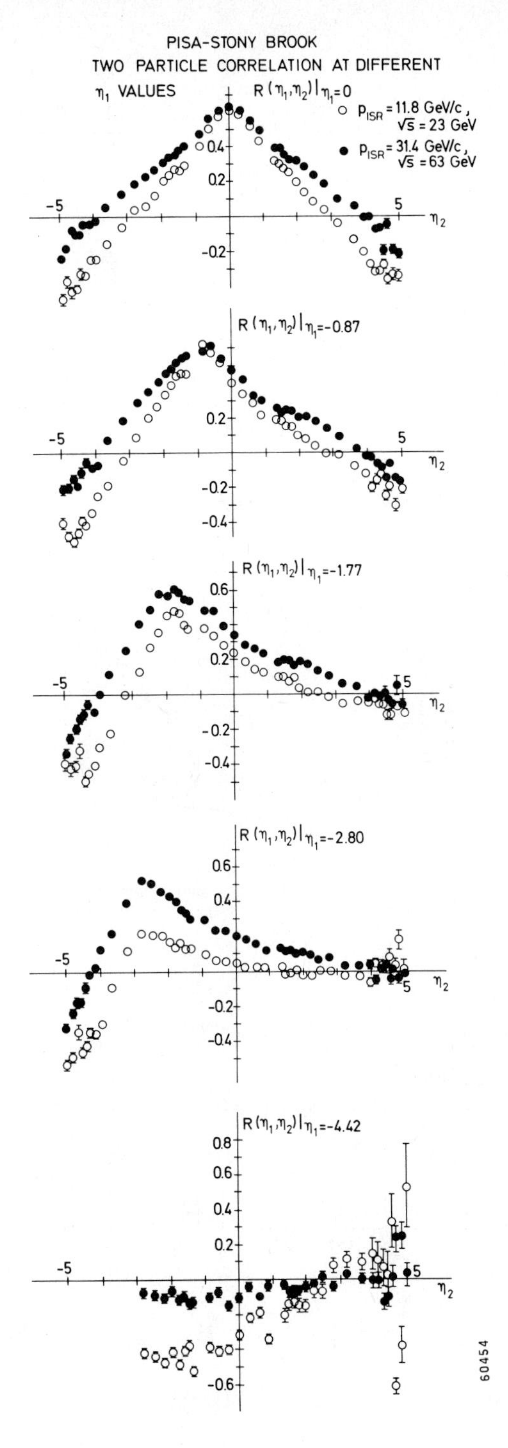

PISA-STONY BROOK
TWO PARTICLE CORRELATION AT DIFFERENT
η_1 VALUES
$R(\eta_1,\eta_2)|_{\eta_1=0}$
$p_{ISR} = 11.8$ GeV/c, $\sqrt{s} = 23$ GeV
$p_{ISR} = 31.4$ GeV/c, $\sqrt{s} = 63$ GeV
$R(\eta_1,\eta_2)|_{\eta_1=-0.87}$
$R(\eta_1,\eta_2)|_{\eta_1=-1.77}$
$R(\eta_1,\eta_2)|_{\eta_1=-2.80}$
$R(\eta_1,\eta_2)|_{\eta_1=-4.42}$
η_2
60454

Fig. 4

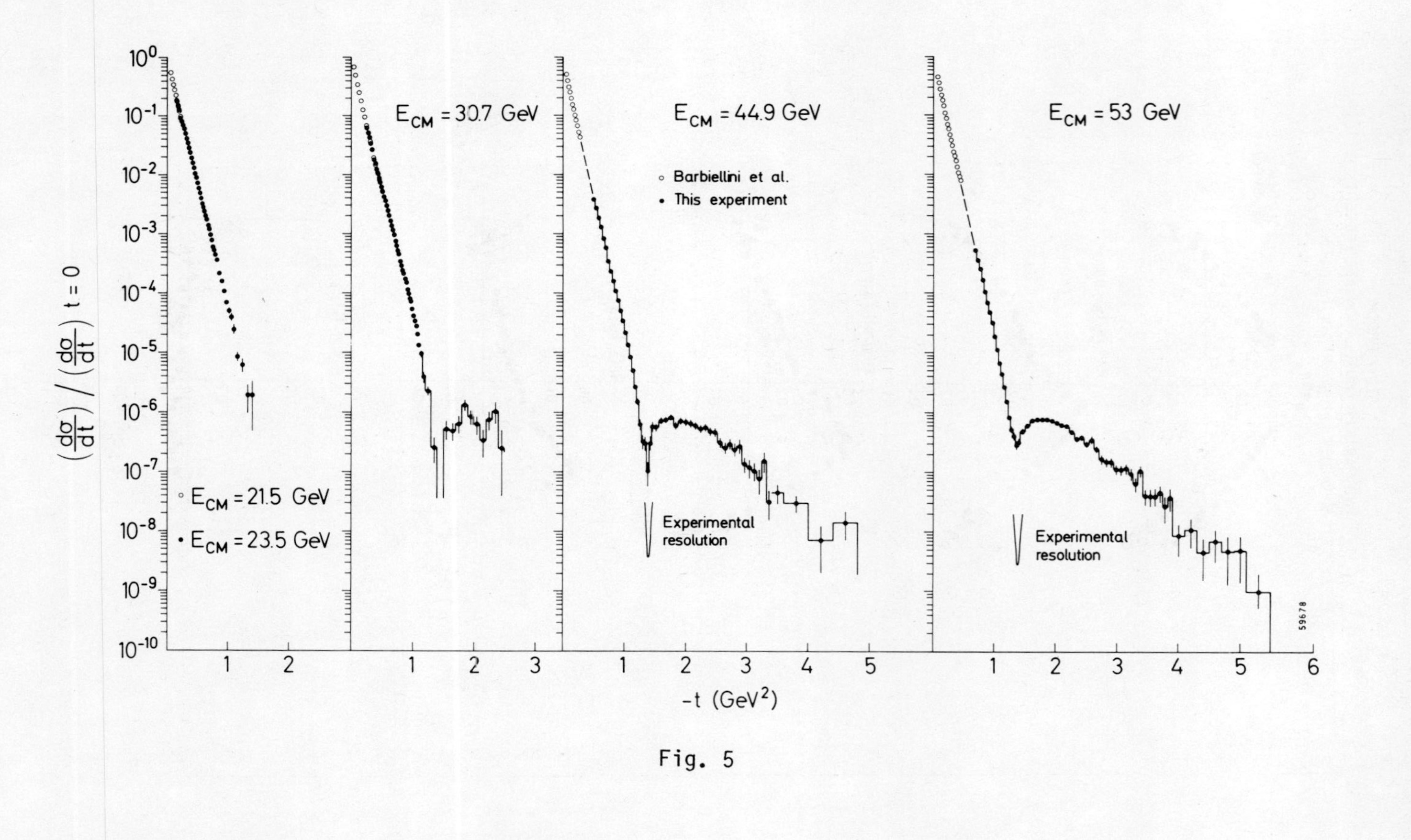

$(\frac{d\sigma}{dt}) / (\frac{d\sigma}{dt})$ t = 0
10^{0}
10^{-1}
10^{-2}
10^{-3}
10^{-4}
10^{-5}
10^{-6}
10^{-7}
10^{-8}
10^{-9}
10^{-10}
∘ E_{CM} = 21.5 GeV
• E_{CM} = 23.5 GeV
E_{CM} = 30.7 GeV
E_{CM} = 44.9 GeV
∘ Barbiellini et al.
• This experiment
Experimental resolution
E_{CM} = 53 GeV
Experimental resolution
-t (GeV^2)
59678

Fig. 5

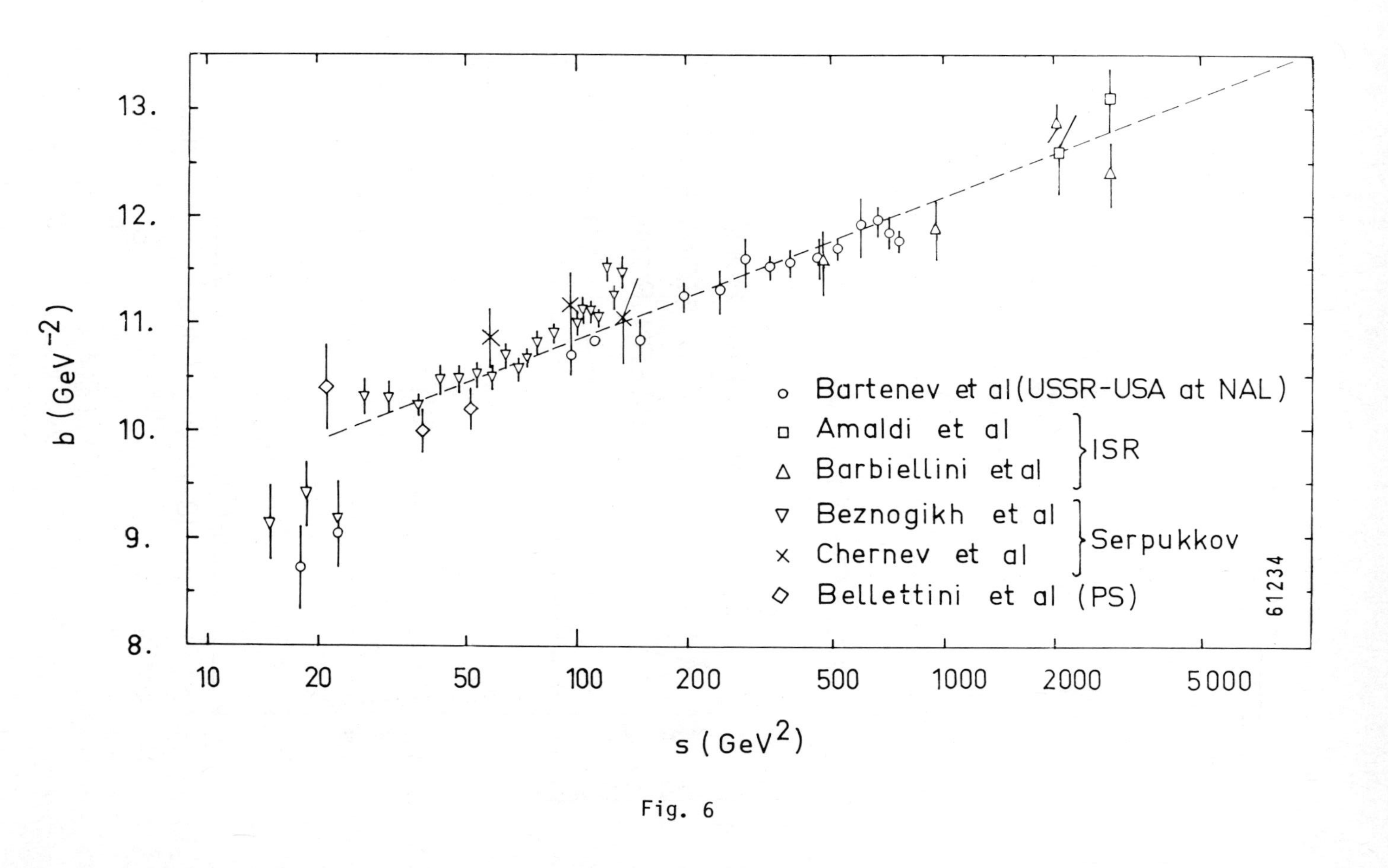

b (GeV⁻²)
13.
12.
11.
10.
9.
8.
10
20
50
100
200
500
1000
2000
5000
s (GeV²)
Bartenev et al (USSR-USA at NAL)
Amaldi et al
Barbiellini et al
ISR
Beznogikh et al
Chernev et al
Serpukkov
Bellettini et al (PS)
61234

Fig. 6

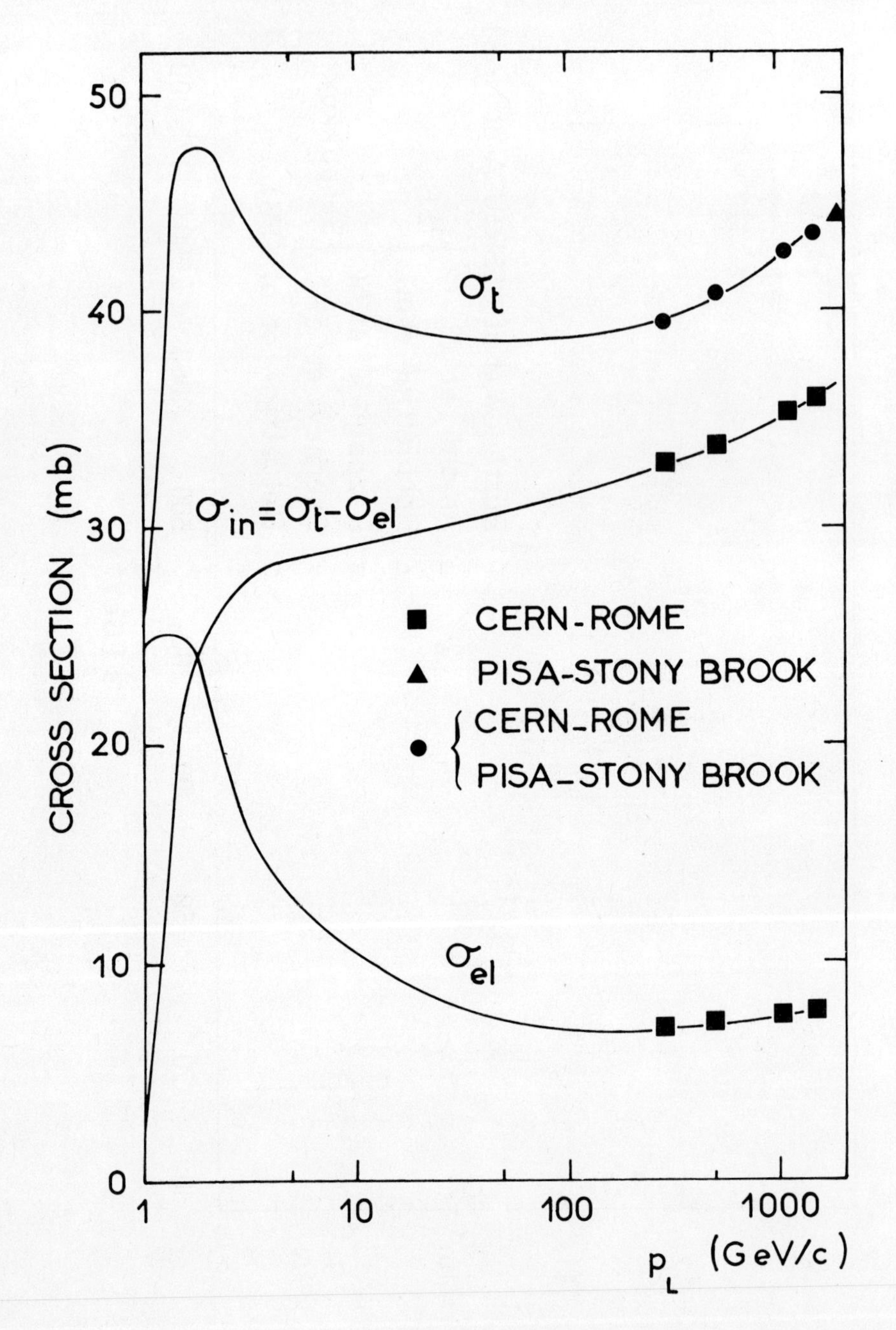

CROSS SECTION (mb)
50
40
30
20
10
0
1
10
100
1000
p_L (GeV/c)
σ_t
$\sigma_{in} = \sigma_t - \sigma_{el}$
σ_{el}
CERN-ROME
PISA-STONY BROOK
CERN-ROME
PISA-STONY BROOK

Fig. 7

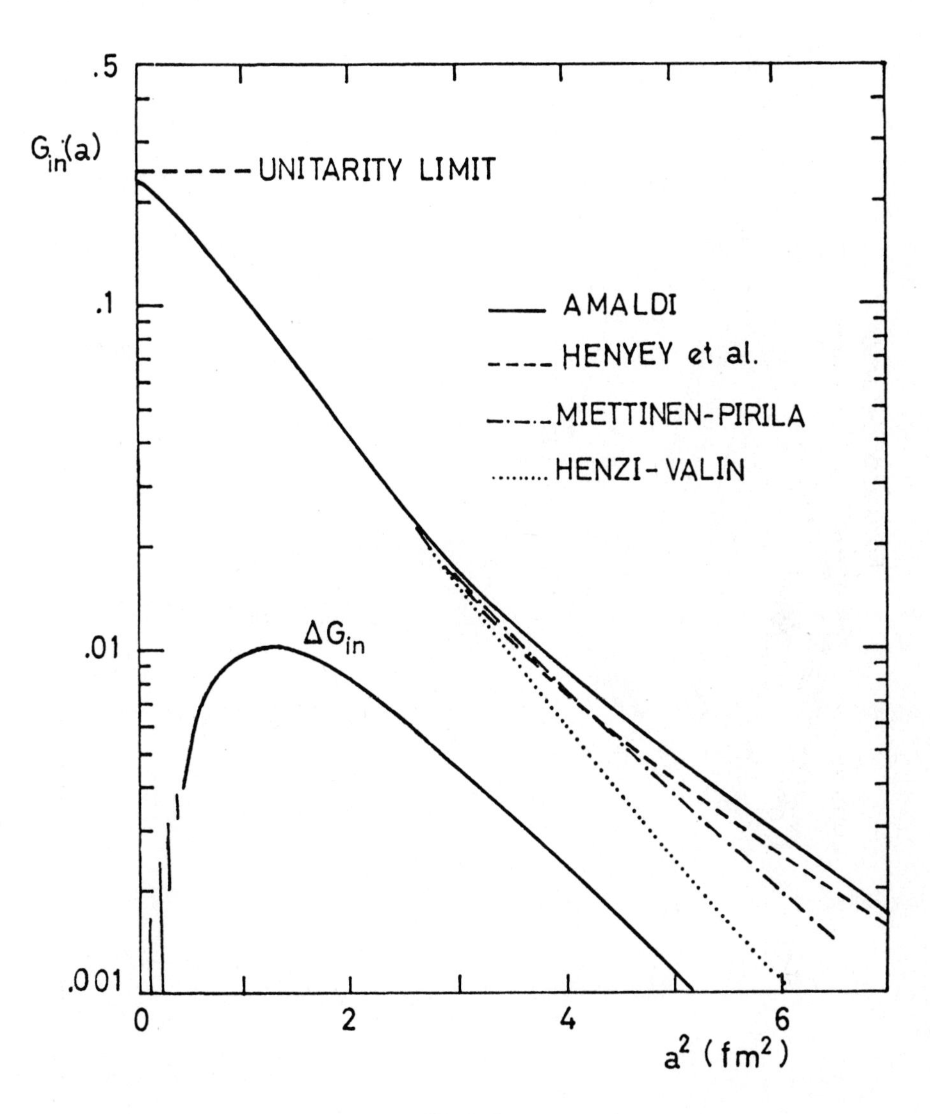

.5
$G_{in}(a)$
UNITARITY LIMIT
.1
AMALDI
HENYEY et al.
MIETTINEN-PIRILA
HENZI-VALIN
.01
ΔG_{in}
.001
0
2
4
6
a^2 (fm^2)

Fig. 8

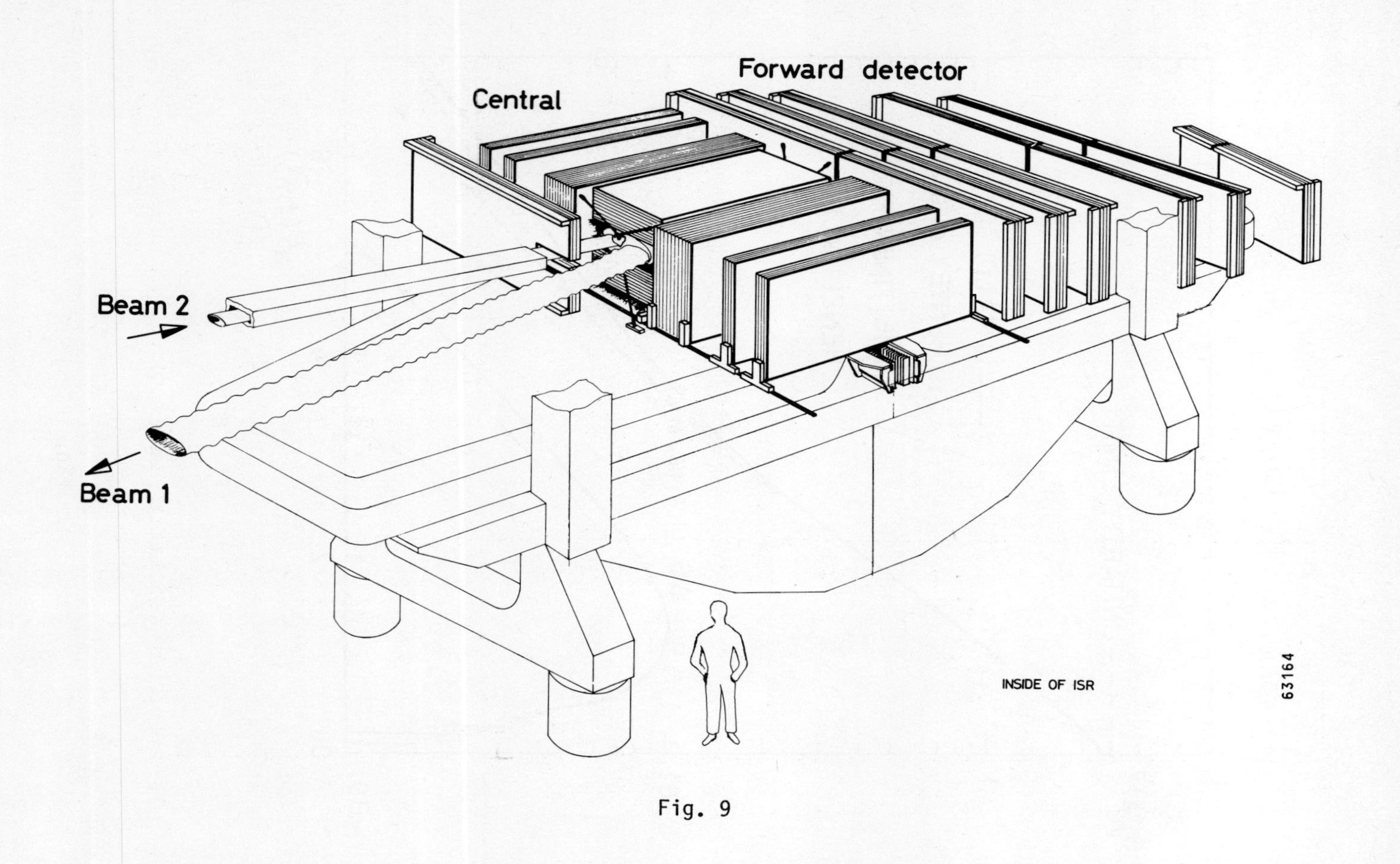

Fig. 9

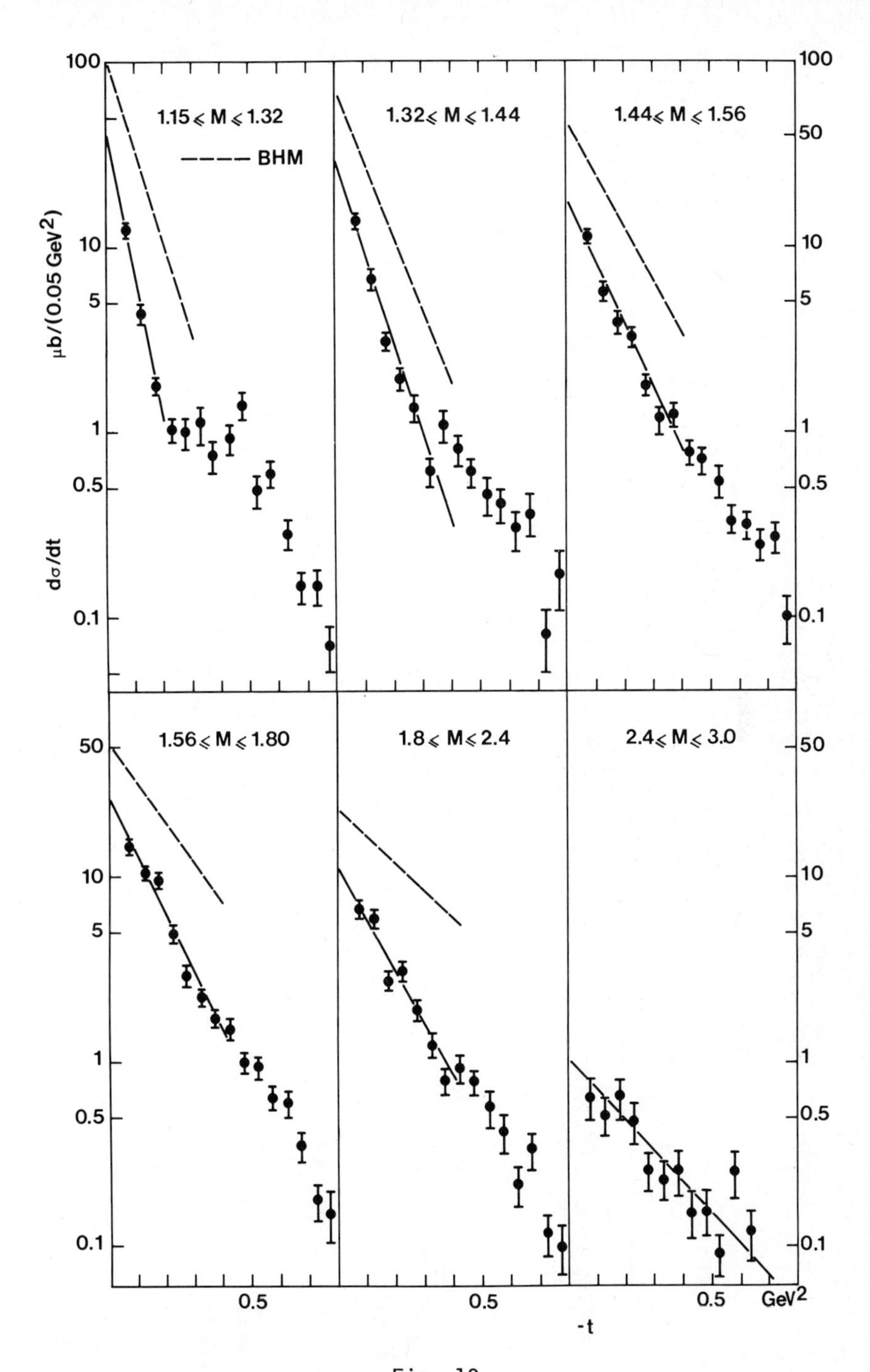

1.15 ⩽ M ⩽ 1.32
BHM
1.32 ⩽ M ⩽ 1.44
1.44 ⩽ M ⩽ 1.56
1.56 ⩽ M ⩽ 1.80
1.8 ⩽ M ⩽ 2.4
2.4 ⩽ M ⩽ 3.0
dσ/dt μb/(0.05 GeV²)
100
50
10
5
1
0.5
0.1
-t
GeV²

Fig. 10

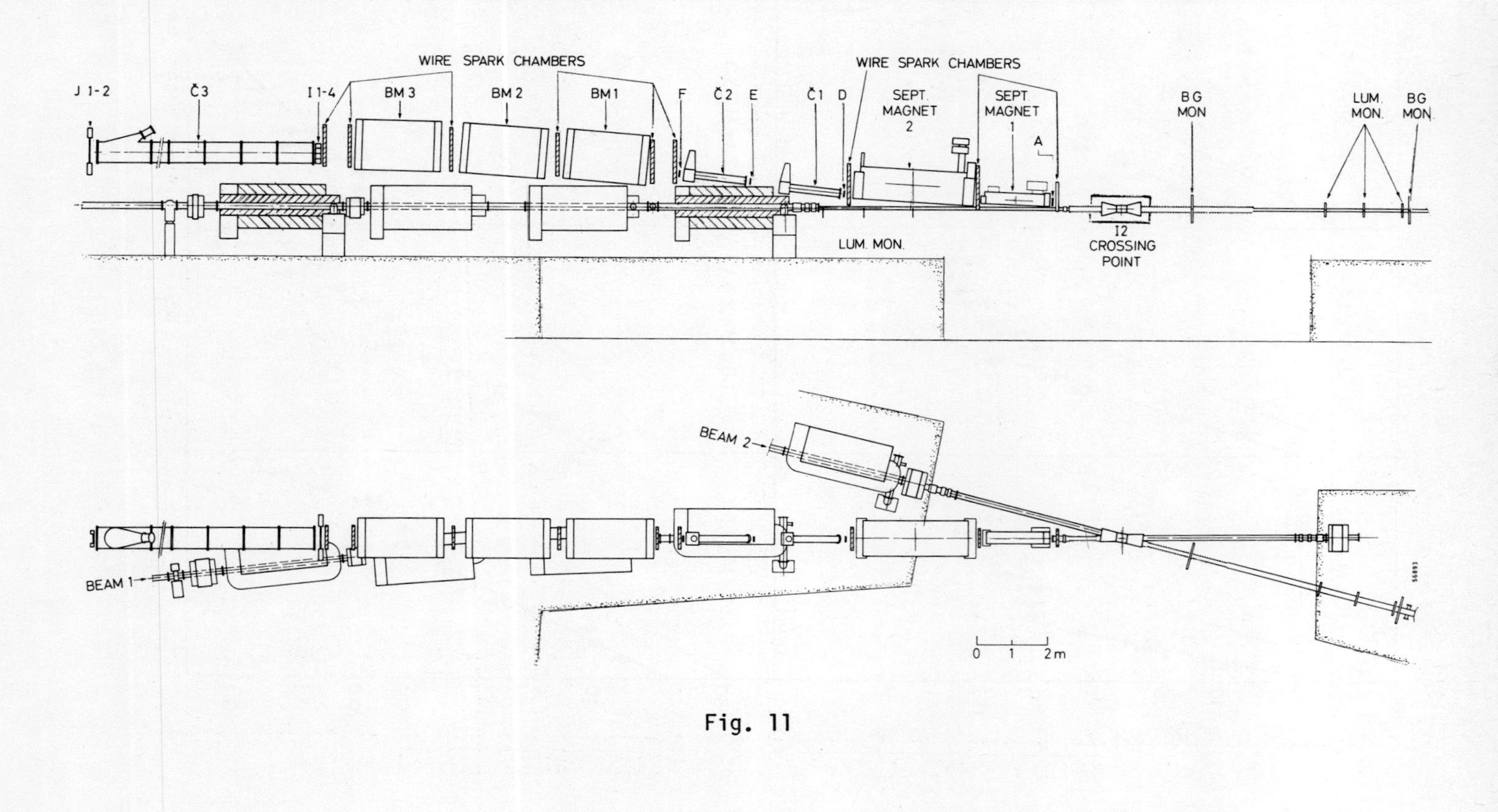

WIRE SPARK CHAMBERS
J 1-2
Č3
I 1-4
BM 3
BM 2
BM 1
F
Č2
E
Č1
D
WIRE SPARK CHAMBERS
SEPT. MAGNET 2
SEPT. MAGNET 1
A
B G MON
LUM. MON.
B G MON.
LUM. MON.
I2 CROSSING POINT
BEAM 2
BEAM 1
0 1 2 m

Fig. 11

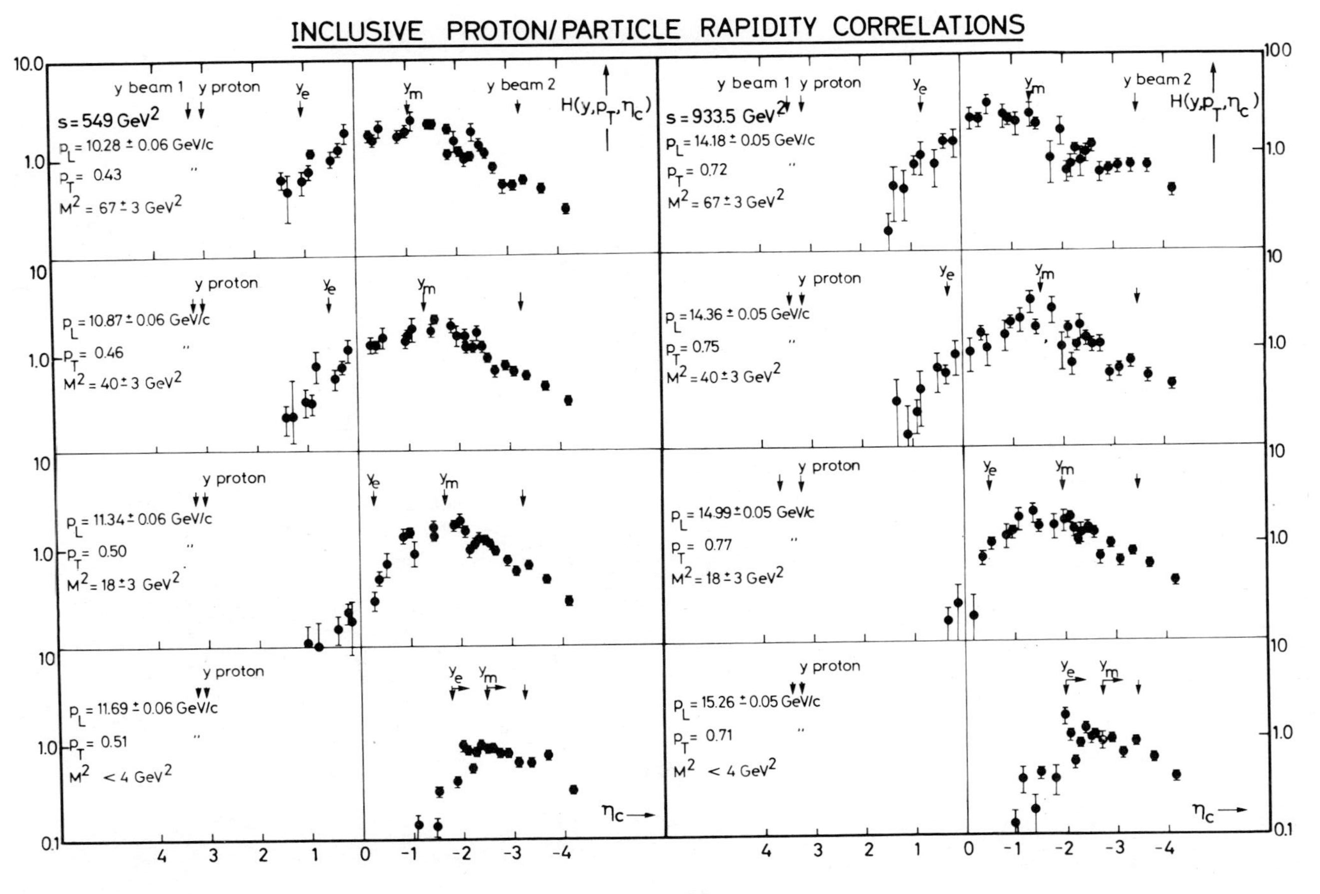
INCLUSIVE PROTON/PARTICLE RAPIDITY CORRELATIONS
y beam 1
y proton
y_e
y_m
y beam 2
$H(y, p_T, \eta_c)$
s = 549 GeV^2
p_L = 10.28 ± 0.06 GeV/c
p_T = 0.43
M^2 = 67 ± 3 GeV^2
p_L = 10.87 ± 0.06 GeV/c
p_T = 0.46
M^2 = 40 ± 3 GeV^2
p_L = 11.34 ± 0.06 GeV/c
p_T = 0.50
M^2 = 18 ± 3 GeV^2
p_L = 11.69 ± 0.06 GeV/c
p_T = 0.51
M^2 < 4 GeV^2
s = 933.5 GeV^2
p_L = 14.18 ± 0.05 GeV/c
p_T = 0.72
M^2 = 67 ± 3 GeV^2
p_L = 14.36 ± 0.05 GeV/c
p_T = 0.75
M^2 = 40 ± 3 GeV^2
p_L = 14.99 ± 0.05 GeV/c
p_T = 0.77
M^2 = 18 ± 3 GeV^2
p_L = 15.26 ± 0.05 GeV/c
p_T = 0.71
M^2 < 4 GeV^2
η_c
10.0
100
10
1.0
0.1
4 3 2 1 0 -1 -2 -3 -4

Fig. 12

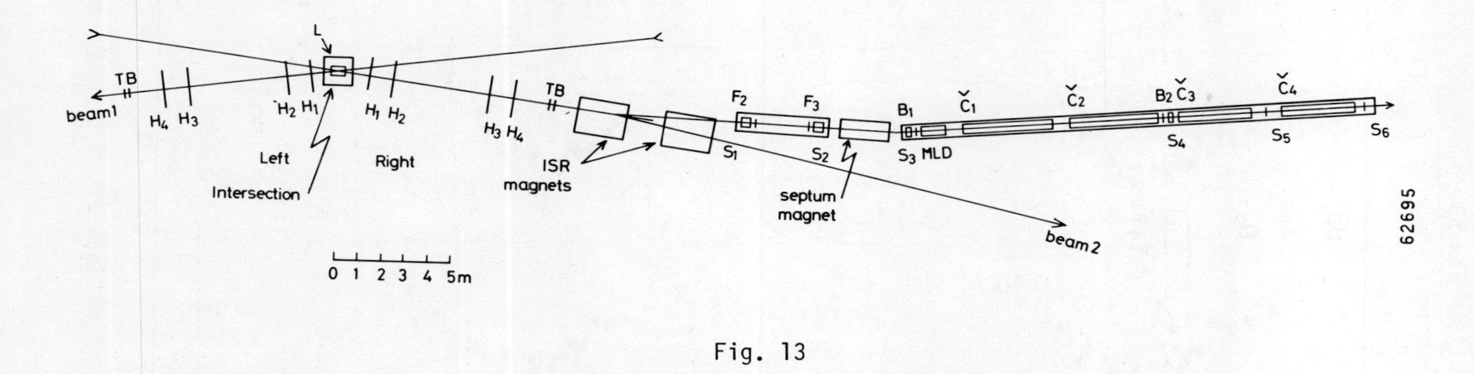

TB
beam1
H4 H3
H2 H1
L
H1 H2
Left
Right
Intersection
H3 H4
TB
ISR
magnets
S1
F2
F3
S2
septum
magnet
B1
S3 MLD
Č1
Č2
B2 Č3
S4
Č4
S5
S6
beam 2
0 1 2 3 4 5m
62695

Fig. 13

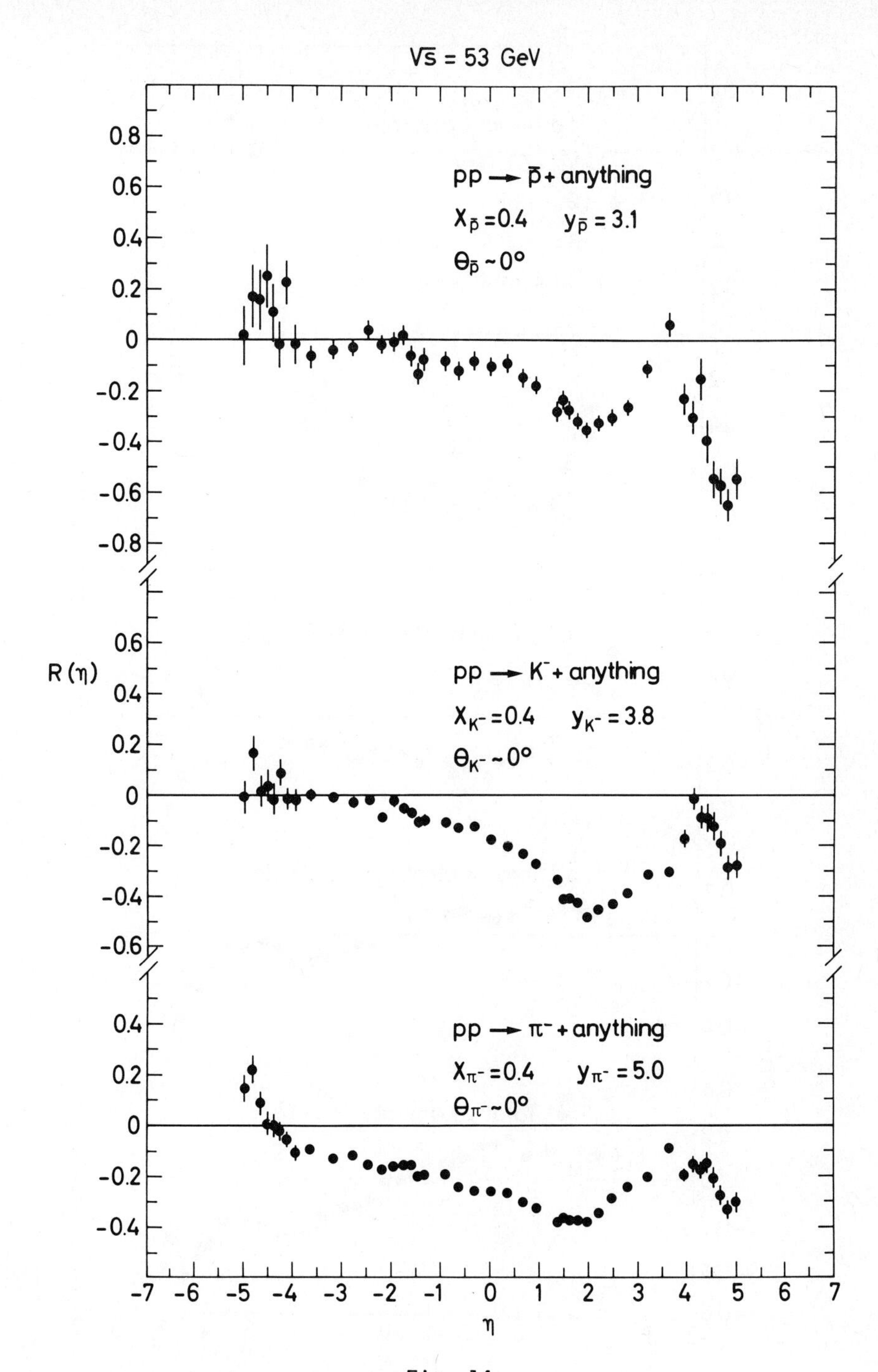

√s = 53 GeV
pp → p̄ + anything
X_p̄ = 0.4   y_p̄ = 3.1
Θ_p̄ ~ 0°
pp → K⁻ + anything
X_K⁻ = 0.4   y_K⁻ = 3.8
Θ_K⁻ ~ 0°
pp → π⁻ + anything
X_π⁻ = 0.4   y_π⁻ = 5.0
Θ_π⁻ ~ 0°
R (η)
η

Fig. 14

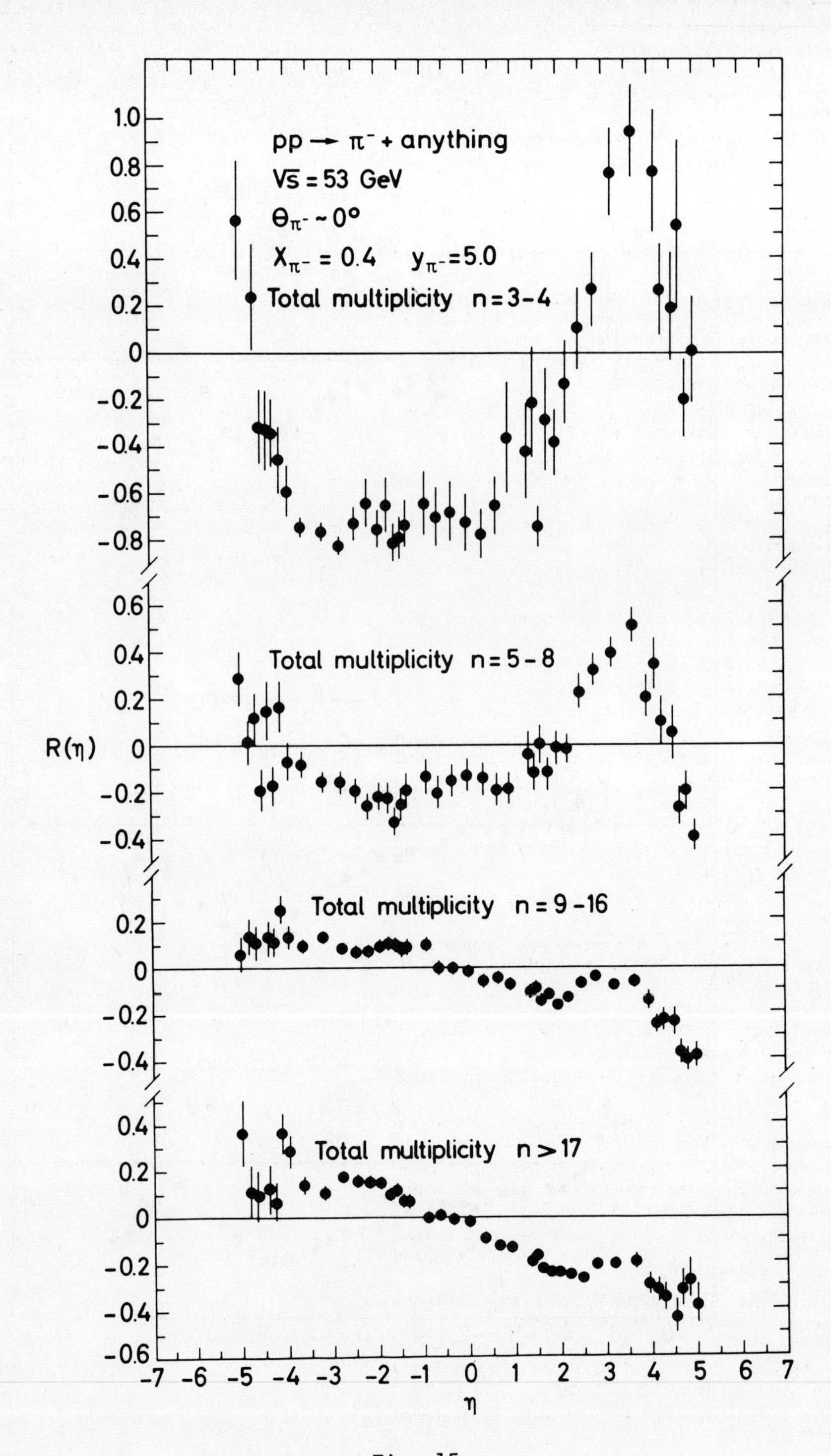
pp → π⁻ + anything
$\sqrt{s}$ = 53 GeV
$\theta_{\pi^-} \sim 0°$
X_{π^-} = 0.4 y_{π^-}=5.0
Total multiplicity n=3-4
Total multiplicity n=5-8
Total multiplicity n=9-16
Total multiplicity n>17
R(η)
η

Fig. 15

Lecture Notes in Physics

This series aims to report new developments in physical research and teaching – quickly, informally and at a high level. The type of material considered for publication includes:

1. Preliminary drafts of original papers and monographs
2. Lectures on a new field, or presenting a new angle on a classical field
3. Seminar work-outs
4. Reports of meetings, provided they are
   a) of exceptional interest and
   b) devoted to a single topic.

Texts which are out of print but still in demand may also be considered if they fall within these categories.

The timeliness of a manuscript is more important than its form, which may be unfinished or tentative. Thus, in some instances, proofs may be merely outlined and results presented which have been or will later be published elsewhere. If possible, a subject index should be included. Publication of Lecture Notes is intended as a service to the international physical community, in that a commercial publisher, Springer-Verlag, can offer a wider distribution to documents which would otherwise have a restricted readership. Once published and copyrighted, they can be documented in the scientific literature.

Manuscripts

Manuscripts should comprise not less than 100 pages.
They are reproduced by a photographic process and therefore must be typed with extreme care. Symbols not on the typewriter should be inserted by hand in indelible black ink. Corrections to the typescript should be made by pasting the amended text over the old one, or by obliterating errors with white correcting fluid. Authors receive 50 free copies and are free to use the material in other publications. The typescript is reduced slightly in size during reproduction; best results will not be obtained unless the text on any one page is kept within the overall limit of 18 x 26.5 cm (7 x 10½ inches). The publishers will be pleased to supply on request special stationery with the typing area outlined.

Manuscripts in English, German or French should be sent to Dr. W. Beiglböck, 69 Heidelberg/Germany, Institut für Angewandte Mathematik, Im Neuenheimerfeld 5, or directly to Springer-Verlag Heidelberg.

Springer-Verlag, D-1000 Berlin 33, Heidelberger Platz 3
Springer-Verlag, D-6900 Heidelberg 1, Neuenheimer Landstraße 28–30
Springer-Verlag, 175 Fifth Avenue, New York, NY 10010/USA

ISBN 3-540-**07160**-1
ISBN 0-387-**07160**-1